Exploration of the Medicinal Potential of Kitchen Ingredients

This book explores the therapeutic potential of everyday kitchen ingredients, blending traditional practices with modern scientific analysis. It provides insights beyond the regular culinary use of vegetables and spices, and presents a comprehensive analysis of their chemistry, nutraceutical roles, and pharmacological applications. The diverse range of topics will offer readers a glimpse into the profound medicinal benefits these ingredients hold.

With contributions from leading researchers at eminent institutions, the book encompasses the therapeutic potential of daily-use kitchen ingredients, their importance in historical and cultural contexts, and their relevance in modern medicine. This volume will help to reveal the true power of kitchen ingredients through scientific evidence.

Key features include:

- A diverse range of topics on the therapeutic roles and nutraceutical significance of vegetables such as garlic, spinach, capsicum, fennels, onion, leafy vegetables, and others.
- A comprehensive insight into the phytochemistry and pharmacology of spices such as clove, cinnamon, ginger, cumin, cardamom, nutmeg, asafoetida, thyme, and more.
- The utilization of different kitchen ingredients for nutraceutical purposes in addressing chronic illnesses such as cancer and diabetes.
- Insight into the pharmacological role of essential oils derived from spices.
- Comprehensive insights into allergenicity, toxicity profiles, and dose-dependent studies.
- A detailed discussion on clinical evidence of assessments done on functional spices.

This book is an essential resource for students, researchers, faculties, and healthcare professionals working in naturopathy. *Exploration of the Medicinal Potential of Kitchen Ingredients* promises to inspire innovative solutions and expand horizons in the fields of pharmacology and natural therapeutics.

Exploration of the Medicinal Potential of Kitchen Ingredients

Edited by
Pankaj Kumar Chaurasia, Shashi Lata Bharati,
Sunita Singh, and Ashutosh Mani

CRC Press
Taylor & Francis Group
Boca Raton London New York

CRC Press is an imprint of the
Taylor & Francis Group, an **informa** business

Designed cover image: Sunita Singh and Pankaj Kumar Chaurasia

First edition published 2026
by CRC Press
2385 NW Executive Center Drive, Suite 320, Boca Raton FL 33431

and by CRC Press
4 Park Square, Milton Park, Abingdon, Oxon, OX14 4RN

CRC Press is an imprint of Taylor & Francis Group, LLC

ISBN: 978-1-032-90806-9 (hbk)
ISBN: 978-1-032-90807-6 (pbk)
ISBN: 978-1-003-55989-4 (ebk)

DOI: 10.1201/9781003559894

Typeset in Times
by Apex CoVantage, LLC

Contents

About the Editors ... viii
Contributors ... x
Preface ... xv

Chapter 1 Different Vegetables and Their Nutraceutical Significances ... 1

Ayush Madan, Pranshul Sethi, Neetu Panwar, Sonali Rastogi, Urvee Dave, and Abhinav Verma

Chapter 2 Garlic through the Ages: Historical Significance, Chemical Composition, and Therapeutic Potential ... 14

Nikhil Rajnani, Ashish Hundani, Heena Bhojwani, and Nalini Kurup

Chapter 3 Spinach: Varieties and Nutraceutical Functions ... 22

Rahul D. Kamble, Deepak B. Shelke, Mahadev R. Chambhare, and Pankaj Kumar Chaurasia

Chapter 4 An Insight into the Chemical Characteristics and Medicinal Properties of Capsicum Species ... 34

Himanko Gogoi, Hriday Kumar Biswas, Yogita Ale, and Mansi Butola

Chapter 5 Fennels: Chemistry and Pharmacology ... 43

Deepika Saini

Chapter 6 Leafy Vegetables and Their Nutraceutical Health Impacts ... 53

Shibashis Das, Partha Mondal, Rajdeep Mohanta, Tanmoy Mondal, Kaushik Kumar Das, and D.K. Murmu

Chapter 7 Onion (*Allium cepa* L.): Chemistry and Pharmacology ... 72

Pankaj Kumar Chaurasia, Shashi Lata Bharati, Payal Joshi, Sunita Singh, Rima Kumari, Alka Singh, Rahul D. Kamble, Deepak B. Shelke, and Vijay Kumar Patel

Chapter 8 Unlocking the Secrets of the King of Spices: Exploring Its Chemistry, Health Benefits, and Medicinal Wonders ... 80

Sukanya Patra, Jyotirmayee, Pradip Paik, and Malkhey Verma

Chapter 9 Clove: Chemistry, Medicinal Application, and Toxicity Profile ... 91

Rahul D. Kamble, Balkrishna A. Shinde, Mahadev R. Chambhare, Deepak B. Shelke, and Pankaj Kumar Chaurasia

Chapter 10 Phytochemistry, Biological Activities, and Therapeutic Potential of *Elettaria cardamomum* and Its Medicinal Perspectives 102

Rima Kumari, Pankaj Kumar Chaurasia, Shweta Yadav, and Nuzhat Parween

Chapter 11 A Comprehensive Insight on Chemistry and Medicinal Values of Fenugreek for Human Health 115

Abhratanu Ganguly, Sayantani Nanda, Nurunnisa Khatun, Kanchana Das, Pritam Maitra, Gopal Biswas, Moutushi Mandi, and Prem Rajak

Chapter 12 Cumin: Varieties and Nutraceutical Function and Applications 128

Sonia Gupta, Dinesh Kumar, Rajni Tanwar, Raman Kumari, and Puja Gulati

Chapter 13 Cinnamon: Chemistry in Action—From Flavour and Aroma to Therapeutic Interventions 143

Anuradha Singh, Javeria, Udit Kumar, Kavya Singh, and Aniqa Tahreem

Chapter 14 Ginger Chemistry and Its Health Impacts with a Toxicity Profile 154

Ayman Younes Allam, Gholamreza Abdi, Avanish Chandra Sharma, Sunita Singh, and Ahmed A. Tayel

Chapter 15 Nutrient Composition and Potential Biomolecule Profile of Black Turmeric: A Comprehensive Review 178

Debashree Kakati, Mukul Barwant, and Himakshi Baishya

Chapter 16 Asafoetida: Phytochemistry and Pharmacological Applications 190

Naveen Kumar, Anoop Kumar, Piyush Sharma, Pankaj Kumar Chaurasia, and Sunita Singh

Chapter 17 Thyme: Chemistry and Pharmacology 205

Alka Singh, Shalu Verma, Kiran Dobhal, and Chetan Singh Chauhan

Chapter 18 Unveiling the Secrets of Nutmeg: A Deep Dive into Its Phytochemistry and Pharmacology 215

Reshma P, Shaik Nazir Basha, and Reddappa JB

Chapter 19 Spices That Heal: From Kitchen to Medicine 226

Nikita Patel, Stephen Oyẹdeji, and Ramar Krishnamurthy

Chapter 20 Pharmacological Chemistry of Essential Oils from Spices 237

Rama Satya Sri Kotipalli, Mani Sharma, Nemala Siva Kumar, and Neeraj Kumar Chouhan

Chapter 21 Clinical Studies and Assessments of Functional Spices 251

Sagar and Jeena Gupta

Chapter 22 Clinical Studies and Assessment of Functional Food (Vegetables) 266

Shubhankar Anand, Bhanu Pratap Singh, R T Thrisha Lakshmi, and V Dhanyalakshmi

Chapter 23 Allergy, Allergens, and Allergenicity Related to Spices 280

Neetu Yadav

Index 285

About the Editors

Dr. Pankaj Kumar Chaurasia, Ph.D., is Assistant Professor in the PG Department of Chemistry, L.S. College, B.R. Ambedkar Bihar University, Muzaffarpur, India. He has more than seven years of experience in teaching at post-graduate and undergraduate levels. He qualified for the National Eligibility Test in 2009 as CSIR-JRF (NET) and has been awarded with SRF-NET in 2012. He received a Ph.D. degree in chemistry. He also worked as guest faculty (2016–2017) in the Department of Chemistry, University of Allahabad, Prayagraj. He was awarded a research associateship by CSIR-New Delhi (India) in 2017 and worked as a CSIR-RA in Motilal Nehru NIT, Allahabad/Prayagraaj (India). In his research career, he has more than 80 publications, including 50 journal articles, 25 book chapters, 9 books, and many more in process. He has guided one Ph.D. student and is also guiding three Ph.D. research scholars. He has worked in different areas of chemistry and has research expertise in biological chemistry, biotechnology, microbiology, enzymology, organometallic chemistry, and environmental chemistry.

Dr. Shashi Lata Bharati (she/her), Ph.D., is Assistant Professor in the Department of Chemistry, North Eastern Regional Institute of Science and Technology, Nirjuli, Arunachal Pradesh (India). She was awarded a UGC-DSA fellowship for meritorious students during her Ph.D. program. She obtained her Ph.D. degree in Chemistry. She was awarded a UGC post-doctoral fellowship for women in 2013 by UGC New Delhi and worked as a post-doctoral fellow in the Department of Chemistry of DDU Gorakhpur University, Gorakhpur (India).

She has more than 55 publications and 8 edited books with national and international journals/publishers of repute. She has guided many M.Sc. project students and one Ph.D. research scholar. She has expertise in the fields of inorganic chemistry, organometallic chemistry, biological chemistry, and enzymology.

Dr. Sunita Singh is Assistant Professor of Chemistry in the Navyug Kanya Mahavidyalaya, University of Lucknow, since 2019. She pursued her Ph.D. in 2015 on the topic "Chemistry, Antioxidant and Antimicrobial Activities of Essential Oils and Oleoresins of Spices" from DDU Gorakhpur University, Gorakhpur. She was awarded with junior and senior research fellowships from the University Grant Commission. She also worked as a research assistant (2010–2011) on the project sponsored by CST: "Chemistry, Antioxidant and Antimicrobial Activities of Oleoresins Extracted from Cardamom, Black Pepper and Caraway". Her work was mainly focused on quantitative and qualitative analyses of essential oils and oleoresins of spices, namely *Piper nigrum, Nigella sativa, Mentha longifolia, Anethum graveolans, Brassica juncea, and Sinapis alba*. Chemistry, antioxidant, and antimicrobial efficacies of essential oil and oleoresins were investigated using different techniques.

She has published 38 research articles in journals of national and international repute. She has authored and coauthored nine book chapters with leading publishing houses, and has edited books with reputed publishers such as Bentham, Nova, and CRC Press. She is also on the review board of journals, namely *Advances in Clinical Toxicology*, *MOJ Food Process & Technology*, and *Pharmaceutical Drug Regulatory Affairs Journal*. She has attended in total 30 conferences and webinars of national and international repute and delivered more than 15 invited talks and presentations. She has presented her work at various global events. She is also an active member of the Association of Chemistry Teachers (India), the International Clinical Aromatherapy Network, and the Global Harmonization Initiative.

Dr. Ashutosh Mani completed an M.Sc. and a Ph.D. in bioinformatics from the University of Allahabad, India. He worked as a visiting fellow at the School of Life Sciences, Jawaharlal Nehru University, New Delhi, India, to complete part of his Ph.D. work. He joined as a MIUR fellow at the University of Cagliari, Italy, and is currently Associate Professor in the Department of Biotechnology, Motilal Nehru National Institute of Technology, Allahabad, India. He has more than 12 years' teaching and research experience in the fields of molecular and structural biology, genomics, proteomics, and computational biology. His areas of interest include herbal drug discovery against cancer, Alzheimer's disease, and diabetes, and plant stress tolerance. Apart from editing two books, Dr. Mani has published more than 50 articles in national and international journals of repute. He has supervised eight Ph.D. and several undergraduate and postgraduate students. He has completed six sponsored research projects and has received research grants from funding agencies such as DST, SERB, CSTUP, and DBT. Dr. Mani has organized several international conferences and workshops in the area of biotechnology. He is a guest editor and reviewer of various international journals in the area of life sciences. Dr. Mani has reviewed research project proposals for leading funding agencies.

Contributors

Gholamreza Abdi
Department of Biotechnology
Persian Gulf Research Institute
Persian Gulf University
Bushehr, Iran

Yogita Ale
Department of Pharmaceutics
Uttaranchal Institute of Pharmaceutical Sciences
Uttaranchal University
Dehradun, Uttarakhand, India

Ayman Younes Allam
Department of Food Science and Technology
Faculty of Agriculture
Menoufia University
Shibin El Kom, Egypt

Shubhankar Anand
Department of Biotechnology
School of Biosciences and Technology
Galgotias University
Greater Noida, Gautam Buddha Nagar, India

Himakshi Baishya
Department of Food Processing Technology
Mangaldai College
Assam, India

Mukul Barwant
Department of Botany
Sanjivani Arts Commerce and Science College
Kopargaon, Maharashtra, India

Shaik Nazir Basha
CSIR-National Institute for Interdisciplinary Science and Technology
Thiruvananthapuram, Kerala, India

Shashi Lata Bharati
Department of Chemistry
North Eastern Regional Institute of Science and Technology
Nirjuli, Itanagar, Arunachal Pradesh, India

Heena Bhojwani
Department of Pharmaceutics
Principal K. M. Kundnani College of Pharmacy
Cuffe Parade, Mumbai, Maharashtra, India

Gopal Biswas
Toxicology Research Unit
Department of Zoology
The University of Burdwan
Purba Bardhaman, West Bengal, India

Hriday Kumar Biswas
Department of Pharmaceutics
Uttaranchal Institute of Pharmaceutical Sciences
Uttaranchal University, Dehradun, Uttarakhand, India

Mansi Butola
Department of Pharmaceutics
Uttaranchal Institute of Pharmaceutical Sciences
Uttaranchal University
Dehradun, Uttarakhand, India

Mahadev R. Chambhare
Amruteshwar Arts
Commerce and Science College
Vinzar, Tal. Velhe, Pune, Maharashtra, India

Chetan Singh Chauhan
Bhupal Noble's University
Udaipur, Rajasthan, India

Pankaj Kumar Chaurasia
P.G. Department of Chemistry
L.S. College
B.R. Ambedkar Bihar University
Muzaffarpur, Bihar, India

Neeraj Kumar Chouhan
Amity Institute of Pharmacy
Amity University
Greater Noida, Uttar Pradesh, India

Kanchana Das
Toxicology Research Unit
Department of Zoology
The University of Burdwan
Purba Bardhaman, West Bengal, India

Kaushik Kumar Das
Department of Entomology
CCS Haryana Agricultural University
Hisar, Haryana, India

Shibashis Das
Department of Vegetable Science
BCKV (Bidhan Chandra Agricultural University)
West Bengal, India

Urvee Dave
Department of Pharmacy
Bhupal Noble's University
Udaipur, Rajasthan, India

V Dhanyalakshmi
Pondicherry University
Kalapet, Puducherry, India

Kiran Dobhal
Shivalik College of Pharmacy
Dehradun, Uttarakhand, India

Abhratanu Ganguly
Toxicology Research Laboratory
Department of Animal Science
Kazi Nazrul University
Asansol, West Bengal, India

Himanko Gogoi
Department of Pharmaceutics
Uttaranchal Institute of Pharmaceutical Sciences
Uttaranchal University
Dehradun, Uttarakhand, India

Puja Gulati
School of Pharmacy
Desh Bhagat University
Mandi Gobindgarh, India

Jeena Gupta
School of Bioengineering and Biosciences
Lovely Professional University
Phagwara, India

Sonia Gupta
Swami Devi Dyal Institute of Pharmacy
Golpura Barwala
Panchkula, India

Ashish Hundani
Department of Pharmaceutics
Principal K. M. Kundnani College of Pharmacy
Cuffe Parade, Mumbai, Maharashtra, India

Javeria
Department of Chemistry
SS Khanna Girls' Degree College
(A Constituent College of the University of Allahabad)
Prayagraj, India

Reddappa JB
Department of Plantation, Spices, Medicinal and Aromatic Crops
College of Agriculture
Kerala Agricultural University
Thiruvananthapuram, India

Payal Joshi
Operations and Method Development
Shefali Research Laboratories
Ambernath, Maharashtra, India

Jyotirmayee
School of Biotechnology
Institute of Science
Banaras Hindu University
Varanasi, Uttar Pradesh, India

Debashree Kakati
Department of Botany
Mangaldai College
Mangaldai, Assam, India

Rahul D. Kamble
Amruteshwar Arts, Commerce and Science College
Vinzar, Tal. Velhe, Pune, Maharashtra, India

Nurunnisa Khatun
Toxicology Research Laboratory
Department of Animal Science
Kazi Nazrul University
Asansol, West Bengal, India

Rama Satya Sri Kotipalli
Jawaharlal Nehru Technological University
School of Pharmacy, Rajeshwari Nagar
Kakinada, Andhra Pradesh, India

Ramar Krishnamurthy
Kishorbhai Institute of Agriculture Sciences and Research Centre
Uka Tarsadia University
Bardoli, Surat, Gujarat, India

Anoop Kumar
Department of Botany
Bahadur Yadav Mahila Mahavidyalay
Deoria, Uttar Pradesh, India

Dinesh Kumar
School of Pharmacy
Desh Bhagat University
Mandi Gobindgarh, Punjab, India

Naveen Kumar
Department of Chemistry
Sanatan Dharm College Muzaffarnagar
Uttar Pradesh, India

Nemala Siva Kumar
Sri Vasavi Institute of Pharmaceutical Sciences
Pedatadepalli, Tadepalligudem
West Godavari, Andhra Pradesh, India

Udit Kumar
Department of Chemistry
SS Khanna Girls' Degree College
(A Constituent College of University of Allahabad)
Prayagraj, India

Raman Kumari
Keshav College of Pharmacy
Matani, Bhiwani, India

Rima Kumari
Department of Botany
L.S. College
Muzaffarpur, Bihar, India

Nalini Kurup
Department of Pharmaceutics
Principal K. M. Kundnani College of Pharmacy
Cuffe Parade, Mumbai, Maharashtra, India

R T Thrisha Lakshmi
Central University of Tamil Nadu
Thiruvarur, Tamilnadu, India

Ayush Madan
Department of Biotechnology
School of Research and Technology
People's University
Bhanpur, Bhopal, Madhya Pradesh, India;
Institute of Tropical Aquaculture and Fisheries (AKUATROP)
Universiti Malaysia Terengganu
Kuala Nerus, Terengganu, Malaysia

Pritam Maitra
Toxicology Research Unit
Department of Zoology
The University of Burdwan
Purba Bardhaman, West Bengal, India

Moutushi Mandi
Toxicology Research Unit
Department of Zoology
The University of Burdwan
Purba Bardhaman, West Bengal, India

Rajdeep Mohanta
Department of Agriculture
Brainware University
Barasat West Bengal, India

Partha Mondal
Department of Agricultural Biochemistry
BCKV (Bidhan Chandra Agricultural University)
West Bengal, India

Tanmoy Mondal
Department of Horticulture
MITS Institute of Professional Studies
Rayagada, Odisha, India

D.K. Murmu
Regional Research Station (Old Alluvial Zone) UBKV
Majhian, West Bengal, India

Sayantani Nanda
Toxicology Research Laboratory
Department of Animal Science
Kazi Nazrul University
Asansol, West Bengal, India

Stephen Oyedeji
Kishorbhai Institute of Agriculture Sciences and Research Centre
Uka Tarsadia University
Bardoli, Surat, Gujarat, India;
Department of Plant Biology
Faculty of Life Sciences
University of Ilorin, Ilorin, Nigeria

Reshma P
Department of Plantation, Spices, Medicinal and Aromatic Crops
College of Agriculture
Kerala Agricultural University
Thiruvananthapuram, India

Pradip Paik
School of Biomedical Engineering
Indian Institute of technology (IIT) BHU
Varanasi, Uttar Pradesh, India

Neetu Panwar
School of Applied Sciences
Shri Venkateshwara University
Gajraula, Uttar Pradesh, India

Nuzhat Parween
Department of Botany
L.S. College
B.R.A. Bihar University
Muzaffarpur, India

Nikita Patel
Kishorbhai Institute of Agriculture Sciences and Research Centre
Uka Tarsadia University
Bardoli, Surat, Gujarat, India

Vijay Kumar Patel
Ram Sharan Roy College of Pharmacy
Panapur, Makkanpur, Vaishali, Bihar, India

Sukanya Patra
School of Biomedical Engineering
Indian Institute of Technology (IIT) BHU
Varanasi, Uttar Pradesh, India

Prem Rajak
Toxicology Research Laboratory
Department of Animal Science
Kazi Nazrul University
Asansol, West Bengal, India

Nikhil Rajnani
Department of Pharmaceutics
Principal K. M. Kundnani College of Pharmacy
Cuffe Parade, Mumbai, Maharashtra, India

Sonali Rastogi
Chitkara College of Pharmacy
Chitkara University
Rajpura, Punjab, India;
Department of Pharmacy
Venkateshwara Group of Institutions
Jatoli, Meerut, India

Sagar
School of Bioengineering and Biosciences
Lovely Professional University
Phagwara, India

Deepika Saini
Amity School of Pharmaceutical Sciences
Amity University
Mohali, Punjab, India

Pranshul Sethi
Department of Pharmacology
College of Pharmacy
Shri Venkateshwara University
Gajraula, Uttar Pradesh, India;
Chitkara College of Pharmacy
Chitkara University
Rajpura, Punjab, India

Avanish Chandra Sharma
Department of Chemistry
Navyug Kanya Mahavidyalaya
University of Lucknow
Lucknow, Uttar Pradesh, India

Mani Sharma
Department of Pharmaceutics
School of Pharmacy, Anurag University
Hyderabad, Telangana, India

Piyush Sharma
Department of Zoology
Sanatan Dharm College Muzaffarnagar
Uttar Pradesh, India

Deepak B. Shelke
Amruteshwar Arts, Commerce and Science College
Vinzar, Tal. Velhe, Pune, Maharashtra, India

Balkrishna A. Shinde
Amruteshwar Arts, Commerce and Science College
Vinzar, Tal. Velhe, Pune, Maharashtra, India

Alka Singh
School of Pharmaceutical Sciences and Technology
Sardar Bhagwan Singh University
Balawala, Dehradun, Uttrakhan, India

Anuradha Singh
Department of Chemistry
SS Khanna Girls' Degree College
(A Constituent College of University of Allahabad)
Prayagraj, India

Bhanu Pratap Singh
Department of Biotechnology
National Institute of Technology Warangal,
Telangana, India

Kavya Singh
Department of Chemistry
SS Khanna Girls' Degree College
(A Constituent College of University of Allahabad)
Prayagraj, India

Sunita Singh
Department of Chemistry
Navyug Kanya Mahavidyalaya
University of Lucknow
Lucknow, Uttar Pradesh, India

Aniqa Tahreem
Department of Chemistry
SS Khanna Girls' Degree College
(A Constituent College of University of Allahabad)
Prayagraj, India

Rajni Tanwar
School of Pharmacy
Desh Bhagat University
Mandi Gobindgarh, Punjab, India

Ahmed A. Tayel
Department of Fish Processing and Biotechnology
Faculty of Aquatic and Fisheries Sciences
Kafrelsheikh University
Kafrelsheikh, Egypt

Abhinav Verma
Department of Pharmaceutics
School of Pharmacy
Shri Venkateshwara University
Gajraula, Uttar Pradesh, India

Malkhey Verma
School of Biotechnology
Institute of Science
Banaras Hindu University
Varanasi, Uttar Pradesh, India

Shalu Verma
Uttaranchal Institute of Pharmaceutical Sciences
Uttaranchal University
Dehradun, Uttarakhand, India

Neetu Yadav
GGIC Khalilabad Sant Kabir Nagar
Uttar Pradesh, India

Shweta Yadav
Department of Botany
MDDM College
B.R.A. Bihar University
Muzaffarpur, India

Preface

Exploration of the Medicinal Potential of Kitchen Ingredients is the collection of 23 chapters contributed by experts in their fields. This book shows the significant medicinal roles of different kitchen ingredients in human life. Each chapter presents useful and promising medicinal information on specific plants or plant sources or materials generally used in the daily food routine by the people of India and/or other countries.

Chapter 1, written by Sethi et al., describes the different vegetables, their potential chemistry, and nutraceutical significances. In **Chapter 2**, Rajnani et al. deal with an informative discussion on the chemical composition, medicinal importance, and historical significances of garlic. In **Chapter 3**, Kamble et al. present the various varieties and nutraceutical benefits of spinach, a commonly used vegetable. Different chemical characteristics and medicinal properties of *Capsicum* species are assessed in **Chapter 4** by Gogoi and co-authors. Detailed chemistry and pharmacology of fennels are described interestingly by Saini in **Chapter 5**. The different health benefits of leafy vegetables are described in **Chapter 6** by Das et al. In **Chapter 7**, Chaurasia et al. evaluate the detailed chemistry and medicinal activities of onion. In **Chapter 8**, Patra et al. explore the chemistry, health benefits, and medicinal properties of the king of the spices: black pepper. In **Chapter 9**, an insightful assessment is provided on the clove for its promising medicinal potential, chemistry, and toxicity profile. In **Chapter 10**, Rima and co-authors offer an inclusive assessment of the phytochemistry, biological activities, and medicinal applications of cardamom. In **Chapter 11**, Ganguly et al. describe the various medicinal aspects of pharmacologically significant fenugreek, along with its chemistry. In **Chapter 12**, Gupta et al. present insightful views on the chemistry, varieties, and nutraceutical applications of cumin. Different therapeutic applications of cinnamon, along with other information, are described by Singh et al. in **Chapter 13**. In **Chapter 14**, Allam et al. present comprehensive discussions on the chemistry, medicinal applications, and toxicity profile of ginger. In **Chapter 15**, Kakati et al. explore the nutrient composition and pharmacological role of black turmeric (*Curcuma caesia*). **Chapter 16**, written by Kumar et al., explores the phytochemistry, traditional uses, and potential pharmacological applications of asafoetida, as well as the bioactive compounds in asafoetida and their future potential in food and medicine. Detailed chemistry and pharmacological roles of thyme are presented in **Chapter 17** by Singh et al. **Chapter 18**, written by Reshma et al., covers the phytochemical and pharmacological properties and toxicity assessment of nutmeg. In **Chapter 19**, Patel et al. highlight the distinctive characteristics, active components, and applications of certain significant spices with their functional and immunomodulatory capabilities. In **Chapter 20**, Kotipalli et al. delve into a comprehensive overview of the chemical composition and pharmacological evaluation of essential oils derived from commonly used spices. Clinical assessments of the functional spices have been elaborated on by Sagar and Gupta in **Chapter 21**. Chapter 22, written by Anand et al., presents the clinical studies and assessment of functional foods (vegetables). The final **Chapter 23**, written by Yadav et al., describes the allergenic effect of some spices.

BOOK EDITORS
Dr. Pankaj Kumar Chaurasia
Dr. Shashi Lata Bharati
Dr. Sunita Singh
Dr. Ashutosh Mani

Preface

1 Different Vegetables and Their Nutraceutical Significances

Ayush Madan, Pranshul Sethi, Neetu Panwar, Sonali Rastogi, Urvee Dave, and Abhinav Verma

1.1. INTRODUCTION

1.1.1. Importance of Vegetables in the Human Diet

Vegetables are a cornerstone of human nutrition, providing essential vitamins, minerals, dietary fibre, and bioactive compounds crucial for maintaining health and preventing diseases. The inclusion of a diverse range of vegetables in daily meals has been shown (see Table 1.1) to reduce the risk of chronic conditions such as cardiovascular disease, diabetes, and certain cancers (Slavin et al., 2012). Vegetables are rich in essential nutrients, including vitamins A, C, and E, potassium, magnesium, and folate, which are necessary for optimal physiological functioning (Craig et al., 2009). Their low caloric content also makes them ideal for weight management while supplying antioxidants that combat oxidative stress (Liu, 2003).

1.1.2. Overview of Nutraceuticals

The term "nutraceutical" refers to food-derived products that offer health benefits beyond basic nutrition. Nutraceuticals bridge the gap between food and medicine, with applications in disease prevention and management. These compounds, derived from vegetables, include phytochemicals, polyphenols, flavonoids, glucosinolates, and carotenoids, which exhibit antioxidative, anti-inflammatory, and antimicrobial properties (Kalra, 2003). Nutraceuticals have gained attention for their potential to modulate metabolic pathways, enhance immune function, and contribute to overall well-being (Diplock et al., 1999), as shown in Figure 1.1. For instance, glucosinolates in cruciferous vegetables like broccoli have been linked to cancer prevention (Higdon et al., 2001).

1.1.3. Role of Phytochemicals in Health

Phytochemicals are bioactive compounds found in vegetables that contribute significantly to their nutraceutical properties. These include carotenoids, polyphenols, alkaloids, and organosulfur compounds, each of which plays a specific role in health promotion. Carotenoids, such as beta-carotene in carrots, function as precursors to vitamin A and are known for their antioxidant capacity, which mitigates cellular damage caused by free radicals (Krinsky et al., 2003). Polyphenols, found in vegetables such as spinach and onions, exhibit anti-inflammatory effects by modulating signaling pathways and reducing oxidative stress (Pandey et al., 2011). Additionally, organosulfur compounds in garlic and onions provide cardiovascular benefits by reducing cholesterol levels and improving blood pressure (Jikah et al., 2024).

The chemopreventive properties of these phytochemicals have been extensively studied. For example, the sulforaphane in broccoli is a potent inducer of phase II detoxification enzymes, aiding in the neutralization of carcinogens (Zhang et al., 1994). Similarly, the flavonoids in leafy greens

DOI: 10.1201/9781003559894-1

TABLE 1.1
Summary of Nutritional Composition and Bioactive Compounds

S.No.	Vegetable Category	Key Nutrients	Bioactive Compounds	Associated Health Benefits	References
1	**Leafy Greens**	**Vitamins**: A, C, K **Minerals**: Magnesium, Folate	**Flavonoids**: Quercetin, Kaempferol **Carotenoids**: Lutein, Beta-carotene	**Antioxidant**: Neutralizes free radicals to prevent oxidative stress **Eye Health**: Reduces risk of macular degeneration **Anti-Inflammatory**: Reduces chronic inflammation **Cardiovascular Health**: Supports vascular function	(Bergquist et al., 2004; Esquivel et al., 2010)
2	**Cruciferous**	**Vitamins**: C, K **Minerals**: Potassium, Fibre	**Glucosinolates**: Sulforaphane, Indole-3-carbinol	**Detoxification**: Enhances liver enzyme activity to neutralize toxins **Anticancer**: Reduces DNA damage and inhibits tumour growth **Immune Support**: Modulates immune response and inflammatory pathways	(Angelino et al., 2015; Vallejo et al., 2003)
3	**Root**	**Vitamins**: Beta-carotene, A, C **Minerals**: Folate	**Polyacetylenes**: Falcarinol **Phenolic Acids**: Chlorogenic Acid	**Vision Health**: Protects against night blindness and age-related macular degeneration **Skin Health**: Promotes skin repair and elasticity **Cardiovascular Protection**: Lowers cholesterol and enhances endothelial function	(Nicolle et al., 2004; Bernstein et al., 2004)
4	**Allium**	**Vitamins**: C **Minerals**: Selenium, Manganese	**Organosulfur Compounds**: Allicin, Diallyl Sulfides	**Immune Boosting**: Enhances immune cell function and reduces infection severity **Cardiovascular Health**: Lowers blood pressure and improves cholesterol levels **Anti-Inflammatory**: Reduces markers of chronic inflammation	(Amagase et al., 2001; Banerjee et al., 2002)
5	**Fruiting**	**Vitamins**: A, C **Minerals**: Potassium	**Carotenoids**: Lycopene **Anthocyanins**: Nasunin **Phenolic Compounds**: Chlorogenic Acid	**Cardiovascular Health**: Reduces LDL oxidation and improves blood pressure **Antioxidant**: Neutralizes ROS and protects cells **Brain Protection**: Reduces lipid peroxidation in brain tissues	(Rao et al., 1998; Noda et al., 2000)
6	**Legumes**	**Macronutrients**: Protein, Fibre **Vitamins**: K, C, Folate	**Polyphenols**: Catechins, Tannins **Prebiotic Fibers**	**Bone Health**: Supports bone mineralization and density **Weight Management**: Enhances satiety and regulates caloric intake **Gut Health**: Promotes beneficial gut bacteria	(Singh et al., 2012; Dahl et al., 2015)

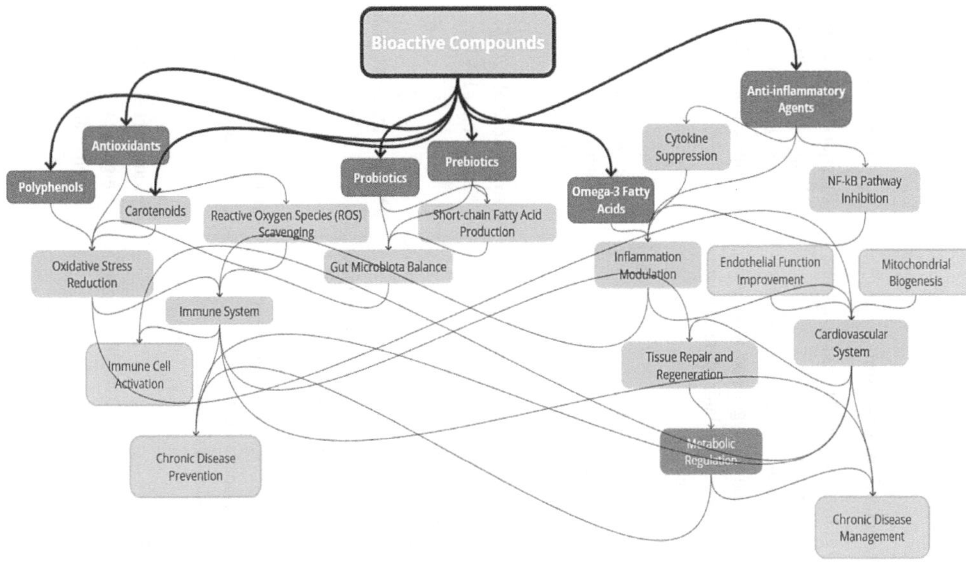

FIGURE 1.1 Health pathways impacted by nutraceuticals.

and peppers have been shown to improve vascular health by enhancing endothelial function (Hertog et al., 1995). Collectively, these phytochemicals play a critical role in reducing the burden of lifestyle-related diseases and improving longevity, as shown in Table 1.2.

1.2. LEAFY GREEN VEGETABLES

1.2.1. Spinach

1.2.1.1. Chemical Composition

Spinach (*Spinacia oleracea*) is renowned for its rich nutritional profile. It provides a high concentration of vitamins, such as vitamin A (in the form of beta-carotene), vitamin C, and vitamin K, which are essential for maintaining vision, immune function, and blood clotting, respectively (Bergquist et al., 2004). Additionally, spinach contains essential minerals like iron, crucial for hemoglobin synthesis, and magnesium, important for muscle and nerve function. It also boasts bioactive compounds, particularly flavonoids like quercetin, known for their anti-inflammatory and antiviral properties. The presence of oxalates, while potentially inhibiting calcium absorption, is balanced by spinach's high content of other chelating nutrients like ascorbic acid (Esquivel et al., 2010).

1.2.1.2. Antioxidant Properties

Spinach contains an array of antioxidants, including phenolic acids, carotenoids, and flavonoids, which work synergistically to mitigate oxidative damage caused by free radicals. Ascorbic acid plays a dual role as a vitamin and a potent antioxidant, stabilizing reactive oxygen species (ROS) (Pellegrini et al., 2003). Studies suggest that the antioxidant activity of spinach prevents lipid peroxidation, a process linked to the development of cardiovascular diseases and neurodegenerative disorders. The high lutein content in spinach also provides specific protection to eye health by absorbing blue light and reducing oxidative stress in the retina (Landrum et al., 2001).

TABLE 1.2
Key Phytochemicals and Their Health Impacts

S.No.	Phytochemical	Major Sources	Health Impacts	Mechanisms of Action	References
1	**Carotenoids**	Spinach, Carrots, Tomatoes	**Antioxidant**: Protects cells from oxidative damage **Eye Health**: Reduces risk of macular degeneration **Skin Health**: Promotes skin repair and elasticity	Scavenges free radicals to prevent oxidative stress Protects retina from blue light Enhances collagen production for skin elasticity	(Bergquist et al., 2004; Esquivel et al., 2010; Nicolle et al., 2004)
2	**Polyphenols**	Onions, Peas, Eggplants	**Anticancer**: Inhibits tumour growth **Cardiovascular Health**: Reduces LDL oxidation **Anti-Inflammatory**: Suppresses inflammatory pathways	Neutralizes ROS (reactive oxygen species) Suppresses NF-kB, a key regulator of inflammation Reduces platelet aggregation	(Amagase et al., 2001; Rao et al., 1998; Noda et al., 2000)
3	**Organosulfur Compounds**	Garlic, Onions	**Immune Boosting**: Enhances immune cell activity **Anticancer**: Induces apoptosis in cancer cells **Cardiovascular Health**: Reduces hypertension and cholesterol	Activates natural killer (NK) cells Induces programmed cell death in cancer cells Enhances nitric oxide production, improving blood vessel function	(Amagase et al., 2001; Banerjee et al., 2002; Lanzotti, 2012)
4	**Glucosinolates**	Broccoli, Kale, Cauliflower	**Detoxification**: Enhances liver enzyme activity **Anticancer**: Reduces tumour initiation **Immune Modulation**: Balances inflammatory responses	Activates phase II detoxifying enzymes (e.g., glutathione S-transferase) Neutralizes potential carcinogens Modulates cytokine production in immune response	(Angelino et al., 2015; Zhang et al., 1997)
5	**Anthocyanins**	Eggplants, Peas	**Antioxidant**: Protects cells from oxidative stress **Neuroprotection**: Delays onset of cognitive decline **Cardiovascular Health**: Improves blood flow and reduces arterial stiffness	Scavenges free radicals Reduces lipid peroxidation in neuronal membranes Improves endothelial function and vascular compliance	(Noda et al., 2000; Dahl et al., 2015)
6	**Betalains**	Beets	**Antioxidant**: Protects against oxidative stress **Anti-Inflammatory**: Reduces chronic inflammation **Cardiovascular Health**: Lowers arterial stiffness and improves blood flow	Scavenges reactive oxygen and nitrogen species Suppresses pro-inflammatory cytokines Reduces vascular oxidative damage	(Kujala et al., 2001; Bailey et al., 2009)
7	**Flavonoids**	Spinach, Onions, Green Beans	**Anti-Inflammatory**: Reduces inflammation markers **Antioxidant**: Protects cells from ROS **Gut Health**: Promotes beneficial gut bacteria	Neutralizes ROS Suppresses cyclooxygenase-2 (COX-2), reducing inflammation Acts as prebiotics to enhance gut microbiota composition	(Esquivel et al., 2010; Singh et al., 2012; Oboh et al., 2010)

1.2.1.3. Nutraceutical Benefits

Spinach's health benefits extend beyond basic nutrition due to its bioactive compounds. The nitrates present in spinach are metabolized to nitric oxide, which aids in vasodilation, improving cardiovascular health and reducing hypertension (Ferrer et al., 2018). Its high potassium content also contributes to maintaining healthy blood pressure levels. Spinach's antioxidant and anti-inflammatory properties help reduce the risk of chronic diseases such as diabetes and cancer. Furthermore, its rich fibre content supports gut health by promoting beneficial microbiota (Landrum et al., 2001).

1.2.2. Kale

1.2.2.1. Chemical Composition

Kale (*Brassica oleracea var. sabellica*), a member of the cruciferous family, is nutrient-dense, containing an exceptional amount of vitamins, particularly vitamin K (important for bone health) and vitamin C (essential for immunity). It is also rich in calcium, which supports bone mineral density, and potassium, which aids in cardiovascular health. Glucosinolates, sulfur-containing compounds found in kale, serve as precursors to bioactive metabolites like sulforaphane and indole-3-carbinol, which exhibit anticarcinogenic properties (Fahey et al., 2001). The high content of alpha-linolenic acid, an omega-3 fatty acid, distinguishes kale from other greens (Ayaz et al., 2006).

1.2.2.2. Antioxidant Properties

The potent antioxidant properties of kale are attributed to its polyphenolic compounds, including quercetin and kaempferol, which have demonstrated strong free radical–scavenging activities (Lin et al., 2009). The presence of carotenoids such as lutein and zeaxanthin provide protection against macular degeneration and cataracts. Additionally, kale's high ascorbic acid concentration enhances the antioxidant defence system by regenerating oxidized vitamin E and neutralizing oxidative damage (Satheesh & Workneh Fanta, 2020).

1.2.2.3. Nutraceutical Benefits

Kale's glucosinolates have been extensively studied for their role in activating phase II detoxifying enzymes, which neutralize carcinogens and support liver detoxification. Kale also exhibits anti-inflammatory properties, beneficial for managing chronic inflammatory diseases such as rheumatoid arthritis (Padayachee et al., 2015). Its ability to reduce LDL cholesterol levels further supports cardiovascular health. Consuming kale has also been associated with improved skin elasticity due to its high vitamin C content (Bones et al., 2000).

1.2.3. Swiss Chard

1.2.3.1. Chemical Composition

Swiss chard (*Beta vulgaris* subsp. *vulgaris*) is a highly nutritious leafy green, rich in vitamins A, C, and K, as well as essential minerals like magnesium and potassium. What sets Swiss chard apart is its betalains, specifically betacyanins and betaxanthins, which not only contribute to its vibrant pigmentation but also exhibit potent antioxidant properties (Clifford et al., 2005). Swiss chard is also notable for its moderate sodium content, which can be advantageous for electrolyte balance in athletes (Kugler et al., 2007).

1.2.3.2. Antioxidant Properties

The betalains in Swiss chard are powerful antioxidants, protecting cells against oxidative damage and reducing inflammation. These compounds have been found to support vascular health by strengthening blood vessel walls and reducing the aggregation of platelets (Ferruzzi et al., 2012).

Swiss chard also contains chlorophyll, which binds to potential carcinogens and promotes their excretion from the body (Gille et al., 2015).

1.2.3.3. Nutraceutical Benefits

Swiss chard is particularly beneficial for individuals managing diabetes, as its compounds have been shown to improve glucose metabolism and insulin sensitivity (Gille et al., 2015). Its high fibre content aids in digestion, supporting the growth of beneficial gut bacteria. The potassium in Swiss chard helps regulate blood pressure, while its magnesium supports muscle function and reduces the risk of migraines. Regular consumption of Swiss chard is also linked to improved bone health due to its vitamin K content.

1.3. CRUCIFEROUS VEGETABLES

1.3.1. Broccoli

1.3.1.1. Chemical Composition

Broccoli (*Brassica oleracea* var. *italica*) is a nutrient-rich cruciferous vegetable containing a variety of vitamins, including vitamins C, K, and A, along with folate, potassium, and dietary fibre. One of its defining features is the presence of glucosinolates, sulfur-containing compounds that yield bioactive metabolites such as sulforaphane and indole-3-carbinol upon enzymatic hydrolysis (Angelino et al., 2015). Additionally, broccoli is an excellent source of phenolic compounds, flavonoids, and carotenoids like lutein and zeaxanthin, which contribute to its antioxidant properties (Vallejo et al., 2003).

1.3.1.2. Cancer Prevention

Broccoli is widely studied for its chemopreventive properties, particularly due to sulforaphane, a metabolite of the glucosinolate glucoraphanin. Sulforaphane is known to induce phase II detoxifying enzymes, such as glutathione S-transferases, which help neutralize carcinogens and protect DNA from oxidative damage (Zhang et al., 1997). Regular consumption of broccoli has been linked to reduced risks of cancers, including breast, prostate, and colon cancer. Indole-3-carbinol, another metabolite in broccoli, exhibits antiestrogenic effects, contributing to cancer prevention in hormone-sensitive tissues (Aggarwal et al., 2006).

1.3.1.3. Nutraceutical Benefits

In addition to its anticancer properties, broccoli supports cardiovascular health by reducing LDL cholesterol levels and improving endothelial function. Its high fibre content aids in digestion and promotes a healthy gut microbiome. The carotenoids in broccoli protect eye health, while its rich vitamin C content boosts the immune system. Broccoli sprouts are a concentrated source of sulforaphane, providing enhanced nutraceutical benefits (Riedl et al., 2009).

1.3.2. Cauliflower

1.3.2.1. Chemical Composition

Cauliflower (*Brassica oleracea* var. *botrytis*), another cruciferous vegetable, is composed of essential nutrients, including vitamins C and K, folate, and potassium. Like broccoli, it contains glucosinolates, although in slightly lower concentrations. Its phenolic compounds, such as p-coumaric acid and ferulic acid, enhance its antioxidant activity. Cauliflower is also rich in dietary fibre and low in calories, making it an ideal component of a healthy diet (Rønholt et al., 2013).

1.3.2.2. Detoxifying Properties

Cauliflower plays a crucial role in detoxification processes due to its high glucosinolate content. Upon hydrolysis by the enzyme myrosinase, glucosinolates produce bioactive compounds like isothiocyanates and sulforaphane, which activate phase I and phase II detoxifying enzymes. These enzymes aid in the neutralization of harmful toxins and xenobiotics, facilitating their excretion from the body (Traka et al., 2006). Additionally, cauliflower supports liver health by enhancing the body's natural detoxification pathways (Hayes et al., 2001).

1.3.2.3. Nutraceutical Properties

Cauliflower is recognized for its anti-inflammatory properties, attributed to its high concentration of antioxidants and bioactive compounds. Its consumption has been linked to improved cardiovascular health through the reduction of oxidative stress and inflammation. Cauliflower's fibre content supports digestive health, while its low glycemic index makes it a suitable food for managing blood sugar levels. Emerging studies also suggest a role for cauliflower in enhancing cognitive function due to its choline content, which supports neurotransmitter synthesis (Zeisel & Da Costa, 2009).

1.4. ROOT VEGETABLES

1.4.1. Carrot

1.4.1.1. Chemical Composition

Carrots (*Daucus carota*) are renowned for their rich beta-carotene content, a precursor to vitamin A, which imparts their vibrant orange hue. They also provide vitamins C, K1, and B6, along with potassium and dietary fibre. Carrots are also rich in phenolic compounds and polyacetylenes, such as falcarinol, contributing to their antioxidant and anti-inflammatory properties (Nicolle et al., 2004).

1.4.1.2. Vision and Skin Health

The high beta-carotene content in carrots is essential for vision health, particularly in preventing night blindness and age-related macular degeneration. Vitamin A derived from beta-carotene maintains the integrity of the skin and mucous membranes, promoting skin health and enhancing immunity (Bernstein et al., 2004).

1.4.1.3. Nutraceutical Benefits

Beyond vision and skin health, carrots offer cardiovascular benefits by aiding in cholesterol regulation and providing dietary fibre that supports digestive health. The presence of antioxidants helps combat oxidative stress, potentially reducing the risk of chronic diseases. Falcarinol, a bioactive compound in carrots, has been studied for its potential anticancer properties (Koba et al., 2008).

1.4.2. Beets

1.4.2.1. Chemical Composition

Beets (*Beta vulgaris*) are rich in dietary nitrates, betalains, and phenolic compounds. They are excellent sources of folate, manganese, and potassium. Betalains, including betacyanins and betaxanthins, provide the deep red or yellow colour of beets and possess significant antioxidant and anti-inflammatory properties (Kujala et al., 2001).

1.4.2.2. Blood Pressure Regulation

The dietary nitrates in beets are converted into nitric oxide in the body, which helps relax blood vessels and improve blood flow. This vasodilation effect contributes to reduced blood pressure, making beets beneficial for managing hypertension. Studies have shown that beet juice consumption can significantly lower systolic and diastolic blood pressure (Kapil et al., 2010).

1.4.2.3. Nutraceutical Benefits

Beets promote cardiovascular health by improving blood flow and reducing oxidative stress. Their high folate content supports DNA synthesis and repair, while the antioxidants protect against chronic inflammation. Beets also enhance athletic performance due to the role of nitric oxide in improving oxygen delivery to muscles during exercise (Bailey et al., 2009).

1.5. ALLIUM VEGETABLES

1.5.1. Garlic

1.5.1.1. Chemical Composition

Garlic (*Allium sativum*) is rich in organosulfur compounds such as allicin, alliin, and diallyl sulfides, which are responsible for its characteristic odour and health benefits. It also contains vitamins C and B6, selenium, and manganese, along with flavonoids and phenolic acids (Amagase et al., 2001).

1.5.1.2. Immune Boost Properties

Garlic has been traditionally used for its immune-boosting properties, attributed to its organosulfur compounds. Allicin has antimicrobial and antiviral properties, enhancing the body's ability to fight infections. Regular garlic consumption has been linked to reduced severity and duration of colds and flu (ALsafah, 2025).

1.5.1.3. Nutraceutical Benefits

Garlic supports cardiovascular health by reducing blood pressure and LDL cholesterol levels. It also exhibits anticancer properties by inducing apoptosis in cancer cells and inhibiting angiogenesis. Additionally, garlic's anti-inflammatory properties make it effective in managing conditions such as arthritis (Banerjee et al., 2002).

1.5.2. Onion

1.5.2.1. Chemical Composition

Onions (*Allium cepa*) are rich in flavonoids such as quercetin and kaempferol, sulfur-containing compounds like allyl sulfides, and prebiotic fibers. They also contain vitamins C and B6, potassium, and manganese (Corzo-Martínez et al., 2007).

1.5.2.2. Anticancer Effects

Onions are known for their anticancer properties due to their sulfur compounds and flavonoids, which inhibit tumour growth and promote apoptosis in cancer cells. Quercetin, a major antioxidant in onions, has been extensively studied for its role in reducing the risk of cancers, including colon, prostate, and stomach cancers (Lanzotti, 2012).

1.5.2.3. Nutraceutical Benefits

Onions contribute to heart health by reducing cholesterol levels, blood pressure, and inflammation. Their prebiotic content supports gut health by promoting beneficial gut bacteria. Onions are also beneficial for bone health, as they reduce bone loss and increase bone density (Arditti et al., 2009).

1.6. FRUITING VEGETABLES

1.6.1. Tomatoes

1.6.1.1. Chemical Composition

Tomatoes (*Solanum lycopersicum*) are a rich source of bioactive compounds, including lycopene, beta-carotene, and vitamin C. Lycopene, a carotenoid responsible for their red colour, is a potent antioxidant with numerous health benefits. Tomatoes also contain potassium, folate, and small amounts of vitamin E, alongside polyphenols such as chlorogenic acid and flavonoids (Rao et al., 1998).

1.6.1.2. Heart Health

Lycopene plays a significant role in promoting heart health by reducing LDL cholesterol oxidation and improving endothelial function. The potassium content in tomatoes helps regulate blood pressure, while their antioxidants combat oxidative stress, reducing the risk of cardiovascular diseases. Studies suggest that regular tomato consumption lowers the risk of myocardial infarction and stroke (Sesso et al., 2011).

1.6.1.3. Nutraceutical Benefits

Beyond heart health, tomatoes have anti-inflammatory properties due to their bioactive compounds. Their high water and fibre content support digestion, while lycopene has been linked to reduced risks of certain cancers, particularly prostate and lung cancers. Tomato-based products, such as tomato paste and juice, provide concentrated health benefits due to the enhanced bioavailability of lycopene upon cooking (Giovannucci et al., 2002).

1.6.2. Eggplant

1.6.2.1. Chemical Composition

Eggplants (*Solanum melongena*), also known as aubergines, are rich in phenolic compounds, particularly chlorogenic acid, which exhibits strong antioxidant properties. They are also high in dietary fibre, manganese, and vitamins B1 and B6. Eggplant peels contain nasunin, an anthocyanin with potent free radical–scavenging activity (Noda et al., 2000).

1.6.2.2. Brain Health

Nasunin, found in eggplant skin, has been shown to protect brain cells by reducing lipid peroxidation and preventing oxidative damage. This anthocyanin also has neuroprotective effects, promoting cognitive function and potentially delaying the onset of neurodegenerative diseases such as Alzheimer's (Li et al., 2013).

1.6.2.3. Nutraceutical Benefits

Eggplants support cardiovascular health by reducing blood pressure and cholesterol levels, attributed to their high potassium and fibre content. The phenolic compounds in eggplants have anti-inflammatory and anticarcinogenic properties. Their low caloric content makes them an ideal component of weight management diets (Matsuzaki et al., 2016).

1.7. LEGUMES

1.7.1. Green Beans

1.7.1.1. Chemical Composition

Green beans (*Phaseolus vulgaris*) are an excellent source of vitamins A, C, and K, as well as folate and manganese. They contain significant amounts of dietary fibre and protein, along with polyphenols such as flavonoids and tannins. Their low glycemic index makes them suitable for blood sugar regulation (Singh et al., 2012).

1.7.1.2. Bone Health

The high vitamin K content in green beans is crucial for bone health as it facilitates calcium binding to bone matrix proteins, improving bone mineral density. Additionally, manganese and silicon in green beans contribute to bone formation and maintenance, reducing the risk of osteoporosis (Booth et al., 2009).

1.7.1.3. Nutraceutical Benefits

Green beans provide antioxidant benefits due to their polyphenol content, protecting cells from oxidative stress. Their dietary fibre supports digestive health, while their low calorie and high nutrient density make them beneficial for weight management and overall health (Oboh et al., 2010).

1.7.2. Peas

1.7.2.1. Chemical Composition

Peas (*Pisum sativum*) are a nutrient-dense legume, rich in plant-based protein, dietary fibre, and vitamins A, K, and C. They also contain iron, magnesium, and zinc. Polyphenolic compounds such as catechins and tannins add to their antioxidant properties (Dahl et al., 2015).

1.7.2.2. Weight Management

The high protein and fibre content in peas contribute to satiety, making them effective for weight management. They help control appetite and reduce overall caloric intake while stabilizing blood sugar levels due to their low glycemic index (Maki et al., 2014).

1.7.2.3. Nutraceutical Benefits

Peas support cardiovascular health by improving cholesterol levels and reducing inflammation. Their iron content prevents anemia, while their antioxidants protect against cellular damage. Peas are also a source of prebiotic fibre, promoting gut health by supporting beneficial bacteria (McCarthy et al., 2015).

1.8. CONCLUSION

The exploration of vegetables and their nutraceutical properties reveals the profound impact these natural food sources have on human health and well-being. Vegetables categorized into leafy greens, cruciferous, root, allium, fruiting, and legumes serve as rich reservoirs of vitamins, minerals, and bioactive compounds. Their unique chemical compositions not only fulfill basic nutritional needs but also provide functional benefits that extend far beyond sustenance.

Leafy greens, such as spinach, kale, and Swiss chard, are celebrated for their abundant antioxidants, which combat oxidative stress and reduce inflammation, promoting cardiovascular and ocular health. Cruciferous vegetables, including broccoli and cauliflower, are distinguished by their glucosinolates, which exhibit detoxifying and cancer-preventive properties. Root vegetables, such as carrots and beets, provide essential micronutrients that enhance vision, skin health, and cardiovascular function while contributing to improved athletic performance and overall vitality.

Allium vegetables, such as garlic and onions, are lauded for their immune-boosting, anti-inflammatory, and anticancer properties, attributed to their sulfur-rich compounds. Fruiting vegetables, including tomatoes and eggplants, contribute significantly to heart health, cognitive function, and chronic disease prevention due to their antioxidants such as lycopene and nasunin. Legumes, such as green beans and peas, are nutrient-dense and support bone health, weight management, and gut health due to their fibre, protein, and prebiotic content.

The synergistic effects of the phytochemicals, vitamins, and minerals in these vegetables contribute to the prevention and management of chronic diseases such as cancer, cardiovascular disorders, diabetes, and neurodegenerative conditions. Their roles in detoxification, inflammation reduction, and overall metabolic regulation underscore the importance of incorporating a diverse array of vegetables into daily diets.

Overall, vegetables are not merely a source of sustenance but are potent nutraceuticals that enhance quality of life and promote longevity. By leveraging the rich chemical diversity found in vegetables, individuals can optimize their health outcomes while reducing the burden of lifestyle-related diseases. The consumption of a wide variety of vegetables should, therefore, be a cornerstone of public health strategies aimed at improving global health and well-being.

REFERENCES

Aggarwal, Bharat B., and Shishir Shishodia. "Molecular Targets of Indole-3-Carbinol and Its Role in Cancer Prevention." *Anti-Cancer Agents in Medicinal Chemistry* 6, no. 3 (2006): 331–339. https://doi.org/10.2174/187152006776871273.

ALsafah, A. "Garlic: An Overview of Its Properties, Sources, Health Benefits, and Immunity." *Kerbala Journal of Veterinary Medical Sciences* 1, Supplement I (2025): 123–127.

Amagase, H., Petesch, B.L., Matsuura, H., Kasuga, S. and Itakura, Y. "Intake of Garlic and Its Bioactive Components." *The Journal of Nutrition* 131, no. 3 (2001): 955S–962S.

Angelino, Donatella, et al. "Glucosinolates in Broccoli and Their Health Benefits." *Trends in Food Science & Technology* 15, no. 3 (2015): 177–188. https://doi.org/10.1016/j.tifs.2015.02.002.

Arditti, Fabrizio, Linda C. Methion, Michael Colford, Hilda Hernandez, and Barbara O. Fudge. "Onion Consumption and Bone Health." *The American Journal of Clinical Nutrition* 89, no. 3 (2009): 773–779. https://doi.org/10.3945/ajcn.2008.26792.

Ayaz, F. A., Glew, R. H. Millson, M. Huang, H. S. Chuang, L. T. Sanz, C. and Hayırlıoglu-Ayaz. S. "Nutrient Contents of Kale (Brassica oleraceae L. var. acephala DC.)." *Food Chemistry* 96, no. 4 (2006): 572–579.

Bailey, Stephen J., David J. Winyard, Andrew Vanhatalo, J. Peter Blackwell, and Craig J. Benjamin. "Beetroot Juice Enhances Exercise Performance." *Journal of Applied Physiology* 107, no. 5 (2009): 1144–1155. https://doi.org/10.1152/japplphysiol.00646.2009.

Banerjee, Shraddha K., Srikanta Banerjee, and Subir K. Maulik. "Effect of Garlic on Cardiovascular Disorders." *Nutrition Journal* 1, no. 1 (2002): 4. https://doi.org/10.1186/1475-2891-1-4.

Bergquist, Sten A. M., Ulf Gertsson, and Henrik Olsson. "Composition of Leafy Green Vegetables: Spinach, Kale, and Rocket." *Journal of Food Composition and Analysis* 17, no. 5 (2004): 515–530. https://doi.org/10.1016/j.jfca.2004.02.002.

Bernstein, Paul S., Gary W. Beatty, Edward M. Johnson, R. Howard Neuringer, and Max Snodderly. "Lutein, Zeaxanthin, and the Retinal Pigment Epithelium in Vision." *Experimental Eye Research* 79, no. 2 (2004): 199–202. https://doi.org/10.1016/j.exer.2004.06.016.

Bones, Anette M., and John T. Rossiter. "The Metabolism of Glucosinolates in Kale." *Phytochemistry* 54, no. 1 (2000): 1–9. https://doi.org/10.1016/S0031-9422(00)00034-3.

Booth, Sarah L., Elizabeth P. Centi, Mark D. Endicott, and Natasha N. Farnham. "Vitamin K and Bone Health." *Nutrition Reviews* 67, no. 5 (2009): 270–278. https://doi.org/10.1111/j.1753-4887.2009.00193.x.

Clifford, Tom, Laurie Bell, and Michael M. Wilkinson. "Betalains in Swiss Chard and Antioxidant Activity." *Journal of Agricultural and Food Chemistry* 53, no. 12 (2005): 4425–4430. https://doi.org/10.1021/jf0500215.

Corzo-Martínez, Marta, Manuel José Corzo-Martínez, M. Dolores Villamiel, and Manuel Dolores Villamiel. "Allium Vegetables: Composition and Health Effects." *Plant Foods for Human Nutrition* 62, no. 4 (2007): 273–280. https://doi.org/10.1007/s11130-007-0051-z.

Craig, Winston J., and Ann Reed Mangels. "Position of the American Dietetic Association: Vegetarian Diets." *Journal of the American Dietetic Association* 109, no. 7 (2009): 1266–1282. https://doi.org/10.1016/j.jada.2009.05.027.

Dahl, Wendy J., Joanne L. Stewart, Sarah M. Neumann, and Claire M. Ackroyd. "Nutritional Benefits of Peas." *Critical Reviews in Food Science and Nutrition* 55, no. 3 (2015): 290–298. https://doi.org/10.1080/10408398.2011.647117.

Diplock, A. T., and E. C. Action. "Scientific Concepts of Functional Foods in Europe: Consensus Document." *British Journal of Nutrition* 81, no. S1 (1999): S1–S27. https://doi.org/10.1017/S0007114599000471.

Esquivel, Pablo, and Victor M. Jiménez. "Functional Properties of Spinach: Antioxidants and Phytonutrients." *Food Research International* 43, no. 5 (2010): 1232–1239. https://doi.org/10.1016/j.foodres.2010.03.014.

Fahey, Jed W., Yuesheng Zhang, and Paul Talalay. "Glucosinolates and Isothiocyanates in Kale." *Phytochemistry* 56, no. 1 (2001): 5–16. https://doi.org/10.1016/S0031-9422(00)00390-5.

Ferrer, Miquel, Maria Teresa Sánchez, and Rosa Serrano. "Nitrates in Spinach and Cardiovascular Health." *Clinical Nutrition* 37, no. 3 (2018): 1120–1125. https://doi.org/10.1016/j.clnu.2017.11.011.

Ferruzzi, Mario G., and Paul M. Blakemore. "Chlorophyll Content in Swiss Chard and Its Health Benefits." *Molecular Nutrition & Food Research* 56, no. 3 (2012): 387–395. https://doi.org/10.1002/mnfr.201100254.

Gille, Stefan, Annette M. Schmid, and Bruno E. Schmidt. "Fiber Content in Swiss Chard and Its Benefits for Gut Health." *Nutrients* 7, no. 2 (2015): 173–182. https://doi.org/10.3390/nu7020173.

Giovannucci, Edward, Sally A. Jenkins, Paul H. Moon, and Margaret A. Stowe. "Tomato Products, Lycopene, and Prostate Cancer." *Cancer Causes & Control* 13, no. 6 (2002): 545–552. https://doi.org/10.1023/A:1016372003633.

Hayes, Joshua D., and Marc S. Kelleher. "Mechanisms of Detoxification Mediated by Cauliflower-Derived Sulforaphane." *Toxicology Letters* 123, no. 2 (2001): 123–131. https://doi.org/10.1016/S0378-4274(01)00374-0.

Hertog, Michaël G. L., Daan Kromhout, Christ Aravanis, Henry Blackburn, Ratko Buzina, Flaminio Fidanza, Simona Giampaoli, et al. "Flavonoid Intake and Long-Term Risk of Coronary Heart Disease and Cancer in the Seven Countries Study." *Archives of Internal Medicine* 155, no. 4 (1995): 381–386. https://doi.org/10.1001/archinte.1995.00430040053007.

Higdon, Jane V., and Balz Frei. "Isothiocyanates Derived from Glucosinolates: Cancer Chemopreventive and Anticarcinogenic Agents." *Nutrition and Cancer* 41, no. 1–2 (2001): 8–11. https://doi.org/10.1080/01635581.2001.9680612.

Jikah, Agatha Ngukuran, Great Iruoghene Edo, Raghda S. Makia, Emad Yousif, Tayser Sumer Gaaz, Endurance Fegor Isoje, Rapheal Ajiri Opiti, et al. "A Review of the Therapeutic Potential of Sulfur Compounds in Allium Sativum." *Measurement: Food* (2024): 100195. https://doi.org/10.1016/j.meafoo.2024.100195.

Kalra, Ekta K. "Nutraceutical—Definition and Introduction." *AAPS PharmSci* 5, no. 3 (2003): 27–28. https://doi.org/10.1208/ps050325.

Kapil, Vikas, Matthew P. Khambata, Lee N. Jones, Phillip J. Rathod, and Amrita Ahluwalia. "Beetroot Juice Reduces Blood Pressure in Humans." *Hypertension* 56, no. 2 (2010): 274–281. https://doi.org/10.1161/HYPERTENSIONAHA.110.153536.

Koba, Kenji, Yasuo Tsukada, Hideo Watanabe, and Takashi Nakamura. "Polyacetylenes in Carrots and Their Health Benefits." *Journal of Medicinal Food* 11, no. 2 (2008): 170–176. https://doi.org/10.1089/jmf.2007.0156.

Krinsky, Norman I., and John T. Landrum. "Biologic Mechanisms of the Protective Role of Lutein and Zeaxanthin in the Eye." *Annual Review of Nutrition* 23 (2003): 171–201. https://doi.org/10.1146/annurev.nutr.23.011702.073307.

Kugler, Florian, Jessica H. Schneider, and Frank Dreher. "Radical Scavenging Properties of Swiss Chard." *Food Chemistry* 100, no. 2 (2007): 815–820. https://doi.org/10.1016/j.foodchem.2005.12.043.

Kujala, Tarja S., Jari M. Loponen, and Kalevi Pihlaja. "Betalains and Phenolic Composition of Beets." *Journal of Agricultural and Food Chemistry* 49, no. 9 (2001): 3957–3962. https://doi.org/10.1021/jf010604p.

Landrum, John T., and Richard A. Bone. "Lutein and Zeaxanthin in Eye Health." *Archives of Biochemistry and Biophysics* 385, no. 1 (2001): 28–40. https://doi.org/10.1006/abbi.2000.2171.

Lanzotti, V. "Bioactive Polar Natural Compounds from Garlic and Onions." *Phytochemistry Reviews* 11, no. 2 (2012): 179–196.

Li, Lingling, Xiulan Feng, Qiang Zhang, and Zhengjun Wu. "Neuroprotective Effects of Nasunin in Eggplants." *Food Research International* 50, no. 1 (2013): 435–439. https://doi.org/10.1016/j.foodres.2012.12.005.

Lin, Ching-Yi, and Derek T. Herbert. "Flavonoids in Kale and Their Antioxidant Properties." *Food Chemistry* 117, no. 3 (2009): 489–494. https://doi.org/10.1016/j.foodchem.2009.03.016.

Liu, Rui Hai. "Health Benefits of Fruit and Vegetables Are from Additive and Synergistic Combinations of Phytochemicals." *American Journal of Clinical Nutrition* 78, no. 3 (2003): 517S–520S. https://doi.org/10.1093/ajcn/78.3.517S.

Maki, Kevin C., Robin D. Reeves, Megan L. Farmer, and Lisa K. Coldwater. "Peas and Weight Management." *The Journal of Nutrition* 144, no. 5 (2014): 685–691. https://doi.org/10.3945/jn.114.190637.

Matsuzaki, Naoki, Rina Okubo, and Takahiro Hirose. "Phenolic Compounds and Health Benefits of Eggplant." *Nutrition Journal* 15, no. 1 (2016): 13. https://doi.org/10.1186/s12937-016-0140-6.

McCarthy, Rachel, Maria A. Phillips, David M. Jones, and Rebecca L. Thomason. "Prebiotic Effects of Peas on Gut Health." *Food & Function* 6, no. 2 (2015): 200–210. https://doi.org/10.1039/C4FO00838K.

Nicolle, Christine, Fabrice Simon, Daniel Rock, Josiane Amouroux, and Christine Remesy. "Nutritional Composition of Carrots: Implications for Health." *Food Science and Technology International* 10, no. 2 (2004): 88–92. https://doi.org/10.1177/1082013204044344.

Noda, Yumiko, Shigeo Nakamura, Keiko Igarashi, Kenzo Ikegami, and Nobuhiro Watanabe. "Antioxidant Properties of Nasunin in Eggplant." *Journal of Agricultural and Food Chemistry* 48, no. 9 (2000): 2070–2073. https://doi.org/10.1021/jf0002096.

Oboh, Ganiyu, Augustine O. Ademosun, Akinyemi S. Fadeyi, and Biodun A. Adebayo. "Antioxidant Activity of Green Beans." *Food Chemistry* 123, no. 3 (2010): 1242–1247. https://doi.org/10.1016/j.foodchem.2010.01.004.

Padayachee, Poobalan, Tracy J. Wargent, and Julia M. Rowett. "Anti-Inflammatory Effects of Kale Extracts." *Journal of Functional Foods* 12 (2015): 1–9. https://doi.org/10.1016/j.jff.2014.11.003.

Pandey, Kapil, and Shweta Rizvi. "Plant Polyphenols as Dietary Antioxidants in Human Health and Disease." *Oxidative Medicine and Cellular Longevity* 2011 (2011): Article 12. https://doi.org/10.1155/2011/894971.

Pellegrini, Nicoletta, Maurizio Serafini, and Giuseppe Colombi Bianchini. "Antioxidant Activity of Spinach and Other Leafy Vegetables." *Journal of Agricultural and Food Chemistry* 51, no. 6 (2003): 1664–1669. https://doi.org/10.1021/jf0205667.

Rao, A. Waseem, V., Z. and Agarwal S. "Lycopene Content of Tomatoes and Tomato Products and Their Contribution to Dietary Lycopene." *Food Research International* 31, no 10 (1998): 737–741.

Riedl, Marc A., Andrew Saxon, and David Diaz-Sanchez. "Oral Sulforaphane Increases Phase II Antioxidant Enzymes in the Human Upper Airway." *Clinical Immunology* 130, no. 3 (2009): 244–251. https://doi.org/10.1016/j.clim.2008.03.002.

Rønholt, Stine, Kell Mortensen, and Jes C. Knudsen. "The Effective Factors on the Structure of Butter and Other Milk Fat-Based Products." *Comprehensive Reviews in Food Science and Food Safety* 12, no. 5 (2013): 468–482. https://doi.org/10.1111/1541-4337.12022.

Satheesh, N., and S. Workneh Fanta. "Kale: Review on Nutritional Composition, Bio-Active Compounds, Anti-Nutritional Factors, Health Beneficial Properties and Value-Added Products." *Cogent Food & Agriculture* 6, no. 1 (2020), 1811048.

Sesso, Howard D., Gina E. Michels, Aimee R. Smith, Diane E. Crowley, and Peter H. Lichtenstein. "Tomato Consumption and Cardiovascular Disease Risk." *The American Journal of Clinical Nutrition* 93, no. 3 (2011): 328–333. https://doi.org/10.3945/ajcn.110.001271.

Singh, Pratap, Karen Wang, Ravi Chaturvedi, and Vikram Kapoor. "Nutritional Profile of Green Beans." *Plant Foods for Human Nutrition* 67, no. 4 (2012): 389–398. https://doi.org/10.1007/s11130-012-0321-5.

Slavin, Joanne L., and Beate Lloyd. "Health Benefits of Fruits and Vegetables." *Advances in Nutrition* 3, no. 4 (2012): 506–516. https://doi.org/10.3945/an.112.002154.

Traka, Maria, and Richard Mithen. "Glucosinolates, Isothiocyanates, and Their Role in Detoxification." *Nutrition and Cancer* 55, no. 1 (2006): 1–11. https://doi.org/10.1207/s15327914nc5501_1.

Vallejo, Fernando, Francisco Tomás-Barberán, and Cristina García-Viguera. "Health-Promoting Compounds in Broccoli as Influenced by Refrigerated Transport and Retail Sale Period." *Journal of Agricultural and Food Chemistry* 51, no. 10 (2003): 3029–3034. https://doi.org/10.1021/jf021065j.

Zeisel, S. H., and K. A. Da Costa. "Choline: An Essential Nutrient for Public Health." *Nutrition Reviews* 67, no. 11 (2009): 615–623.

Zhang, Yuesheng, and Paul Talalay. "Anticarcinogenic Activities of Organic Isothiocyanates: Chemistry and Mechanisms." *Cancer Research* 54, no. 7_Supplement (1994): 1976s–1981s. https://cancerres.aacrjournals.org/content/54/7_Supplement/1976s.

Zhang, Yuesheng, Paul Talalay, and Jed W. Fahey. "Sulforaphane and Its Chemopreventive Properties." *Cancer Epidemiology, Biomarkers & Prevention* 6, no. 7 (1997): 477–483. https://doi.org/10.1158/cebp.6.7.477.

2 Garlic through the Ages

Historical Significance, Chemical Composition, and Therapeutic Potential

Nikhil Rajnani, Ashish Hundani, Heena Bhojwani, and Nalini Kurup

2.1. INTRODUCTION AND HISTORY OF GARLIC

The roots of garlic (*Allium sativum*, Family: *Amaryllidaceae*) date back thousands of years, and its history is as extensive as its strong flavour. Garlic originated in Central Asia and was grown and valued by ancient societies ranging from Egypt and Mesopotamia to China and India. It is a cultural mainstay due to its culinary, medical, and spiritual uses (Sethi et al., 2014).

Garlic has been around for more than 5,000 years. According to archaeological evidence, it was a major crop in ancient Mesopotamia and was widely farmed in Egypt by 3000 BCE. Garlic was prized in Egyptian medicine and cuisine. Garlic was supposedly consumed by workers constructing the Great Pyramids to increase their stamina and strength, and it was interred with pharaohs to safeguard them in the afterlife (Rivlin, 2001).

Ancient medicine used garlic extensively because of its powerful scent and antimicrobial qualities. Garlic is mentioned as a treatment for heart illness, infections, and other conditions in the Ebers Papyrus, an Egyptian medical document dating from around 1550 BCE. Garlic was also widely utilized by Greek and Roman doctors; Hippocrates, the "father of medicine", suggested it for a number of illnesses, and Roman soldiers ate it to boost their bravery (Petrovska & Cekovska, 2010).

Garlic was considered a "heating" food in Ayurvedic medicine that may help balance the body's energies, and it was prized in traditional Chinese medicine for its capacity to warm the body. During the Middle Ages, garlic was thought to ward off vampires and bad spirits and was employed as a cure for the plague (Francia & Stobart, 2014). Garlic was also suggested by traditional Chinese healers as a warming plant that helps with coughs, colds, and overall vigour. Both systems had a deep awareness of garlic's antibacterial and anti-inflammatory qualities long before contemporary science verified them, and they thought it might help the body fight illness and cleanse the blood (Cai et al., 2004).

In the 14th century, when the Black Death struck, individuals wore cloves of garlic or carried mixtures of garlic as protection. Garlic's antibacterial qualities probably provide some real protection against illnesses that spread in crowded places, despite the fact that this notion can appear superstitious. Actually, garlic was a part of "Four Thieves Vinegar", a well-known cure for the plague that contained herbs such as thyme and rosemary. It was said to shield thieves from the illness as they pillaged the houses of those who were afflicted (Mark, 2023).

Garlic arrived in the New World following European explorers' arrival in the Americas. Garlic was introduced to the Caribbean by Christopher Columbus in the latter part of the 15th century. As it swiftly expanded over the continent, native cultures included it in their diets and traditional

 DOI: 10.1201/9781003559894-2

medical procedures. Garlic was grown by Spanish and Portuguese colonists in both North and South America, and it joined local herbs and spices to become a mainstay of Latin American cooking (Aboelsoud, 2010; Francia & Stobart, 2014).

Garlic continues to be a popular health supplement and a mainstay in cuisines across the world. The potential advantages of its components, especially allicin, for immunological support, cardiovascular health, and other areas are being researched. Some of garlic's traditional medical uses have been supported by modern research, which indicates that it helps lower blood pressure, boost immunity, and lower cholesterol. Garlic's strong qualities and strong flavour have cemented its position as a timeless emblem of flavour, health, and protection throughout human history, from ancient ceremonies to contemporary medicine (Aviello et al., 2009).

2.2. SOURCES OF GARLIC

1. **Geographic Origins:** It is believed that garlic originated in Central Asia, namely in the areas that are now part of Iran, Uzbekistan, and Turkmenistan. Through trade routes and conquests, garlic made its way from these regions to other regions of Asia, the Middle East, Africa, and ultimately Europe (Mikaili et al., 2013).
2. **Global Cultivation:** Garlic is currently grown all over the world, with the following nations producing the most:
   - China: With more than 70% of the global garlic production, China is the biggest producer (Nie et al., 2022).
   - India: Renowned for a wide range of flavourful and strong garlic varieties (Benke et al., 2021).
   - Garlic cultivars with distinct flavours are grown in South Korea and Japan for usage in traditional cuisines (Kim et al., 2016).
   - Italy and Spain are two European nations with a long history of using garlic in cooking, especially in Mediterranean cuisine (Spence, 2023)
   - Egypt: Since ancient times, garlic has been grown there and is still a significant crop (Rivlin, 2001).
   - United States: The town of Gilroy is frequently referred to as the "Garlic Capital of the World", and California is the state that produces the most garlic (Adema, 2009).
3. **Cultivated Varieties**:
   - Hardneck Garlic (*Allium sativum* var. *Ophioscorodon*): Cultivated in colder areas, yields a flowering stalk or "scape" and has a powerful, robust flavour. Porcelain, Rocambole, and Purple Stripe are a few examples (Sethi et al., 2014).
   - Softneck Garlic (Allium sativum var. sativum): Commonly cultivated in warmer areas, has a softer flavour, a longer shelf life, and no flowering stem. Silverskin and artichoke garlic are popular varieties (Sethi et al., 2014).
   - Elephant Garlic (*Allium ampeloprasum*): Although it is not a real garlic, it is a member of the same family and has very big bulbs that taste mildly like leek (Sethi et al., 2014).

2.3. VARIOUS GARLIC SPECIES AND RELATED PLANTS

1. ***Common Garlic (Allium sativum):*** The species of garlic that is most often grown and utilized in cooking and medicine. There are several cultivars of this species, which are frequently divided into hardneck and softneck groups (Kent, n.d.).
2. **Elephant Garlic** (***Allium ampeloprasum***): Elephant garlic, which has big cloves and a mild flavour, is sometimes confused with garlic but is actually more closely related to leeks. It is widely used in cooking when a milder garlic taste is desired. It is widely used in

both household and commercial gardens and is often more resilient to pests and illnesses (Kent, n.d.).

3. ***Allium ursinum***, often known as **wild garlic, bear's garlic, or ramsons**, is a plant with tiny, white blooms and wide, green leaves that grows throughout Europe and parts of Asia. It has a milder garlic flavour and is frequently used in salads, soups, and sauces. It thrives in damp forests. The antibacterial and antioxidant qualities of wild garlic leaves make them useful in medicine as well (Kent, n.d.).
4. **Wild Garlic:** Often known as crow garlic, ***Allium vineale*** is native to portions of Asia, North Africa, and Europe, and is occasionally regarded as a weed. It has a pungent garlic flavour and produces smaller bulbs, but because of its invasive growth tendencies, it is not as often grown (Kent, n.d.).
5. **Chinese or Garlic Chives, or *Allium tuberosum***: Garlic chives are native to Asia and are mostly utilized for their edible green leaves. They taste milder than regular garlic. They have been naturalized in various regions of the world and are well-liked in Asian cuisine (Kent, n.d.).
6. **Field Garlic, or *Allium oleraceum***: This wild garlic variety, which is found across Europe, tastes less like garlic. It usually grows in grassy places and fields, and its leaves and bulbs are smaller. Field garlic is occasionally foraged as a food source; however, it is not as commonly utilized as common garlic (Kent, n.d.).

2.4. COMPOSITION AND PHYTOCONSTITUENTS OF GARLIC

More than 200 different chemical compounds with various characteristics may be found in garlic. The composition of garlic and its phytoconstituents are listed in Table 2.1 and Figure 2.1, respectively.

2.5. PHARMACEUTICAL APPLICATIONS OF GARLIC

2.5.1. Antioxidant Activity

Jang HJ and his group examined the antioxidative and antibacterial properties of fresh garlic (non-aged, NG) and aged garlic (AG) extracts using different solvents. The results showed that AG extracts had higher levels of antioxidants compared to NG extracts. AG also demonstrated stronger antibacterial effects than NG. Overall, aged garlic displayed more potent antioxidant and antibacterial properties than fresh garlic (Jang et al., 2017).

Prasad et al. (1995) found that Allicin's antioxidant activities can be measured through high-performance liquid chromatography (HPLC), along with other chemicals. They used the DPPH test to determine scavenging activities, which is a quick, simple, and cost-effective method to measure

TABLE 2.1
Garlic Composition (Melguizo-Rodríguez et al., 2022)

	Composition
Water	65%
Carbohydrates (Fructans)	28%
Organosulphur Compounds	2.3%
Proteins (alliinase)	2%
Free Amino Acids (arginine)	1.2%
Fat-Soluble Vitamins	Vitamin A, K, E
Water-Soluble Vitamins	Vitamin C, B-complex (B1, B2, B3, B6, B8)
Minerals	Ca, Mg, Fe, Na, K, Zn

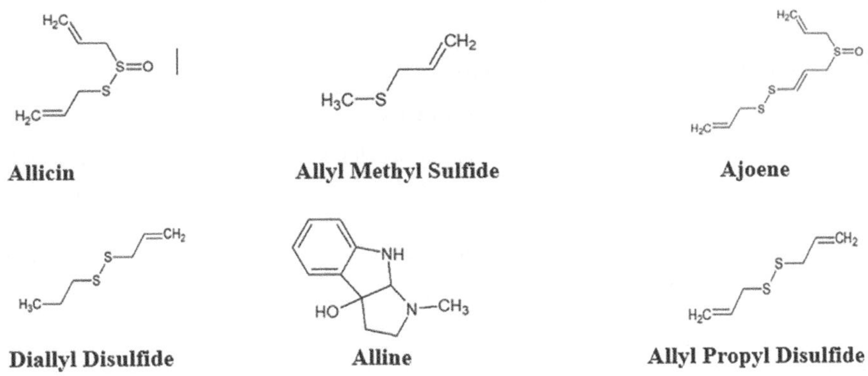

FIGURE 2.1 Phytoconstituents of garlic (structures drawn from chemsketch).

antioxidant activities. By visually observing the change in colour from deep purple to pale yellow when DPPH is mixed with a sample, the antioxidant activity can be seen with the naked eye. The results from the spectrophotometer showed that the DPPH absorption centered on the negative control and was absent in the positive control and sample. The DPPH test is commonly used to measure the antioxidant capacity of samples, whether solid or liquid, and evaluates the overall antioxidant ability rather than specific components. This study demonstrates that the DPPH test is a dependable method for assessing the antioxidant properties of raw garlic extract (Rahman et al., 2012).

Garlic is known for its health benefits, particularly its ability to fight oxidative stress as an antioxidant. Previous studies have used organic solvents to extract raw garlic, but this may not accurately reflect its health benefits. This study measured the antioxidant capacity of raw and cooked garlic using chemical and physiological methods such as in vitro gastrointestinal digestion. Results showed that in vitro digested extracts of raw garlic had the highest antioxidant capacity compared to other extracts. Cooking garlic led to a loss of phenolic compounds and antioxidant potential. Therefore, the physiological method of antioxidant extraction is more reliable and applicable than conventional chemical methods (Bhatt & Patel, 2013).

2.5.2. Anticancer Activity

Many laboratory studies have demonstrated that garlic has important effects in killing cancer cells. The anticancer properties of garlic are mainly linked to: (1) reducing mutations, (2) neutralizing free radicals, (3) controlling enzyme activities, (4) blocking protein folding in the endoplasmic reticulum, and (5) preventing cancer cell behaviours such as growth, resistance to cell death, and avoidance of immune detection (Corzo-Martínez et al., 2007).

Garlic contains powerful compounds that can help prevent cancer by affecting various stages of cancer cell development. It interferes with cell cycle regulation, inhibits signaling pathways, and induces apoptosis and autophagy (Shang et al., 2019). Garlic contains sulfur compounds that fight cancer by reducing oxidative stress, metabolizing carcinogens, and enhancing immune response.

Table 2.2 summarizes how various garlic compounds affect specific cancer traits (Talib et al., 2024).

2.5.3. Cardiovascular Diseases

Garlic has been shown to have cardioprotective effects on primary cultured cardiac myocytes, fibroblasts, and endothelial cells in a number of in vitro studies by lowering ROS production and inhibiting ROS-dependent extracellular signal-regulated kinase (ERK)1/2, JNK1/2, AKT, NF-κB, and SMADS receptors (Liu et al., 2010; Arreola et al., 2015).

TABLE 2.2
Garlic Phytoconstituents on Cancer Markers

Compounds	Cancer Markers	References
Allicin	Tolerance activity inhibition and inhibitor for gastric cancer	(Sun & Wang, 2003)
Z-ajoene	Decrease in the amounts of TP1 and hTRT mRNA linked to telomerase	(Ye et al., 2005)
S-allylcysteine (SAC)	• Cell cycle progression inhibition • Wee1 activity enhancement • CDK inhibitor accumulation	(Talib et al., 2024)

Atherosclerosis is a complicated disease marked by an intense inflammatory response to damage in blood vessel linings, involving various cell types such as smooth muscle cells and macrophages. Studies suggest that garlic might help improve heart health by lowering LDL cholesterol and blood pressure, reducing platelet clumping, and slowing atherosclerosis. However, some garlic compounds do not affect NF-κB activity in endothelial cells, indicating they may influence inflammation differently. More research is required to identify the specific benefits of various garlic compounds (Arreola et al., 2015; Budoff et al., 2009; Ried et al., 2010).

2.5.4. Anti-Inflammatory Action

Allicin can help control the immune system and lower TNF-α and IL-8 levels. A study found that in mice with ankylosing spondylitis, allicin improved spinal issues, decreased IL-6, IL-8, and TNF-α release, and blocked HLA-B2704 protein expression. Furthermore, some types of waves and microwave radiation were seen to boost the anti-inflammatory effects of allicin (Bose, 2024; Kang et al., 2024).

Allicin has shown potential anti-inflammatory and antiapoptotic effects in cerebral ischemia/reperfusion injury. It reduced cerebral infarction, brain water content, neuronal apoptosis, and MPO activity in serum, indicating its protective role for the brain (Zhang et al., 2015).

2.5.5. Hepatoprotective Activity

Allicin has significant liver-protective benefits, as shown in various studies. Its strong SH-modifying and antioxidant qualities help safeguard the liver from damage caused by oxidative stress and lipid peroxidation. In a model of liver injury induced by acetaminophen, taking allicin orally increased levels of Bcl-2 and Ki-67, highlighting its protective role. This protection is linked to allicin's reduction of oxidative stress and its ability to inhibit the inflammasome pathway. Additionally, in another study, allicin improved liver enzyme levels and reduced inflammation, while also addressing liver cell damage by blocking stress related to the TLR4/NF-κB pathway (Ipsen et al., 2018).

In studies with rat models of liver damage from carbon tetrachloride, allicin showed protective effects by preventing harmful changes in liver tissue, without indicating bile duct adenocarcinoma or tumour cells. Additionally, pairing allicin with omega-3 fatty acids improved liver health in a model of liver damage caused by paracetamol. Allicin pretreatment also boosted the antitumour effects of tamoxifen and helped protect the liver by lowering oxidative stress, lipid peroxidation, and inflammation, while increasing antioxidant enzyme activities (Suddek, 2014).

2.5.6. Antidiabetic Activity

Various studies have shown that allicin has properties that lower blood sugar, especially in animal tests. In experiments with diabetic rabbits treated with alloxan, combining allicin with tolbutamide significantly reduced blood sugar levels, suggesting it might be useful for this purpose. Allicin has also been noted to reduce blood pressure, insulin, and triglyceride levels. Its antidiabetic effects may involve a specific mechanism related to the opening of SUR 2 (Dubey et al., 2012).

Additionally, allicin may help prevent insulin-dependent diabetes mellitus (IDDM) by influencing insulin and blood sugar levels and protecting pancreatic tissue. It could also help manage blood lipids, potentially preventing formation of harmful substances. Studies on diabetic rats have shown that allicin offers benefits for heart injuries related to diabetes, protecting against heart dysfunction and the progression of fibrosis by affecting proteins related to these processes (Liu et al., 2012).

2.5.7. Anti-Alzheimer Action

Allicin has been shown to improve brain function in APP/PS1 double-transgenic mice. It does this by lowering Aβ levels, reducing oxidative stress, and enhancing mitochondrial function. Its effects are partly due to the blocking of the p-JNK/p-c-Jun signaling pathway. Additionally, allicin can inhibit acetylcholinesterase and butyrylcholinesterase enzymes, which break down acetylcholine, a key neurotransmitter for cognition. In a mouse model of Alzheimer's disease, a dose of 180 mg/kg of allicin improved cognitive abilities, likely due to increased superoxide dismutase (SOD) activity and lower reactive oxygen species (ROS) levels (Li et al., 2012; Zhu et al., 2015).

2.6. CONCLUSION

Garlic (*Allium sativum*) has a rich history that spans thousands of years, originating in Central Asia and being valued in various ancient cultures for its culinary, medicinal, and spiritual properties. Archaeological findings show that garlic was a significant crop in ancient Mesopotamia and Egypt, where it was believed to enhance the stamina of pyramid workers and was buried with pharaohs for protection in the afterlife. The strong scent and antimicrobial qualities of garlic made it a common treatment in ancient medicine, noted in documents like the Ebers Papyrus, and used by notable figures such as Hippocrates for various ailments.

In the Ayurvedic and traditional Chinese medicinal practices, garlic was recognized for its warming effects on the body and was employed to treat ailments like coughs and colds. During the Middle Ages, garlic was thought to ward off evil and was used against the plague, with people wearing it for protection. The "Four Thieves Vinegar", a mix containing garlic, was believed to protect against the disease when looting homes of the infected.

Garlic made its way to the New World with European explorers, notably introduced to the Caribbean by Columbus, and was incorporated into the diets and medicinal practices of Native cultures across the Americas. Today, garlic remains a popular health supplement, with research supporting its benefits for immunity, heart health, and various ailments.

Garlic is now cultivated worldwide, with China as the largest producer, followed by India and several European countries that embrace its use in cooking, particularly Mediterranean cuisine. Different types of garlic exist, including hardneck, softneck, and elephant garlic, each with unique flavors and characteristics.

In terms of health benefits, garlic contains over 200 chemical compounds, with allicin being particularly noted for its antioxidant, anticancer, cardiovascular protective, anti-inflammatory, hepatoprotective, antidiabetic, and anti-Alzheimer properties. Studies have shown that garlic can effectively fight oxidative stress, inhibit cancer cell growth, promote heart health by lowering cholesterol and blood pressure, and reduce inflammation. Additionally, allicin has demonstrated protective effects on the liver and potential benefits in managing blood sugar levels in diabetes.

Garlic's ability to improve cognitive function and reduce neurodegeneration has also been documented, with findings suggesting it enhances mitochondrial function and reduces harmful substances in the brain. Overall, garlic has established itself as a powerful component in both traditional and modern medicine, with continued research validating its wide-ranging health benefits.

REFERENCES

Aboelsoud, N. H. 2010. "Herbal Medicine in Ancient Egypt." Journal of Medicinal Plants Research 4 (2): 82–86. https://doi.org/10.5897/jmpr09.013.

Adema, Pauline. 2009. Garlic Capital of the World. https://doi.org/10.14325/mississippi/9781604731200.001.0001.

Arreola, R., S. Quintero-Fabián, R. I. López-Roa, E. O. Flores-Gutiérrez, J. P. Reyes-Grajeda, L. Carrera-Quintanar, and D. Ortuño-Sahagún. 2015. "Immunomodulation and Anti-Inflammatory Effects of Garlic Compounds." Journal of Immunology Research 2015: 401630. https://doi.org/10.1155/2015/401630.

Aviello, Gabriella, Ludovico Abenavoli, Francesca Borrelli, Raffaele Capasso, Angelo Antonio Izzo, Francesca Lembo, Barbara Romano, and Francesco Capasso. 2009. "Garlic: Empiricism or Science?" Natural Product Communications 4 (12). https://doi.org/10.1177/1934578x0900401231.

Benke, A. P., Krishna, R., Mahajan, V., Ansari, W. A., Gupta, A. J., Khar, A., ... & Manjunathagowda, D. C. (2021). "Genetic diversity of Indian garlic core germplasm using agro-biochemical traits and SRAP markers." Saudi journal of biological sciences 28 (8), 4833–4844.

Bhatt, Anjali, and V. H. Patel. 2013. "Antioxidant Activity of Garlic Using Conventional Extraction and in Vitro Gastrointestinal Digestion." Free Radicals and Antioxidants 3 (1) (apirilak 1): 30–34. https://doi.org/10.1016/J.FRA.2013.03.003. https://www.sciencedirect.com/science/article/abs/pii/S2231253613000040.

Bose, Sankhadip. 2024. "Anti-Inflammatory Activity of Isolated Allicin from Garlic with Post-Acoustic Waves and Microwave Radiation." Accessed November 27, 2024.

Budoff, Matthew J., Naser Ahmadi, Khawar M. Gul, Sandy T. Liu, Ferdinand R. Flores, Jima Tiano, Junichiro Takasu, Elizabeth Miller, and Sotirios Tsimikas. 2009. "Aged Garlic Extract Supplemented with B Vitamins, Folic Acid and L-Arginine Retards the Progression of Subclinical Atherosclerosis: A Randomized Clinical Trial." Preventive Medicine 49 (2–3): 101–107. https://doi.org/10.1016/j.ypmed.2009.06.018.

Cai, Yizhong, Qiong Luo, Mei Sun, and Harold Corke. 2004. "Antioxidant Activity and Phenolic Compounds of 112 Traditional Chinese Medicinal Plants Associated with Anticancer." Life Sciences 74 (17): 2157–2184. https://doi.org/10.1016/j.lfs.2003.09.047.

Corzo-Martínez, Marta, Nieves Corzo, and Mar Villamiel. 2007. "Biological Properties of Onions and Garlic." Trends in Food Science and Technology 18 (12) (December 1): 609–625. https://doi.org/10.1016/J.TIFS.2007.07.011.

Dubey, Harikesh, Anamika Singh, Angad Mohanrao Patole, Chandrashekhar Ramdas Tenpe, and Balu Vinayak Ghule. 2012. "Allicin, a SUR2 Opener: Possible Mechanism for the Treatment of Diabetic Hypertension in Rats." Revista Brasileira de Farmacognosia 22 (5): 1053–1059. https://doi.org/10.1590/s0102-695x2012005000046.

Francia, Susan, and Anne Stobart. 2014. "Critical Approaches to the History of Western Herbal Medicine: From Classical Antiquity to the Early Modern Period." Choice Reviews Online 52 (4): 52–2014. https://doi.org/10.5860/choice.185515.

Ipsen, David Højland, Jens Lykkesfeldt, and Pernille Tveden-Nyborg. 2018. "Molecular Mechanisms of Hepatic Lipid Accumulation in Non-Alcoholic Fatty Liver Disease." Cellular and Molecular Life Sciences 75 (18): 3313–3327. https://doi.org/10.1007/s00018-018-2860-6.

Jang, Hyun-Joo, Hyun-Jin Lee, Dong-Kyu Yoon, Da-Som Ji, Ji-Han Kim, and Chi-Ho Lee. 2017. "Antioxidant and Antimicrobial Activities of Fresh Garlic and Aged Garlic by-Products Extracted with Different Solvents." Food Science and Biotechnology 27 (1) (December 12): 219–225. https://doi.org/10.1007/s10068-017-0246-4.

Kang, Moon Sung, Hee Jeong Jang, Jeong Min Kim, Hyo Jung Jo, Kyung Min Park, Young-Hwa Chung, and Dong-Wook Han. 2024. "Evaluation of Anti-Inflammatory Activity of Garlic Extracts in 3D Bioprinted Skin Equivalents." Advanced NanoBiomed Research (May). https://doi.org/10.1002/anbr.202400007.

Kent, Laura. n.d. "Garlic Types and Varieties." Gardeningtheme.Com. https://www.gardeningtheme.com/theme/garlic-types-and-varieties.

Kim, Soon Hee, Myung Sunny Kim, Myoung Sook Lee, Yong Soon Park, Hae Jeong Lee, Soon-Ah Kang, Hyun Sook Lee, et al. 2016. "Korean Diet: Characteristics and Historical Background." Journal of Ethnic Foods 3 (1): 26–31. https://doi.org/10.1016/j.jef.2016.03.002.

Li, Xian-Hui, Chun-Yan Li, Jiang-Ming Lu, Rong-Bo Tian, and Jia Wei. 2012. "Allicin Ameliorates Cognitive Deficits Ageing-Induced Learning and Memory Deficits through Enhancing of Nrf2 Antioxidant Signaling Pathways." Neuroscience Letters 514 (1): 46–50. https://doi.org/10.1016/j.neulet.2012.02.054.

Liu, C., F. Cao, Q. Z. Tang, et al. 2010. "Allicin Protects against Cardiac Hypertrophy and Fibrosis Via Attenuating Reactive Oxygen Species-Dependent Signaling Pathways." Journal of Nutritional Biochemistry 21 (12):1238–1250. https://doi.org/10.1016/j.jnutbio.2009.11.001.

Liu, Yang, Hanping Qi, Ye Wang, Mingli Wu, Yonggang Cao, Wei Huang, Lei Li, Zhong Ji, and Hongli Sun. 2012. "Allicin Protects against Myocardial Apoptosis and Fibrosis in Streptozotocin-Induced Diabetic Rats." Phytomedicine 19 (8–9): 693–698. https://doi.org/10.1016/j.phymed.2012.04.007.

Mark, Joshua J. 2023. "Medieval Cures for the Black Death." World History Encyclopedia, February. https://www.worldhistory.org/article/1540/medieval-cures-for-the-black-death/.

Melguizo-Rodríguez, Lucía, Enrique García-Recio, Concepción Ruiz, Elvira De Luna-Bertos, Rebeca Illescas-Montes, and Víctor J. Costela-Ruiz. 2022. "Biological Properties and Therapeutic Applications of Garlic and Its Components." Food & Function 13 (5): 2415–2426. https://doi.org/10.1039/d1fo03180e.

Mikaili, Peyman, Surush Maadirad, Milad Moloudizargari, Shahin Aghajanshakeri, and Shadi Sarahroodi. 2013. "Therapeutic Uses and Pharmacological Properties of Garlic, Shallot, and Their Biologically Active Compounds." PubMed 16 (10): 1031–1048. https://pubmed.ncbi.nlm.nih.gov/24379960.

Nie, Jing, Rui Weng, Chunlin Li, Xiuhua Liu, Fang Wang, Karyne M. Rogers, Yongzhong Qian, Yongzhi Zhang, and Yuwei Yuan. 2022. "Chemometric Origin Classification of Chinese Garlic Using Sulfur-Containing Compounds, Assisted by Stable Isotopes and Bioelements." Food Chemistry 394 (June): 133557. https://https://doi.org/10.1016/j.foodchem.2022.133557.

Petrovska, Biljana Bauer, and Svetlana Cekovska. 2010. "Extracts from the History and Medical Properties of Garlic." Pharmacognosy Reviews/Bioinformatics Trends/Pharmacognosy Review 4 (7): 106. https://doi.org/10.4103/0973-7847.65321.

Prasad, Kailash, et al. 1995. "Antioxidant Activity of Allicin, an Active Principle in Garlic." Molecular and Cellular Biochemistry 148 (2) (uztailak 19): 183–189. https://doi.org/10.1007/BF00928155. https://link.springer.com/article/10.1007%2FBF00928155.

Rahman, Md. Mokhlesur, V. Fazlic, and N. W. Saad. 2012. "Antioxidant Properties of Raw Garlic (Allium Sativum) Extract." International Food Research Journal 19 (2) (martxoak 31): 589–591. https://www.cabdirect.org/cabdirect/abstract/20133162320.

Ried, Karin, Oliver R. Frank, and Nigel P. Stocks. 2010. "Aged Garlic Extract Lowers Blood Pressure in Patients with Treated But Uncontrolled Hypertension: A Randomised Controlled Trial." Maturitas 67 (2): 144–150. https://doi.org/10.1016/j.maturitas.2010.06.001.

Rivlin, Richard S. 2001. "Historical Perspective on the Use of Garlic." Journal of Nutrition 131 (3): 951S–954S. https://doi.org/10.1093/jn/131.3.951s.

Sethi, N., S. Kaura, N. Dilbaghi, M. Parle, and M. Pal. (2014). "Garlic: A Pungent Wonder from Nature." International Research Journal of Pharmacy 5 (7): 523–529. https://doi.org/10.7897/2230-8407.0507106.

Shang, Ao, et al. 2019. "Bioactive Compounds and Biological Functions of Garlic (Allium sativum L.)." Foods 8 (7) (uztailak 5): 246. https://doi.org/10.3390/FOODS8070246. https://www.ncbi.nlm.nih.gov/pmc/articles/PMC6678835/.

Spence, Charles. 2023. "Garlic: Tracing Its Changing Popularity in British Cuisine." International Journal of Gastronomy and Food Science 35 (December): 100854. https://https://doi.org/10.1016/j.ijgfs.2023.100854.

Suddek, Ghada M. 2014. "Allicin Enhances Chemotherapeutic Response and Ameliorates Tamoxifen-Induced Liver Injury in Experimental Animals." Pharmaceutical Biology 52 (8): 1009–1014. https://doi.org/10.3109/13880209.2013.876053.

Sun, L., and X. Wang. 2003. "Effects of Allicin on Both Telomerase Activity and Apoptosis in Gastric Cancer SGC-7901 Cells." World Journal of Gastroenterology 9 (9): 1930–1934. https://doi.org/10.3748/wjg.v9.i9.1930.

Talib, W. H., S. Atawneh, A. N. Shakhatreh, G. N. Shakhatreh, I. S. Rasheed Aljarrah, R. Hamed, D. A. Banyyounes, and I. H. Al-Yasari, 2024. "Anticancer Potential of Garlic Bioactive Constituents: Allicin, Z-Ajoene, and Organosulfur Compounds." Pharmacia 71, 1–23. https://doi.org/10.3897/pharmacia.71.e114556.

Ye Y, Yang HY, Wu J, Li M, Min JM, Cui JR. [Z-ajoene causes cell cycle arrest at G2/M and decrease of telomerase activity in HL-60 cells]. Zhonghua Zhong Liu Za Zhi. 2005 Sep; 27 (9): 516–520. Chinese. PMID: 16438845.

Zhang, Benping, Feng Li, Weijiang Zhao, Jiebing Li, Qingsong Li, and Weizhi Wang. 2015. "Protective Effects of Allicin against Ischemic Stroke in a Rat Model of Middle Cerebral Artery Occlusion." Molecular Medicine Reports 12 (3): 3734–3738. https://doi.org/10.3892/mmr.2015.3883.

Zhu, Yao-Feng, Xian-Hui Li, Zhi-Peng Yuan, Chun-Yan Li, Rong-Bo Tian, Wei Jia, and Zhu-Ping Xiao. 2015. "Allicin Improves Endoplasmic Reticulum Stress-Related Cognitive Deficits via PERK/Nrf2 Antioxidative Signaling Pathway." European Journal of Pharmacology 762 (September): 239–246. https://doi.org/10.1016/j.ejphar.2015.06.002.

3 Spinach
Varieties and Nutraceutical Functions

Rahul D. Kamble, Deepak B. Shelke, Mahadev R. Chambhare, and Pankaj Kumar Chaurasia

3.1. INTRODUCTION

India has a legacy of more than 8,000 years of culinary tradition. The food culture of India has a vibrant history filled with tasty twists and turns. It is an affirmation that Indian food culture was solely determined by the attributive property of Guna that is an oversimplification and a Hindu philosophical perception referring to quality. Sattva, rajas, and tamas, the three 'Gunas', were believed to have manifestation in the form of vegetarian, spicy, and nonvegetarian. In the kitchens of India, housewives or cooks try to prepare the food that complements Guna (Antani and Mahapatra, 2022). Among the diverse kitchen ingredients, leafy green vegetables acquired a unique position in the culinary matrix due to their wide nutraceutical functions (Achaya, 2003). Spinach (*Spinacia oleracea* L.) is a widely cultivated green leafy vegetable popular for its nutritional values and diverse health benefits.

China is the leading country among the top ten countries that are dedicated for the cultivation of spinach. The annual production of spinach in China in 2022 was more than 30 million tons, contributing alone to 93% of total global production, followed by the USA, Japan, Turkey, and Indonesia, with less than about a half million tons. Other countries among the top ten producers of spinach are Iran, France, Pakistan, Italy, and South Korea (UN Food and Agriculture Organization, 2024; Wikipedia/spinach, 2024). Spinach is also cultivated across India. The states with the most annual production of spinach are Andhra Pradesh, Telangana, Kerala, Tamil Nadu, Uttar Pradesh, Karnataka, Maharashtra, West Bengal, and Gujrat.

Spinach is known for its nutritional and health benefits. Spinach is an excellent natural source of minerals, such as iron, copper, calcium, phosphorus, zinc, selenium, vitamin B complex including niacin and folic acid, ascorbic acid, β-carotene, lutein, xanthophylls, polyphenols, flavonoids, coumarins, apocynins, and omega-3 fatty acids (Roberts and Moreau, 2016). Spinach is medicinally important and is used as traditional medicine for numerous therapeutic effects due to the presence of polyphenols such as tannins and other phytochemicals such as alkaloids, flavonoids, terpenoides, glycosides, and steroids (Ekor, 2014). Spinach is also extensively used in the treatment of diabetes (Park et al., 2012), leprosy, urinary infections, inflammation, multiple organ dysfunction (Kapoor et al., 2017), and diseases related to the heart and brain. Spinach and spinach-based phytochemicals are able to scavenge reactive oxygen species and prevent macromolecular damage, activity of genes involved in the metabolism, proliferation, inflammation (Gupta et al., 2017), antioxidant defence (Gupta et al., 2017; Sun et al., 2018), and prompting secretion of satiety hormone to restrain intake of food (Abolhassanzadeh et al., 2016; Akasaka et al., 2016). The ability of green leafy vegetables to scavenge reactive oxygen species designated them as "nature's antiaging wonders" (Gupta and Prakash, 2008).

 DOI: 10.1201/9781003559894-3

3.2. VARIETIES OF SPINACH

Plant variety and nutritional quality of vegetable crops are mainly essential for improving food and nutrition safety worldwide (Rashid et al., 2022). Besides, it could be an imperative aspect for plant breeders and crop growers to select improved varieties for further development (Grant et al., 2007; Rana and Thakur, 2017). The progressive purpose of current spinach (*Spinacia oleracea* L.) breeding programs includes identifying varieties low in nitrate and oxalate content as well as high in bioactive metabolites with health-promoting actions (Koh et al., 2012; Mou, 2019). To comprehend these relationships, more studies encompassing a wider variety of crops are still deserved because different species and varieties do not respond to agronomic and environmental stresses optimistically (Serpolay et al., 2011; Bhattarai and Shi, 2021). Hence, the objective of the present chapter is to explore the improved levels of ascorbic acid, vitamin C, flavonoids, nitrate, oxalic acid, nutrients, and morphological traits in different spinach varieties; to assess the effect of certified organic and conventional cultivation methods on these components; and to identify varietal or cultivation permutations that may lead to an overall improvement in the quality of spinach vegetable (Morelock and Correll, 2007; Koh et al., 2012). For instance, some of the spinach varieties comprise 'Indian Summer', 'Supergreen', 'Pelican', 'Bloomsdale', 'Crocodile', 'Califlay', 'Fallgreen', 'Tyee', 'Red Cardinal', 'Regiment', 'Red Kitten', 'Gaudry', 'Viroflay', 'Monarch Long', 'Nobel', 'Punjab Green', 'Punjab all green', 'Pusa jyoti', 'Pusa palak', 'Pusa harit', 'Pusa bharti', 'Red spinach', 'Rembrandt', 'Spiros', 'Sathi', 'Sp75', 'Spinat-2', 'Spinat-5', 'Diamond', 'Resistoflay', 'Winter Giant', 'Giant 157', 'Flamingo', 'Spinoza', and 'Spargo' (Serpolay et al., 2011). The 'Fallgreen' variety is open-pollinated resistant and attained high resistance against white rust used by the spinach breeding industries, and it was developed by the University of Arkansas (Morelock and Correll, 2007). Shohag et al. (2011) have described 'PI 499372', 'PI 261787', 'NSL 6095', and 'TOT7337-B' spinach varieties, which contain enriched folate content (150 μg/100 g of fresh weight). 'Indian Summer' is a flattened, smooth-leaved, good choice for all-season cultivation, slow-bolting and desirable spinach variety for summer season. Ribera et al. (2020) have reported that 'Bloomsdale', 'Califlay', 'Géant d'Hiver', 'Viroflay', and 'Resistoflay' are promising spinach varieties for CMV-resistant, blight resistance, and Downey mildew tolerance. 'Alwaro', 'Supergreen', and 'Verbeterde Hollandia' spinach varieties have been improved with longer petioles and leaves, dark-green leaves, more erect leaves, and more flowering. The 'Monarch Long' variety exhibited improved characteristics with less bolting and shorter petioles (Serpolay et al., 2011). The 'Red Cardinal' variety has red veins in the leaves, deep red stems, must be harvested young, and is a beautiful addition to a salad. 'Balady' was significantly improved for higher chlorophyll and carotene contents (0.0023 and 0.0018 g/g FW, respectively) (Alessa et al., 2017). 'Punjab all green' spinach was potentially grown by the farmers for their increased yield (10.44 t/ha) (Nath et al., 2020). 'Red spinach' was developed for its rich content of betacyanin and increased antioxidants (Mou, 2019). The tallest plant height (25.52 cm), maximum number of leaves (9.6), higher leaf length (15.52 cm), and highest total yield (18.08 t/ha) was noticed in variety 'Sathi' (Nath et al., 2020). The 'Virofly' spinach variety significantly increased the average stem diameter (7.43 mm), leaf area (236.9 cm^2), and fresh weight (Alessa et al., 2017). Likewise, several spinach varieties were improved for morphological, nutritional, antioxidant, disease-resistance, phytoremediation, and agronomic yield traits, as described in the Table 3.1.

3.3. NUTRITIONAL COMPOSITION

Spinach is a leafy green vegetable consumed both raw such as in salads, as well as in cooked forms such as steamed, soups, etc. The nutritional composition of spinach is mainly water (approximately 91.4%), along with other nutritional components in small quantity, including 2.9% protein, 3.6% carbohydrate, and 0.4% fats. The lipid content primarily possessed mono- and polyunsaturated fatty acids (PUFA) such as α-linoleic acid, linoleic acid, and oleic acid, while saturated fatty acids

TABLE 3.1
Commercially Improved Spinach Varieties

Spinach Variety	Improved Traits	References
Alwaro	Longer petioles and leaves, more greener leaves, more erect leaves, more flowering	(Serpolay et al., 2011)
ASR 157, ACX 3665 and DMC 09	Significant improvement in marketable yields	(Leskovar and Piccinni, 2005)
Avenger	Improved resistance against downy mildew and white rust	(Correll et al., 2010)
Balady	Significantly higher chlorophyll and carotene contents (0.0023 and 0.0018 g/g FW)	(Alessa et al., 2017)
Bloomsdale	CMV-resistant variety	(Ribera et al., 2020)
Califlay	Promising variety for blight resistance	(Ribera et al., 2020)
Campania	Improved resistance against downy mildew and white rust	(Correll et al., 2010)
Clemont and Dolphin	Improved resistance against downy mildew and white rust	(Correll et al., 2010)
Fallgreen	Open-pollinated resistant and white rust resistant variety	(Morelock and Correll, 2007)
Geant d'Hiver	Winter-hardy variety	(Ribera et al., 2020)
Gaudry and Monstrueux de Viroflay	Smooth-leaved variety	(Ribera et al., 2020)
Hybrid 178' and 'WRG 70–5'	White rust resistance	(Morelock and Correll, 2007)
Lazio	Improved resistance against downy mildew and white rust	(Correll et al., 2010)
Monarch Long	Shorter petioles, less bolting	(Serpolay et al., 2011)
Nobel and Viking	Late-bolting summer variety	(Ribera et al., 2020)
Ozaka and Greenvalley	Disease-resistant varieties	(Bowers and Goode, 1981)
Punjab all green	Improved total yield (10.44 t/ha)	(Nath et al., 2020)
Punjab Green	Take 30 days for harvest, average yield of 125 q/ha	(Nath et al., 2020)
PI 499372, PI 261787, NSL 6095, and TOT7337-B	Enriched folate content over 150 µg/100 g of fresh weight	(Shohag et al., 2011)
Red spinach	Rich in betacyanin, increased antioxidant content	(Mou, 2019)
Rembrandt and Spiros	Improvement in spinach yield (18.6–44.8 t/ha).	(Edward et al., 2010)
Sathi	Improved height (25.52 cm), significant increment in number of leaves (average 9.6) and length of leaves (15.52 cm), and highest total yield (18.08 t/ha)	(Nath et al., 2020)
Sp75	Improved heat tolerance	(Li et al., 2019)
Supergreen	Narrower leaves, more erect leaves, more flowering	(Serpolay et al., 2011)
Spinat-2, Spinat-5, RANCHO, and Diamond	Improved agronomic traits	(Meng et al., 2017)
Tyee	Resistant to downy mildew	(Rana and Thakur, 2017)
Virofly	Significantly higher stem diameter (7.43 mm), leaf area (236.96 cm^2), and fresh weight	(Alessa et al., 2017)
Verbeterde Hollandia	More yellow and green leaves, more flowering	(Serpolay et al., 2011)
Viroflay and Resistoflay	Improved resistance to downy mildews	(Ribera et al., 2020)
Viking	More greener and lighter leaves, less erect leaves	(Serpolay et al., 2011)
Viking-Matador	Longer leaves	(Serpolay et al., 2011)
Whale	Improved resistance against downy mildew and white rust	(Correll et al., 2010)
501	Improved morphology	(Koh et al., 2012)

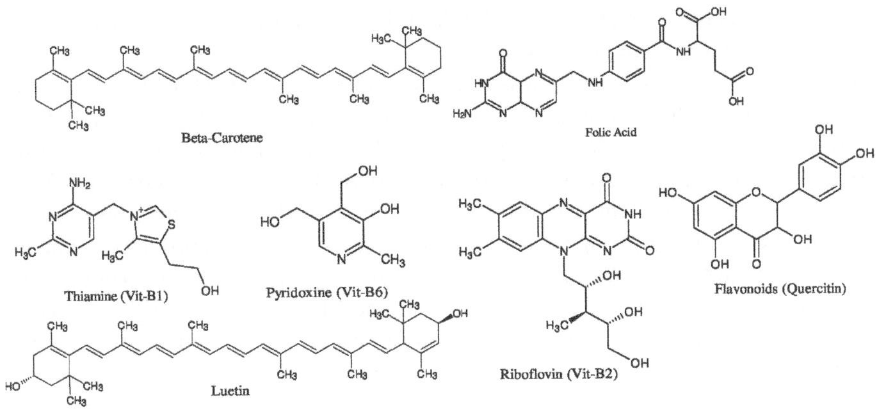

FIGURE 3.1 Nutritional composition of spinach.

including stearic, capric, and myristic acids are present in trace amounts (Rahati et al., 2023; Wolf et al., 1962; Singh et al., 2001).

Spinach is rich in fibre content (2.2 g/100 g) and satisfies the recommended daily allowance (RDA) for a 2,000-kcal diet. Spinach has diverse mineral and vitamin composition compared with other green leafy vegetables. Spinach leaves comprise higher level of potassium (16%), magnesium (20%), and iron (15%), respectively, that convene its RDA per 100 g serving (Roberts and Moreau, 2016). Spinach leaves are an enormous source of vitamin A precursor (β-carotene), C, K (phylloquinone), E, $B_{2,}$ B_6, folate, niacin, lutein, and quercetin with other bioflavonoids (Gaikwad et al., 2010; Roughani et al., 2017) (Figure 3.1).

3.4. HEALTH BENEFITS

A change in lifestyle, lack of physical activities, malnourishment, etc. increases the incidence of chronic diseases (diabetics and cancer) and cardiovascular diseases (Roberts and Moreau, 2016; Ward and Schiller, 2013). A clear link between diet and physiological function has been proved, and it has been observed that the use of functional foods becomes a cost-effective strategy for prevention and disease treatment (Roberts and Moreau, 2016; Sofi, et al., 2013). The spinach leaves and its leaf extracts acquire a leading position and play an important role to protect against chronic diseases. Spinach exhibits various functional properties, including antioxidant, antidiabetic, hypoglycemic, antiobesity, and lipid-lowering ability, which brought it to the centre of the therapeutic horizon (Roberts and Moreau, 2016).

3.4.1. Antioxidant Properties

A diet rich in leafy green vegetables protects against chronic diseases, due to the presence of phytoconstituents, and vitamins C and E in spinach exhibit free radical–scavenging potential (Roberts and Moreau, 2016). Rao et al. (2015) studied the influence of drying on the phytochemical constituents of the spinach leaf. There was no significant change in phytochemical composition of dried and fresh spinach leaf. The antioxidant potential of spinach extract was studied *in vitro* using ABTS [2,2′-Azino-bis(3-ethylbenzothiazoline-6-sulfonic acid)], peroxyl, and hydroxyl radicals, Fe^{2+}, peroxynitrite, and superoxide anion (O_2^-) (Roberts and Moreau, 2016; Chu et al., 2002; Zhou and Yu, 2005). The total oxygen radical–scavenging potential of spinach extract is higher than that of other green vegetables. The aqueous extract of spinach also exhibits remarkable antioxidant potential. The aqueous extract of spinach primarily constitutes aromatic polyphenols, which includes flavonoids, p-coumaric acid derivatives, and other phyrophilic compounds. The *in vitro* antioxidant study of aqueous extract of spinach reveals that it inhibits the formation of peroxide and malondialdehyde by about 35% (Bergman et al., 2001).

The antioxidant potential of spinach was also tested in Sprague-Dawley rats. The oxidative stress studied by examining production of the 2,7-dichloroflurocien by providing diets fortified with spinach leaf extract (0.64%) for a period of eight months led to a remarkable decline in the formation of stress indicator 2,7-dichloroflurocien as compared with the control (Roberts and Moreau, 2016; Joseph et al., 1998). The human intervention experiments have also advocated the antioxidant potential of spinach. The volunteer group with an average age of 66.9 years provided a particular single meal (294 g) of spinach facilitate 25% increase in the oxygen radical absorption capacity, 24% enhancement in ferric-reducing potential of plasma, and 21% increased level of Trolox-equivalent antioxidant capacity four hours after consumption of spinach compared with volunteers not consuming spinach (Roberts and Moreau, 2016; Cao et al., 1998).

3.4.2. Antidiabetic Property

Spinach contains flavonoids such as luteolin, kaempferol, quercetin, apigenin, and qyricetin, which may contribute to its antidiabetic action. Medicinal herbs with hypoglycemic and antidiabetic activities often include flavonoids as active ingredients. The aqueous and ethanol extracts of spinach leaves were examined orally at the concentrations of 200 and 400 mg/kg for antidiabetic effects in alloxan-induced diabetic rats and in controls (Roberts and Moreau, 2016). The influence of aqueous and ethanol extract therapy in a diabetic rat's body weight, serum cholesterol, and triglyceride levels were compared to those of diabetic control and normal animals. Histopathological findings were also investigated after 12 days of treatment. The aqueous and ethanol extracts of spinach leaves significantly reduced blood sugars in control and alloxan-induced diabetic mice (Roberts and Moreau, 2016). Diabetic rats treated with ethanolic or aqueous solutions showed substantial variations in blood lipid profiles, including cholesterol and triglycerides, as well as body weight alterations. Nuutila et al. (2002) found that extracts could regenerate pancreatic tissues that had previously been damaged by alloxan. Diabetic individuals continue to face a significant risk of chronic ulcers. Diabetes often leads to delayed or disrupted ulcer regeneration. Spinach extract's antioxidative and anti-inflammatory properties may help cure diabetic ulcers faster.

Significant changes were detected in the macroscopic inspection of inflammation of the control and aqueous extract of spinach groups among 7 to 21 days as compared to the diabetes group. On microscopic examination, the spinach aqueous extract group and nondiabetic group had significantly better epithelial tissue creation, granulation tissue production, and new blood vessel formation than did the diabetes group. On days 3 and 7, there were substantial variations in vascular endothelial growth factor levels across groups. Spinach aqueous extract can effectively regenerate diabetic ulcers, improving their pace and structure (Rahati et al., 2015). Spinach leaf extract (50 mg/kg BW) was studied for its potential to protect against systemic lipopolysaccharide (LPS) (5mg/kg BW)-induced toxicity. Toxological parameters integrated enlarged oxidative load in body organs due to cooperated antioxidant defences (glutathione, MDA levels, and catalase activity), blood glucose, and cytokine ranges. Glucose ranges were observed to be considerably lower in LPS-exposed rats than in controls, although SLE normalized the glucose level. In reaction to LPS, the liver produces cytokines IL-1 and TNFα, which affect glucose metabolism and can cause hypoglycemia. Spinach co-administration normalized blood glucose levels in LPS-treated rats by declining the synthesis of pro-wounding cytokines such as TNF-α and IL-1β.

Spinach was evaluated for nutritional value, antioxidant capacity, and antidiabetic properties in vitro. The methanol extract of spinach leaves was prepared using Soxhlet apparatus and tested for OH-scavenging, 2,2-diphenyl-1-picrylhydrazyl, α-amylase inhibitory actions, nutritional value, and physicochemical parameters. The findings revealed that overall crude fibre, proteins, vitamins A and C, oils, carbohydrates, and fats were all high in nutritional value. This extract of spinach leaves demonstrated antidiabetic and free radical–scavenging actions, with inhibitory concentrations of

3.03 μg/ml for OH-scavenging, 6.03 μg/ml for DPPH inhibition, and 3.046 μg/ml for α-amylase inhibition (Kapoor et al., 2017).

3.4.3. Cardiovascular Effects

Spinach is an excellent source of nitrates, compounds that have been shown to enhance the health of blood vessels and improve circulation (Elena et al., 2015). Once ingested, nitrates are converted to nitric oxide, a molecule that dilates blood vessels, leading to improved blood flow and reduced blood pressure (Liu et al., 2013). Nitric oxide helps in relaxing and widening the blood vessels (vasodilation) (Sobko et al., 2009), which reduces the workload on the heart and decreases the risk of conditions like hypertension (high blood pressure), a major risk factor for heart attacks and strokes.

The potassium content in spinach helps balance out the negative effects of sodium, which is crucial for maintaining healthy blood pressure levels (Ferreira et al., 2020). Potassium aids in the dilation of blood vessels and promotes better blood flow, which can prevent or manage hypertension. Magnesium and other minerals in spinach are positively implicated in regulating blood pressure by relaxing the walls of the blood vessels, further promoting cardiovascular health.

Spinach has high amounts of fibre, particularly soluble fibre, which helps in reducing levels of LDL (low-density lipoprotein) cholesterol, the "bad" cholesterol (Iritani and Nogi, 1972). Soluble fibre binds to bile acids in the digestive system, promoting the excretion of cholesterol from the body rather than being absorbed into the bloodstream (Hafez et al., 2017). This reduction in LDL cholesterol helps in preventing plaque formation in the arteries, a condition referred to as atherosclerosis that is a significant cause of heart attacks and strokes.

Spinach has been shown to improve overall lipid profiles (the balance of cholesterol and triglycerides in the blood), contributing to better cardiovascular health (Djoussé et al., 2004). The combination of antioxidants, fibre, and nitrates helps reduce bad cholesterol while supporting healthy HDL (high-density lipoprotein) levels, which protect the heart by removing excess cholesterol from the bloodstream.

Spinach is an excellent source of folate (vitamin B9), which plays an important role in maintaining heart health (Shohag et al., 2011). Folate helps regulate levels of homocysteine, an amino acid present in the blood; at increased levels, homocysteine can injure the inside layer of blood vessels and increase the risk of cardiovascular diseases. High homocysteine levels are associated with a greater risk of atherosclerosis and blood clots, but adequate folate from spinach helps convert homocysteine into less harmful compounds, thereby reducing these risks.

Spinach is loaded with vitamin K that serves a critical function in maintaining the health of blood vessels. It helps in preventing the calcification of arteries, a process in which calcium deposits form in the blood vessel walls, leading to stiffness and impaired blood flow. By promoting vascular health, vitamin K assists to diminish the threat of arterial blockages that might direct heart attacks or strokes (Mladenka et al., 2021).

Keeping blood sugar levels stable is important for heart health, especially for people with diabetes, which is a significant risk factor for cardiovascular diseases. The fibre and alpha-lipoic acid found in spinach help regulate blood sugar, preventing spikes that can damage blood vessels and the heart over time.

Although spinach is not as plentiful in omega-3 fatty acids in comparison to fish, it does contain small amounts of plant-based omega-3s in the form of α-linolenic acid. Omega-3s are essential for reducing inflammation, lowering triglycerides, and preventing heart arrhythmias, contributing to overall cardiovascular protection. Spinach is low in calories but high in fibre, which can aid in weight management. Sustaining an optimized body weight is essential for the healthy heart, as overweight is a foremost threat to hypertension, heart failure, heart disease, and strokes. The fibre in spinach helps control appetite, promotes fullness, and aids in the digestion of food, supporting a heart-healthy diet.

3.4.4. Anti-Inflammatory Effects

Wounding is an innate immune feedback to infections, which if untreated may lead to development of serious chronic diseases including cancer, cardiovascular disease, and type 2 diabetes. The existing pharmacological remedies to treat inflammation are effective but associated with some adverse side effects, such as vomiting, dizziness, constipation, nausea, etc. Thus, an alternative solution could be functional food for the prevention and treatment of wounding or inflammation and wound-arbitrated diseases.

Spinach contains several types of phytonutrients (plant-based compounds) that have anti-inflammatory effects, including flavonoids, carotenoids, and phenolic compounds (Marinaccio et al., 2023). Spinach is rich in flavonoids such as kaempferol, quercetin, and myricetin, which have potent anti-inflammatory properties (Park et al., 2023). These compounds help modulate inflammatory pathways by reducing the activity of enzymes and cytokines that promote inflammation. Spinach is particularly high in lutein and beta-carotene, two carotenoids known for their ability to reduce inflammation. They work by scavenging free radicals and inhibiting pro-inflammatory molecules in the body (Park et al., 2023). Spinach and its extract are rich in the phenolic compounds. These include ferulic acid and caffeic acid, which are powerful antioxidants that prevent oxidative damage and reduce inflammatory responses at the cellular level.

Cytokines are small proteins released by cells, and they play a crucial role in signaling inflammatory responses. Spinach helps regulate pro-wounding cytokines such as interleukin-6 (IL-6), tumour necrosis factor-alpha (TNF-α) (Lee et al., 2022), and C-reactive protein (CRP) (Babu et al., 2018). Quercetin, found in spinach, is particularly effective in inhibiting the production of TNF-α, a major pro-inflammatory molecule (Amirinejad et al., 2021). By lowering the levels of these cytokines, spinach can help reduce systemic inflammation and protect against chronic inflammatory diseases such as arthritis and cardiovascular diseases.

One of the key mechanisms by which inflammation occurs in the body is through the activation of the nuclear factor-kappa B (NF-kB) route. NF-kB is a protein composite that manages the expression of genes involved in inflammation. When activated, it leads to the production of pro-inflammatory molecules. The flavonoids and carotenoids in spinach inhibit the NF-kB pathway (Gutierrez et al., 2019), reducing the expression of these inflammatory genes. This action helps lower the body's inflammatory response and can prevent the development of chronic inflammation-related diseases.

Cyclooxygenase (COX) enzymes, particularly COX-1 and 2, are implicated in the production of prostaglandins, which are lipid compounds that promote inflammation. Nonsteroidal anti-inflammatory drugs (NSAIDs) like aspirin work by inhibiting these enzymes, and spinach appears to have a similar but natural effect. Some of the compounds in spinach, such as lutein and quercetin, have been found to inhibit the activity of COX-2 enzymes, thereby reducing the synthesis of pro-inflammatory prostaglandins (Lomnitski et al., 2000). This makes spinach a natural option for reducing pain and inflammation caused by conditions such as arthritis.

3.4.5. Anticancer Property

Cancer is a primary cause of mortality globally, with an estimated 20 million new diagnosed cancer cases and 9.7 million cancer-linked deaths recorded in 2023 (Siegel et al., 2023). The projected 70% rise in cancer incidence over the upcoming two decades emphasizes the need for secure, safe, and efficient avoidance and management options. Diet has a well-established impact on hyperplasia, among other aspects involved with carcinogenesis. Vegetable-rich diets, particularly dark green leafy vegetables, are associated with a decreased vulnerability of several malignancies.

Spinach plays a significant role in cancer prevention due to its abundance of bioactive compounds, antioxidants, and anti-inflammatory properties. Many of these nutrients and phytochemicals in spinach have been shown to combat cancer development at multiple stages, from preventing DNA damage to inhibiting tumour growth and promoting the elimination of harmful carcinogens. According to epidemiological research, consumption of spinach or spinach-added food products may protect from colon, esophageal, and breast cancers. A case-control study of 6,888 cases and 9,428 controls found that females who consumed rawer spinach leaves (greater than 52 servings annually) have about 45% lesser occurrence of breast cancer (Longnecker et al., 1997). A carotenoid lutein was found to be associated in lowering the threat of breast cancer (Freudenheim et al., 1996). A pooled examination of 14 potential group investigations from Europe and North America found that consuming one 73 g serving of spinach weekly to one-half serving a day lowered the incidence of colon cancer by 11% (Roberts and Moreau, 2016; Koushik et al., 2007). Among the 12 vegetables investigated in this study, spinach was the only one that showed a substantial reduction in colon cancer risk (Koushik et al., 2007).

Spinach leaves have more glycolipids than green garlic and onion, sweet pepper, chive, and carrot. Spinach glycolipids comprise sulfoquinovosyl diacylglycerol (SQDG), monogalactosyl diacylglycerol (MGDG), and digalactosyl diacylglycerol (DGDG). These glycolipids are significantly suppressed in the proliferation of cancer cells, comprising HeLa cell line of cervix cancer, NUGC-3 of gastric cancer, HL-60 of leukaemia, A549 cell line of lung cancer, Colon-26 of colon cancer, BALL-1 of B-cell acute lymphoblastoid leukaemia, and Molt-4 cell line of T-cell acute lymphoblastic leukaemia (Roberts and Moreau, 2016; Kuriyama et al., 2005; Maeda et al., 2008).

Spinach may reduce the incidence of breast, esophageal, and colon cancers in people, according to epidemiologic data. However, most studies on its cancer-protective benefits have used rat and cell models. There is an urgent need of human involvement research to determine the anticancer effects of *Spinacia oleracea* and its active metabolites.

3.5. CONCLUSION

Spinach is a popular green leafy vegetable rich in vitamins, minerals, phytochemicals, and bioactive substances. Spinach has a high fibre content. Consuming 100 g of spinach gives a high concentration of minerals and vitamins, including folates, vitamins K, A, and C, iron, magnesium, and potassium, with minimal calories. Spinach, which contains vitamin C and fibre, lowers the risk of cardiovascular disease. Spinach attributes include high fibre composition, complex carbohydrates, and lower fat ranges, which can augment blood glucose management, reduce insulin provisions (magnesium), and promote satiety in diabetic patients. Spinach saponins can decrease blood pressure and cholesterol. Consuming spinach can boost NO levels, improve lipid balance, and enhance endothelial function. Spinach contains vitamin K, folic acid, lutein, and β-carotenes, which help prevent oxidative stress caused by ageing and improve cognitive function skills and mental ability. Spinach contains both nutritional and non-nutritional antioxidants, including phenolic and flavonoids, as well as antioxidant enzymes, which may help prevent cancer. Spinach can help alleviate chronic diseases caused by oxidative stress and inflammation by quenching ROS, reducing oxidative injuries, and hampering NF-κB commencement and inflammatory indicators.

Spinach, when consumed, has hypocaloric and functional effects, as well as antioxidant properties that promote health. Including spinach in daily meals and salads might boost exposure to the phytochemicals and bioactive compounds discussed herein, perhaps leading to good health consequences.

REFERENCES

Abolhassanzadeh, Zohreh, Elham Aflaki, Gholamhossein Yousefi, and Abdolali Mohagheghzadeh. "Medicinal Plants for Joint Pain in Traditional Persian Medicine," June 1, 2016. https://tips.sums.ac.ir/article_42189.html.

Achaya, K. T. *The Story of Our Food*, 2003. Universities Press.

Akasaka, Hiroaki, Yoshiyuki Mizushina, Kenji Yoshida, Yasuo Ejima, Naritoshi Mukumoto, Tianyuan Wang, Sachiko Inubushi, Masao Nakayama, Yuki Wakahara, and Ryohei Sasaki. "MGDG Extracted from Spinach Enhances the Cytotoxicity of Radiation in Pancreatic Cancer Cells." *Radiation Oncology* 11, no. 1 (November 22, 2016). https://doi.org/10.1186/s13014-016-0729-0.

Alessa, Ola, Safaa Najla, and Ramzi Murshed. "Improvement of Yield and Quality of Two Spinacia Oleracea L. Varieties by Using Different Fertilizing Approaches." *Physiology and Molecular Biology of Plants* 23, no. 3 (June 9, 2017): 693–702. https://doi.org/10.1007/s12298-017-0453-8.

Amirinejad, Ali, Ali Saneei Totmaj, Farzaneh Mardali, Azita Hekmatdoost, Hadi Emamat, Majid Safa, and Farzad Shidfar. "Administration of Hydro-Alcoholic Extract of Spinach Improves Oxidative Stress and Inflammation in High-Fat Diet-Induced NAFLD Rats." *BMC Complementary Medicine and Therapies* 21, no. 1 (September 3, 2021). https://doi.org/10.1186/s12906-021-03396-x.

Antani, Vishu, and Santosh Mahapatra. "Evolution of Indian Cuisine: A Socio-Historical Review." *Journal of Ethnic Foods* 9, no. 1 (April 28, 2022). https://doi.org/10.1186/s42779-022-00129-4.

Babu, Ng Ramesh, J. Divakar, U. Lal Krishna, and C. Vigneshwaran. "Study of Antimicrobial, Antioxidant, Anti-Inflammatory Activities and Phytochemical Analysis of Cooked and Uncooked Different Spinach Leaves." *Journal of Pharmacognosy and Phytochemistry* 7, no. 5 (January 1, 2018): 1798–1803. https://www.phytojournal.com/archives/2018/vol7issue5/PartAE/7-4-667-208.pdf.

Bergman, Margalit, Lucy Varshavsky, Hugo E. Gottlieb, and Shlomo Grossman. "The Antioxidant Activity of Aqueous Spinach Extract: Chemical Identification of Active Fractions." *Phytochemistry* 58, no. 1 (September 1, 2001): 143–152. https://doi.org/10.1016/s0031-9422(01)00137-6.

Bhattarai, Gehendra, and Ainong Shi. "Research Advances and Prospects of Spinach Breeding, Genetics, and Genomics." *Vegetable Research* 1, no. 1 (January 1, 2021): 1–18. https://doi.org/10.48130/vr-2021-0009.

Bowers, J. L., and M. J. Goode. "Ozarka and Greenvalley: New Disease Resistant Spinach Cultivars." *Arkansas Farm Research* 6 (1981): 1198.

Cao, Guohua, Robert M. Russell, Neal Lischner, and Ronald L. Prior. "Serum Antioxidant Capacity Is Increased by Consumption of Strawberries, Spinach, Red Wine or Vitamin C in Elderly Women." *Journal of Nutrition* 128, no. 12 (December 1, 1998): 2383–2390. https://doi.org/10.1093/jn/128.12.2383.

Chu, Yi-Fang, Jie Sun, Xianzhong Wu, and Rui Hai Liu. "Antioxidant and Antiproliferative Activities of Common Vegetables." *Journal of Agricultural and Food Chemistry* 50, no. 23 (October 10, 2002): 6910–6916. https://doi.org/10.1021/jf020665f.

Correll, J. C., B. H. Bluhm, C. Feng, K. Lamour, L. J. Du Toit, and S. T. Koike. "Spinach: Better Management of Downy Mildew and White Rust through Genomics." In *Springer eBooks*, 61–73, 2010. https://doi.org/10.1007/978-94-007-1281-2_5.

Djoussé, Luc, Donna K. Arnett, Hilary Coon, Michael A, Province, Lynn L, Moore, and R. Curtis Ellison. "Fruit and Vegetable Consumption and LDL Cholesterol: The National Heart, Lung, and Blood Institute Family Heart Study." *American Journal of Clinical Nutrition* 79, no. 2 (February 1, 2004): 213–217. https://doi.org/10.1093/ajcn/79.2.213.

Ekor, Martins. "The Growing Use of Herbal Medicines: Issues Relating to Adverse Reactions and Challenges in Monitoring Safety." *Frontiers in Pharmacology* 4 (January 1, 2014). https://doi.org/10.3389/fphar.2013.00177.

Ferreira, Jorge F. S., Jaime Barros Da Silva Filho, Xuan Liu, and Devinder Sandhu. "Spinach Plants Favor the Absorption of K+ over Na+ Regardless of Salinity, and May Benefit from Na+ When K+ Is Deficient in the Soil." *Plants* 9, no. 4 (April 15, 2020): 507. https://doi.org/10.3390/plants9040507.

Freudenheim, Jo L., James R. Marshall, John E. Vena, Rosemary Laughlin, John R. Brasure, Mya K. Swanson, Takuma Nemoto, and Saxon Graham. "Premenopausal Breast Cancer Risk and Intake of Vegetables, Fruits, and Related Nutrients." *JNCI Journal of the National Cancer Institute* 88, no. 6 (March 20, 1996): 340–348. https://doi.org/10.1093/jnci/88.6.340.

Gaikwad, Priyanka Subhash, Rajkumar Virbhadrappa Shete, and Kishor Vasant Otari. "Spinacia Oleracea Linn: A Pharmacognostic and Pharmacological Overview." *International Journal of Research in Ayurveda & Pharmacy* 1, no. 1 (2010): 78–84. https://www.ijrap.net.

Grant, C. A., J. M. Clarke, S. Duguid, and R. L. Chaney. "Selection and Breeding of Plant Cultivars to Minimize Cadmium Accumulation." *The Science of the Total Environment* 390, no. 2–3 (December 1, 2007): 301–310. https://doi.org/10.1016/j.scitotenv.2007.10.038.

Gupta, Charu, Consuelo Pacheco, and Dhan Prakash. "Anti-Inflammatory Functional Foods." In *Advances in Human Services and Public Health (AHSPH) Book Series*, 48–78, 2017. https://doi.org/10.4018/978-1-5225-2970-5.ch003.

Gupta, Sheetal, and Jamuna Prakash. "Studies on Indian Green Leafy Vegetables for Their Antioxidant Activity." *Plant Foods for Human Nutrition* 64, no. 1 (November 4, 2008): 39–45. https://doi.org/10.1007/s11130-008-0096-6.

Gutierrez, Rosa Martha Perez, Elisa Gutierrez Velazquez, and Silvia Patricia Paredes Carrera. "Spinacia Oleracea Linn Considered as One of the Most Perfect Foods: A Pharmacological and Phytochemical Review." *Mini-Reviews in Medicinal Chemistry* 19, no. 20 (June 3, 2019): 1666–1680. https://doi.org/10.2174/1389557519666190603090347.

Hafez, Hetta M., A. S. Moawad, M. Abdel-Aziz Hamed, and A. I. Sabri "*In-vitro* and *In-vivo* Hypolipidemic Activity of Spinach Roots and Flowers." *Iran Journal of Pharm Research* 16, no. 4 (Fall, 2017): 1509–1519.

Iritani, N., and J. Nogi. "Effect of Spinach and Wakame on Cholesterol Turnover in the Rat." *Atherosclerosis* 15, no. 1 (January 1, 1972): 87–92. https://doi.org/10.1016/0021-9150(72)90041-x.

Joseph, J. A., B. Shukitt-Hale, N. A. Denisova, R. L. Prior, G. Cao, A. Martin, G. Taglialatela, and P. C. Bickford. "Long-Term Dietary Strawberry, Spinach, or Vitamin E Supplementation Retards the Onset of Age-Related Neuronal Signal-Transduction and Cognitive Behavioral Deficits." *Journal of Neuroscience* 18, no. 19 (October 1, 1998): 8047–8055. https://doi.org/10.1523/jneurosci.18-19-08047.1998.

Jovanovski, Elena, Laura Bosco, Kashif Khan, Fei Au-Yeung, Hoang Ho, Andreea Zurbau, Alexandra L. Jenkins, and Vladimir Vuksan. "Effect of Spinach, a High Dietary Nitrate Source, on Arterial Stiffness and Related Hemodynamic Measures: A Randomized, Controlled Trial in Healthy Adults." *Clinical Nutrition Research* 4, no. 3 (January 1, 2015): 160. https://doi.org/10.7762/cnr.2015.4.3.160.

Kapoor, Monika, Neha Sharma, and Bimla Nehru. "Spinach Leaf Extract Ameliorates Multiple Organ Dysfunction Induced by Lipopolysaccharide Exposure in Animal Model of Parkinson's Disease." *Alzheimers & Parkinsons Disease Open Access* 1, no. 1 (August 18, 2017): 1–9.

Koh, Eunmi, Suthawan Charoenprasert, and Alyson E. Mitchell. "Effect of Organic and Conventional Cropping Systems on Ascorbic Acid, Vitamin C, Flavonoids, Nitrate, and Oxalate in 27 Varieties of Spinach (Spinacia Oleracea L.)." *Journal of Agricultural and Food Chemistry* 60, no. 12 (March 6, 2012): 3144–3150. https://doi.org/10.1021/jf300051f.

Koushik, A., D. J. Hunter, D. Spiegelman, W. L. Beeson, P. A. Van Den Brandt, J. E. Buring, and E. E. Calle. "Fruits, Vegetables, and Colon Cancer Risk in a Pooled Analysis of 14 Cohort Studies." *JNCI Journal of the National Cancer Institute* 99, no. 19 (September 25, 2007): 1471–1483. https://doi.org/10.1093/jnci/djm155.

Kunicki, Edward, Aneta Grabowska, Agnieszka Sękara, and Renata Wojciechowska. "The Effect of Cultivar Type, Time of Cultivation, and Biostimulant Treatment on the Yield of Spinach (Spinacia Oleracea L.)." *Folia Horticulturae* 22, no. 2 (December 1, 2010): 9–13. https://doi.org/10.2478/fhort-2013-0153.

Kuriyama, Isoko, Keiichi Musumi, Yuko Yonezawa, Masaharu Takemura, Naoki Maeda, Hiroshi Iijima, Takahiko Hada, Hiromi Yoshida, and Yoshiyuki Mizushina. "Inhibitory Effects of Glycolipids Fraction from Spinach on Mammalian DNA Polymerase Activity and Human Cancer Cell Proliferation." *The Journal of Nutritional Biochemistry* 16, no. 10 (August 4, 2005): 594–601. https://doi.org/10.1016/j.jnutbio.2005.02.007.

Lee, Sang-Hee, Ah-Ram Han, Byoung-Mok Kim, Mi Sung, and Sun-Mee Hong. "Lactococcus Lactis--Fermented Spinach Juice Suppresses LPS---Induced Expression of Adhesion Molecules and Inflammatory Cytokines -through the NF--κB Pathway in HUVECs.-" *Experimental and Therapeutic Medicine* 23, no. 6 (April 12, 2022). https://doi.org/10.3892/etm.2022.11317.

Leskovar, Daniel I., and Giovanni Piccinni. "Yield and Leaf Quality of Processing Spinach under Deficit Irrigation." *HortScience* 40, no. 6 (October 1, 2005): 1868–1870. https://doi.org/10.21273/hortsci.40.6.1868.

Li, Shanshan, Juanjuan Yu, Ying Li, Heng Zhang, Xuesong Bao, Jiayi Bian, Chenxi Xu, "Heat-Responsive Proteomics of a Heat-Sensitive Spinach Variety." *International Journal of Molecular Sciences* 20, no. 16 (August 8, 2019): 3872. https://doi.org/10.3390/ijms20163872.

Liu, Alex H., Catherine P. Bondonno, Kevin D. Croft, Ian B. Puddey, Richard J. Woodman, Lisa Rich, Natalie C. Ward, Joseph A. Vita, and Jonathan M. Hodgson. "Effects of a Nitrate-Rich Meal on Arterial Stiffness and Blood Pressure in Healthy Volunteers." *Nitric Oxide* 35 (October 10, 2013): 123–130. https://doi.org/10.1016/j.niox.2013.10.001.

Lomnitski, Liat, Julie F. Foley, Shlomo Grossman, Varda Ben Shaul, Robert R. Maronpot, Cindy R. Moomaw, Michela Carbonatto, and Abraham Nyska. "Effects of Apocynin and Natural Antioxidant from Spinach on Inducible Nitric Oxide Synthase and Cyclooxygenase-2 Induction in Lipopolysaccharide-Induced Hepatic Injury in Rat." *Pharmacology & Toxicology* 87, no. 1 (July 1, 2000): 18–25. https://doi.org/10.1111/j.0901-9928.2000.870104.x.

Longnecker, M. P., P. A. Newcomb, R. Mittendorf, E. R. Greenberg, and W. C. Willett. "Intake of Carrots, Spinach, and Supplements Containing Vitamin A in Relation to Risk of Breast Cancer." *Cancer Epidemiology, Biomarkers and Prevention* 6 no. 11 (November, 1997): 887–892.

Maeda, Naoki, Yasuo Kokai, Seiji Ohtani, Hiroeki Sahara, Yuko Kumamoto-Yonezawa, Isoko Kuriyama, Takahiko Hada, Noriyuki Sato, Hiromi Yoshida, and Yoshiyuki Mizushina. "Anti-Tumor Effect of Orally Administered Spinach Glycolipid Fraction on Implanted Cancer Cells, Colon-26, in Mice." *Lipids* 43, no. 8 (July 1, 2008). https://doi.org/10.1007/s11745-008-3202-5.

Marinaccio, Lorenza, Gokhan Zengin, Stefano Pieretti, Paola Minosi, Edina Szucs, Sandor Benyhe, Ettore Novellino, Domiziana Masci, Azzurra Stefanucci, and Adriano Mollica. "Food-Inspired Peptides from Spinach Rubisco Endowed with Antioxidant, Antinociceptive and Anti-Inflammatory Properties." *Food Chemistry X* 18 (March 13, 2023): 100640. https://doi.org/10.1016/j.fochx.2023.100640.

Meng, Shuchun, Changjiang Liu, Xiuping Xu, Songquan Song, Shunhua Song, Zhenxian Zhang, and Ling Liu. "Comparison of Morphological Features of Fruits and Seeds for Identifying Two Taxonomic Varieties of Spinacia Oleracea L." *Canadian Journal of Plant Science* (September 12, 2017). https://doi.org/10.1139/cjps-2017-0119.

Mladenka, Premysl, Katerina Macakova, Lenka Kujovská Krcmova, Lenka Javorska, Kristyna Mrstna, Alejandro Carazo, Michele Protti, Fernando Remiao, and Lucie Novakova. "Vitamin K—Sources, Physiological Role, Kinetics, Deficiency, Detection, Therapeutic Use, and Toxicity." *Nutrition Reviews* 80, no. 4 (July 9, 2021): 677–698. https://doi.org/10.1093/nutrit/nuab061.

Morelock, Teddy E., and James C. Correll. "Spinach." In *Springer eBooks*, 189–218, 2007. https://doi.org/10.1007/978-0-387-30443-4_6.

Mou, Beiquan. "'USDA Red' Spinach." *HortScience* 54, no. 11 (November 1, 2019): 2070–2072. https://doi.org/10.21273/hortsci14308-19.

Nath, Chumki Rani, Methela Nusrat Jahan, Ruhi Roksana Aftab, and Islam Mohammad Shafiqul "Varietal Performances of Spinach (Spinacia Oleracea L.) at Coastal Region in Bangladesh." *American-Eurasian Journal of Agronomy* 13 (2020). https://doi.org/10.5829/idosi.aeja.2020.39.45.

Nuutila, Anna-Maria, Kari Kammiovirta, and Kirsi-Marja Oksman-Caldentey. "Comparison of Methods for the Hydrolysis of Flavonoids and Phenolic Acids from Onion and Spinach for HPLC Analysis." *Food Chemistry* 76, no. 4 (April 1, 2002): 519–525. https://doi.org/10.1016/s0308-8146(01)00305-3.

Park, Hyewon, Jin Seok Lee, Nuri Lee, Kisang Kwon, Jung Boon Kim, Suk Bin Kim, Hyeong-Geug Kim, and Dong Woon Kim. "Red Stem of Spinach Promotes Antioxidant and Anti-Inflammatory Chondroprotection in a Rat Model of Osteoarthritis." *Journal of Functional Foods* 109 (September 14, 2023): 105789. https://doi.org/10.1016/j.jff.2023.105789.

Park, Jae-Hee, Rae-Young Kim, and Eunju Park. "Antidiabetic Activity of Fruits and Vegetables Commonly Consumed in Korea: Inhibitory Potential Against A-glucosidase and Insulin-like Action in Vitro." *Food Science and Biotechnology* 21, no. 4 (August 1, 2012): 1187–1193. https://doi.org/10.1007/s10068-012-0155-5.

Rahati, Sara, Mohammadreza Eshraghian, Abdolali Ebrahimi, and Hamideh Pishva. "Effect of Spinach Aqueous Extract on Wound Healing in Experimental Model Diabetic Rats with Streptozotocin." *Journal of the Science of Food and Agriculture* 96, no. 7 (July 23, 2015): 2337–2343. https://doi.org/10.1002/jsfa.7349.

Rahati, Sara, Mohammad Kamalinezhad, Abdolali Ebrahimi, Mohammadreza Eshraghian, and Hamideh Pishva. "Accelerated Wound Healing Induced by Spinach Extract in Experimental Model Diabetic Rats with Streptozotocin." *Scientific Reports* 13, no. 1 (September 11, 2023). https://doi.org/10.1038/s41598-023-42033-0.

Rana, Mahesh Kumar, and Bhupinder Singh Thakur. *Vegetable Crops Science*, 1st ed., 2017. https://www.taylorfrancis.com/chapters/edit/10.1201/9781315116204-27/spinach-bhupinder-singh-thakur.

Rao, K. N. V., Bushra Tabassum, S. Raghu Babu, Alagara Raja, David Banji, and Nalanda College of Pharmacy. "Preliminary Phytochemical Screening of Spinacia Oleracea L." *World Journal of Pharmacy and Pharmaceutical Sciences* 4 (2015).

Rashid, Madiha, Zubaida Yousaf, Ahmad Din, Muneeb Munawar, Arusa Aftab, Nadia Riaz, Afifa Younas, Ibrahim A. Alaraidh, Mohammad K. Okla, and Hamada AbdElgawad. "Assessment of Mineral Nutrient Efficiency in Genetically Diverse Spinach Accessions by Biochemical and Functional Marker Strategies." *Frontiers in Plant Science* 13 (May 30, 2022). https://doi.org/10.3389/fpls.2022.889604.

Ribera, Arnau, Yuling Bai, Anne-Marie A. Wolters, Rob Van Treuren, and Chris Kik. "A Review on the Genetic Resources, Domestication and Breeding History of Spinach (Spinacia Oleracea L.)." *Euphytica* 216, no. 3 (February 27, 2020). https://doi.org/10.1007/s10681-020-02585-y.

Roberts, Joseph L., and Régis Moreau. "Functional Properties of Spinach (Spinacia Oleracea L.) Phytochemicals and Bioactives." *Food & Function* 7, no. 8 (January 1, 2016): 3337–3353. https://doi.org/10.1039/c6fo00051g.

Roughani, Afra, Seied Mehdi Miri, Abdol Karim Kashi, and Behnam Naserian Khiabani. "Increasing the Ploidy Level In Spinach (Spinacia Oleracea L.) Using Mitotic Inhibitors." *Plant Cell Biotechnology and Molecular Biology* 18, no. 3–4 (March, 2017), 124–130.

Serpolay, Estelle, Nicolas Schermann, Julie Dawson, Edith T. Lammerts Van Bueren, Isabelle Goldringer, and Véronique Chable. "Phenotypic Changes in Different Spinach Varieties Grown and Selected under Organic Conditions." *Sustainability* 3, no. 9 (September 23, 2011): 1616–1636. https://doi.org/10.3390/su3091616.

Shohag, M. J. I., Yan-Yan Wei, Ning Yu, Jie Zhang, Kai Wang, Johan Patring, Zhen-Li He, and Xiao-E Yang. "Natural Variation of Folate Content and Composition in Spinach (Spinacia Oleracea) Germplasm." *Journal of Agricultural and Food Chemistry* 59, no. 23 (October 18, 2011): 12520–12526. https://doi.org/10.1021/jf203442h.

Siegel, Rebecca L., Kimberly D. Miller, Nikita Sandeep Wagle, and Ahmedin Jemal. "Cancer Statistics, 2023." *CA: A Cancer Journal for Clinicians* 73, no. 1 (January 1, 2023): 17–48. https://doi.org/10.3322/caac.21763.

Singh, G., Asha Kawatra, and S. Sehgal. "Nutritional Composition of Selected Green Leafy Vegetables, Herbs and Carrots." *Plant Foods for Human Nutrition* 56, no. 4 (January 1, 2001): 359–364. https://doi.org/10.1023/a:1011873119620.

Sobko, Tanja, Claude Marcus, Mirco Govoni, and Shigeru Kamiya. "Dietary Nitrate in Japanese Traditional Foods Lowers Diastolic Blood Pressure in Healthy Volunteers." *Nitric Oxide* 22, no. 2 (November 3, 2009): 136–140. https://doi.org/10.1016/j.niox.2009.10.007.

Sofi, Francesco, Rosanna Abbate, Gian Franco Gensini, and Alessandro Casini. "Importance of Diet on Disease Prevention." *International Journal of Medicine and Medical Sciences* 5, no. 2 (February 28, 2013): 55–59. https://doi.org/10.5897/ijmms12.107.

Sun, Feilong, Yaru Yan, and Long Lin. "The Evaluation of Antioxidant Properties and Stability of Polyphenols from Spinacia Oleracea." *Journal of Biotech Research* 9 (2018): 8–13.

UN Food and Agriculture Organization. "Spinach production in 2022; Crops/Regions/World/Production Quantity/Year from pick lists." UN Food and Agriculture Organization, 2024. Retrieved May 7, 2024. https://en.wikipedia.org/wiki/Spinach/08/20/2024.

Ward, Brian W., and Jeannine S. Schiller. "Prevalence of Multiple Chronic Conditions Among US Adults: Estimates from the National Health Interview Survey, 2010." *Preven Preventing Chronic Disease* 10 (April 25, 2013): E65. https://doi.org/10.5888/pcd10.120203.

Wikipedia Contributors. "Spinach." *Wikipedia*, October 29, 2024. https://en.wikipedia.org/wiki/Spinach.

Wolf, Frederick T., John G. Coniglio, and James T. Davis. "Fatty Acids of Spinach Chloroplasts." *Plant Physiology* 37, no. 1 (January 1, 1962): 83–85. https://doi.org/10.1104/pp.37.1.83.

Zhou, Kequan, and Liangli Yu. "Total Phenolic Contents and Antioxidant Properties of Commonly Consumed Vegetables Grown in Colorado." *LWT* 39, no. 10 (September 22, 2005): 1155–1162. https://doi.org/10.1016/j.lwt.2005.07.015.

4 An Insight into the Chemical Characteristics and Medicinal Properties of Capsicum Species

Himanko Gogoi, Hriday Kumar Biswas, Yogita Ale, and Mansi Butola

4.1. INTRODUCTION

Chilli peppers originated in Mexico following the Columbian Exchange; several varieties disseminated globally, utilized in cuisine and traditional old medicine. Pepper (Capsicum spp.) is a crop that has been grown from a long time back and is also one of the most utilized crops. Its utilization commenced from considerably long ago and is believed to originate from America (Musfiroh et al. 2013). The genus Capsicum consists of more than 200 species, bearing fruits varying significantly in flavour and pungency. Five primary Capsicum species identified in scholarly literature are: *C. baccatum*, *C. annuum*, *C. frutescens*, *C. chinense*, and *C. pubescens*. These species of Capsicum peppers are indigenous to the humid and tropical regions of Central America and South America and include economically important varieties (Masud Parvez 2017). They are typically utilized as condiments or nutrients, offering extensive medicinal applications in Chinese, Indian, and American practices for acne treatment, joint inflammation, and dyspepsia. Pepper is classified within the Capsicum genus, which is part of the family of Solanaceae. The Capsicum genus has approximately 31 species, with the principal five farmed species being *C. frutescens*, *C. annuum*, *C. chinense*, *C. baccatum*, and *C. pubescens*. *C. chinense* and *C. frutescens* are currently extensively grown (Patel and Patel 2023).

Pepper is widely used as seasoning worldwide and is highly esteemed because of its spiciness, enhancing the flavour of different cuisines worldwide. Historically, it was predominantly employed for flavouring and medical purposes; however, its contemporary applications now include preparation of veggies, crisps, and flavourings. It is cultivated as an ornamental plant and utilized in the production of extracts for various pharmaceutical and cosmetic industries.

The cultivation and utilization of Capsicum fruits have proliferated globally since the era of maritime exploration, when various civilizations began to consume as well as cultivate them, adapting to diverse environmental circumstances through hybridization. The rapid surge in their popularity can be attributed to its remarkable properties; for example, it has been documented that prior to 1600 BCE, the fruits were utilized as colours, culinary spices, and in the alleviation of gastrointestinal ailments and headaches (Snitker et al. 2009; Zhang et al. 2024). The application of pepper in culinary and medicinal contexts has prompted numerous researchers to investigate the chemical qualities associated with peppers. Concurrently, the use of Capsicum fruits expanded and garnered other applications, primarily in medicinal contexts. Subsequently, study advancements within the genus species attributed these qualities to the plentiful presence of flavonoids, capsaicinoids, along with carotenoids in the peppers. Furthermore, these three categories of chemicals have already been

DOI: 10.1201/9781003559894-4

identified as bioactive compounds. The initial two metabolites contribute to the appealing colours of peppers, while capsaicinoids impart a characteristic pungent flavour, which results in the global popularity of Capsicum (Antonio, Wiedemann, and Veiga Junior 2018). Pungency involves stimulation of the terminal nerves, resulting in sensations of pain and heat, hence serving as an effective means in culinary applications to eliminate blandness and impart new flavours to dishes. Recently, these fruits have been utilized across multiple industries as food colourants, spices, analgesics, antioxidants, anti-inflammatories, anticancer agents, and antitumour agents (Batiha et al. 2020; Ishimov, Ziyavitdinov, and Sagdiev 2011).

4.2. CHEMICAL CHARACTERISTICS OF CAPSICUM SPECIES

The chemical characteristics of pepper change according to the species, environmental factors, circumstances, seasonality, and the plant's life cycle. Selecting the economically viable Capsicum species and cultivars is tailored to a certain aim. Consequently, understanding the composition, properties, and factors influencing the chemical profile of Capsicum fruits may facilitate the selection of cultivars and methods of cultivation, enhance the industrial application of pepper-based products, boost productivity, and promote the advancement of technology applicable after harvesting (Batiha et al. 2020). In addition to adding flavour to food, Capsicum peppers are also rich sources of phytochemicals that provide possible health benefits. It contains various phytoconstituents, including capsaicinoids, carotenoids, vitamins, capsiate, flavonoids, etc., as shown in Figure 4.1. These phytoconstituents possess several pharmacological benefits, such as antioxidant, anti-inflammatory, antimicrobial, neuroprotective, etc. (Yadav et al. 2009; Antonio, Wiedemann, and Veiga Junior 2018).

4.2.1. Capsaicinoids

Capsaicinoids (CAPS) are a series of compounds characterized by an alkaloid-like chemical structure, exclusively synthesized within the Capsicum species and responsible for creating pungency. This trait, along with their bioactivity, stems from the chemical structure that enables distinct interactions with nerve receptors and cellular membranes. The fundamental chemical structure of capsaicinoids consists of a group of vanillyl linked to an alkyl chain and an amide. The diversity in molecular level of capsaicinoids arises from the chain of carbon, which can be classified as: (1) unsaturated and branched, characterized by capsaicin; (2) branched and saturated, represented by dihydrocapsaicin; or (3) linear and saturated, in nonivamide. This molecular structure imparts an amphiphilic characteristic to the CAPS, exhibiting a lack of basicity, which is uncharacteristic to the common behaviour reported in alkaloids (Hudáková et al. 2024; Naves et al. 2019). Among the chemicals classified as capsaicinoids, only four are typically examined due to their elevated in situ concentration. The principal chemicals are capsaicin and dihydrocapsaicin, whereas the secondary compounds are homocapsaicin and nordihydrocapsaicin. The CAPS molecular structure exhibits resistance to ionizing radiations; yet, it decomposes by more than half of its actual concentration when subjected to more than 80°C temperatures. The degeneration by heat is mostly ascribed to the breakage of the nitrogen-carbon bond, resulting in the formation of the acyl chain and a vanillyl group. Capsaicinoids exhibit solubility in solvents with low to medium polarity, including ethanol, alcohol, methanol, and acetonitrile; the latter has been utilized most in capsaicin extraction due to its superior solubility for capsaicinoids (Hudáková et al. 2024; Quinones-Seglie, Burns, and Villalon 1989; Ishimov, Ziyavitdinov, and Sagdiev 2011).

4.2.2. Carotenoids

Carotenoids comprise around 750 tetraterpenoid chemicals, extensively found in plants and recognized for their chromogenic qualities that impart vibrant orange, red, and yellow hues to organisms.

This category of colours can be produced in certain nonphotosynthetic animals and in all photosynthetic organisms, serving a crucial function in stabilization of free radicals, photosynthesis, and photoprotection. Carotenoids are significant for the nutrition and health of animals, although not produced by them due to their anticarcinogenic and antioxidant properties, and they serve as precursors to vital molecules such as vitamin A and retinoic acid (Antonio, Wiedemann, and Veiga Junior 2018; Batiha et al. 2020; Gómez-García and Ochoa-Alejo 2013).

The diversity of colours in the Capsicum genus is primarily influenced by the array of carotenoids it contains, particularly the oxygenated types, whose composition can vary depending on the species, maturity stage, and cultivars, highlighting the intricacy of carotenoid synthesis. Carotenoids exhibit a less polar structure of the molecule founded on a tetraterpenoid framework, characterized by predominant structures containing 40 carbon atoms produced from units of isoprene. It can be grouped into two primary categories depending on their chemical structure: xanthophylls and carotenes (Guzman et al. 2021). Carotenes consist of cyclized or linear hydrocarbons, exemplified by alpha-carotene and beta-carotene, whereas xanthophylls encompass carotenoids with oxygenated functionalities, including methoxy-, hydroxy-, keto-, carbonyl, and epoxy groups, as found in violaxanthin, lutein, neoxanthin, and zeaxanthin. A further classification is based on the chromogenic characteristics of carotenoids (Gómez-García and Ochoa-Alejo 2013), categorized into two factions: red and yellow. Each proportion corresponds to a certain chemical structural property that enables the substance to interact variably with visible radiation. Consequently, their conjugation of double bonds, oxygenated and length functionalities enable them to exhibit a diverse range of absolute values in the visible spectrum. The yellow portion consists of carotenoids that exhibit orange or yellow hues as perceived by the human eye, which include hydroxylated carotenoids (β-cryptoxanthin and zeaxanthin), carotenoids based on hydrocarbon (β-carotene and α-carotene), and epoxidized carotenoids (cucurbitaxanthin A and violaxanthin). The red part structurally varies from the yellow due to the replacement of the 3-hydroxy b rings with a 3-hydroxyacylcyclopentane ring, which imparts enhanced stability. Within this portion, ketocarotenoids are present, which are predominantly found in the Capsicum genus, including capsanthin 5,6-epoxide, capsanthin, and capsorubin (Hassan et al. 2019).

4.2.3. Capsinoid

Capsaicinoids have been thoroughly investigated by the food industry owing to their bioactive and pungent characteristics. However, their application is constrained by the pungency and toxicological effects that even little amounts can exert on humans. Analogous compounds, known as capsinoids, is a viable alternative to CAPS for industrial use, distinguished mostly by their reduced pungency. Capsinoids are known as a category of compounds within the class of CAPS that are synthesized solely within the genus Capsicum, encompassing both nonpungent and pungent cultivars. Capsinoids typically occur in modest concentrations in fruit, roughly 300 mg per gram. Their molecular structure diverges from CAPS due to the presence of a functional group ester linking the vanillyl group to the carbon chain. Moreover, the limited research on the pharmaceutical usage of capsinoids predominantly centres on capsiate. Capsiate has been shown to have qualities akin to capsaicin regarding analgesia, anticarcinogenic effects, and the prevention of cardiovascular and gastrointestinal illnesses, underscoring its potential as a less hazardous alternative to capsaicin (Snitker et al. 2009; Quinones-Seglie, Burns, and Villalon 1989; Naves et al. 2019).

4.2.4. Flavonoids

Flavonoids are a category of chemical substances comprising more than 7,000 secondary plant metabolites, characterized by a structure containing 15 carbon atoms. These compounds gained prominence due to their bioactive qualities in mammals, which cannot synthesize them, primarily because of their antioxidant activity. They are defensive against hypertension, inflammatory processes, arthritis, and AIDS. These qualities have rendered them highly appealing to various industrial sectors, including food, pharmaceuticals, and apparel. Flavonoids are present in the Capsicum species, primarily concentrated in the peel, and are represented as aglycone flavonoids. In peppers, they may exist as O-glycosides and C-glycoside conjugation derivatives. In Capsicum, luteolin and quercetin are identified as the predominant flavonoids, constituting roughly 41% of the total content of flavonoid, often in their hydrolyzed form. The flavonoid concentration in pepper is typically represented as the aggregate of the aforementioned compounds, owing to the challenges associated in conducting acid hydrolysis on several additional conjugated flavonoids (Oulaï et al. 2018; Ishimov, Ziyavitdinov, and Sagdiev 2011).

4.2.5. Vitamins

Vitamins, particularly vitamin C (ascorbic acid), are abundant in Capsicum peppers. They also contain a good concentration of vitamin B6, vitamin E, folate, and vitamin A (from beta-carotene). Vitamin C is necessary for immune system function, collagen synthesis, and wound healing. Vitamin A helps in maintaining healthy skin, good eyesight, and a healthy immune system. Vitamin E functions as an antioxidant to shield cell membranes from oxidative damage, whereas B vitamins are involved in energy metabolism (Marín et al. 2004). Table 4.1 summarizes the various chemical constituents present in the Capsicum species.

4.3. PHARMACOLOGICAL ACTIVITY OF CAPSICUM SPECIES

Capsicum is rich in bioactive chemicals, which give it a wide range of pharmacological actions. These compounds, which include minerals, vitamins, flavonoids, and capsaicinoids, contribute to the numerous health benefits of consuming Capsicum. Capsicum posses analgesic, anti-inflammatory, antioxidant, gastrointestinal, cardiovascular, metabolic, neuroprotective, antimicrobial, and potentially anticancer activities. However, further research is warranted to explore the therapeutic potential of Capsicum and its constituents in clinical settings and to elucidate their mechanisms of action (Khan et al. 2014; Masud Parvez 2017).

4.3.1. Analgesic Activity

One of Capsicum's most well-known pharmacological effects is analgesia. Capsaicin, the major capsaicinoid found in the Capsicum, is known for its ability to bind to and activate the transient receptor potential vanilloid 1 (TRPV1) receptors on nociceptive neurons. This stimulation initially produces a burning sensation but gradually desensitizes the nerve terminals, reducing the transmission of pain. For the treatment of neuropathic pain, arthritis, and musculoskeletal discomfort, capsaicin is frequently found in topical creams and patches. Deal et al. (1991) found that people using capsaicin of the Capsicum spp., formulated in the form of cream for topical application, get comparatively more relief against pain caused by neuropathic pain disorders, osteoarthritis, and rheumatoid arthritis when compared to a placebo. Although the most common side effect found is a mild to moderate burning sensation, long-term usage is safe. Another research conducted by Anand and Bley (2011) found that when high-concentration capsaicin patches (8%) were used, they gave substantial pain relief after only one 60-minute treatment, lasting up to 12 weeks. According to their findings, these patches desensitize sensory neurons to reduce discomfort and hence have shown effectiveness in diseases such as HIV-related neuropathy and postherpetic neuralgia (Jawad et al. 2017; Kalpana and Vrinda 2024).

4.3.2. Anti-Inflammatory Properties

The chemical constituent found in Capsicum has the ability to modify inflammatory pathways and prevent the synthesis of pro-inflammatory cytokines such as TNF-α and IL−6. The therapy

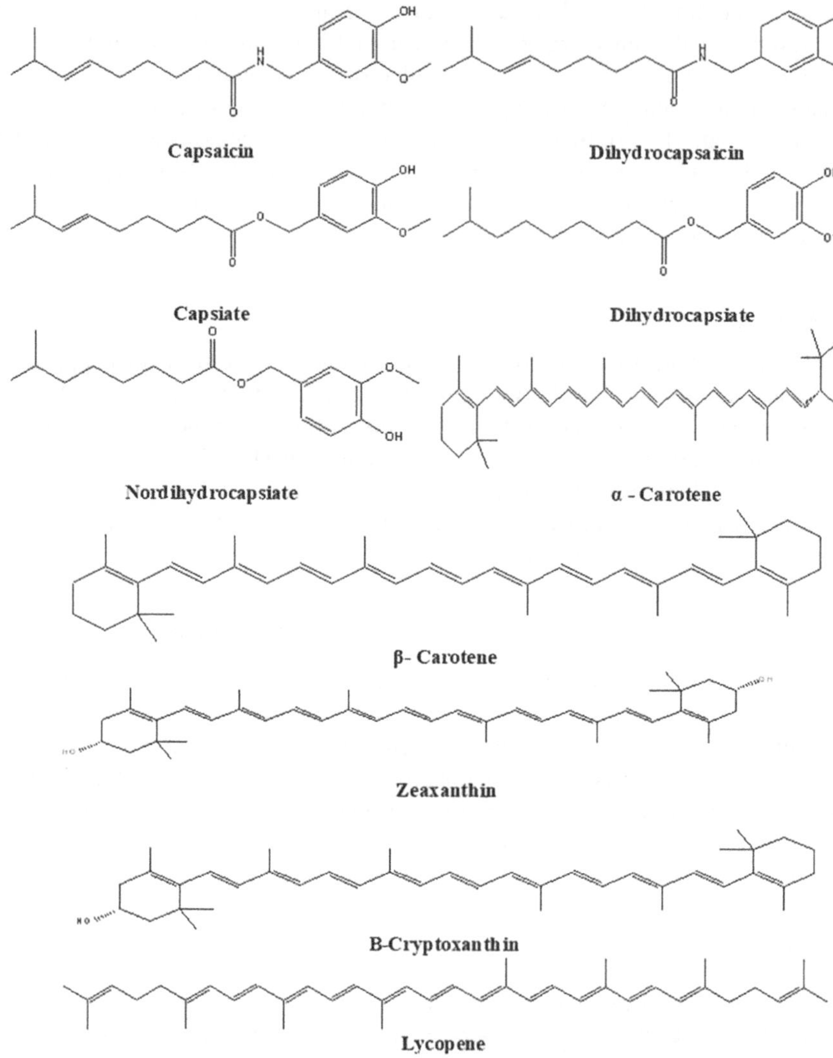

FIGURE 4.1 Chemical constituent present in *Capsicum annuum*.

TABLE 4.1
Various Phytochemicals Present in Capsicum Species

Chemical Constituents	Function/Properties	Concentration Range	References
Capsaicinoids	Responsible for spiciness; potential health benefits includes relief of pain and boost of metabolism	6.78 to 32.5 mg/100g	(Zaki et al. 2017)
Carotenoids (e.g., β-carotene, lutein, zeaxanthin)	Antioxidant properties and support eye health conditions	0.8 – 3.2 (mainly beta-carotene, lutein, zeaxanthin)	(Musfiroh et al. 2013)
Flavonoids (e.g., quercetin, luteolin)	Antioxidant and anti-inflammatory properties	17.17 to 85.49	(Hudáková et al. 2024)
Vitamin C (ascorbic acid)	Antioxidant; supports functions of immune system	170–280	(Badia, Spina, and Vassalotti 2018)
Vitamin E (tocopherols)	Antioxidant; supports skin health	0.1–0.5	(Musfiroh et al. 2013)

of inflammatory illnesses such rheumatoid arthritis, osteoarthritis, and inflammatory bowel disorders may be modified by this activity. Research investigation found out that capsaicin nasal spray decreases cluster headaches and nonallergic rhinitis symptoms by desensitizing nasal sensory neurons, thereby reducing nasal congestion and pain (Li et al. 2023; Kim et al. 2003).

4.3.3. Antioxidant Activity

Capsicum extracts are rich in antioxidants, such as vitamins C and E, carotenoids (such as beta-carotene), and flavonoids (quercetin, luteolin). These antioxidants support general cellular health by scavenging free radicals and preventing oxidative damage to cells and tissues. The antioxidative properties of Capsicum can help prevent illnesses linked to oxidative stress and slow down the aging process (Marín et al. 2004; Oulaï et al. 2018).

4.3.4. Gastrointestinal Effect

Several research studies have revealed that Capsicum improves digestion and strengthens the mucosal barriers in the stomach. It can promote the secretion of protective mucosal compounds and improve the flow of blood to the stomach mucosa, both of which may help in the prevention and treatment of gastric ulcers (Zhang et al. 2024).

4.3.5. Cardiovascular Effect

According to certain research, extracts from Capsicum may be beneficial for the heart. Capsaicin activates TRPV1 receptors in endothelial cells, which leads to vasodilation and improved blood flow. It may also help in reducing cholesterol levels and preventing the formation of blood clots, thereby reducing the risk of cardiovascular diseases (Khan et al. 2014; Masud Parvez 2017).

4.3.6. Metabolic Effects

Consuming Capsicum has been associated to increased fat oxidation and higher energy expenditure. The production of heat and an increase in caloric expenditure are a result of brown adipose tissue's TRPV1 receptor activation, which is responsible for this thermogenic effect. Supplements containing capsaicin are occasionally used to enhance metabolic health and help with weight management. Snitker et al. (2009) showed in their findings that capsaicin capsules help in controlling weight and improving metabolism by promoting fat oxidation and thermogenesis. They assigned 40 women and 40 men to either the capsinoid or placebo group with a mean (+/- SD) age of 42 +/- 8 years and a BMI of 30.4 +/– 2.4. Mean (+/- SD) weight change was 0.9 +/– 3.1 and 0.5 +/– 2.4 kg in the capsinoid and placebo groups, respectively ($P = 0.86$). The study found that there was a higher correlation between the body weight change and the decrease in adiposity of abdomen in the capsinoid group (–1.11 +/– 1.83%) compared to the placebo group (–0.18 +/– 1.94%) ($P = 0.049$).

4.3.7. Neuroprotective Effect

In several preclinical investigations, capsaicin has demonstrated potential benefits on neuroprotection. It has the ability to increase the synthesis of nerve growth factor (NGF), which is necessary for the survival, development, and maintenance of neurons. This neurotrophic activity may be useful to treat neurodegenerative diseases such as Alzheimer's and Parkinson's (Abdel-Salam et al. 2018; Byun et al. 2018).

4.3.8. Antimicrobial Activity

Research has shown that extracts from Capsicum have the ability to resist a variety of pathogens, such as bacteria, fungus, and even certain viruses. Capsaicin and other bioactive chemicals may

TABLE 4.2
Pharmacological Activities of Capsicum Formulations Along with Disease Cured

Formulations	Pharmacological Activity	Disease Cure	Key Findings	Reference
Capsaicin topical cream	Analgesic activity	Arthritis, neuropathy, psoriasis	Long-term use is safe, moderate burning sensations.	(Deal et al. 1991)
High concentration capsaicin patch	Analgesic activity	Cutaneous hypersensitivity, HIV-associated neuropathy	Reduce pain by defunctionalization of nociceptor fibres.	(Anand and Bley 2011)
Capsaicin capsules	Metabolic effect	Reduces obesity	Increase in caloric expenditure due to brown adipose tissue's TRPV1 receptor activation.	(Snitker et al. 2009)
Capsaicin gel	Analgesic activity	Relief of musculoskeletal pain	Effective against pain from conditions such as back pain, muscle strains, and fibromyalgia.	(Frerick et al. 2003)
Capsaicin nasal spray	Anti-inflammatory activity	Relief from nonallergic rhinitis and cluster headaches	Desensitizes nasal sensory neurons, thereby reducing nasal congestion and pain.	(Li et al. 2023)

rupture microbial membranes, interfere with enzyme function, and interfere with microbial DNA replication, making Capsicum a promising natural antibacterial agent (kom, Tamokou, and Kuete 2021; Koffi-Nevry et al. 2012).

4.3.9. Anticancer Activity

Studies indicate that extracts from Capsicum plants may have anticancer effects. In lab experiments, capsaicin has shown to cause apoptosis, or programmed death of cells, and to stop the growth of certain cancer cells. To better understand the processes behind its anticancer properties and its possible uses in cancer therapy, more investigation is necessary (Al-Samydai et al. 2021). Table 4.2 summarizes the pharmacological activities of Capsicum formulations along with diseases.

4.4. CONCLUSION AND FUTURE PERSPECTIVE

The active ingredient in Capsicum species, capsaicin, is already utilized in topical pain relievers. More sophisticated topical gels and transdermal patches are examples of future innovations that might improve patient compliance and efficacy. Capsaicinoids have been demonstrated to increase the expenditure of energy and oxidation of fat, which speeds up metabolism and helps with weight reduction. Future studies may concentrate on developing more effective dietary supplements and functional foods that utilize these properties. Capsaicin has shown potential in preventing the development of cancer cells and triggering apoptosis in several cancer forms. Future studies might result in the creation of novel adjuvant treatments or anticancer medications that contain capsaicin or its analogs. Extract from Capsicum species are used in nasal sprays to alleviate sinus problems and congestion. Future interactions could lessen adverse effects and increase efficacy, offering better alternatives for people with long-term respiratory disorders. Since capsaicin has anti-inflammatory qualities, it may be used to treat inflammatory illnesses like arthritis. Investigating more specialized delivery methods may improve its medicinal potential. Capsicum species have the potential to offer health advantages beyond just nourishment when added to functional meals and drinks. With everything considered, Capsicum species have many health advantages that make them a crucial part of consumer products for wellness and health in the future. Further investigation and advancement are expected to broaden its scope of use and enhance its efficacy across diverse therapeutic fields.

REFERENCES

Abdel-Salam, Omar M. E., Amany A. Sleem, Eman R. Youness, Noha N. Yassen, Nermeen Shaffie, and Sayed A. El-Toumy. 2018. "Capsicum Protects against Rotenone-Induced Toxicity in Mice Brain Via Reduced Oxidative Stress and 5-Lipoxygenase Activation." *Journal of Pharmacy and Pharmacology Research* 2 (3). https://doi.org/10.26502/jppr.0011.

Al-Samydai, Ali, Yasser Khalil Bustanji, Hanan Azzam, Farah Al-Mamoori, Deniz M. Al-Tawalbe, Bashaer Abu-Irmaileh, and Talal Aburjai. 2021. "Promising Cytotoxicity and Anticancer Activity of Capsicum Annuum." *Indian Journal of Forensic Medicine & Toxicology* 15 (3). https://doi.org/10.37506/ijfmt.v15i3.16360.

Anand, P., and K. Bley. 2011. "Topical Capsaicin for Pain Management: Therapeutic Potential and Mechanisms of Action of the New High-Concentration Capsaicin 8 Patch." *British Journal of Anaesthesia*. https://doi.org/10.1093/bja/aer260.

Antonio, A. S., L. S. M. Wiedemann, and V. F. Veiga Junior. 2018. "The Genus: Capsicum: A Phytochemical Review of Bioactive Secondary Metabolites." *RSC Advances*. https://doi.org/10.1039/c8ra02067a.

Badia, Antonella Della, Anna Antonella Spina, and Giuseppe Vassalotti. 2018. "Capsicum Annuum L.: An Overview of Biological Activities and Potential Nutraceutical Properties in Humans and Animals." *Journal of Nutritional Ecology and Food Research* 4 (2). https://doi.org/10.1166/jnef.2017.1163.

Batiha, Gaber El Saber, Ali Alqahtani, Oluwafemi Adeleke Ojo, Hazem M. Shaheen, Lamiaa Wasef, Mahmoud Elzeiny, Mahmoud Ismail, et al. 2020. "Biological Properties, Bioactive Constituents, and Pharmacokinetics of Some Capsicum Spp. and Capsaicinoids." *International Journal of Molecular Sciences*. https://doi.org/10.3390/ijms21155179.

Byun, Eui Baek, Woo Young Park, Woo Sik Kim, Ha Yeon Song, Nak Yun Sung, and Eui Hong Byun. 2018. "Neuroprotective Effect of Capsicum Annuum Var: Abbreviatum against Hydrogen Peroxide-Induced Oxidative Stress in HT22 Hippocampus Cells." *Bioscience, Biotechnology and Biochemistry* 82 (12). https://doi.org/10.1080/09168451.2018.1514248.

Deal, C. L., T. J. Schnitzer, E. Lipstein, J. R. Seibold, R. M. Stevens, M. D. Levy, D. Albert, and F. Renold. 1991. "Treatment of Arthritis with Topical Capsaicin: A Double-Blind Trial." *Clinical Therapeutics* 13 (3).

Ekom, Steve Endeguele, Jean De Dieu Tamokou, and Victor Kuete. 2021. "Antibacterial and Therapeutic Potentials of the Capsicum Annuum Extract against Infected Wound in a Rat Model with Its Mechanisms of Antibacterial Action." *BioMed Research International* 2021. https://doi.org/10.1155/2021/4303902.

Frerick, Helmut, Wolfgang Keitel, Ulrike Kuhn, Stefan Schmidt, Angelika Bredehorst, and Maike Kuhlmann. 2003. "Topical Treatment of Chronic Low Back Pain with a Capsicum Plaster." *Pain* 106 (1–2). https://doi.org/10.1016/S0304-3959(03)00278-1.

Gómez-García, María del Rocío, and Neftalí Ochoa-Alejo. 2013. "Biochemistry and Molecular Biology of Carotenoid Biosynthesis in Chili Peppers (Capsicum Spp.)." *International Journal of Molecular Sciences*. https://doi.org/10.3390/ijms140919025.

Guzman, Ivette, Krystal Vargas, Francisco Chacon, Calen McKenzie, and Paul W. Bosland. 2021. "Health-Promoting Carotenoids and Phenolics in 31 Capsicum Accessions." *HortScience* 56 (1). https://doi.org/10.21273/HORTSCI15039-20.

Hassan, Norazian Mohd, Nurul Asyiqin Yusof, Amirah Fareeza Yahaya, Nurul Nasyitah Mohd Rozali, and Rashidi Othman. 2019. "Carotenoids of Capsicum Fruits: Pigment Profile and Health-Promoting Functional Attributes." *Antioxidants*. https://doi.org/10.3390/antiox8100469.

Hudáková, Terézia, Martina Šemeláková, Peter Očenáš, Mária Kožurková, Kristína Krochtová, Simona Sovová, Zuzana Tóthová, Zuzana Guľášová, Peter Popelka, and Peter Solár. 2024. "Chili Pepper Extracts, Capsaicin, and Dihydrocapsaicin as Potential Anticancer Agents Targeting Topoisomerases." *BMC Complementary Medicine and Therapies* 24 (1). https://doi.org/10.1186/s12906-024-04394-5.

Ishimov, U. Zh, Zh F. Ziyavitdinov, and N. Zh Sagdiev. 2011. "Minor Constituents of Total Capsaicinoids from Capsicum Annuum." *Chemistry of Natural Compounds* 46 (6). https://doi.org/10.1007/s10600-011-9813-8.

Jawad, Muhammad, Haroon Khan, Samreen Pervaiz, Saud S. Bawazeer, Tareq Abu-Izneid, Muhammad Saeed, and Mohammad Amjad Kamal. 2017. "Pharmacological Validation of the Anxiolytic, Muscle Relaxant and Sedative Like Activities of Capsicum Annuum in Animal Model." *Bangladesh Journal of Pharmacology* 12 (4). https://doi.org/10.3329/bjp.v12i4.33182.

Kalpana, C. V., and K. C. Vrinda. 2024. "Comparative Study on Morphological Characteristics, Ethno Botanical and Ethno Pharmacological Importance of Capsicum Annuum L. Var. Annuum and Capsicum Annuum L. Var. Frutescens." *Agricultural Science Digest* 44 (3). https://doi.org/10.18805/ag.D-5650.

Khan, Farhan A., Tariq Mahmood, Muhammad Ali, Abdul Saeed, and Aneela Maalik. 2014. "Pharmacological Importance of an Ethnobotanical Plant: Capsicum Annuum L." *Natural Product Research*. https://doi.org/10.1080/14786419.2014.895723.

Kim, Chu Sook, Teruo Kawada, Byung Sam Kim, In Seob Han, Suck Young Choe, Tadao Kurata, and Rina Yu. 2003. "Capsaicin Exhibits Anti-Inflammatory Property by Inhibiting IkB-a Degradation in LPS-Stimulated Peritoneal Macrophages." *Cellular Signalling* 15 (3). https://doi.org/10.1016/S0898-6568(02)00086-4.

Koffi-Nevry, Rose, Kouassi Clément Kouassi, Zinzerdof Yesse Nanga, Marina Koussémon, and Guillaume Yao Loukou. 2012. "Antibacterial Activity of Two Bell Pepper Extracts: Capsicum Annuum L. and Capsicum Frutescens." *International Journal of Food Properties* 15 (5). https://doi.org/10.1080/10942912.2010.509896.

Li, Jingshuang, Hui Wang, Lili Zhang, Ni An, Wan Ni, Quanqi Gao, and Yang Yu. 2023. "Capsaicin Affects Macrophage Anti-Inflammatory Activity via the MAPK and NF-KB Signaling Pathways." *International Journal for Vitamin and Nutrition Research* 93 (4). https://doi.org/10.1024/0300-9831/a000721.

Marín, Alicia, Federico Ferreres, Francisco A. Tomás-Barberán, and María I. Gil. 2004. "Characterization and Quantitation of Antioxidant Constituents of Sweet Pepper (Capsicum Annuum L.)." *Journal of Agricultural and Food Chemistry* 52 (12). https://doi.org/10.1021/jf0497915.

Masud Parvez, G. M. 2017. "Current Advances in Pharmacological Activity and Toxic Effetcs of Various Capsicum Species." *International Journal of Pharmaceutical Sciences and Research* 8 (5).

Musfiroh, I. D. A., Mutakin Mutakin, Treesye Angelina, and Muchtaridi Muchtaridi. 2013. "Capsaicin Level of Various Capsicum Fruits." *International Journal of Pharmacy and Pharmaceutical Sciences* 5 (1).

Naves, Emmanuel Rezende, Lucas de Ávila Silva, Ronan Sulpice, Wagner L. Araújo, Adriano Nunes-Nesi, Lázaro E. P. Peres, and Agustin Zsögön. 2019. "Capsaicinoids: Pungency beyond Capsicum." *Trends in Plant Science*. https://doi.org/10.1016/j.tplants.2018.11.001.

Oulaï, Ayamaé Casimir, Kouakou Martin Djè, Krou Philippe Eba, Amissa Augustin Adima, and Eugène Jean Parfait Kouadio. 2018. "Chemical Composition, Antioxidant and Antimicrobial Activities of Capsicum Annuum Var: Annuum Concentrated Extract Obtained by Reverse Osmosis." *GSC Biological and Pharmaceutical Sciences* 5 (2). https://doi.org/10.30574/gscbps.2018.5.2.0123.

Patel, Kanika, and Dinesh Kumar Patel. 2023. "Biological Importance, Pharmacological Activities, and Nutraceutical Potential of Capsanthin: A Review of Capsicum Plant Capsaicinoids." *Current Drug Research Reviews* 16 (1). https://doi.org/10.2174/2589977515666230331093712.

Quinones-Seglie, Celia R., Edward E. Burns, and BenignoVillalon. 1989. "Capsaicinoids and Pungency in Various Capsicums." *LWT—Food Science and Technology* 22 (4).

Snitker, Soren, Yoshiyuki Fujishima, Haiqing Shen, Sandy Ott, Xavier Pi-Sunyer, Yasufumi Furuhata, Hitoshi Sato, and Michio Takahashi. 2009. "Effects of Novel Capsinoid Treatment on Fatness and Energy Metabolism in Humans: Possible Pharmacogenetic Implications." *American Journal of Clinical Nutrition* 89 (1). https://doi.org/10.3945/ajcn.2008.26561.

Yadav, Mukesh, Shalini Jain, Aarti Bhardwaj, Ravinder Nagpal, Monica Puniya, Radha Tomar, Vinod Singh, et al. 2009. "Biological and Medicinal Properties of Grapes and Their Bioactive Constituents: An Update." *Journal of Medicinal Food* 12 (3). https://doi.org/10.1089/jmf.2008.0096.

Zaki, N., A. Hasib, K. C. Eddine, F. Dehbi, H. El Batal, and A. Ouatmane. 2017. "Comparative Evaluation of the Phytochemical Constituents and the Antioxidant Activities of Five Moroccan Pepper Varieties (Capsicum Annuum L.)." *Journal of Chemical, Biological and Physical Sciences* 7 (4). https://doi.org/10.24214/jcbps.b.7.4.129406.

Zhang, Wei, Yu Zhang, Jinke Fan, Zhiguo Feng, and Xinqiang Song. 2024. "Pharmacological Activity of Capsaicin: Mechanisms and Controversies (Review)." *Molecular Medicine Reports* 29 (3). https://doi.org/10.3892/mmr.2024.13162.

5 Fennels
Chemistry and Pharmacology

Deepika Saini

5.1. INTRODUCTION

5.1.1. Overview of Fennel

Fennel (*Foeniculum vulgare* Mill.), widely recognized as fennel, can be consumed through leaves, seeds, and bulbs. It is widely used in cooking as a kitchen spice and an ingredient for making pickles, tea, and dietary supplements. It is a plant species that belongs to the carrot family (Apiaceae). Fennel is a renowned staple in kitchens worldwide, characterized by a distinct anise-like flavour and aromatic properties.

It is a perennial herb that can grow up to 2.5 meters in height with yellow flowers (umbrella-shaped clusters), strongly flavoured feathery leaves, and fruits. It is widely distributed to many parts of the world such as Asia, Europe, parts of North America, and Australia, and grown on the coasts of the Mediterranean regions (Abebie et al. 2017). The most prominent species of fennel are *F. vulgare*, var. azoricum, *Foeniculum vulgare* var. dulce, and *Foeniculum vulgare* var. vulgare. Other related species exist within the genus *Foeniculum*, but they are not as widely recognized or utilized (Shahat et al. 2011).

Small quantities of dry fennel seeds are commonly added as a spice in cooking. The reported data revealed that approximately 100 g of fennel seeds deliver 350 kilocalories of nutritional energy. Fennel consists of a rich blend of dietary minerals, proteins, fibres, vitamins, and dietary minerals, especially calcium, iron, magnesium, and manganese. The seeds of fennel contain 52% carbohydrates (including 40% dietary fibre), 15% fat, 16% protein, and 9% water (Dheebisha et al. 2020).

5.1.2. Historical Overview

In ancient civilizations, the fennel herb was known as *foeniculum,* from its Latin word "fragrant hay", which was claimed for its eyesight improvement property. The plant parts, e.g., stalks, were utilized as food, and seeds have been used as an antiflatulent. The fennel plant, because of its flavouring characteristics, has been used as a flavouring agent in toffies, cordials, and medications, and the oil was used as an antifungal ingredient for storing fruits and vegetables. Fennel is reported to be used as an antidote against the effects of toxic herbs, edible mushrooms, and reptile venoms (Loewenfeld and Back 1974), and it has been used to enhance digestion, promote lactation in nursing mothers, and alleviate respiratory ailments (Meyer 1934). Its decoction in the form of crushed fennel seeds has served as an eyewash.

5.1.3. Importance in Traditional Medicinal System

Fennel has held a prominent place in traditional medicine systems, such as Ayurveda and Traditional Chinese Medicine (TCM). Fennel has been used to alleviate digestive issues, such as bloating, gas, and colic, particularly in infants because of its carminative properties that help to relax the gastrointestinal tract, facilitating digestion and reducing discomfort (Rather et al. 2016). The plant also helps to treat kidney and bladder disorders, owing to its diuretic properties. It also works as an antiemetic and

DOI: 10.1201/9781003559894-5

removes nausea-like feelings. The fennel plant is also responsible for treating chronic fever conditions, cataracts, and obstructions in the hepatic, respiratory, endocrine, and reproductive systems. Moreover, it is also consumed during breastfeeding for stimulating milk production by women (Jadid et al. 2023).

5.2. PHARMACOLOGICAL APPLICATIONS OF FENNEL

5.2.1. Antibacterial Properties

It was reported that the methanolic extract of fennel seeds showed antibacterial activity against *S. pyrogenes* and *P. aeroginosa* strains (Kaveh et al. 2023). The essential oils isolated from fennel's sub-varieties demonstrated 100% mortality at 200μg/ml concentration against gram-positive and gram-negative bacterial strains. These essential oils resulted in higher antimicrobial activity due to inhibition of biofilm formation at very low concentrations (Napoli et al. 2022). Fennel essential oil showed antimicrobial activity with notable higher area of inhibition zones against the bacterial strains of *C. albicans* (45.3 mm) and *B. subtilis* (24.0 mm) (Okorska et al. 2023).

5.2.2. Antifungal Activity

Fennel's essential oil displayed a higher level of antifungal activity against biofilm-forming fluconazole-resistant *Candida albicans* strain, with 50% inhibition concentrations of 3.12% and 6.25% with 90% inhibition concentrations (Bassyouni et al. 2019). Fennel's essential oil showed a wide spectrum of bio-fungicidal activity by reducing the extent of infection with a 500-ppm concentration against the growth of *F. culmorum* and other species in pea plants (Okorska et al. 2023).

5.2.3. Antioxidant Activity

Reported findings demonstrated significant DPPH activity between fennel essential oils and their extracts. Ethanolic extract and essential oils of fennel showed strong antioxidant activity with 50% inhibition concentrations 6.34 and 7.17 mg/g, respectively. The Chinese variety of fennel essential oil revealed higher radical scavenging activity than the Egyptian variety (IC50 values 15.66 and 141.82 mg/g, respectively) (Ahmed et al. 2019). However, in other investigated reports, the antioxidant activity was evaluated for the essential oils from stems and leaves of fennel, which displayed higher activity in fennel stem oil than fennel leaves oil (EC_{50} value of 2.58 and 6.91 mg/mL, respectively) (Milenković et al. 2022).

5.2.4. Inflammatory Activity

The findings on volatile oils of *F. vulgare* fruits displayed considerable inhibition of the neutrophilic inflammation via deactivation of Ca and MAPKs-mediated human neutrophils activation with respiratory burst and the degranulation induced by receptor agonists fMLF/CB and MMK1 with notable inhibition concentration of 3.8–17.2 μg/ml in human neutrophils. Moreover, fennel oils also inhibited the phosphorylation of mitogen-activated protein kinase expression as well as recovery time in the entry of calcium ions (Korinek et al. 2021). The oral administration of *F. vulgare* extract (0.8 ml)/kg/day displayed decreased anti-inflammatory activity through bowel damage and apoptosis, therefore preventing the reduction of intestinal tissue damage and protein, DNA damage, and lipid peroxidation, as well as in an experimental necrotizing enterocolitis (NEC) model, as compared to the control group (Yakut et al. 2020).

5.2.5. Gastroprotective Activity

A combination of polyherbal extract demonstrated significant gastroprotection against ulcers induced via aspirin and ethanol in a mice model. The seed extract of fennel successfully prevented ulcer and mucosal lesions by 65% (250 mg/kg) and 75% (500 mg/kg), respectively, associated with the standard drug omeprazole (Reddy et al. 2013).

5.2.6. Estrogenic Effects

Fennel is noted for its estrogen-like properties, primarily attributed to compounds such as *t*-anethole and estragole, which are structurally related to synthetic estrogen diethylstilbesterol (Albert-Puleo 1980). Due to this structural relatability, the vast research studies were mainly focused on the anti-fertility impact and its estrogenic effects. This led to an investigation on the seed extract of fennel that proved significant effects in the reduction of hormone imbalances and improvement in estrogen levels, which have been utilized locally earlier to improve hormonal disbalances. It was discovered that the oral administration of fennel extract increased the concentrations of nucleic acids, proteins, and organ weights for 10 days. This lowers the inflammation and metabolic abnormalities in polycystic ovarian syndrome (PCOS) (Meena et al. 2019).

Fennel extract also displayed estrogenic effects concerning the weight of female reproductive organs. One study that investigated the enhanced production of milk also validates its galactogenic activity, which resulted in supporting childbirth and initiating the menstrual cycle (Albert-Puleo 1980; Rather et al. 2016). These studies suggested that fennel may help alleviate symptoms of menopause such as heat surges and emotional fluctuations, by acting on estrogen receptors. Fennel contains phytoestrogens that may help balance hormonal levels, particularly in women experiencing menopausal symptoms. This hormonal modulation is an area of growing interest in phytotherapy (Rahimikian et al., 2017). A controlled, double-blind randomized trial was carried out involving 60 postmenopausal women aged 45 to 65 years, which demonstrated the effects of fennel (5 g/day for 8 weeks) and resulted in the reduction of atrophic vaginitis symptoms in women's post-climacteric phase with no adverse effects (Yaralizadeh et al. 2016).

5.2.7. Cardiovascular and Lipid Activity

Currently, it is reported that the fennel seed extract prepared in ethanol solvent and its isolated anethole compound displayed protective effects against hypoxia/reoxygenation injury model of H9C2 heart myoblast cells. The extract also inhibits the reactive oxygen species (ROS) generation, DNA double-strand break damage. Moreover, anethole is responsible for the treatment of heart damage by hypoxia (Seo et al. 2024).

5.2.8. Antidiabetic Activity

Studies demonstrated that the aqueous extract of fennel exhibited reduction in blood glucose levels when administered to diabetic rat models with IC_{50} at 43 ± 1.19 µg/ml (El-Ouady et al. 2020). Furthermore, the combination of *F. vulgare* and black cumin (2.5 g each) significantly causes reduction in the body fat percentage and lipoprotein level in women patients (Abdelwahab et al. 2021). Another finding demonstrated that the encapsulated nano-emulsion prepared from the essential oil of fennel revealed noticeable decrease in the glycemic levels (Zolkepli et al. 2022).

5.2.9. Anticancer Activity

The crude extracts and fractions of fennel seeds displayed marked antineoplastic activity towards breast carcinoma cell lines. Thus, fennel proved its potential as chemo-preventive and anti-mutagenic agents in cancer treatment (Megeressa et al. 2020). Furthermore, simulation molecular docking studies analyzed the strong binding energy of α-pinene and D-limonene (each displayed −6 and −5.9 K/cal/mol) via simulation of strong binding energy towards breast cancer cell target (Kaur et al. 2022). Moreover, the ethanol extract of fennel seed demonstrated antihepatocellular carcinoma activity against hepatocellular carcinoma via potential inhibition of oncogene and inhibitor of apoptosis (Ke et al. 2021).

5.2.10. Hepatoprotective Activity

The extract of fennel plant (400 mg/Kg) evaluated on hyperglycemia-induced, liver-injured rats resulted in reduction of lipid peroxidation and serum aminotransferases such as ALT and AST markers (Samadi-Noshahr et al. 2021).

5.3. CHEMICAL COMPOSITION OF FENNEL

Fennel (*Foeniculum vulgare*) is characterized by a complex chemical composition that consists of a rich library of essential oils, flavonoids, phenolic compounds, and dietary fibre. The essential oil of fennel, extracted predominantly from the seeds, contains a variety of volatile compounds, primarily trans-anethole, which is responsible for its characteristic flavour and aroma. Other significant constituents include fenchone, estragole, and α-pinene, each contributing to fennel's unique sensory profile (Alam et al. 2019; Mehra et al. 2022).

5.3.1. Major Chemical Compounds in Fennel

5.3.1.1. Essential Oils

Fennel seeds are particularly rich in essential oils, comprising about 3–6% of their weight. The major components include:

1. **Trans-Anethole:** Responsible for the characteristic aroma, with documented anti-inflammatory and antimicrobial properties. The most prominent compound in fennel essential oil, trans-anethole can constitute 50–60% of its composition. This phenylpropene compound imparts a sweet, anise-like flavour. Its chemical structure, featuring a methoxy group and a double bond, is crucial for its biological activities, which include anti-inflammatory and antioxidant effects (Shahat et al. 2011).
2. **Fenchone:** Representing 10–20% of the essential oil, fenchone is a bicyclic monoterpene ketone with a camphor-like aroma. It has potential neuroprotective and antispasmodic properties, contributing to fennel's medicinal applications (Badgujar et al. 2014).
3. **Estragole:** This methoxy-substituted allylbenzene is another significant component, present in smaller quantities. Although estragole has been associated with potential carcinogenic effects in high doses, it also exhibits notable antioxidant properties (Topal et al. 2007; Fang et al. 2006).
4. **α-Pinene:** A bicyclic monoterpene, α-pinene adds to fennel's aromatic profile and has demonstrated various pharmacological activities, including anti-inflammatory and bronchodilator effects (Salehi et al. 2019).

5.3.1.2. Phenolic Compounds

Phenolic compounds found in fennel contribute to its overall health benefits, including their ability to modulate inflammation and support digestive health. The compounds, such as caffeic acid, chlorogenic acid, and ferulic acid, also contribute to fennel's health benefits, particularly through their anti-inflammatory and antimicrobial properties (Rather et al. 2016).

5.3.1.3. Flavonoids

Fennel contains various flavonoids, such as quercetin, kaempferol, and apigenin, which are known for their antioxidant and anti-inflammatory properties. These compounds perform a significant task in reducing oxidative stress in the body. Furthermore, fennel is rich in dietary fibre, particularly inulin, which is beneficial for the gastrointestinal system. Inulin acts as a digestive supportive aid in the growth of intestinal microbiome and enhancing digestive health. This multifaceted composition underscores fennel's potential as a functional food and therapeutic agent (Niness 1999).

5.3.1.4. Other Bioactive Components

Fennel contains vitamins (such as vitamins C and B) and minerals, further enhancing its nutritional profile.

5.3.2. Chemical Moieties and Their Therapeutic Significance

The therapeutic effects of fennel's compounds can often be traced to specific chemical moieties:

- **Phenolic Moieties:** Compounds with phenolic groups, such as trans-anethole and flavonoids, play a crucial role in their antioxidant capacity. This is particularly important in the context of drug development, as oxidative stress is linked to numerous health issues, including cancer and neurodegenerative diseases.
- **Alkenes and Aromatic Structures:** Compounds such as estragole and trans-anethole, with their alkenes and aromatic rings, enhance lipophilicity, allowing better membrane permeability. This property is essential for their effectiveness as therapeutic agents.
- **Ketones:** The ketone functional group found in fenchone contributes not only to its distinctive scent but also to its biological activities, such as promoting muscle relaxation and neuroprotection.

5.3.3. Reported Therapeutic Applications of Isolated Compounds

1. **Anti-Inflammatory Agents:** Trans-anethole (250 mg/kg) and various flavonoids have shown promise in reducing inflammation, making them potential candidates for new anti-inflammatory therapies. Their mechanisms often involve inhibiting pro-inflammatory cytokines and enzymes (Kim et al. 2017; da Rocha et al. 2023; Tong et al. 2022; da Rocha et al. 2017).
2. **Antioxidants:** The wide array of phenolic compounds and flavonoids contribute to fennel's strong antioxidant properties, which are essential in formulating dietary supplements and pharmaceuticals that target oxidative stress-related conditions (Badgujar et al. 2014). A current study determined that the antioxidant effect of anethole (31.25–250 mg/kg dose) against acetic acid–induced colitis in mice diminished injury to the epithelium, penetration of inflammatory cells, and edema, which results in significant antioxidant properties (Choi and Hwang 2004).
3. **Digestive Health:** Traditionally used as a carminative, fennel is effective in alleviating gastrointestinal discomfort. Compounds such as fenchone (Araruna et al. 2024) and trans-anethole (Kwiatkowski et al. 2019) relax the gastrointestinal tract, making them beneficial for digestive health. *F. vulgare* essential oil and anethole (100 mg kg–1 oral administration) provided significant protection towards ethanol-induced gastric lesions in rats (Kwiatkowski et al. 2019).
4. **Hormonal Regulation:** Certain compounds in fennel, particularly trans-anethole, exhibit estrogen-like properties that may help manage hormone-related conditions, including symptoms of menopause (Chen et al. 2012). This characteristic is being explored for developing natural alternatives to hormone replacement therapies (Albert-Puleo 1980).
5. **Antimicrobial Agents:** The essential oil, rich in trans-anethole and other volatile compounds, displays significant antimicrobial activity (Kubo et al. 2008) against various pathogens, positioning fennel as a candidate for natural preservatives and therapeutic agents. Anethole from *F. vulgare* and hexane extract of Florence fennel exerted antimicrobial effects against probiotic bacteria. The antimicrobial activity was higher in gram-positive bacteria (Senatore et al. 2013).
6. **Antithrombotic Activity:** Both *F. vulgare* essential oil and its major component anethole (30 mg kg–1 day–1 for 5 days) were tested against guinea pig plasma, which inhibited

arachidonic acid, collagen, ADP, and U46619-induced aggregation, with an inhibitory concentration between 4 to 147 μg ml–1 (Tognolini et al. 2007). It was found as a significant and safe antithrombotic agent in both in vitro and in vivo tested models due to a wide range of antiplatelet activity, thrombus subverting effect, and vasodilation (Badgujar et al. 2014).

7. **Mosquito-Repellent Activity:** Isolated compounds (+)-fenchone and E-9-octadecenoic acid and fennel oil from methanol or hexane extracts of *Foeniculum vulgare* fruits exhibited insect-repellent activity (> 80% repellency) by a patch test; however, they do not show any toxicity to people (Ahn et al. 2007).
8. **Galactogogue Activity:** It was reported that an acetone extract of *F. vulgare* seeds worked as a stimulant to promote lactation. The seeds showed estrogenic effects on the male and female reproductive organs of rat models. The key estrogenic element of fennel was originally considered as anethole, but it is found to be a polymer of anethole, e.g., dianethole (Albert-Puleo 1980).
9. **Antibacterial Activity:** Trans-anethole with muprocin demonstrated inhibition of *S. aureus* (isolated from asymptomatic carriers) growth at concentrations > 4.0% with MIC value 0.064 with 86% (Kwiatkowski et al. 2019).
10. **Lung-Protective Activity:** The major component of fennel, i.e., trans-anethole, revealed protection against lung injury induced in mice models by LPS through regulation of Th17/Treg function and also decreased the number of inflammatory cells, hence promote the healing effect on injured lungs (Zhang et al. 2018).

5.4. SAFETY AND TOXICOLOGY

As with any herbal remedy, understanding fennel's safety profile is crucial, especially regarding its estrogenic effects and potential toxicity when consumed at elevated doses. Concerns about the safety of fennel also extend to its potential toxicity, particularly at elevated doses. Some studies have reported adverse effects associated with high consumption of fennel essential oil. For instance, high doses of estragole have been linked to hepatotoxicity in animal studies, raising concerns about its carcinogenic potential when consumed in large quantities. The studies showed that people gave good feedback for reducing menstrual cramps (dysmenorrhea), and it was also helpful in colic and indigestion problems of infants after consuming fennel oil and extract by mouth compared to the effects of Ibuprofen and mefenamic acid (https://www.webmd.com/vitamins/ai/ingredient-mono-311/fennel; Rather et al. 2016).

5.4.1. Dosage and Administration

The seeds of fennel and their isolated oil have been utilized as an appetite enhancer and stimulants within doses of 5–7 g and 0.1–0.6 mL, respectively (Blumenthal et al. 2000). Oral essential fennel oil (30%) capsules and fennel extract vaginal cream (5%) have been used to improve symptoms in postmenopausal women at 200 mg/day and 5 g/day, respectively (Yaralizadeh et al. 2016). Fennel is generally regarded as safe when consumed in culinary amounts. However, concentrated extracts should be used with caution, and standardized dosages should be followed in supplement form.

5.4.2. Potential Toxicity and Side Effects

It is essential to prepare any herbal product from initial step to the final product without compromising its quality standards and with a permissible amount of toxicity and negative effects. The most important and critical constituent of fennel is estragole. This compound is reported to be a prohibited substance because it develops malignant tumours in animals (Sekizawa et al. 1982), but estragole has not proven to have any carcinogenic effects on people, and it does cause formation of tumours in rats. Fennel oil was reported as a mutagenic against *B. subtilis* in the DNA-repair test (Ostad et al. 2004). Moreover, research findings noticed that fennel may cause some photo reactions such as photo-dermatitis, contact dermatitis, and cross reactions (Mohamad et al. 2011). Furthermore, the fennel oil reported to induce neuropsychological effects such as deliriums and epileptic symptoms. Specific investigated reports indicated untimely development of mammary glands in females that has been observed after the application of fennel oil (Türkyilmaz et al. 2008). Fennel is well tolerated, except an excessive amount of fennel seed consumption causes gastrointestinal disturbances, and some individuals also respond with allergic reactions (Badgujar et al. 2014). It was also mentioned that fennel looks similar to poison hemlock, so hemlock's misinterpretation as fennel led to the release of a hazardous nicotine coniine-like substance that resulted in vomiting, paralysis, and death (Colombo et al. 2009).

5.4.3. Drug Dynamics

A study demonstrated that an in vitro test on 5-methoxypsoralen isolated from fennel caused inhibition to the liver substrate cytochrome P450 3A4. Therefore, fennel may interact with certain medications, particularly those affecting hormonal balance due to having this substrate of an isoenzyme (Subehan et al. 2007). Consultation with a healthcare provider is advised before incorporating fennel supplements, especially for individuals on medications.

5.5. CLINICAL TRIAL STUDIES ON FENNEL

A recent clinical trial of fennel seed extract capsule (200 mg, oral) on women patients with knee osteoarthritis showed a significant percentage change in the fennel-tested group compared to the placebo-treated group (Alazadeh et al. 2020). The results revealed that use of fennel extract for knee pain is an alternate treatment option as compared to complementary therapy in knee-related problems.

A systematic review and meta-analysis of randomized controlled clinical trial studies found fennel as an effective remedy for reducing pain in primary dysmenorrhea (Lee et al. 2020). Fennel has been investigated for its potential benefits in managing menopausal symptoms. A randomized controlled trial was conducted on women experiencing menopausal symptoms, who were given fennel extract for 12 weeks. The results showed a significant decrease in hot flashes and improvements in overall quality of life related to the control group. This suggests that fennel may serve as a natural alternative for managing menopausal discomfort (Rahimikian et al. 2017).

One more study evaluated the bioefficacy of fennel seed–fortified crackers, which significantly lowered the basal metabolic index (BMI) 38.48 to 36.95 kg/mg in 45 days; however, they were also increasing the lipid profiles against cholesterol-lowering medicines and other herbal therapies for obesity (Anayatullah et al. 2023). A single-centre prospective randomized trial conducted in China described the effects of heated fennel therapy improved the recovery of gastrointestinal functions in patients after abdominal surgery (Chen et al. 2020).

In a different spectrum of crossover clinical trial studies, 20 patients with chronic constipation were selected randomly, individually, and for controlled study to analyze the effect of an herbal tea containing fennel, elderberry, senna, and anise herbs. After trial study, the results showed significant improvements in the frequency of colon and gastric evacuations, with the latter effects beginning during the second day of therapy (Tognolini et al. 2007). One of the most common uses of fennel is for alleviating digestive issues. A clinical trial on fennel examined the efficacy of fennel in treating irritable bowel syndrome (IBS). In this randomized, double-blind study, participants who received fennel extract reported significant reductions in abdominal pain and bloating compared to the placebo group. The study concluded that fennel might be an effective natural treatment for IBS, highlighting its carminative properties (Portincasa et al. 2016).

In a controlled trial of a sporadic group, women who received fennel extract reported less severe menstrual cramps compared to those given a placebo. The study highlighted fennel's potential as a nonpharmacological option for managing dysmenorrhea (Omidvar et al. 2012). Fennel is often used to enhance lactation in breastfeeding mothers. A clinical trial explored the effects of fennel tea on breast milk production. The study involved postpartum women who consumed fennel tea daily for a specified period. The findings indicated an increase in milk volume among the participants, supporting the traditional use of fennel as a galactagogue (Turkyılmaz et al. 2011).

5.6. FUTURE DIRECTIONS AND RESEARCH OPPORTUNITIES

Despite its historical use, extensive research findings are essential to discover the pharmacokinetics and consistent actions of fennel and its constituents. Ongoing studies are exploring fennel's potential in treating metabolic syndrome, cancer, and its gastrointestinal tract effects. Future research should focus on clinical trials to validate traditional uses and explore new therapeutic applications, alongside investigations into optimal extraction methods to enhance bioavailability.

5.7. CONCLUSIONS

Fennel is not merely a culinary herb but a complex plant with significant pharmacological properties. Its rich chemical composition, featuring a diverse array of bioactive compounds, contributes to its therapeutic potential across various health domains. The exploration of fennel's chemistry and pharmacology offers promising insights into its role in promoting health and wellness. Ongoing studies are crucial for gaining a complete understanding of the processes driving fennel's positive effects and for incorporating this knowledge. Clinical trials on fennel have provided promising insights into its therapeutic potential, particularly for digestive health, menopausal symptoms, lactation support, and antioxidant properties. Although these studies suggest fennel can be beneficial, further research is necessary to establish standardized dosages and long-term safety profiles.

REFERENCES

Abdelwahab, M. et al. 2021. Effect of low caloric diet supplemented by fennel (*Foeniculum vulgare*) seeds or black cumin (*Nigella sativa*) seeds and its mixture on obese adult female patients. Afr J Biol Sci. 17:137–147.

Abebie, B. et al. 2017. Phytochemical screening, antioxidant and antimicrobial activities of seeds of *Foeniculum vulgare* (ensilal). *World J Pharm Res*. 5(3):198–208.

Ahmed, A. F. et al. 2019. Comparative analysis of antioxidant activities of essential oils and extracts of fennel (*Foeniculum vulgare* Mill.) seeds from Egypt and China. *Food Sci Hum Wellness* 8(1):67–72.

Ahn, et al. 2007. *Insect Repellent Isolated from the Fruit of Foeniculum Vulgare* (U.S. Patent No. US 7,179,479 B1U.S. Patent and Trademark Office. https://patents.google.com/patent/KR10-2000-0038336A.

Alam, P. et al. 2019. Chemical composition of fennel seed extract and determination of fenchone in commercial formulations by GC-MS method. *J Food Sci Technol*. 56:2395–2403.

Alazadeh, M. F. et al. 2020. Effect of sweet fennel seed extract capsule on knee pain in women with knee osteoarthritis. *Complement Ther Clin Pract*. 40:101219.

Albert-Puleo, M. 1980. Fennel and anise as estrogenic agents. *J Ethnopharmacol*. 2(4):337–344. https://doi.org/10.1016/S0378-8741(80)81015-4.

Anayatullah, S. et al. 2023. Effect of fennel seeds fortified crackers on various obesity biomarkers. J Agr Food Res, 14:100906. https://doi.org/10.1016/j.jafr.2023.100906

Araruna, M. E. C. et al. 2024. (-)-Fenchone prevents cysteamine-induced duodenal ulcers and accelerates healing promoting re-epithelialization of gastric ulcers in rats via antioxidant and immunomodulatory mechanisms. *Pharmaceuticals (Basel, Switzerland)*. 17(5):641.

Badgujar, S. et al. 2014. *Foeniculum vulgare* Mill: A review of its botany, phytochemistry, pharmacology, contemporary application, and toxicology. *BioMed Res Int*. 84:2674.

Bassyouni, R. H. et al. 2019. Fennel oil: A promising antifungal agent against biofilm forming fluconazole resistant *Candida albicans* causing vulvovaginal candidiasis. *J Herb Med*. 15:100227.

Blumenthal, M. et al. 2000. *Herbal Medicine: Expanded Commission E Monographs*. Newton, MA: Integrative Medicine Communications.

Chen, B. et al. 2020. Heated fennel therapy promotes the recovery of gastrointestinal function in patients after complex abdominal surgery: A single-center prospective randomized controlled trial in China. *Surgery.* 168:793–799.

Chen, C. H. et al. 2012. Anethole suppressed cell survival and induced apoptosis in human breast cancer cells independent of estrogen receptor status. *Phytomedicine.* 19:763–767.

Choi, E. M. and deGraffenried, Linda A. 2004. Antiinflammatory, analgesic and antioxidant activities of the fruit of *Foeniculum vulgare. Fitoterapia.* 75:557–565.

Colombo, M. et al. 2009. Hemlock poisoning due to plant misidentification. *J Pharm Sci Res.* 1(3):71–75.

da Rocha, B. A. et al. 2017. Acetaminophen-induced hepatotoxicity: Preventive effect of trans anethole. *Biomed Pharmacother.* 86: 213–220.

da Rocha, E. M. T. et al. 2023. Development and characterization of trans-anethole-containing solid lipid microparticles: Antiinflammatory and gastroprotective effects in experimental inflammation. *Naunyn-Schmiedeberg's Arch Pharmacol.* 396:469–484.

Dheebisha, C. et al. 2020. Advances in cultivation of fennel. *J Pharmacogn Phytochem.* 9:1295–1300.

Di Napoli, M. et al. 2022. Antimicrobial, antibiofilm, and antioxidant properties of essential oil of *Foeniculum vulgare* Mill: Leaves. *Plants.* 11:3573.

El-Ouady, F. et al. 2020. Antihyperglycemic effect of the aqueous extract of *Foeniculum vulgare* in normal and streptozotocin-induced diabetic rats. *Cardiovasc Hematol Disord Drug Targets.* 20:54–63.

Fang, L. et al. 2006. Headspace solvent microextraction-gas chromatography-mass spectrometry for the analysis of volatile compounds from *Foeniculum vulgare* Mill. *J Pharm Biomed Anal.* 41:791–797.

Jadid, N. et al. 2023. The medicinal Umbelliferae plant fennel (*Foeniculum vulgare* Mill): Cultivation, traditional uses, phytopharmacological properties, and application in animal husbandry. *Arab J Chem.* 16:104541.

Kaur, B. et al. 2022. An in silico investigation to explore anti-cancer potential of *Foeniculum vulgare* Mill: Phytoconstituents for the management of human breast cancer. *Molecules.* 27:4077.

Kaveh, R. et al. 2023. Antibacterial and antioxidant activities and anticancer effects of fennel seeds (*Foeniculum vulgare*) against Lung Cancer cells. *Proc Natl Acad Sci India Sect B Biol Sci.* 93:311–316.

Ke, W. et al. 2021. *Foeniculum vulgare* seed extract exerts anti-cancer effects on hepatocellular carcinoma. *Food Funct.* 12:1482–1497.

Kim, K. Yt. et al. 2017. Anti-inflammatory effects of trans-anethole in a mouse model of chronic obstructive pulmonary disease. *Biomed Pharmacother.* 91:925–930.

Korinek, M. et al. 2021 Anti-inflammatory and antimicrobial volatile oils: Fennel and cumin inhibit neutrophilic inflammation via regulating calcium and MAPKs. *Front Pharmacol.* 12. https://doi.org/10.3389/fphar.2021.674095.

Kubo, I. et al. 2008. Antimicrobial activity of anethole and related compounds from aniseed. *J Sci Food Agric.* 88:242–247.

Kwiatkowski, P. et al. 2019. The effect of fennel essential oil and trans-anethole on antibacterial activity of mupirocin against *Staphylococcus aureus* isolated from asymptomatic carriers. *Postepy Dermatol Alergol.* 36:308–314.

Lee, H. W. et al. 2020. Fennel for reducing pain in primary dysmenorrhea: A systematic review and meta-analysis of randomized controlled trials. *Nutrients.* 12:3438.

Loewenfeld, C. and Back, P. 1974. *The Complete Book of Herbs and Spices.* New York, NY: Putnam.

Meena, J. P. et al. 2019. Therapeutic potential of Foeniculum *vulgare* on PCOS: Effects on nucleic acids, proteins, and organ weights. *World J Pharm Res.* 8(10):348–349.

Megeressa, M. et al. 2020. Cytotoxic activity of proteins extracted from fennel (*Foeniculum vulgare*) seeds against human cancer cell lines. *FASEB J.* 34:1.

Mehra, N. et al. 2022. Phytochemical screening and in vitro antioxidant assays in *Foeniculum vulgare* Mill:(Fennel) seeds collected from tarai region in the Uttarakhand. *Indian J Nat Prod Resour.* 2:213–222.

Meyer, J. E. 1934. *The Herbalist.* Hammond, IN: Hammond Book Co.

Milenković, A. et al. 2022. Essential oil yield, composition, antioxidant and microbial activity of wild fennel (*Foeniculum vulgare* Mill.) from Monte Negro Coast. *Horticulturae.* 8:1015.

Mohamad, R. H. et al. 2011. Antioxidant and anticarcinogenic effects of methanolic extract and volatile oil of fennel seeds (*Foeniculum vulgare*). *J Med Food.* 14:986–1001.

Niness, K. R. 1999. Inulin and oligofructose: What are they? *J Nutr* 129(7 Suppl):1402S–1406S. https://doi.org/10.1093/jn/129.7.1402S

Okorska, S. B. et al. 2023. The fungicidal effect of essential oils of fennel and hops against fusarium disease of pea. *Appl Sci.* 13:6282.

Omidvar, S., et al. 2012. Effect of *Foeniculum vulgare* (fennel) on pain intensity in dysmenorrhea: A placebo-controlled trial. *Iran J Obstet Gynecol Infertil* 15(6):1–5.

Ostad, S. N. et al. 2004. Evaluation of the teratogenicity of fennel essential oil (FEO) on the rat embryo limb buds culture. *Toxicol in Vitro*.18:623–627.

Portincasa, P. et al. 2016. Curcumin and fennel essential oil improve symptoms and quality of life in patients with irritable bowel syndrome. *J Gastrointestin Liver Dis*. 25(2):151–157. https://doi.org/10.15403/jgld.2014.1121.252.ccm

Rahimikian, F. et al. 2017. Effect of *Foeniculum vulgare* Mill. (fennel) on menopausal symptoms in postmenopausal women: A randomized, triple-blind, placebo-controlled trial. *Menopause*. 24:1017–1021.

Rather, M. A. et al. 2016. *Foeniculum vulgare*: A comprehensive review of its traditional use, phytochemistry, pharmacology, and safety. *Arab J Chem*. 9:S1574–S1583.

Reddy, G. R. S. et al. 2013. Investigation of poly-herbal aqueous extract for potential antiulcer activity. *Int J Pharmacol*. 3:53–58.

Samadi-Noshahr, Z. et al. 2021. The hepatoprotective effects of fennel seeds extract and trans-anethole in streptozotocin-induced liver injury in rats. *Food Sci Nutr*. 9:1121–1131.

Sekizawa, J. et al 1982. Genotoxicity of safrole-related chemicals in microbial test systems. *Mutat Res*. 101:127–140.

Senatore, F. et al. 2013. Chemical composition, antimicrobial and antioxidant activities of anethole-rich oil from leaves of selected varieties of fennel [*Foeniculum vulgare* Mill. ssp. *vulgare* var. *azoricum* (Mill.) Thell]. *Fitoterapia*. 90:214–219.

Seo, J. W. et al. 2024. Protective effects of anethole in *Foeniculum vulgare* Mill. seed ethanol extract on hypoxia/reoxygenation injury in H9C2 heart myoblast cells. *Antioxidants*. 13:1161.

Salehi, B. et al. 2019. Therapeutic potential of α- and β-pinene: A miracle gift of nature. Biomolecules, 9(11):738. https://doi.org/10.3390/biom9110738

Shahat, A. A. et al. 2011. Chemical composition, antimicrobial and antioxidant activities of essential oils from organically cultivated fennel cultivars. *Molecules*. 16:1366–1377.

Subehan, Syed F. H. Zaidi, Shigetoshi Kadota, Yasuhiro Tezuka. 2007. Inhibition on human liver cytochrome P450 3A4 by constituents of fennel (*Foeniciulum vulgare*): Identification and characterization of a mechanism-based inactivator. *J Agric Food Chem*. 55:10162–10167.

Tognolini, M. et al. 2007. Protective effect of *Foeniculum vulgare* essential oil and anethole in an experimental model of thrombosis. *Pharmacol Res*. 56:254–260.

Tong, Y. et al. 2022. Trans-anethole ameliorates lipopolysaccharide-induced acute liver inflammation in broilers via inhibiting NF-κB signaling pathway. *Poult Sci*. 101:101962.

Topal, U. et al. 2007. Chemical compositions and antioxidant properties of essential oils from nine species of Turkish plants obtained by supercritical carbon dioxide extraction and steam distillation. *Int J Food Sci Nutr*. 18:1–16.

Türkyilmaz, Z. et al. 2008. A striking and frequent cause of premature thelarche in children: *Foeniculum vulgare*. *J Pediatr Surg*. 43:2109–2111.

Turkyılmaz, C. et al. 2011. The effect of galactagogue herbal tea on breast milk production and short-term catch-up of birth weight in the first week of life. J Altern Complement Med. 17(2):139–142. https://doi.org/10.1089/acm.2009.0207

Yakut, H. I. et al. 2020. Preventative and therapeutic effects of fennel (*Foeniculum vulgare*) seed extracts against necrotizing enterocolitis. *J Food Biochem*. 44:e13284.

Yaralizadeh, M. et al. 2016. Effect of *Foeniculum vulgare* (fennel) vaginal cream on vaginal atrophy in postmenopausal women: A double-blind randomized placebo-controlled trial. *Maturitas*. 84:75–80.

Zhang, S. et al. 2018. Fennel main constituent, trans-anethole treatment against LPS-induced acute lung injury by regulation of Th17/treg function. *Mol Med Rep*. 18:1369–1376.

Zolkepli, H. et al. 2022. A review on the delivery of plant-based antidiabetic agents using nanocarriers: Current status and their role in combatting hyperglycaemia. *Polymers*. 14:2991.

6 Leafy Vegetables and Their Nutraceutical Health Impacts

Shibashis Das, Partha Mondal, Rajdeep Mohanta, Tanmoy Mondal, Kaushik Kumar Das, and D.K. Murmu

6.1. INTRODUCTION

Nutrition is a fundamental and essential need for maintaining a healthy livelihood. A balanced diet is crucial at every stage of life, from fetal development to old age. Vegetables are commonly referred to as "protective foods" in human diets due to their numerous health benefits. According to the dietary guidelines of the U.S. Department of Agriculture (USDA), consuming five servings of vegetables daily is recommended for a 2,000-calorie diet (HHS/USDA, 2015). Among these, one serving should ideally include green leafy vegetables (GLVs). GLVs are increasingly recognized as an integral part of the food pyramid, valued for their rich content of trace elements and bioactive compounds (Gibson and Hotz, 2001). Their nutritional profile includes ascorbic acid, calcium, copper, beta carotene, iron, manganese, magnesium, potassium, and zinc. The significance of GLVs extends beyond their nutritional value, as traditional knowledge often highlights their medicinal properties (Ogunsuyi et al., 2018). Growing consumer awareness of the health and nutraceutical benefits of GLVs has led to increasing research on active metabolites derived from these vegetables. Many of these metabolites are classified as bioactive compounds. In vitro studies have revealed that GLVs possess anticancer, antioxidant, and anti-inflammatory properties (Chester et al., 2017; Vithya and Jayshree, 2017). These properties are attributed to their bioactive metabolites, including phenolic acids, flavonoids, glucosinolates, and vitamins. Such compounds have demonstrated potential in preventing noncommunicable diseases (NCDs).

Despite their numerous benefits, GLVs remain underutilized, undervalued, and often neglected (Ebert, 2014). They face several challenges, particularly in their integration into mainstream food systems. These challenges include limited research on their bioactivity, the effects of processing and digestion on their bioactive compounds, and the bioavailability of their health-promoting metabolites. This chapter highlights the potential of GLVs as sources of nutraceutical benefits, focusing on their anticancer, antioxidant, and anti-inflammatory properties. It aims to identify knowledge gaps, address existing challenges, and ultimately advocate for increased cultivation and consumption of GLVs.

6.2. BOTANY

Botanical descriptions including taxonomy, edible parts, habitat, etc. are mentioned in Table 6.1.

6.3. NUTRITIONAL COMPONENTS IN GREEN LEAFY VEGETABLES

6.3.1. Proteins

Proteins are vital for daily sustenance, metabolic functions, and cellular activities. Green leafy vegetables are among the most economical and excellent sources of protein. The essential amino

DOI: 10.1201/9781003559894-6

TABLE 6.1
Botanical Description of Leafy Vegetables

Sl No	Common Name	Family	Chromosome Number (2n)	Botanical Name	Edible Part	Habitat
1	Lal shak	Amaranthaceae	34	*Amaranthus tricolor* L.	Leaf, whole plant	Terrestrial plant
2	Notay shak	Amaranthaceae	34	*Amaranthus viridis* L.	Leaf	Terrestrial plant
3	Shanchi shak	Amaranthaceae	66–100	*Alternanthera sessilis*	Whole plant	Aquatic plant
4	Endive	Asteraceae	18	*Cichorium endivia* L.	Leaf	Terrestrial plant
5	Chicory	Asteraceae	18	*Cichorium intybus* L.	Leaf	Terrestrial plant
6	Butterhead lettuce	Asteraceae	18	*Lactuca sativa* var. *capitata* L	Leaf	Heliophytes plants
7	Crisphead lettuce	Asteraceae	18	*L. sativa* var. *capitata* L.	Leaf	Terrestrial plant
8	Cutting lettuce	Asteraceae	18	*L. sativa* var. *acephala*	Leaf	Terrestrial plant
9	Pui shak	Basellaceae	44	Basella alba L.	Leaf, leaf stalk, stem	Mesophyte type
10	Abyssian mustard	Brassicaceae	34	*Brassica carinata*	Leaf	Heliophytes plant
11	Mustard	Brassicaceae	36	*Brassica juncea* L.	Leaf	Terrestrial plant
12	Black mustard	Brassicaceae	16	*Brassica nigra* L.	Leaf	Terrestrial plant
13	Kale, collards	Brassicaceae	18	*Brassica oleracea* L. Acephala group DC.	Leaf	Mesophyte plant
14	Cabbage	Brassicaceae	18	*Brassica oleracea* var Capitata L.	Leaf	Terrestrial plant
15	Savoy cabbage	Brassicaceae	18	*Brassica oleracea* var Sabauda L.	Leaf	Terrestrial plant
16	Pak choi	Brassicaceae	20	*Brassica* Chinensis	Leaf	Terrestrial plant
17	Chinese cabbage	Brassicaceae	20	*Brassica campestris*	Leaf	Mesophyte plant
18	Garden cress	Brassicaceae	16, 24, 32	*Lepidium sativum* L.	Leaf	Dumping grounds
19	Watercress	Brassicaceae	32	*Nasturtium officinale* R.Br.	Leaf	Sun and water-loving perennial
20	Swiss chard	Chenopodiaceae	18	*Beta vulgaris* L. cicla group	Leaf	Terrestrial plant
21	Spinach	Chenopodiaceae	12	*Spinacia oleracea* L.	Leaf	Terrestrial plant
22	Water Spinach	Convolvulaceae	30	*Ipomoea aquatica* Forsk.	Leaf, stem	Aquatic plant
23	Sorrel	Polygonaceae	14, 15	*Rumex* spp.	Leaf	Grasslands and roadside verges
24	Bathua	Amaranthaceae	36	*Chenopodium album* L.	Leaf, stem	Terrestrial plant
25	Moringa	Moringaceae	28	*Moringa Olifera*	Leaf	Terrestrial plant
26	Asparagus lettuce, celtuce	Asteraceae	18	*Lactuca sativa* var. *angustana* L.	Stem and leaf	Terrestrial plant
27	Cardoon	Asteraceae	26	*Cynara cardunculus* L.	Main nervation	Mesophyte plant

28	Celery	Apiaceae	22	*Apium graveolens* L.	Prevalently petiole	Mesophyte plant
29	Fennel	Apiaceae	22	*Foeniculum vulgare* Mill.	Thickened base of the petiole	Terrestrial plant
30	Purslane	Portulacaceae	18, 36, 54	*Portulaca oleracea* L.	Leaf and young shoot	Along roadways, and disturbed environments.
31	Rhubarb, pieplant	Polygonaceae	44	*Rheum rhabarbarum* L.	Petiole	Roadside, waste areas
32	Vegetable rape	Brassicaceae	38	*Brassica napus* L. Napus group	Leaf and early flower stalk	Human-made or disturbed habitats
33	Parsley	Apiaceae	22	*Petroselinum crispum*	Leaf	Cliffs and rocks
34	Chervil	Apiaceae	14, 22	*Anthriscus cerefolium* L.	Leaf	Terrestrial plant
35	Spearmint	Lamiaceae	48	*Mentha spicata* L	Leaf and inflorescence	Moist fields and pond or lake margins
36	Common basil,	Lamiaceae	48	*Ocimum basilicum* L.	Leaf	Terrestrial plant
37	Oregano	Lamiaceae	30	*Origanum vulgare* L.	Leaf	Terrestrial plant
38	Perilla	Lamiaceae	40	*Perilla frutescens* L.	Leaf	Hills and mountain, sunny and fertile soil,
39	Rosemary	Lamiaceae	24	*Rosmarinus officinalis* L.	Leaf	Dry scrublands, rocky cliffs

(*Source:* Alvino and Barbieri, 2015; Siddiq and Uebersax, 2018)

acids include lysine, isoleucine, leucine, phenylalanine, threonine, tryptophan, valine, histidine, and methionine. Significantly, *Chenopodium album* supplies all of the essential amino acids necessary for optimal protein quality (Guerrero and Isasa, 2009). Ribulose-1,5-bisphosphate carboxylase/oxygenase (RUBISCO), a protein found in leaf chloroplasts, constitutes approximately 50% of the total protein in leaf cells (Kawashima and Wildman, 1970). All green leafy vegetables contain RUBISCO, albeit the amino acid composition varies significantly across species. Recent study indicates that green leafy vegetables surpassing Food and Agricultural Organization (FAO) nutritional standards, such as spinach (*Spinacia oleracea*), broccoli (*Brassica oleracea* var. Italica), and duckweed (*Lemna perpusilla*), encompass all necessary amino acids (Edelman and Colt, 2016). The amino acid composition of cassava (*Manihot esculenta*) leaves resembles that of legumes and dairy products, but with a reduced methionine concentration (Fasuyi, 2005).

According to research, several leafy vegetables are protein-rich, such as cassava leaves and septic weed (*Senna occidentalis*), and have about 7g of protein per 100g of fresh weight. Agricultural practices and environmental factors might affect the amount of protein in leafy greens (Nwanekezie and Obiakor-Okeke, 2014).

6.3.2. Vitamins

Spinach is a great source of vitamins, just like other LGVs. The vitamin A content of half a cup of boiled spinach is 573 mcg, or 229% of the dietary requirement. The vitamin C content of young, fresh spinach leaves is higher than that of mature plants. In a similar vein, cabbage and lettuce have more vitamins in their green outer leaves than in their white interior ones. Particularly noteworthy is the high vitamin C content of amaranth species. Provitamin A carotenoids, such as beta-carotene (around 25–30%), alpha-carotene, beta-cryptoxanthin, and gamma-carotene, make up the majority of vitamin A found in leafy greens. Non-provitamin A carotenoids, such as lutein (about 45%), violaxanthin (about 15%), and neoxanthin (roughly 15%), are also significant (Raju et al., 2007). The USDA National Nutrient Database states that when compared to cereals and pulses, leafy vegetables such as spinach and kale have remarkably high quantities of vitamins A and K1 (phylloquinone). Duckweed (*Lemna perpusilla*) is unique among leafy greens in that it has a higher vitamin E concentration than many other leafy vegetables and seeds, with 45.7 mg per 100 g. Folate is a water-soluble vitamin B9 group molecule that is mostly found in plants, particularly in GLVs. Folate levels in staple diets based on cereal grains and tubers are often low, but they can be raised by including GLVs. For instance, Indian diets contain a lot of folate from the leaves of colocasia (*Colocasia esculenta*). Nonetheless, it is estimated that cooking eliminates more than 50% of the folate in food. Consequently, it is advised to drink the cooking water in order to preserve the folate level (Kala and Prakash, 2004).

6.3.3. Minerals

If micronutrient deficits are not addressed concurrently with protein-energy status, optimal development and functionality will not be achieved. Because malnutrition and micronutrient status are closely related, the World Health Organization stressed that the two must always be taken into account together (Kumar et al., 2020). For instance, sustaining a robust immune system associated with Zinc, and magnesium is involved in more than 300 bodily functions. *Chenopodium album* possesses significant amounts of vital elements, such as phosphorus, nitrogen, potassium, calcium, magnesium, iron, and manganese, according to two years of field research by Shahi (1977). But as the plant ages, its nutritious content declines. Zinc shortage affects the gastrointestinal and immunological systems, while iron deficiency causes anemia, especially in women and children (Nath et al., 2004). Minerals are abundant in green leafy plants. For example, spinach has the highest concentrations of sodium (827 mg/100 g), calcium (1036 mg/100 g), magnesium (827 mg/100 g), and iron (28.4 mg/100 g). Zinc content in duckweed (*Lemna perpusilla*) is very high (15 mg/100 g). Iron (3490 mg/100 g), phosphorus (83 mg/100 g), and calcium (397 mg/100 g) are all present in fresh amaranth leaves (Kumar et al., 2020). According to Abbas et al. (2018), moringa leaves also contain a substantial amount of minerals, such as calcium, iron, magnesium, manganese, phosphorus, potassium, sodium, and zinc. Compared to vitamins and proteins, minerals are more stable during food processing, which makes them more resistant to nutrient loss (Kala and Prakash, 2004).

6.3.4. Carbohydrates and Dietary Fibre

Leafy vegetables are naturally low in carbohydrates, typically containing about 2–4 grams per 100 grams of fresh weight. Most of these carbohydrates are derived from dietary fibre, which aids digestion, supports gut health, and provides various health benefits without significantly affecting blood sugar levels. For example, spinach and *Bathua* contain 2.90 g, cabbage 4.6 g, and moringa 4.1 g of carbohydrates per 100 grams (Gopalan et al., 2004). Their carbohydrate content, combined with high water levels and essential nutrients, contributes to overall well-being, promoting heart and gut health. It is often known that GLVs are a great source of dietary fibre. According to research, soluble dietary fibre is especially abundant in Indian GLVs such as basella, fenugreek, coriander, spinach, and lettuce. Increased consumption of vegetable fibre lowers the risk of colon cancer and cardiovascular diseases, maintains a healthy body weight, and supports a healthy digestive system. Additionally, dietary fibre plays a significant role in alleviating issues such as obesity, constipation, diabetes, and diverticulitis (Spiller, 2001).

6.3.5. Essential Fatty Acids

Omega-3 is a necessary nutrient that must be obtained through diet because those could not be synthesized by the human body. Among its many health benefits, it is extremely useful. A precursor to omega-3 fatty acids, alpha-linolenic acid (ALA) is found in GLVs and is linked to several health benefits. Studies show that compared to farmed veggies, wild plants have comparatively higher quantities of omega-3 fatty acids. The same study found that several leafy greens, including spinach, mustard greens, red leaf lettuce, and butter crunch lettuce, had significant levels (0.6–1.7 mg/g) of alpha-linolenic acid (Hamazaki and Okuyama, 2001). Particularly high in omega-3 fatty acids is spinach, which has about 370 mg of omega-3 per 100 g.

6.4. PHYTOCHEMISTRY AND NUTRACEUTICAL COMPONENTS IN LEAFY VEGETABLES

Phytochemicals are basically compounds that are non-nutritional in nature found in plants in trace quantities. Phytochemicals affect the secondary metabolism of plants including vegetables and may considerably influence human health after consumption. Bioactive molecules in plants can be classified as per their chemical nature, such as terpene derivatives, phytoestrogens, antioxidant vitamins, polyphenols, sulfur compounds, minerals, polyunsaturated fatty acids, and dietary fibre (Barba et al., 2014). Bioactive chemicals obtained from leafy greens have many biological effects, including antioxidant and antibacterial properties associated with the antiaging process. Nutraceuticals refer to naturally occurring dietary components in foods that are consumed orally and are believed to offer medical or health benefits (Chauhan et al., 2013). Leafy vegetables are gaining recognition as important nutraceuticals due to their abundance of bioactive compounds, essential nutrients, and health-enhancing properties. Including them in the diet can help combat nutritional deficiencies and improve overall health.

Leafy vegetables are rich in different vitamins such as vitamins A, C, and B and minerals. Moreover, as photosynthetic tissues, leafy vegetables have vitamin K in more ample amounts than other fruits and vegetables because vitamin K (phylloquinone) is directly involved in photosynthesis. Some GLVs are medicinally significant, having few or no toxic substances and the ability to produce several secondary metabolites with complex structures that demonstrate antibacterial characteristics. GLVs have compounds with antidiabetic, anticarcinogenic, antihistaminic, and hypolipidemic properties, providing preventive or therapeutic benefits against several lifestyle disorders, including cardiovascular disease, aging, obesity, hypertension, and insomnia (Bhat and Al-Daihan, 2014). Numerous bioactive compounds are found to have significant antioxidant properties, evidenced by their capacity to neutralize free radicals, especially reactive oxygen species. Moreover, numerous of them, such as carotenoids, anthocyanin, etc., influence the coloration of certain leaves and other parts. They are being investigated in order to study cancer prevention, cardiovascular disease, and other conditions. Bioactive compounds encompass lycopene, resveratrol, lignans, tannins, and indoles. The principal chemical groups of bioactive compounds are outlined here (Barba et al., 2014).

6.4.1. Carotenoids

Carotenoids comprise around 700 compounds that provide different colours of fruits and vegetables, such as orange, yellow, and red hues, to many fruits and vegetables. Carotenoids are categorized into two groups: carotenes and xanthophyll. Carotenes are only hydrocarbons lacking oxygen, and xanthophylls incorporate oxygen (Lakshminarayana et al., 2005). Carotenoids work as colour markers and are recognized as essential ingredients of a healthy diet, with some varieties demonstrating vitamin A activity. Thus, β-carotene and lutein (xanthophyll) are well-documented as two principal carotenoids found in green vegetables. Furthermore, alpha-carotene, zeaxanthin, beta-cryptoxanthin, lutein 5,6-epoxide, and antheraxanthin are often present as minor constituents. Leafy vegetables can effectively increase dietary levels of zeaxanthin and/or lutein (Barba et al., 2014). Thus, leafy vegetables may diminish the incidence of ocular disorders such as age-related macular degeneration and cataract formation, since they serve as a valuable source of zeaxanthin and/or lutein in the diet (Barba et al., 2014).

6.4.2. Polyphenols

Polyphenols constitute an essential category of chemical compounds considered as plant secondary metabolites, characterized by diverse chemical structures and functions. Chemically, phenolic compounds are molecules characterized by an aromatic ring each containing hydroxyl group(s), along with functional derivatives such as esters, methylesters, and glycosides (Craft et al., 2012). The predominant phenolics in green leafy vegetables are chlorogenic acid and quercetin, succeeded by ferulic acid, kaempferol, and, in certain products, caffeic acid. Chlorogenic acid is present in several green vegetables, including lettuce, broccoli, and endive, which contain 3.78 mg/100 g fresh weight, 1.78 mg/100 g fresh weight, and 101 mg/100 g fresh weight, respectively. Caffeic acid is found in fresh herbs, including rosemary, sage, thyme, and oregano, with concentrations ranging from 2.08 to 11.7 mg per 100 g. Additionally, ferulic acid is present in other popular leafy vegetables, including cauliflower leaves, chayote, hoary basil, oregano, and rosemary, ranging from 0.13 to 2.88 mg/100 g (Barba et al., 2014).

6.4.3. Ascorbic Acid

Ascorbic acid possesses a carbonyl group with an enediol structure in the lactone ring. The acidic property is due to the presence of two enolic hydrogen atoms, which supply the electrons necessary for its role as an antioxidant. Ascorbic acid is converted into dehydroascorbic acid in the presence of oxygen, which retains equivalent vitamin activity. Ascorbic acid is regarded as one of the most potent natural antioxidants and is essential for the organism's regular functioning due to its involvement in several physiological processes (Hamrick and Counts, 2008). Leafy vegetables are an excellent source of ascorbic acid. Ascorbic acid is present in several commonly consumed leafy vegetables, including broccoli (*Brassica oleracea* var. italica) at 96.7 mg/100 g, kale (*Brassica oleracea* var. sabellica) at 52–67 mg/100 g, spinach (*Spinacia oleracea*) at 35.67–75.00 mg/100 g, curry (*Murraya koenigii*) at 140.50 mg/100 g, mint leaf (*Mentha longifolia*) at 52.75 mg/100 g, Stevia (*Stevia rebaudiana* Bertoni) at 14.98 mg/100 g, turnip greens (*Brassica rapa* subsp. rapa) at 27 mg/100 g, water lettuce (*Pistia stratiotes*) at 21.02 mg/100 g, and mustard greens (*Brassica juncea*) at 36.20 mg/100 g. Consequently, these products can serve as an effective means to fulfill the necessary daily consumption for adults, which is 90 mg (Barba et al., 2014).

6.4.4. Phytosterols

Plants contain a certain number of sterols, specifically phytosterols. Both cholesterol and phytosterols have a tetra-cyclic ring structure along with a C17 side chain; however, the side chain in phytosterols is alkylated at C24 with a methyl/ethyl group (e.g., sitosterol), and several phytosterols, such as stigmasterol, also include double bonds. Sterol content varies throughout plants, especially among families. Plant leaves and other tissues contain different sterol and acyl sterol glycosides where a C-3 hydroxyl group is attached with sugar via a glycosidic bond. ß-sitosterol is the primary phytosterol found in leafy greens, followed by campesterol, stigmasterol, and sitostanol (Barba et al., 2014). Diverse leafy greens possess different varieties. Spinach (*Spinacia oleracea*) possesses ß-sitosterol, campesterol, stigmasterol, ß-sitostanol, and campestanol; Romaine lettuce (*Lactuca sativa* L. var. longifolia) includes ß-sitosterol, campesterol, stigmasterol, ß-sitostanol, and campestanol; endive lettuce (*Cichorium endivia*) contains campesterol, ß-sitosterol, stigmasterol, and ß-sitostanol (Ching and Mohamed, 2001; Barba et al., 2014).

6.4.5. Tocopherols and Tocotrienols

The term "vitamin E" pertains to a category of lipid-soluble antioxidant vitamin in plants, collectively identified as tocopherols and tocotrienols. These molecules possess a phenolic-chromanol ring structure linked to an isoprenoid side chain, which is saturated in tocopherols and unsaturated in tocotrienols. Vitamin E comprises four principal types of tocopherols and tocotrienols: α, β, γ, and δ. Among them, α-tocopherol is the predominant type in leafy vegetables and is also the most physiologically active (Bjørneboe et al., 1990). Research indicates that α-tocopherol is found in some leafy vegetables, including celery (*Apium graveolens*), which has 13.64 mg per 100 g, and amaranth (*Amaranthus* spp.), which contains 1.16–1.57 mg per 100 g (Ching and Mohamed, 2001).

6.4.6. Melatonin

Animals, bacteria, unicellular creatures, and plants all contain the extremely conserved chemical melatonin. It is thought that melatonin, an antioxidant found in plants, protects against oxidative stress caused by both internal and external factors. Certain herbs have melatonin concentrations higher than 1 μg/g, and these plants have long been used to treat ailments associated with damage from free radicals. Studies analyzing melatonin content in various leafy vegetables have identified Chinese sage (*Salvia miltiorrhiza Bge*) and *Chenopodium rubrum* leaves as significant sources of melatonin (Chen et al., 2003).

6.4.7. Sulfur Compounds

Phytochemicals containing sulfur are frequently present in plants of the Cruciferae family. Organosulfur compounds comprise sulfur atoms linked to a cyanate group or additional carbon atoms in either cyclic or acyclic configurations. The bioactive components of foods with organosulfur compounds are released just when the vegetable is sliced, masticated, or pulverized, resulting in the disruption of its tissue. Isothiocyanates, including phenethyl isothiocyanate, sulforaphane, and benzyl isothiocyanate, are generated from glucosinolates in cruciferous vegetables by the enzymatic activity of myrosinase. Allicin is synthesized from alliin in Allium species and then transformed into diallyl sulfide, diallyl disulfide, or diallyl trisulfide by the enzymatic activity of allinase. The hydrolytic degradation products of cruciferous and allium vegetables are bioactive dietary constituents that enhance health. Brassicaceae has around 50 unique glucosinolates, but more than 100 glucosinolates have been found globally across plants. Cruciferous species comprise around 20 glucosinolates. Cabbage has 14 glucosinolates, cauliflower has 12, Brussels sprouts hold 11, while both broccoli and Chinese cabbage each include 7. Glucoiberin was recognized as the predominant sulfur compound in Brussels sprouts, broccoli, cabbage, and cauliflower, while glucobrassicin and 4-Methylthio-3-butenyl glucosinolate were the prominent sulfur compounds in kale and radish, respectively (Ciska et al., 2000; Barba et al., 2014).

The immense significance of all phytochemicals from leafy vegetables is mentioned in Table 6.2.

6.5. ANTIOXIDANT ACTIVITY

Reactive oxygen species (ROS) induce lipid oxidation and membrane degradation in living organisms. The production of ROS includes hydroxyl radicals, superoxide anions, hydrogen peroxide, and singlet oxygen, which are linked to several degenerative human diseases, including cancer and cerebrovascular disorders, through multiple mechanisms. The growing emphasis on natural foods and food-derived antioxidants has become more evident due to their chemical protection against oxidative damage. Plants are a vital source of bioactive natural chemicals, exhibiting considerable structural and biological diversity. Recent studies have linked the protection of cancer and cardiovascular disorders to the consumption of fresh fruits, vegetables, or teas abundant in natural antioxidants. The defensive benefits of plant products stem from several components with distinct modes of action, including enzymes, proteins, and low-molecular-weight chemicals such as vitamins, carotenoids, flavonoids, anthocyanins, and other phenolic substances (Kim et al., 2011). Ascorbic acid is considered one of the most powerful natural antioxidants and is crucial for the organism's healthy functioning owing to its role in several physiological processes. Studies suggest that a daily intake of 90–100 mg of vitamin C reduces the risk of chronic diseases among nonsmokers. Ascorbic acid has shown effectiveness in therapeutically reducing blood cholesterol levels and several cancer types (Barba et al., 2014). Ingesting leafy greens is essential for the provision of dietary antioxidants. The benefits of consuming lettuce depend on its makeup, particularly its antioxidant levels, which may function as nutrients. The health advantages are contingent upon their biochemical effects upon entering the bloodstream. Specific components may be released from the food matrix and altered inside the digestive tract. The bioaccessibility of lettuce measures the quantity of these chemicals released from the food matrix during digestion, which is essential for health-promoting attributes. Natural antioxidant molecules originating from plants, such as those found in lettuce, have gained significance due to the growing awareness of the necessity for a healthy diet and the limitations on dangerous synthetic antioxidants. Anthocyanins in iceberg lettuce augment antioxidant activity by blocking enzymes and sequestering trace components that facilitate free radical production (Shi et al., 2022). Polyphenols from leafy greens can reduce oxidative reactions in several food products, including dairy and fruit/vegetable juices (Barba et al., 2014). Oregano extracts have been recognized as a significant beneficial substitute for omega-3 fatty acids in dairy beverage formulations, owing to the vulnerability of omega-3 fatty acids to oxidation. The oregano extract has around 269 mg of gallic acid equivalents per gram of phenols and exhibits considerable antioxidant activity, as demonstrated by the DPPH radical assay (Boroski et al., 2012).

TABLE 6.2
Phytochemical and Health Benefits of Leafy Vegetables

Sl. No.	Leafy Vegetable	Botanical Name	Family	Phytochemicals	Health Benefits, Use	Reference
1.	Lettuce	*Lactuca sativa L.*	Asteraceae	Phenolics, glycosylated flavonoids, sesquiterpene lactones, etc.	Reduce risk of cancer and cardiovascular disease; heal Alzheimer's disease and diabetes	(Shi et al., 2022)
2.	Amaranthus	*Amaranthus cruentus*	Amaranthaceae	Kaempferol-3-rutinoside, caffeoylglucaric isomer, coumaroyl glucaric isomer, feruloyl isocitrat isomer, feruloylglucaric isomer, quercetin-3-rutinoside, etc.	Useful in treatment of fever, body pain, and asthma	(Moyo et al., 2021)
3.	**Jute leaf**	*Corchorus olitorius*	Tiliaceae	Quinic acid, monogalactosyldiacylglycerol, chlorogenic acid, quercetin-3-O-galactoside, etc.	Used as laxatives, in creams for skin care, pain management, dysentery, enteritis, fever	(Islam, 2013; Moyo et al., 2021)
4.	*Stevia*	*Stevia rebaudiana*	Asteraceae	Diterpene glycoside, stevioside, Quercetin, catechin and Pyrogallol	Stevioside (about 300 times sweeter than sucrose, served as noncaloric sweetener, substitutes for sucrose for diabetes mellitus patients	(Kim et al., 2011)
5.	Moringa	*Moringa pterygosperma G.*	Moringaceae	Chlorogenic acid, ferulic acid, luteolin, quercetin, kaempferol, rhamnosylated glucosinolates	Show antiasthmatic activity, control undesirable skin sebum secretions, act as an inhibitor of human platelet aggregation	(Fahey et al., 2018; Barba et al., 2014)
6.	Oregano leaves	*Origanum vulgare*		Carnosol, carnosic acid, rosmarinic acid	Significant healthy alternative to omega-3 fatty acids in dairy beverages formulation	(Neveu et al., 2010)
7.	Kale	*Brassica oleracea*	Brassicaceae	Glucoiberin, progoitrin, sinigrin, glucoraphanin	Protective in coronary artery disease, anti-inflammatory activity, gastroprotective activity, antigenotoxic ability, inhibition of the carcinogenic compounds formation, positive to gut microbes	(Barba et al., 2014; Satheesh and Fanta, 2020)
8.	**Radish**	*Raphanus sativus*	Brassicaceae	Raphasatin, progoitrin, glucoraphanin, glucoraphanin, glucoapoleiferin, glucoalyssin, gluconapin	Improve the organoleptic characteristics of the dishes, used in treatment of infectious and noncommunicable diseases	(Barba et al., 2014)

6.6. LEAFY VEGETABLES IN HEALTH BENEFITS

Leafy vegetables are protective foods as they are a rich source of nutraceuticals. Consumption of leafy vegetables offers the following preventive roles.

6.6.1. Antidiabetic Properties

Diabetes mellitus is a prevalent noncommunicable illness found in most countries worldwide. The primary classifications are type 1 (juvenile-onset or insulin-dependent) diabetes and the more common type 2 (adult-onset or non-insulin-dependent) diabetes. Shaw et al. (2010) reported that the global prevalence of type 2 diabetes was 6.4% in 2010, projected to rise to 7.7% by 2030. Patients with type 2 diabetes frequently exhibit oxidative damage to numerous biomolecules, resulting in an increase in lipid peroxidation. Consequently, alleviating oxidative damage is seen as essential in the management of diabetes and its associated problems, frequently related to lipid peroxidation (Stanely and Menon, 2001; Giugliano et al., 1996). Numerous green leafy vegetables, including spinach, kale, and lettuce, are abundant in antioxidants, vitamins, carbohydrates, phytochemicals, and minerals, enhancing their protective function against oxidative stress. These chemicals are essential in mitigating diabetes risk by affecting many biological systems. A prospective therapeutic approach for diabetes focuses on inhibiting amylase enzymes, which catalyze the hydrolysis of starch into glucose that subsequently enters the circulation (Kabir et al., 2016; Rahman et al., 2022). Consequently, research has concentrated on natural bioactive substances possessing antidiabetic characteristics. Lettuce leaves are regarded as a phytonutrient powerhouse, abundant in carotenoids such as lutein, lactucaxanthin, and β-carotene (Llorach et al., 2008). Notably, lactucaxanthin, a special carotenoid from lettuce that is rare in other plants, inhibits amylase activity. This inhibition helps reduce postprandial hyperglycemia by controlling starch breakdown during digestion. Additionally, plants rich in polyphenols exhibit antidiabetic effects and possess ferric-reducing properties. These properties are particularly important, as ferric ion accumulation can lead to beta-cell destruction, a key factor in diabetes mellitus development (Pulido et al., 2000).

6.6.2. Anticancer/Antiproliferative Properties

Uncontrolled cellular proliferation can result in various forms of cancer, including carcinomas, leukemias, sarcomas, and lymphomas (Basli et al., 2017). Green leafy vegetables (GLVs) are a rich source of anticancer compounds, particularly phenolic molecules (Kaur et al., 2011), by causing cell cycle arrest and blocking oncogenic signalling pathways that control proliferation, angiogenesis, and apoptosis. These chemicals can also increase the activity of tumour suppressor proteins like p53, modify the levels of reactive oxygen species, and improve the differentiation and conversion of aberrant cells back into normal ones. Because the gastrointestinal tract (GIT) is in direct contact with ingested food, it serves as the primary site for the action of many phenolic compounds. The anticancer potential of various GLVs in the GIT was found to be significant in cancer prevention. The effectiveness of *Amaranthus cruentus* in inhibiting the growth of HCT-116 colon cancer cells has been compared to the anticancer drug 5-fluorouracil because IC50 value of 5-fluorouracil (32.14 μg/mL) was found to be comparable to that of *A. cruentus* ethanolic extract (138 μg/mL) (Vithya and Jayshree, 2017).

Lettuce has potential cancer-preventive biomolecules, such as β-carotene and ascorbic acid, known to reduce the risk of colorectal cancer. Secondly, the chlorophyll pigment, particularly in green, dark-colored varieties, provides a protective effect against cancers such as colon and liver cancers. Thirdly, lettuce is a significant source of phenolic compounds, including dicaffeoyl tartaric acid and chlorogenic acid. These phenols are highly valued in plant-based therapeutics for their roles in cancer prevention and chemotherapy.

Apigenin, a flavonoid derived from leafy vegetables, has demonstrated significant anticancer potential in various experimental and biological studies by modulating multiple signalling pathways, especially by activating the extrinsic caspase-dependent pathway. It could inhibit cell proliferation and promote apoptosis in several tumour types, including lung, breast, skin, liver, blood, and prostate cancers.

6.6.3. Anti-Inflammatory Properties

A variety of agents, including toxic chemicals, viruses, reactive nitrogen and oxygen species, and bacterial lipopolysaccharides, can induce inflammation in the human body. These factors enhance the production and secretion of pro-inflammatory cytokines. A variety of polyphenol groups have exhibited anti-inflammatory properties. Flavonoids are recognized as natural anti-inflammatory agents that can inhibit certain pro-inflammatory proteins (Salaritabar et al., 2017). Likewise, by inhibiting IKK-κB kinase activity, phenolic acids such as ferulic acid might diminish inflammation (Villela-Castrejón et al., 2017). Dietary phenolics are predominantly concentrated in the stomach and oropharyngeal tract, with smaller concentrations in the intestine before absorption, metabolism, and excretion (Metere and Giacomelli, 2017). This is particularly significant as these phenolics might mitigate inflammation in the GIT. Inflammatory bowel disease (IBD) refers to a group of chronic conditions marked by persistent inflammation of the intestinal mucosa (Nunes et al., 2018). Although historically associated with Western nations, IBD is increasingly prevalent across Asia, Africa, South America, and Eastern Europe (Ng et al., 2017).

Amaranthus dubius (pigweed or amaranth) has been shown to decrease the production of the inflammatory cytokine TNF-α in human acute monocytic THP-1 cells stimulated by lipopolysaccharide (LPS) (Tufts et al., 2015). This inhibition reduces tissue damage and inflammation induced by interleukin-6 (IL-6) and interleukin-1β (IL-1β). Extracts from *Adansonia digitata* (baobab leaves) have anti-inflammatory properties via inhibiting the activation of nuclear factor kappa B (NF-κB) (Ma et al., 2017).

6.6.4. Cardiovascular Disease and Leafy Vegetables

Cardiovascular diseases (CVDs), a category of chronic noncommunicable disorders, represent a considerable public health concern and are linked to multifaceted risk factors, such as hyperlipidemia, hypertension, diabetes, and metabolic syndrome (Praveen et al., 2013). Epidemiological studies indicate that the intake of vegetables, particularly GLVs, has advantageous effects on the management of CVDs due to their fibre content, antioxidant capabilities, and phytochemicals. The cholesterol-reducing properties of GLVs are mostly ascribed to their fibre composition. Soluble fibers, including pectin, are essential in lipid metabolism, decreasing cholesterol absorption in humans and animals (Nishimura et al., 2000). This impact is mainly accomplished by inhibiting micelle production and impeding cholesterol transfer across the unstirred layer (Stedronsky, 1994). Moreover, dietary fibre facilitates the elimination of steroids, including bile acids, via feces, so further decreasing cholesterol levels. Fermentation of fibre in the large intestine generates short-chain fatty acids, such as propionate, which is an effective cholesterol-lowering compound, hence contributing to the reduction of plasma cholesterol levels (Nicolle et al., 2004). Additionally, green leafy vegetables, such as lettuce, are abundant in antioxidants, including vitamins E and C, carotenoids, and polyphenols, which provide cardiovascular protection (Proteggente et al., 2002). These antioxidants mitigate lipid peroxidation in the heart, a primary target tissue for ROS, hence diminishing oxidative stress and offering protective benefits against CVDs.

6.6.5. Fertility and Human Fetus

GLVs are abundant in important minerals, vitamins, and antioxidants. These vegetables may be ingested either raw or cooked; however, consuming them raw or minimally cooked is advisable, since cooking might diminish their concentrations of vitamins and antioxidants, especially zinc, vitamin C, and folate, which are essential for fertility. Vegetables such as broccoli, Brussels sprouts, and cabbage are distinguished by their diindolylmethane concentration, a chemical that assists the body in expelling surplus "bad" estrogen (Randhawa et al., 2015). Women with inadequate iron levels may encounter anovulation and diminished oocyte quality, elevating the probability of conception difficulties by up to 60% relative to those with appropriate iron reserves. GLVs, including spinach and asparagus, are outstanding providers of dietary iron. Vitamin C enhances iron absorption and protects cells from damage inflicted by free radicals, among its several essential roles.

Preserving a healthy acid-base equilibrium is crucial for proper bodily function. Contemporary diets, frequently characterized by high consumption of meat, sugar, and processed foods, are typically too acidic. Integrating more GLVs can aid in reestablishing this equilibrium. Spinach, regarded as a superfood, offers an abundance of vital vitamins and minerals required for fertility and gestation. It is very abundant in iron and folic acid, both essential for reproductive health and embryonic development (Randhawa et al., 2015). Folic acid, scientifically termed pteroylmonoglutamic acid, is collectively known as "folate." These folates comprise pteroylmonoglutamic acid molecules with two to eight connected glutamic acid groups. It is an essential element of the vitamin B complex and a water-soluble vitamin (Cossins, 2000). GLVs such as spinach, broccoli, lettuce, and asparagus are abundant in folic acid, which is vital for proper development of the fetus and required for the production of amino acids and nucleic acids (Antony, 2007). The incidence of neural tube abnormalities, including anencephaly, encephalocele, and spina bifida, is significantly reduced when women adequately intake folic acid before conception and throughout the early stages of pregnancy.

6.7. ANTINUTRITIONAL FACTORS IN VEGETABLES

6.7.1. Phenolic Compounds

Polyphenols are important compounds, exhibiting greater antioxidant activity than carotenoids and vitamins (Gardner et al., 2000). They include tannins, flavonoids, lignans, and phenolic acids. GLVs contain flavonoids such as myricetin, quercetin, and luteolin. Tannin content in African leafy vegetables (ALVs) varies from about 650 mg/100 g in *Xanthosoma* species to 1200 mg/100 g in *Euphorbia hirta*, potentially reducing nutritional value by interacting with proteins and enzymes (Serrano et al., 2009). Polyphenol-rich extracts correlate with high antioxidant activity, although this varies with methodology and vegetable type. GLVs such as *Portulaca oleracea* and *Xanthosoma mafaffa* exhibit strong antioxidant activity. Factors such as plant age, growth conditions, and processing influence activity; for instance, amaranth harvested after 60 days shows higher antioxidant activity (Modi, 2007). Cooking may increase or decrease flavonoid content due to leaching or structural changes.

6.7.2. Oxalates

Oxalates are organic compounds formed by oxalic acid and its salts or minerals, commonly found in many GLVs such as *Xanthosomas* sp., *Euphorbia hirta*, *Amaranthus* sp., *Ipomoea involucrata*, *Celosia argentea*, *Manihot esculenta*, *Vernonia* sp., and *Telfairia occidentalis* (Champ, 2002). These can occur as soluble salts (e.g., sodium and potassium oxalates) or insoluble salts (e.g., calcium, magnesium, and iron oxalates), with the type and concentration varying among plant species. Soluble oxalates form strong chelates with dietary calcium, reducing its bioavailability and potentially leading to deficiencies. High soluble oxalate intake can result in kidney stone formation (renal calculi or nephrolithiasis) due to the accumulation of insoluble calcium oxalate crystals in the kidneys. It has been shown that dietary supplementation with divalent minerals, such as calcium, can mitigate these effects by reducing the intestinal absorption of soluble oxalates (Radek and Savage, 2008). Leaves generally contain higher concentrations of oxalates than do seeds or stems, as seen in rhubarb, spinach, amaranth, and beet leaves.

6.7.3. Protease Inhibitors

GLVs may contain significant amounts of chymotrypsin and trypsin inhibitors, which interfere with protein and amino acid utilization by binding to proteolytic enzymes, thereby preventing their digestion (Glew et al., 2005). Protease enzyme inhibitors were identified in species such as *Boscia senegalensis*, *Maerua crassifolia*, and *Brassica* sp. The effectiveness of these inhibitors is significantly influenced by temperature, water availability, and heating duration. Conventional blanching is more effective than microwave blanching in reducing the inhibitory activities of chymotrypsin and trypsin inhibitors. However, studies have shown that in many leafy vegetables, the inhibitory activity of trypsin remains stable to heat even after 5 minutes of boiling, potentially impairing protein utilization in humans (Vanderjagt et al., 2000).

6.7.4. Nitrate

Nitrate, a naturally occurring compound in vegetables, significantly influences their quality. Leafy vegetables such as spinach and lettuce tend to accumulate higher nitrate levels than do root or fruit vegetables, driven by agronomic practices (e.g., nitrogen fertilizer use) and environmental factors (Santamaria, 2006). Low light intensity during harvesting (≤ 200 µmol m^{-2} s^{-1}) further increases nitrate accumulation in leafy vegetables. Although dietary nitrate is nontoxic, its metabolites—nitrite, nitric oxide, and N-nitrous compounds—pose potential health risks, including methemoglobinemia ("blue baby syndrome") in vulnerable populations such as infants and individuals with gastroenteritis (Majumdar, 2003). The WHO recommends a daily intake limit of 0.06 mg/kg body weight for nitrites and 3.7 mg/kg body weight for nitrates. European Union regulations set nitrate limits for vegetables, such as 2500–3500 mg/kg in spinach and 3000–5000 mg/kg in lettuce (Colonna et al., 2016).

6.7.5. Tannins

Tannins, first identified by Seguin in 1976, are polyphenolic compounds (up to 20,000 Da) capable of forming complexes with proteins and carbohydrates in aqueous solutions (Khanbabaee and Van Ree, 2001). Found across a wide range of plant parts, including bark, wood, fruits, leaves, roots, and galls, tannins are significant as normal metabolic products of plants and not artifacts of chemical transformations. Hydrolyzable tannins consist of a central glucose molecule esterified with phenolic acids, making them prone to breakdown by acids, bases, or enzymes. In contrast, condensed tannins, composed of proanthocyanidins, are more resistant to hydrolysis and are found predominantly in higher plants (Salunkhe and Chavan, 1989). Tannins have dual roles in human health and food systems. Although they contribute to flavour and preservation in some foods, they can produce undesirable effects when present in excessive quantities. Their ability to bind proteins through covalent linkages and hydrogen bonding reduces protein bioavailability for absorption. High levels of dietary tannins have been implicated in conditions such as esophageal cancer (Harborne, 1975).

6.7.6. Phytic Acid

Phytic acid, or myo-inositol hexaphosphoric acid, a compound that functions as a primary phosphorus storage, form in leafy vegetables (Reddy and Shridhar, 2001). In plant tissues, phytic acid is primarily associated with cations such as magnesium, potassium, and calcium. These phosphate groups form strong complexes with essential minerals in the human body, making these nutrients less bioavailable for absorption. For instance, it chelates zinc, impacting zinc homeostasis. Additionally, phytates can reduce the activity of different digestive enzymes by chelating cofactors or through protein interactions, either directly or via cation bridges. Phytic acid is remarkably heat stable, and traditional cooking or boiling does not significantly reduce its levels in food. However, alternative food preparation methods, such as fermentation of lactic acid and acidic media soaking, have proven to be more effective at lowering phytic acid content.

6.7.7. Cyanogenic Glycosides

These natural metabolites are present in more than 2,500 species of leafy plants, including many economically significant crops. These compounds, derived from amino acids, are glycosides of α-hydroxynitriles. Upon enzymatic hydrolysis, cyanogenic glycosides release hydrogen cyanide (HCN), a compound with potential toxicity. Primarily produced by plants as a defence mechanism, cyanogenic glycosides are present in crops such as cassava (*Manihot esculenta*), durian (*Durio zibethinus*), and white clover (*Trifolium repens*). Toxic effects in humans include interference with protein absorption and inhibition of enzymes such as Na-K-ATPase, leading to electrolyte imbalances and potassium depletion, as observed with linamarin, a cyanogenic glycoside in cassava (Kamalu, 1993). Chronic exposure is associated with conditions such as tropical ataxic neuropathy and goiter, highlighting the need for proper food preparation to mitigate health risks.

6.8. PRECAUTIONS WITH LEAFY VEGETABLE CONSUMPTION

Antinutrients are a major concern when it comes to the consumption of green leafy vegetables, as they can pose potential health risks. Compounds such as nitrates, oxalates, phytates, cyanogenic glycosides, and tannins present in these vegetables can interfere with the absorption of micronutrients, rendering them unavailable for the body. Additionally, calcium oxalate, a common component of kidney stones, is often linked to their formation. Therefore, individuals with kidney stones should limit their intake of these vegetables. It is crucial to employ various methods to reduce antinutritional factors in order to minimize associated health risks.

6.9. METAL TOXICITY IN LEAFY VEGETABLES

Leafy vegetables are widely valued across the globe for their numerous health benefits. However, contamination with heavy metals poses a significant ecological concern worldwide. Heavy metals can build up in the tissues of these vegetables without causing any obvious toxicity signs (Manzoor et al., 2018). Some leafy plants, such as lettuce, have been identified as possible heavy metal hyperaccumulators (Ramos et al., 2012). Different sections of the same plant may have varying concentrations of heavy metals. The buildup of heavy metals is generally greater in the leaves than in the roots, according to Xu et al. (2019).

Vegetables are essential to the ecosystem because they are effective in absorbing metal elements from the soil. However, plant absorption of soil pH, which influences metal solubility, increases health hazards, especially for leafy greens that have a tendency to collect high amounts of heavy metals (Ahmad et al., 2019). Industrial pollution and tainted irrigation water are the main sources of heavy metals that humans transfer into agricultural soils through practices such as pesticide application and wastewater irrigation (Alam et al., 2018). There is increased concern about toxic metal pollution in soils due to the widespread use of wastewater for irrigation, particularly in developing nations. The uptake of heavy metals by leafy vegetables is greatly influenced by a number of factors, including soil type, concentration, atmospheric deposition, and plant development. Heavy metal concentrations in vegetables are also increased by transportation-related air pollution and irrigation with untreated water (Ashraf et al., 2021a, 2021b).

The body requires essential heavy metals for various biological processes. However, excessive levels of these metals can lead to health complications. Cadmium, lead, arsenic, and mercury are highly toxic and pose significant health risks (Khan et al., 2016a, 2016b). Prolonged exposure to cadmium can weaken the immune system and cause hepatic and respiratory issues (Latif et al., 2018). Dietary intake of chromium and nickel has been linked to symptoms such as fatigue, headaches, skin rashes, and cardiac problems. Arsenic exposure can lead to skin cancer, excessive zinc can impair immune function, and high levels of copper can harm the stomach and liver. Heavy metals have both direct and indirect effects on health. Direct impacts occur through the consumption of contaminated leafy vegetables, ingestion, or dermal exposure, potentially causing cancers (e.g., skin, colon, lung) and organ damage. Indirect impacts include the reduction of nutritional content in contaminated crops, which can lead to deficiencies of essential nutrients in plants and, consequently, in human diets.

A focused study on the toxic effects of consuming green leafy vegetables contaminated with heavy metals in humans is essential. Additionally, strategies and policies should be developed to regulate and minimize heavy metal contamination in vegetables. It is equally important to identify vegetable species that are hyperaccumulators of heavy metals and raise public awareness regarding their consumption risks.

6.10. REDUCTION OF ANTINUTRITIONAL FACTORS THROUGH PROCESSING

Heat treatment has been identified as one of the most effective methods for reducing antinutritional factors in green leafy vegetables (Mosha and Gaga, 1999). Processes such as cooking and blanching help remove antinutrients by breaking down the plant cell walls, allowing soluble compounds to leach into the blanching medium. These methods have been shown to significantly reduce levels of phytic acid and oxalic acid (Oboh, 2005). However, such practices may also lead to the loss of essential nutrients through leaching (Mepba et al., 2007). Studies on different green leafy vegetables reveal that blanching for 10–15 minutes can significantly reduce phytic acid content. Oxalic acid levels are effectively reduced through both cooking and blanching, whereas drying and storage methods have minimal impact on reducing antinutritional factors (Yadav and Sehgal, 2003; Ramulu and Rao, 2003). Further research is necessary to explore different vegetable varieties and to develop agronomic practices that minimize antinutrient content while enhancing the nutritional value of green leafy vegetables.

6.11. CONCLUSION

GLVs are an excellent source of critical nutrients, including vitamins, minerals, amino acids, essential fatty acids, and dietary fibre, which are crucial for human health and well-being, regardless of dietary categories. Growing leafy vegetables in home gardens could be the source of secondary income, mostly by women. However, leafy vegetable cultivation has become commercial because of growing awareness of their suitable health impacts. Traditional leafy vegetables are a good nutritional source, especially in rural areas, because of their phytochemical and antioxidant richness. Those are the cheapest source of vitamins and micronutrients, which could be useful for nutrient deficiencies. Again, unique phytochemical constituents made some selective leafy vegetables herbally and medicinally significant. The antinutritional factors such as phytates, nitrates, oxalates, cyanogenic glycosides, and tannins sometimes limit the excellent utility of leafy vegetables, mostly by impairing micronutrient absorption. Thermal processing through boiling, cooking, and blanching leafy vegetables could reduce the antinutrient level. There is a lot of scope for identification of unexplored leafy vegetables, their phytochemical profile, and health benefits towards inclusion in the agri-food system, and it is necessary to consider the conservation strategy, too.

REFERENCES

Abbas, R., Elsharbasy, F. S., and Fadlelmula, A. A. 2018. Nutritional values of moringa oleifera, total protein, amino acid, vitamins, minerals, carbohydrates, total fat and crude fiber, under the semi-arid conditions of Sudan. *Journal* of *Microbial and Biochemical Technology*: 56–58.

Ahmad, K., Wajid, K., Khan, Z. I., Ugulu, I., Memoona, H., and Sana, M. 2019. Evaluation of potential toxic metals accumulation in wheat irrigated with wastewater. *Bulletin of Environmental Contamination and Toxicology* 102(6):822–828.

Alam, M., Khan, M., Khan, A., Zeb, S., Khan, M. A., and Amin, N. U. 2018. Concentrations, dietary exposure, and human health risk assessment of heavy metals in market vegetables of Peshawar, Pakistan. *Environmental Monitoring and Assessment* 190(9):505.

Alvino, A., and Barbieri, G. 2015. Vegetables of temperate climates: Leafy vegetables. In: *Encyclopedia of Food and Health* (pp. 393–400). Oxford: Academic Press, Elsevier Inc.

Antony, A. C. 2007. In utero physiology: Role of folic acid in nutrient delivery and fetal development. *American Journal* of *Clinical Nutrition* 85:598–603.

Ashraf, I., Ahmad, F., Sharif, A., Altaf, A. R., and Teng, H. 2021a. Heavy metals assessment in water, soil, vegetables and their associated health risks via consumption of vegetables, District Kasur, Pakistan. *SN Applied Sciences* 3(4):552.

Ashraf, M. A., Maah, M. J., and Yusof, I. 2021b. Heavy metals accumulation in plants growing on landfll and their potential health risks. *International Journal of Environmental Science and Technology* 18(6):1439–1450.

Barba, F. J., Esteve, M. J., and Frígola, A.2014. Bioactive components from leaf vegetable products. *Studies in Natural Products Chemistry* 41:321–346.

Basli, A., Belkacem, N., and Amrani, I. 2017. Health benefits of phenolic compounds against cancers. In: Soto-Hernández, M., Tenango, M. P., and García-Mateos, R. (Eds.), *Phenolic Compounds-Biological Activity* (pp. 195–210). London: Intech.

Bhat, R. S., and Al-Daihan, S. 2014. Phytochemical constituents and antibacterial activity of some green leafy vegetables. *Asian Pacific Journal of Tropical Biomedicine* 4(3):189–193.

Bjørneboe, A., Bjørneboe, G. E., and Drevon, C. A. 1990. Absorption, transport and distribution of vitamin. *European Journal of Nutrition* 120(3):233–242.

Boroski, M., Giroux, H. J., Sabik, H., Petit, H. V., Visentainer, J. V., Matumoto-Pintro, P. T., and Britten, M. 2012. Use of oregano extract and oregano essential oil as antioxidants in functional dairy beverage formulations. *LWT:Food Science and Technology* 47:167–174.

Champ, M. M. J. 2002. Non-nutrient bioactive substances of pulses. *British Journal of Nutrition* 88(S3): S307–S319.

Chauhan, B., Kumar, G., Kalam, N., and Ansari, S. H. 2013. Current concepts and prospects of herbal nutraceutical: A review. *Journal of Advanced Pharmaceutical Technology and Research* 4:4–8.

Chen, G., Huo, Y., Tan, D. X., Liang, Z., Zhang, W., and Zhang, Y. 2003. Melatonin in Chinese medicinal herbs. *Life Science* 73:19–26.

Chester, K., Paliwal, S., Khan, W., Ahmad, S. 2017. UPLC-ESI-MS/MS and HPTLC method for quantitative estimation of cytotoxic glycosides and aglycone in bioactivity guided fractions of *Solanum nigrum* L. *Frontiers in Pharmacology* 8:1–15.

Ching, L. S., and Mohamed, S. 2001. Alpha-tocopherol content in 62 edible tropical plants. *Journal of Agricultural and Food Chemistry* 49(6):3101–3105.

Ciska, E., Martyniak-Przybyszewska, B., and Kozlowska, H. 2000. Content of glucosinolates in cruciferous vegetables grown at the same site for two years under different climatic conditions. *Journal of Agricultural and Food Chemistry* 48:2862–2867.

Colonna, E., Rouphael, Y., Barbieri, G., et al. 2016. Nutritional quality of ten leafy vegetables harvested at two light intensities. *Food Chemistry* 199:702–710.

Cossins, E. 2000. The fascinating world of folate and one-carbon metabolism. *Canadian Journal of Botany* 78:691–708.

Craft, B. D., Kerrihard, A. L., Amarowicz, R., and Pegg, R. B. 2012. Phenol-based antioxidants and the in vitro methods used for their assessment. *Comprehensive Reviews in Food Science and Food Safety* 11(2): 148–173.

Ebert, A. W. 2014. Potential of underutilized traditional vegetables and legume crops to contribute to food and nutritional security, income and more sustainable production systems. *Sustainability* 6, 319–335.

Edelman, M., and Colt, M. 2016. Nutrient value of leaf vs. seed. *Frontiers in Chemistry* 4:1–32.

Fahey, J. W., Olson, M. E., Stephenson, K. K., Wade, K. L., Chodur, G. M., Odee, D., Nouman, W., Massiah, M., Alt, J., Egner, P. A., and Hubbard, W. C., 2018. The diversity of chemoprotective glucosinolates in Moringaceae (Moringa spp.). *Scientific Reports* 8(1):7994.

Fasuyi, A. O. 2005. Nutrient composition and processing effects on cassava leaf (*Manihot esculenta* crantz) anti nutrients. *Pakistan Journal of Nutrition* 4(1):37–42.

Gardner, P. T., White, T. A. C., McPhail, D. B., and Duthie, G. G. 2000. The relative contributions of vitamin C, carotenoid, and phenolics to the antioxidant potential of fruit juices. *Food Chemistry* 68:471–474.

Gibson, R. S., and Hotz, C. 2001. Dietary diversification/modification strategies to enhance micronutrient content and bioavailability of diets in developing countries. *British Journal of Nutrition* 85:159–166.

Giugliano, D., Ceriello, A., and Paolisso, G. 1996. Oxidative stress and diabetic vascular complications. *Diabetes Care* 19:257–267.

Glew, R. S., Vanderjagt, D. J., Bosse, R., Huang, Y.-S., Chuang, L.-T., and Glew, R. H. 2005. The nutrient content of three edible plants of the Republic of Niger. *Journal of Food Composition and Analysis* 18: 15–27.

Gopalan, C., Sastri, R., and Balasubramanian, S. C. 2004. *Nutritive Value of Indian Foods* (p. 52). Hyderabad: National Institute of Nutrition, ICMR.

Guerrero, J. L. G., and Isasa, M. E. T. 2009. Nutritional composition of leaves of *Chenopodium* species (*C. album* L., *C. murale* L. and *C. opulifolium* Shraeder). *International Journal of Food Sciences and Nutrition* 5:321–327.

Hamazaki, T., and Okuyama, H. 2001. Fatty acids and lipids–new findings. *Karger Medical and Scientific Publishers, Japan*: 1–53.

Hamrick, I., and Counts, S. H. 2008. Vitamin and mineral supplements. *Primary Care: Clinics in Office Practice* 35(4):729–747.

Harborne, J. 1975. The biochemical systematics of flavonoids. In: Harborne, J. B., et al. (Eds.), *The Flavonoids* (pp. 1056–1095). USA: Springer.

Islam, M. M. 2013. Biochemistry, medicinal and food values of jute (Corchorus capsularis L. and C. olitorius L.) leaf: A review. *International Journal of Enhanced Research in Science, Technology, and Engineering* 2(11):135–144.

Kabir, M. S. H., Hossain, M. M., Kabir, M. I., Rahman, M. M., Hasanat, A., Emran, T. B., and Rahman, M. A. 2016. Phytochemical screening, antioxidant, thrombolytic, alpha-amylase inhibition and cytotoxic activities of ethanol extract of *Steudnera colocasiifolia* K. Koch leaves. *Journal of Young Pharmacists* 8:391.

Kala, A., and Prakash, J. 2004. Nutrient composition and sensory profile of differently cooked green leafy vegetables. *International Journal of Food Properties* 7(3):659–669.

Kamalu, B. P. 1993. Cassava (Manihot esculenta Crantz) in the etiology of kwashiorkor. *Nutrition Research Reviews* 6(1):121–135.

Kaur, R., Kapoor, K., and Kaur, H. 2011. Plants as a source of anticancer agents. *International Journal of Drug Research and Technology* 1:119–124.

Kawashima, N., and Wildman, S. G. 1970. Fraction I protein. *Annual Review of Plant Physiology* 21(1):325–358.

Khan, M. U., Muhammad, S., Malika, R. N., Khand, S. A., and Tariq, M. 2016a. Heavy metals potential health risk assessment through consumption of wastewater irrigated wild plants: A case study. *Human and Ecological Risk Assessment* 22:141–152.

Khan, S., Cao, Q., Zheng, Y. M., Huang, Y. Z., and Zhu, T. G. 2016b. Health risks of heavy metals in contaminated soils and food crops irrigated with wastewater in Beijing, China. *Environmental Pollution* 152(3):686–692.

Khanbabaee, K., and Van Ree, T. 2001. Tannins: Classification and definition. *Natural Product Reports* 18(6):641–649.

Kim, I. S., Yang, M., Lee, O. H., and Kang, S. N. 2011. The antioxidant activity and the bioactive compound content of Stevia rebaudiana water extracts. *LWT-Food Science and Technology* 44(5):1328–1332.

Kumar, D., Kumar, S., and Shekhar, C. 2020. Nutritional components in green leafy vegetables: A review. *Journal of Pharmacognosy and Phytochemistry* 9(5): 2498–2502.

Lakshminarayana, R., Raju, M., Krishnakantha, T. P., and Baskaran, V. 2005. Determination of major carotenoids in a few Indian leafy vegetables by high-performance liquid chromatography. *Journal of Agricultural and Food Chemistry* 53(8):2838–2842.

Latif, A., Bilal, M., Asghar, W., Azeem, M., Ahmad, M. I., Abbas, A., Ahmad, M. Z., and Shahzad, T. 2018. Heavy metal accumulation in vegetables and assessment of their potential health risk. *Journal of Environmental Analytical Chemistry* 5:234.

Llorach, R., Martínez-Sánchez, A., Tomás-Barberán, F. A., Gil, M. I., and Ferreres, F. 2008. Characterisation of polyphenols and antioxidant properties of five lettuce varieties and escarole. *Food Chemistry* 108:1028–1038.

Ma, T., Wang, Z., Zhang, Y. M., Luo, J. G., Kong, L. Y., Choi, D. K., and Ganesan, P. 2017. Bioassayguided isolation of anti-inflammatory components from the bulbs of *Lilium brownii* var. viridulum and Identifying the underlying mechanism through acting on the NF-KB/MAPKs pathway. *Molecules* 22:506.

Majumdar, D. 2003. The blue baby syndrome: Nitrate poisoning in humans. *Resonance* 8(10):20–30.

Manzoor, J., Sharma, M., and Wanu, A. K. 2018. Heavy metals in vegetables and their impact on the nutrient quality of vegetables: A review. *Journal of Plant Nutrition* 41(3):1744–1763.

Mepba, H., Eboh, L., and Banigo, D. 2007. Effects of processing treatments on the nutritive composition and consumer acceptance of some Nigerian edible leafy vegetables. *African Journal of Food, Agriculture, Nutrition and Development* 7(1).

Metere, A., and Giacomelli, L. 2017. Absorption, metabolism and protective role of fruits and vegetables polyphenols against gastric cancer. *European Review for Medical and Pharmacological Sciences* 21:5820–5858.

Modi, A. T. 2007. Growth temperature and plant age influence on nutritional quality of Amaranthus leaves and seed germination capacity. *Water SA* 33:369–375.

Mosha, T., and Gaga, H. 1999. Nutritive value and effect of blanching on the trypsin and chymotrypsin inhibitor activities of selected leafy vegetables. *Plant Foods for Human Nutrition* 54(3):271–283.

Moyo, S. M., Serem, J. C., Bester, M. J., Mavumengwana, V., and Kayitesi, E. 2021. African green leafy vegetables health benefits beyond nutrition. *Food Reviews International* 37(6):601–618.

Nath, P., Gaddagimath, P. B., and Dutta, O. 2004. *Food and Nutrition Security and Vegetables: A Global Perspective* (pp. 1–412). India: Prem Nath Agricultural Science Foundation.

Neveu, V., Perez-Jiménez, J., Vos, F., Crespy, V., du Chaffaut, L., Mennen, L., Knox, C., Eisner, R., Cruz, J., Wishart, D., and Scalbert, A. 2010. Phenol-Explorer: An online comprehensive database on polyphenol contents in foods. *Database* 2010.

Ng, S. C., Shi, H. Y., Hamidi, N., Underwood, F. E., Tang, W., Benchimol, E. I., Panaccione, R., Ghosh, S., Wu, J. C. Y., and Chan, F. K. L. 2017. Worldwide incidence and prevalence of inflammatory bowel disease in the 21st century: A systematic review of population-based studies. *Lancet* 390:2769–2778.

Nicolle, C., Cardinault, N., Gueux, E., Jaffrelo, L., Rock, E., Mazur, A., Amouroux, P., and Rémésy, C. 2004. Health effect of vegetable-based diet: Lettuce consumption improves cholesterol metabolism and antioxidant status in the rat. *Clinical Nutrition* 23:605–614.

Nishimura, N., Taniguchi, Y., and Kiriyama, S. 2000. Plasma cholesterol-lowering effect on rats of dietary fiber extracted from immature plants. *Bioscience, Biotechnology, and Biochemistry* 64:2543–2551.

Nunes, S., Danesi, F., Del Rio, D., and Silva, P. 2018. Resveratrol and inflammatory bowel disease: The evidence so far. *Nutrition Research Reviews* 31:85–97.

Nwanekezie, E. C., and Obiakor-Okeke, P. N. 2014. Mineral content of five tropical leafy vegetables and effect of holding methods and time. *American Journal of Experimental Agriculture* 4(12):1708–1717.

Oboh, G. 2005. Effect of blanching on the antioxidant properties of some tropical green leafy vegetables. *LWT–Food Science and Technology* 38(5):513–517.

Ogunsuyi, O. B., Ademiluyi, A. O., Oboh, G., Oyeleye, S. I., Dada, A. F. 2018. Green leafy vegetables from two solanum spp. (*Solanum nigrum* L and *Solanum nacrocarpon* L) ameliorate scopolamine-induced cognitive and neurochemical impairments in rats. *Food Science & Nutrition*: 860–870.

Praveen, P. A., Roy, A., and Prabhakaran, D. 2013. Cardiovascular disease risk factors: A childhood perspective. *Indian Journal* of *Pediatrics* 80:3–12.

Proteggente, A. R., Pannala, A. S., Paganga, G., Buren, L. V., Wagner, E., Wiseman, S., Put, F. V. D., Dacombe, C., and Rice-Evans, C. A. 2002. The antioxidant activity of regularly consumed fruit and vegetables reflects their phenolic and vitamin C composition. *Free Radical* Research 36:217–233.

Pulido, R., Bravo, L., and Saura-Calixto, F. 2000. Antioxidant activity of dietary polyphenols as determined by a modified ferric reducing antioxidant power assay. *Journal of Agricultural and Food Chemistry* 48:3396–3402.

Radek, M., and Savage, G. P. 2008. Oxalates in some Indian green leafy vegetables. *International Journal of Food Sciences and Nutrition* 59(3):246–260.

Rahman, M. M., Islam, M. R., Shohag, S., Hossain, M. E., Rahaman, M. S., Islam, F., Ahmed, M., Mitra, S., Khandaker, M. U., and Idris, A. M. 2022. The multifunctional role of herbal products in the management of diabetes and obesity: A comprehensive review. *Molecules* 27:1713.

Raju, M., Varakumar, S., Lakshminarayana, R., Krishnakantha, T. P., and Baskaran, V. 2007. Carotenoid composition and vitamin A activity of medicinally important green leafy vegetables. *Food Chemistry* 101(4):1598–1605.

Ramos, I., Esteban, E., Lucena, J. J., and Gárate, A. 2012 Cadmium uptake and sub cellular distribution in plants of Lactuca sp. Cd-Mn interaction. *Plant Science* 162:761–767.

Ramulu, P., and Rao, P. U. 2003. Total, insoluble and soluble dietary fiber contents of Indian fruits. *Journal of Food Composition and Analysis* 16(6):677–685.

Randhawa, M. A., Khan, A. A., Javed, M. S., and Sajid, M. W. 2015. Green leafy vegetables: A health promoting source. In: *Handbook of Fertility*. USA: Academic Press.

Reddy, N. R., and Shridhar Sathe, K. 2001. *Occurrence, Distribution, Content, and Dietary Intake of phytate* (1st edition). USA: CRC Press.

Salaritabar, A., Darvishi, B., Hadjiakhoondi, F., Manayi, A., Sureda, A., Nabavi, S. F., Fitzpatrick, L. R., Nabavi, S. M., and Bishayee, A. 2017. Therapeutic potential of flavonoids in inflammatory bowel disease: A comprehensive review. *World Journal of Gastroenterology* 23, 5097–5114.

Salunkhe, D. K., and Chavan, J. 1989. *Dietary Tannins: Consequences and Remedies* (pp. 1–208). USA: CRC Press.

Santamaria, P. 2006. Nitrate in vegetables: Toxicity, content, intake, and EC regulation. *Journal of the Science of Food and Agriculture* 86(1):10–17.

Satheesh, N., and Workneh Fanta, S. 2020. Kale: Review on nutritional composition, bio-active compounds, anti-nutritional factors, health beneficial properties and value-added products. *Cogent Food & Agriculture* 6(1):1811048.

Serrano, J., Pupponen-Pimiä, R., Dauer, A., Aura, A.-M., & Saura-Calixto, F. 2009. Tannins: Current knowledge of food sources, intake, bioavailability, and biological effects. *Molecular Nutrition and Food Research* 53(S2):S310–S329.

Shahi, H. N. 1977. Studies on chemical composition of Chenopodium album L. *Plant and Soil* 46:271–273.

Shaw, J. E., Sicree, R. A., and Zimmet, P. Z. 2010. Global estimates of the prevalence of diabetes for 2010 and 2030. *Diabetes Research and Clinical Practice* 87:4–14.

Shi, M., Gu, J., Wu, H., Rauf, A., Emran, T. B., Khan, Z., Mitra, S., Aljohani, A. S., Alhumaydhi, F. A., Al-Awthan, Y. S., and Bahattab, O. 2022. Phytochemicals, nutrition, metabolism, bioavailability, and health benefits in lettuce—a comprehensive review. *Antioxidants* 11(6):1158.

Siddiq, M., and Uebersax, M. A. 2018. *Handbook of Vegetables and Vegetable Processing*. USA: Wiley Blackwell.

Spiller, G. 2001. Dietary fiber in prevention and treatment of disease. In: *CRC Handbook of Dietary Fiber in Human Nutrition* (pp. 363–431). USA: CRC Press LLC.

Stanely, M. P., and Menon, V. P. 2001. Antioxidant action of *Tinospora cordifolia* root extract in alloxan diabetic rats. *Phytotherapy Research* 15:213–218.

Stedronsky, E. R. 1994. Interaction of bile acids and cholesterol with non-systemic agents having hypocholesterolemic properties. *Biochimica et Biophysica Acta (BBA)—Lipids and Lipid Metabolism* 1210:255–287.

Tufts, H. R., Harris, C. S., Bukania, Z. N., and Johns, T. 2015. Antioxidant and anti-inflammatory activities of Kenyan leafy green vegetables, wild fruits, and medicinal plants with potential relevance for kwashiorkor. *Evidence-Based Complementary* and *Alternative Medicine*: 807158.

Vanderjagt, D. J., Freiberger, C., Vu, H.-T. N., Mounkaila, G., Glew, R. S., and Glew, R. H. 2000. The trypsin inhibitor content of 61 wild edible plant foods of Niger. *Plant Foods for Human Nutrition* 55:335–346.

Villela-Castrejón, J., Antunes-Ricardo, M., and Gutiérrez-Uribe, J. A. 2017. Bioavailability and anti-inflammatory activity of phenolic acids found in spray-dried nejayote after its in vitro digestion. *Journal of Functional Foods* 39:37–43.

Vithya, S., and Jayshree, N. 2017. Phytochemical analysis and in vitro anticancer study of ethanolic extract of leaves of *Amaranthus cruentus* Linn. against colon cancer cell line (HCT-116). *World Journal of Pharmacy and Pharmaceutical Sciences* 6:1644–1658.

Xu, D., Chen, Z., Sun, K., Yan, D., Kang, M., and Zhao, Y. 2019. Effect of cadmium on the physiological parameters and the subcellular cadmium localization in the potato (*Solanum tuberosum* L.). *Ecotoxicology and Environmental Safety* 97:147–153.

Yadav, S. K., and Sehgal, S. 2003. Effect of domestic processing and cooking on selected anti nutrient contents of some green leafy vegetables. *Plant Foods for Human Nutrition* 58(3):1–11.

7 Onion (*Allium cepa* L.) *Chemistry and Pharmacology*

Pankaj Kumar Chaurasia, Shashi Lata Bharati, Payal Joshi, Sunita Singh, Rima Kumari, Alka Singh, Rahul D. Kamble, Deepak B. Shelke, and Vijay Kumar Patel

7.1. INTRODUCTION

Plants have a significant role in human life due to their extensive chemistry and pharmacologically significant activities (Chaurasia et al., 2024a; Singh et al., 2023). Out of many types of plants, several vegetables are renowned for their unique nutraceutical properties, and onion is one of them. *Allium cepa* L. is an edible herbaceous plant commonly known as 'onion' that belongs to the Alliaceae family. Its genus is *Allium*, which has several species, such as *Allium cepa* L. (onion), *Allium ascalonicum* L. (shallot), *Allium schoenoprasum* L. (chive), *Allium fistulosum* L. (Japanese bunching onion/green onions), *Allium sativum* L. (garlic) (Chaurasia et al., 2024b, 2025), *Allium ampeloprasum* L. var. porrum (leek), and *Allium wakegi* (scallion) (Ochar and Kim, 2023). Onion's growth, cultivation, and production need fertile soils, cool weather, and proper caring. It is widely cultivated worldwide specifically for food purposes and may be used as fresh salad or in cooked form or in various processed products (Manohar et al., 2017).

Onion has a strong place in the human diet due to their novel flavour, health-activating qualities, unique taste, and culinary versatility (Ochar and Kim, 2023). It is cultivated in more than 140 countries, and China, India, the United States, Turkey, Mali, Angola, and Japan produce onions and shallots (green) in large amount. Out of 4,970,615 tonnes production of onions and shallots (green) in 2022, China contributed 864,488 tonnes while other countries such as Mali, Angola, and Japan's contributions were 610,576 tonnes, 558,480 tonnes, and 510,462 tonnes, respectively (United Nations Food and Agriculture Organization, Statistics Division (FAOSTAT)).

Onion and its whole plant are treasured for its culinary as well as several medicinal and nutritional properties (Ochar and Kim, 2023). Anthocyanins are the molecules in red onion responsible for their colour in the epidermal cells of the bulbs' scale leaves, and cyanidin 3-glucoside is reported as the main anthocyanin (Lee et al., 2015; Fossen and Andersen, 2003; Fossen et al., 1996). Anthocyanins are chiefly made of cyanidin or peonidin glucosides that are nonacylated or acylated with malonic acid (Mazza and Miniati, 1993). Anthocyanins' content is almost 10% of the content of total flavonoids (Slimestad et al., 2007), and anthocyanins have also been shown as an antioxidant source (Geetha et al., 2011).

Onions show excellent antioxidant properties due to the presence of phenolic components (Sidhu et al., 2019). In an antioxidant study on the 15 onions (short-day) with yellow, white, and red colour, it was observed that white onions have a low level of activity whereas red onions have a high level (Lee et al., 2015). Lesser contents of flavonoid are found in white onions than colored onions (Herrmann, 1976). Regular onion consumption has been shown to reduce the risks of cancer, cataracts, DNA damage, and heart-related diseases (Arung et al., 2011; Jimenez et al., 2011; Hamauzu et al., 2011). Carcinogenic heterocyclic amines as well as azaarenes' formation was found to be reduced due to the presence of onions and garlic in pan-fried meat (Janoszka, 2010). Onions are a great source of bioactive chemical molecules, which have several health benefits. Therefore,

 DOI: 10.1201/9781003559894-7

considering the versatile applicability of this important vegetable, this chapter has been prepared to present the chemistry and pharmacology of this nutritious vegetable in concise form.

7.2. CHEMISTRY OF ONIONS

Onion is rich in several important phytoconstituents, such as carbohydrates, vitamins, proteins, and minerals, along with a few flavonoids and polyphenolic constituents (Sami et al., 2021). A high amount of sugar contents and dietary fibre of about 90% water has been found in onion. It has sodium in low concentration while Vitamin B6, folic acid, Mg (magnesium), Ca (calcium), K (potassium), and P (phosphorus) are present in high concentrations (Chakraborty et al., 2022). It has a low level of lipid, and in the amino acids, only arginine and glutamic acid are exceptional (Suleria et al., 2015), except for sulphur-containing components and flavonoids. Lectins, pectin, prostaglandins, adenosine, fructan, vitamins (e.g., B1, B2, B6, C, and E), fatty acids, biotic, phospholipids, nicotinic acid, glycolipids, and many essential amino acids are other constituents of intact onion and are studied for different biological applications (Pareek et al., 2017). Phenolic components such as gallic acid, ferulic acid, quercetin, and protocatechuic acid found in different onion varieties varied considerably according to the variety of onions (Chakraborty et al., 2022; Prakash et al., 2007).

Different onion varieties contain different types of flavonoids, such as quercetin-40-monoglucoside (Huma et al., 2011), quercetin aglycon, isorhamnetin 3,40-diglucoside, quercetin-3-monoglucoside, delphinidin 3,5-diglycosides, quercetin-3,40-diglucoside, quercetin 3-glycosides (Zhang et al., 2016), quercetin 3,7,40-triglucoside, quercetin 7,40-diglucoside, quercetin 3,40-diglucoside (Pérez-Gregorio et al., 2010), and many more. Onion's pungency is due to the presence of a number of sulphur components (Chakraborty et al., 2022). Ideally, low pungency and high sugar make the onion sweet (Vågen and Slimestad, 2008). Dipropyl disulphide, propenyl disulfides, and dipropenyl sulphide are the chief components of onion volatiles, while many other constituents, such as dipropenyl sulfide and dipropyl sulfide, are recognized among them (Munday and Munday, 2001). Many sulphur-containing flavour precursors have been quantitatively determined in onion, which were S-alk(en)yl-L-cysteine derivatives, methiin, alliin, cycloalliin, isoalliin, deoxyalliin, N-(γ-glutamyl)-S-methyl-L-cysteine, N-(γ-glutamyl)-S-(*E*-1-propenyl)-L-cysteine, N-(γ-glutamyl)-S-(2-propenyl)-L-cysteine, N-(γ-glutamyl)-S-(2-propenyl)-L-cysteine sulfoxide, S-(2-carboxy-propyl) glutathione, and N-(γ-glutamyl)-S-(*E*-1-propenyl)-L-cysteine sulfoxide (Chakraborty et al., 2022; Yamazaki et al., 2011).

Along with several biologically active phytochemicals, a number of phenolic compounds and flavonoids are reported in onion bulbs by different techniques of their identification (Sagar et al., 2022). Amongst organic acids, many compounds are reported in onion bulbs, such as folic acid, ferulic acid, triethyl citrate, pivmecillinam, pyruvic acid, ureidoisobutyric acid, and valaciclovir acid in different concentrations (Zhou et al., 2020). Amongst alkaloids, ergoloid, galantamine, harman, 3-hydroxyquinine, and morphine-6-glucuronide are also found in onion bulbs in various concentrations (Zhou et al., 2020). In this way, it can be seen that onion is a rich source of various medicinally and nutritionally significant organic compounds that have noteworthy biological activities (Sagar et al., 2022).

7.3. ANTIMICROBIAL PROPERTIES

Microbial contamination of food materials is a serious concern because it causes the damage of a large quantity of food materials. Several synthetic preservatives, such as benzoate, butylated hydroxyl toluene, butylated hydroxyl anisole, sulphites, and propionate (Soubra et al., 2007), which inhibit the growth of microbes due to their strong antimicrobial properties, are used to save spoilage of food materials. Although synthetic antimicrobial preservatives are effective, they are associated with several health concerns (Balogun et al., 2023). Therefore, nowadays, natural preservatives as potent antimicrobial agents are preferred over synthetic preservatives. It was observed that onion has high phenolic as well as flavonoid contents and have the tendency to reduce the growth of several studied bacteria such as *Staphylococcus aureus and Bacillus cereus* (two gram-positive) and *Pseudomonas aeruginosa, Escherichia coli,* and *Salmonella spp* (three gram-negative). Susceptibility of gram-negative bacteria was lower than gram-positive bacteria for onion. A study showed the onion's antimicrobial potential in beef and chicken in the course of cold storage and like butylated hydroxyl toluene (Balogun et al., 2023).

A study on polyphenolic extracts of different cultivars' onion skin with different colour (pink, red, dark red, and white) were performed to know their potential as antimicrobial agents (Sagar and Pareek, 2020). Pink colour onion skin's extract was very effective against studied pathogenic bacteria, followed by the extracts of red onion skin and dark red onion skin while extract of white onion was least effective on bacterial growth. This study shows that skin waste may be utilized as a natural antimicrobial preservative in the field of food and health. Antibacterial properties of *Allium cepa* were studied on the bacterial isolates obtained from diabetic patients' wounds. Greater inhibitory effect was shown by the ethanolic extract of the onion in comparison with aqueous extract against bacterial isolates (Oyawoye et al., 2022). A study by Mounir et al. (2023) suggests the potential role of onion peel extract in wound treatment by decreasing infection caused by microbes, decreasing inflammation, and tissue regeneration activation. Besides these references, a number of literature sources also prove the capability of onion as an effective antimicrobial agent and food preservative (Joković et al., 2024; Harivananthan and Razali, 2024; Yash et al., 2024).

7.4. ANTIOXIDANT PROPERTIES

Onion is a rich source of natural antioxidants. Natural antioxidants are generally preferred over synthetic antioxidants because of no side effects or negative effects. Flavonoids, phenolics, flavanols, anthocyanins, ferulic acid, vanillic acid, and tannins are identified by various studies in the extracts of onion peel (Kumar et al., 2022). Flavonoid quercetin is found in onion peels as the most common compound showing antioxidant properties, which may be present in free aglycon form or in mono- and diglycoside forms (Stoica et al., 2023). Onion's chemical composition is mainly dependent on the variety and cultivar, and better antioxidant activities are reported in red onion peels in comparison with yellow and white peels (Albishi et al., 2013). Joković et al. (2024) studied the antioxidant potential of four types of extracts of onion peels based on ethyl acetate, acetone, methanol, and ethanol. Quercetin was found as the principal component in all the extracts. Acetone extract showed the maximum total phenols and flavonoids while the maximum antioxidant activity was shown by the methanol extract. Mardani et al. (2023) reported the antioxidant properties of internal and external onion's layers on 10 Iranian onions landrace and observed the highest phenolic and flavonoid contents and also antioxidant activities in the red onions like landraces from Nowshahr and Azarshahr. In this study, antioxidant capacity range was 10.34–40.54%. In a study, ethanol-based extract of onion peel showed higher DPPH scavenging activity and higher antioxidant activity measured by ferric thiocyanate method in comparison with extracts based on hot water extraction and subcritical water extraction at temperature 165°C (Lee et al., 2014). At 110°C, antioxidant activity of extract by subcritical water extraction was alike to that of ethanol extraction. Aqueous extract of onion bulbs showed a good level of antioxidant activities in term of DPPH radical scavenging and reducing power assay (Dimitry et al., 2021). A complete onion's ethanolic extract has also shown a high level of antioxidant effects based on the study using DPPH free radical and ABTS radical scavenging methods (Kim et al., 2022).

7.5. ANTICANCER ACTIVITIES

Onion has a tendency to inhibit the cell growth of various types of cancer. In study, *A. cepa*'s quercetin-rich extracts show a cytotoxicity effect on adrenocortical carcinoma (Veiga et al., 2022).

Mitotane is an approved drug for the treatment of this rare subtype of cancer. Fractions rich in quercetin have been observed to show better results on H295R (adrenocortical carcinoma cell line). The effectiveness of quercetin *vs.* mitotane was compared using H295R and SW-13 cell lines, and greater cytotoxicity response was shown by quercetin on these cell lines after the concentration of 10 μM, whereas mitotane showed only after the concentration of 30 μM. Thus, Veiga et al. (2022) conclude that quercetin-dominant onions may be a natural anticancer source for this type of cancer. Murayyan et al. (2017) studied the anticancer properties of Ontario-grown onions. These onions have shown antiproliferative activity and thus, dietary supplementation may have anticancer benefits. In another study, potential cytotoxic effects have been shown by the extracts of onion skin against cancer cell lines, which indicates the potential of extracts as a natural source of anticancer agents (Bozinou et al., 2023).

7.6. ANTI-INFLAMMATORY PROPERTIES

Onion has been found to show protective influence against inflammation and pain (Dimitry et al., 2021). The different peel extract of *A. cepa* was utilized to determine its anti-inflammatory effect by inhibition of the JAK (Janus kinase)-STAT (signaling transducer and activator of transcription) pathway in RAW264.7 cells stimulated by lipopolysaccharide (Lee et al., 2023). The 50% ethanol extract of *A. cepa* peel (AP50E) was observed as the maximum effective at preventing lipopolysaccharide-induced nitric oxide and inducible nitric oxide synthase. The levels of pro-inflammation cytokines interleukin (IL)-1α, IL-1β, IL-6, and IL-27 were noticeably reduced, and the JAK-STAT pathway was also directly inhibited by AP50E (Lee et al., 2023). This study indicates the potential of this extract in the prevention of inflammatory illness. Whole onion ethanolic extract lowered the secretion of nitric oxide and inflammatory cytokines and IL-1β and IL-12's mRNA expressions (Kim et al., 2022). Results were similar to quercetin, which shows good antioxidant and anti-inflammatory properties and, therefore, use of whole onion as potential anti-inflammatory food material may be recommended to inhibit health-related issues such as arthritis, cardiovascular disease, atopic dermatitis, and others (Kim et al., 2022). Many reviews explore the anti-inflammatory potential of onion and present descriptive information regarding it (Cordeiro et al., 2023; Marefati et al., 2021).

7.7. ANTIDIABETIC AND ANTIHYPERGLYCEMIC PROPERTIES

Diabetes mellitus is a type of metabolic disorder in which carbohydrate, protein, and fat metabolism gets disturbed (Cherian et al., 2023). Pancreatic beta cells destruction is responsible for type 1 diabetes while type 2 diabetes is linked with insulin resistance and lowered production of insulin, along with gestational diabetes in the course of pregnancy and other forms. Approximately 90–95% of cases are related to type 2 diabetes (Hu et al., 2024). Onion is an important herb, which may play a favourable role in diabetes control due to their antidiabetic ability. Dried onion peel–based ethanolic extracts inhibit alpha-glucosidase, causing lowered level of blood glucose. Moreover, an enhancement in the production of insulin was also prompted by this inhibition (Esakki et al., 2024). This study shows the potential of *A. cepa* leaf's ethanolic extract because of hypoglycemic and antidiabetic properties. Another study on diabetic rats showed the potential role of onion supplementation in blood glucose lowering, where the level of blood glucose was significantly reduced by white bread supplemented with onion peel extract (1% and 3%) and onion powder (7%) in treated rats (Masood et al., 2021). Type 2 insulin resistance and glucose response may be improved by the onion peel extract by alleviating the free fatty acid metabolic dysregulation, oxidative stress suppression, glucose uptake upregulation, and/or inflammatory gene expression downregulation in the liver (Jung et al., 2011). During the study, Jung et al. (2011) observed the higher potency of onion peel extract in comparison with pure quercetin equivalent. In type 2 diabetes, onion peel may be used in order to improve insulin insensitivity.

7.8. OTHER POTENTIAL APPLICATIONS

Due to enormous numbers of biologically active molecules, especially sulphur-containing components, onion shows a number of biological activities besides its specific tastes, aroma, and flavour. Onion has been used since ancient time to treat various human illnesses, such as stomach illness, infection of the throat, and hepatitis (Akash et al., 2014). *Allium cepa* tea has been used in Chinese medicine to provide relief in fever, cholera, cold, headache, dysentery, and arthritis (Corzo-Martínez et al., 2007). Besides these medicinal roles, this herb also acts as an antispasmodic, antiasthmatic, antidiabetic, antithrombotic, antihyperlipidemic, antiplatelet, and antiarthritic agent. It is also very helpful in neurodegenerative diseases, heart-related problems, kidney problems, anthelmintic, carminative, vertigo, migraine, and many other human health-related issues (Marefati et al., 2021).

7.9. CONCLUSION

This discussion shows that onion has an extensive chemistry that makes it an important nutraceutically significant vegetable. It has different types of medicinal properties and may be used for relaxation in various types of health issues. Thus, its use as an effective supplement in our daily food may improve human health by exerting its health-promoting activities. Its multiple medicinal properties, such as antioxidant, antimicrobial, anti-inflammatory, anticancer, antidiabetic, antihyperlipidemic, antiarthritic, and many others, make it an excellent, must-eat food material.

7.10. ACKNOWLEDGEMENT

All authors are thankful to their respective department/institutions for any type of support provided during their work.

REFERENCES

Akash, M.S.H., Rehman, K., and Chen, S. 2014. Spice plant Allium cepa: Dietary supplement for treatment of type 2 diabetes mellitus. *Nutrition*, 30(10):1128–1137. https://doi.org/10.1016/j.nut.2014.02.011

Albishi, T., John, J.A., Al-Khalifa, A.S., and Shahidi, F. 2013. Antioxidative phenolic constituents of skins of onion varieties and their activities. *J. Funct. Foods*, 5:1191–1203. https://doi.org/10.1016/j.jff.2013.04.002

Arung, E.T., Furuta, S., Ishikawa, H., Kusuma, I.W., Shimizu, K., and Kondo, R. 2011. Anti-melanogenesis properties of quercetin-and its derivative-rich extract from *Allium cepa*. *Food Chem.*, 124(3):1024–1028. https://doi.org/10.1016/j.foodchem.2010.07.067

Balogun, M.A., Sobande, O.S, and Oyeyinka, S.A. 2023. Antimicrobial properties of onion and garlic extracts in beef and chicken, *Food Chem. Adv.*, 3:100519. https://doi.org/10.1016/j.focha.2023.100519.

Bozinou, E., Pappas, I.S., Patergiannakis, I.-S., Chatzimitakos, T., Palaiogiannis, D., Athanasiadis, V., Lalas, S.I., Chatzilazarou, A., and Makris, D.P. 2023. Evaluation of antioxidant, antimicrobial, and anticancer properties of onion skin extracts. *Sustainability*, 15:11599. https://doi.org/10.3390/su151511599

Chakraborty, A.J., Uddin, T.M., Matin Zidan, B.M.R., Mitra, S., Das, R., Nainu, F., Dhama, K., Roy, A., Hossain, M.J., Khusro, A., and Emran, T.B. 2022. *Allium cepa*: A treasure of bioactive phytochemicals with prospective health benefits. In *Evidence-Based Complementary and Alternative Medicine*, 4586318, 27 pages. https://doi.org/10.1155/2022/4586318

Chaurasia, P.K., Bharati, S.L., and Singh, S. 2024a. The Chemistry inside Spices and Herbs: Research and Developments. Bentham Science Publisher, Volume 4, 1–217. https://doi.org/10.2174/97898151968321240401

Chaurasia, P.K., Bharati, S.L., and Singh, S. 2024b. Garlic against heart-related ailments: Chemistry, pharmacology, and future perspective. *Mini Rev. Med. Chem.* 24(5):521–530. http://dx.doi.org/10.2174/1389557523666230821102512

Chaurasia, P.K., Yadav, M., Bharati, S.L., and Patel, V.K. 2025. *Allium sativum* L. a potent functional food with potential anticancer properties. *Curr. Nutr. Food Sci.* 21(8):926–929. http://dx.doi.org/10.2174/0115734013386242250427163856

Cherian, E., Goyal, M., Mittal, N., Yesodharan, V., Ramadoss, R., and Thomas, C. 2023. Assessment of therapeutic bio-activity of Cinnamoyl Sulfonamide Hydroxamate in squamous cell carcinoma. *Cureus*, 15(8):e43949. https://doi.org/10.7759/cureus.43949

Cordeiro, G.S., Santos, L.S., Vieira, G.P., Espírito-Santo, D.A., Teixeira, R.S., Matos, R.J.B., Costa, C.A.S., Deiró, T.C.B.J., and Barreto-Medeiros, J.M. 2023. Antioxidant, anti-inflammatory and anti-obesity effects of onion and its by-products in high-fat fed rodents: A systematic review. *Braz. J. Biol.*, 83. https://doi.org/10.1590/1519-6984.266108

Corzo-Martínez, M., Corzo, N., and Villamiel, M. 2007. Biological properties of onions and garlic. *Trends Food Sci. Technol.*, 18(12):609–625. https://doi.org/10.1016/j.tifs.2007.07.011

Dimitry, M.Y., Emmanuel, P.A., Edith, D.M.J., Armand, A.B., and Nicolas, N.Y. 2021. Antioxidant property, anti-inflammatory and analgesic effects of aqueous extracts of two onion bulbs varieties (*Allium cepa L.*). *Clin. Phytosci.*, 7:49. https://doi.org/10.1186/s40816-021-00284-2

Esakki, A., Ramadoss, R., Ananthapadmanabhan, L., Sundar, S., Panneerselvam, S., and Ramani, P. 2024. Quantification of the anti-diabetic effect of *Allium cepa. Cureus*, 16(4):e59174. https://doi.org/10.7759/cureus.59174

Fossen, T., and Andersen, O.M. 2003. Anthocyanins from red onions, *Allium cepa*, with novel aglycone. *Phytochem.*, 62:1217–1220. https://doi.org/10.1016/S0031-9422(02)00746-X

Fossen, T., Andersen, O.M., Ovstedal, D.O., Pedersen, A.T., and Raknes, A. 1996. Characteristic anthocyanin pattern from onions and other *Allium* spp. *J. Food Sci.*, 61(4):703–706. https://doi.org/10.1111/j.1365-2621.1996.tb12185.x

Geetha, M., Ponmozhi, P., Saravanakumar, M., and Suganyadevi, P. 2011. Extraction of anthocyanin and analyzing its antioxidant properties from different onion (Allium cepa) varieties. *Int. J. Pharm. Sci. Res.*, 2, 497–506.

Hamauzu, Y., Nosaka, T., Ito, F., Suzuki, T., Torisu, S., Hashida, M., Fukuzawa, A., Ohguchi, M., and Yamanaka, S. 2011. Physicochemical characteristics of rapidly dried onion powder and its anti-atherogenic effect on rats fed high-fat diet. *Food Chem.*, 129(3):810–815. https://doi.org/10.1016/j.foodchem.2011.05.027

Harivananthan, S., and Razali, N.F. 2024. Evaluation of antioxidant and antibacterial activity of red onion (*A. cepa* L.) and garlic (*A. sativum* L.) wastes on homemade tofu. *Prog. Eng. Appl. Tech.*, 5(1):709–716.

Herrmann, K. 1976. Flavonols and flavones in food plants: A review. *J. Food Technol.*, 11:433–448.

Hu, J., Hao, G., Xu, J., Wang, X., and Chen, M. 2024. Deep learning-based coronary artery calcium score to predict coronary artery disease in type 2 diabetes mellitus. *Heliyon*, 10(6):e27937. https://doi.org/10.1016/j.heliyon.2024.e27937

Huma, Z.E., Vian, M.A., Fabiano-Tixier, A.-S., Elmaataoui, M., Dangles, O., and Chemat, F. 2011. A remarkable influence of microwave extraction: Enhancement of antioxidant activity of extracted onion varieties, *Food Chem.*, 127(4):1472–1480. https://doi.org/10.1016/j.foodchem.2011.01.112

Janoszka, B. 2010. Heterocyclic amines and azaarenes in pan-fried meat and its gravy fried without additives and in the presence of onion and garlic. *Food Chem.*, 120(2):463–473. https://doi.org/10.1016/j.foodchem.2009.10.039

Jimenez, L., Alarcon, E., Trevithick-Sutton, C., Gandhi, N., and Scaiano, J.C. 2011. Effect of γ-radiation on green onion DNA integrity: Role of ascorbic acid and polyphenols against nucleic acid damage. *Food Chem.*, 128(3):735–741. https://doi.org/10.1016/j.foodchem.2011.03.098

Joković, N., Matejić, J., Zvezdanović, J., Stojanović-Radić, Z., Stanković, N., Mihajilov-Krstev, T., and Bernstein, N. 2024. Onion peel as a potential source of antioxidants and antimicrobial agents. *Agronomy*, 14:453. https://doi.org/10.3390/agronomy14030453

Jung, J.Y., Lim, Y., Moon, M.S., Kim, J.Y., and Kwon, O. 2011. Onion peel extracts ameliorate hyperglycemia and insulin resistance in high fat diet/streptozotocin-induced diabetic rats. *Nutr. Metab. (Lond)*, 8:18. https://doi.org/10.1186/1743-7075-8-18

Kim, J.-H., Kim, J.-S., Kim, S.-H., Jeong, S.-H., Jeong, U.-Y., Jung, J.-E., Lee, S.-K., and Lee, S.-H. 2022. Antioxidant and anti-inflammatory effects of ethanol extract from whole onion (*Allium cepa* L.) with leaves. *Agriculture*, 12:963. https://doi.org/10.3390/agriculture12070963

Kumar, M., Barbhai, M.D., Hasan, M., Punia, S., Dhumal, S., Rais, N., Chandran, D., Pandiselvam, R., Kothakota, A., and Tomar, M. 2022. Onion (*Allium cepa* L.) peels: A review on bioactive compounds and biomedical activities. *Biomed. Pharmacother.*, 146:112498. https://doi.org/10.1016/j.biopha.2021.112498

Lee, E.J., Patil, B.S., and Yoo, K.S. 2015. Antioxidants of 15 onions with white, yellow, and red colors and their relationship with pungency, anthocyanin, and quercetin. *LWT Food Sci. Technol.*, 63(1):108–114. https://doi.org/10.1016/j.lwt.2015.03.028

Lee, H.-S., Kwon, Y.-J., Seo, E.-B., Kim, S.K., Lee, H., Lee, J.T., Chang, P.S., Choi, Y.J., Lee, S.H., and Ye, S.K. 2023. Anti-inflammatory effects of Allium cepa L. peel extracts via inhibition of JAK-STAT pathway in LPS-stimulated RAW264.7 cells. *J. Ethnopharmacol.*, 317:116851. https://doi.org/10.1016/j.jep.2023.116851

Lee, K.A., Kim, KT., Kim, H.J., Chung, M.S., Chang, P.S., Park, H., and Pai, H.D. 2014. Antioxidant activities of onion (*Allium cepa* L.) peel extracts produced by ethanol, hot water, and subcritical water extraction. *Food Sci. Biotechnol.*, 23:615–621. https://doi.org/10.1007/s10068-014-0084-6

Manohar, C.M., Xua, J., Murayyan, A., Neethirajan, S., and Shi, J. 2017. Antioxidant activity of polyphenols from Ontario grown onion varieties using pressurized low polarity water technology. *J. Funct. Foods.*, 31:52–62. https://doi.org/10.1016/j.jff.2017.01.037

Mardani, N., Jahadi, M., Sadeghian, M., Keighobadi, K., and Khosravi-Darani, K. 2023. Antimicrobial activities, phenolic and flavonoid contents, antioxidant and DNA protection of the internal and outer layers of *Allium cepa* L. from Iran. *NFS Journal*, 31:93–101. https://doi.org/10.1016/j.nfs.2023.03.003.

Marefati, N., Ghorani, V., Shakeri, F., Boskabady, M., Kianian, F., Rezaee, R., and Boskabady, M.H. 2021. A review of anti-inflammatory, antioxidant, and immunomodulatory effects of *Allium cepa* and its main constituents. *Pharm. Biol.*, 59(1):287–302. https://doi.org/10.1080/13880209.2021.1874028

Masood, S., Rehman, A.U., Bashir, S., Shazly, M.E., Imran, M., Khalil, P., Ifthikar, F., Jaffar, H.M., and Khursheed, T. 2021. Investigation of the anti-hyperglycemic and antioxidant effects of wheat bread supplemented with onion peel extract and onion powder in diabetic rats. *J. Diabetes Metab. Disord.,* 20:485–495. https://doi.org/10.1007/s40200-021-00770-x

Mazza, G., and Miniati, E. 1993. Anthocyanins in Fruits, Vegetables and Grain, CRC Press, Inc. Boca Raton, FL, pp. 1–28.

Mounir, R., Alshareef, W.A., El Gebaly, E.A., El-Haddad, A.E., Ahmed, A.M.S., Mohamed, O.G., Enan, E.T., Mosallam, S., Tripathi, A., Selim, H.M.R.M., Bukhari, S.I., Alfaraj, R., Ragab, G.M., El-Gazar, A.A., and El-Emam, S.Z. 2023. Unlocking the power of onion peel extracts: Antimicrobial and anti-inflammatory effects improve wound healing through repressing Notch-1/NLRP3/Caspase-1 signaling. *Pharmaceuticals*, 16:1379. https://doi.org/10.3390/ph16101379

Munday, R., and Munday, C.M. 2001. Relative activities of organosulfur compounds derived from onions and garlic in increasing tissue activities of quinone reductase and glutathione transferase in rat tissues. *Nutrition and Cancer*, 40(2):205–210, https://doi.org/10.1207/s15327914nc402_18

Murayyan, A.I., Manohar, C.M., Hayward, G., and Neethirajan, S. 2017. Antiproliferative activity of Ontario grown onions against colorectal adenocarcinoma cells. *Food Res. Int.*, 96:12–18. https://doi.org/10.1016/j.foodres.2017.03.017.

Ochar, K., and Kim, S.-H. 2023. Conservation and global distribution of onion (*Allium cepa* L.) germplasm for agricultural sustainability. *Plants*, 12:3294. https://doi.org/10.3390/plants12183294

Oyawoye, O.M., Olotu, T.M., Nzekwe, S.C., Idowu, J.A., Abdullahi, T.A., Babatunde, S.O., Ridwan, I.A., Batiha, G.E., Idowu, N., Alorabi, M., and Faidah, H. 2022. Antioxidant potential and antibacterial activities of *Allium cepa* (onion) and *Allium sativum* (garlic) against the multidrug resistance bacteria. *Bull. Natl. Res. Cent.*, 46:214. https://doi.org/10.1186/s42269-022-00908-8

Pareek, S., Sagar, N.A., Sharma, S., and Kumar, V. 2017. Onion (*Allium cepa* L.). In Fruit and Vegetable Phytochemicals, E.M. Yahia (Ed.). https://doi.org/10.1002/9781119158042.ch58

Pérez-Gregorio, R.M., García-Falcón, M.S., Simal-Gándara, J., Rodrigues, A.S., and Almeida D.P.F. 2010. Identification and quantification of flavonoids in traditional cultivars of red and white onions at harvest. *J. Food Comp. Anal.*, 23(6):592–598. https://doi.org/10.1016/j.jfca.2009.08.013

Prakash, D., Singh, B.N., and Upadhyay, G. 2007. Antioxidant and free radical scavenging activities of phenols from onion (*Allium cepa*). *Food Chem.*, 102(4):1389–1393. https://doi.org/10.1016/j.foodchem.2006.06.063

Sagar, N.A., and Pareek, S. 2020. Antimicrobial assessment of polyphenolic extracts from onion (*Allium cepa* L.) skin of fifteen cultivars by sonication-assisted extraction method. *Heliyon*, 6(11):e05478. https://doi.org/10.1016/j.heliyon.2020.e05478

Sagar, N.A., Pareek, S., Benkeblia, N., and Xiao, J. 2022. Onion (*Allium cepa* L.) bioactives: Chemistry, pharmacotherapeutic functions, and industrial applications. *Food Front.*, 3:380–412. https://doi.org/10.1002/fft2.135

Sami, R., Elhakem, A., Alharbi, M., Benajiba, N., Almatrafi, M., and Helal, M. 2021. Nutritional values of onion bulbs with some essential structural parameters for packaging process. *Appl. Sci.*, 11:2317. https://doi.org/10.3390/app11052317

Sidhu, J.S., Ali, M., Al-Rashdan, A., and Ahmed, N. 2019. Onion (*Allium cepa* L.) is potentially a good source of important antioxidants. *J. Food Sci. Technol.*, 56(4):1811–1819. https://doi.org/10.1007/s13197-019-03625-9

Singh, S., Chaurasia, P.K., and Bharati, S.L. 2023. Pharmacological Aspects of Essential Oils: Current and Future Trends, CRC Press, Taylor & Francis, 2023, ISBN–9781003389774. https://doi.org/10.1201/9781003389774

Slimestad, R., Fossen, T., and Vågen, I.M. 2007. Onions: A source of unique dietary flavonoids. *J. Agric. Food Chem.*, 55:10067–10080.

Soubra, L., Sarkis, D., Hilan, C., and Verger, P.H. 2007. Dietary exposure of children and teenagers to benzoates, sulphites, butylhydroxyanisol (BHA) and butyl hydroxyl toluene (BHT) in Beirut (Lebanon). *Reg. Toxicol. Pharmacol.*, 47:68–77. https://doi.org/10.1016/j.yrtph.2006.07.005

Stoica, F., Raţu, R.N., Veleşcu, I.D., Stănciuc, N., and Râpeanu, G. 2023. A comprehensive review on bioactive compounds, health benefits, and potential food applications of onion (Allium cepa L.) skin waste. *Trends Food Sci. Technol.*, 141:104173. https://doi.org/10.1016/j.tifs.2023.104173

Suleria, H.A.R., Butt, M.S., Anjum, F.M., Saeed F., and Khalid, N. 2015. Onion: Nature protection against physiological threats. *Crit. Rev. Food Sci. Nutr.*, 55(1):50–66. https://doi.org/10.1080/10408398.2011.646364.

United Nations Food and Agriculture Organization, Statistics Division (FAOSTAT). 2024. Production of onions and shallots (green) in 2022: Crops/World Regions/Production Quantity/Year from pick lists. http://www.fao.org/faostat/en/#data/QC

Vågen, I.M., and Slimestad, R. 2008. Amount of characteristic compounds in 15 cultivars of onion (*Allium cepa* L.) in controlled field trials. *J. Sci. Food and Agric.*, 88(3):404–411. https://doi.org/10.1002/jsfa.3100

Veiga, A.A., Irioda, A.C., Mogharbel, B.F., Bonatto, S.J.R., and Souza, L.M. 2022. Quercetin-rich extracts from onions (*Allium cepa*) play potent cytotoxicity on adrenocortical carcinoma cell lines, and quercetin induces important anticancer properties. *Pharmaceuticals*, 15:754. https://doi.org/10.3390/ph15060754

Yamazaki, Y., Iwasaki, K., Mikami, M., and Yagihashi, A. 2011. Distribution of eleven flavor precursors, S-Alk(en)yl-L-cysteine derivatives, in seven allium vegetables. *Food Sci. Technol. Res.*, 17(1):55–62. https://doi.org/10.3136/fstr.17.55

Yash, S., Gourav, T., and Sandhya, T. (2024). Antimicrobial potential activity of onion (*Allium Cepa*) waste extracts prepared in different solvent systems. *J. Multidiscipl. Res.*, 4(2):14–27. https://doi.org/10.37022/tjmdr.v4i2.621

Zhang, S.-l., Deng, P., Xu, Y.-c., Lü, S.-w., and Wang J.-j. 2016. Quantification and analysis of anthocyanin and flavonoids compositions, and antioxidant activities in onions with three different colors. *J. Integr. Agric.*, 15(9):2175–2181. https://doi.org/10.1016/s2095-3119(16)61385-0

Zhou, Y., Li, C., Feng, B., Chen, B., Jin, L., and Shen, Y. 2020. UPLC-ESI-MS/MS based identification and antioxidant, antibacterial, cytotoxic activities of aqueous extracts from storey onion (*Allium cepa* L. var. proliferum Regel). *Food Res. Int.*, 130:108969. https://doi.org/10.1016/j.foodres.2019.108969

8 Unlocking the Secrets of the King of Spices

Exploring Its Chemistry, Health Benefits, and Medicinal Wonders

Sukanya Patra, Jyotirmayee, Pradip Paik, and Malkhey Verma

8.1. INTRODUCTION

Originating from the Malabar Coast of India, black pepper is a globally traded commodity with major production centres in India, Indonesia, Vietnam, and Brazil. Vietnam is the largest producer and exporter of black pepper and contributes almost 40% of global supply (Newerli-Guz et al., 2023). Today, the black pepper market is a multimillion-dollar industry, driven by culinary traditions, medicine, and industrial application. This is a key ingredient in Indian and European cuisine, used as a common seasoning, along with garam masala and rasam, and has significant cultural impact on literature, art, and folklore, demonstrating its global appeal (Jaffrey, 2020).

Black pepper is gaining attention within the scientific community due to its chemical composition and pharmacological properties (Figure 8.1). The primary bioactive compound, piperine, has been the focus of extensive research highlighting its potential health benefits, including antioxidant, anti-inflammatory, and antimicrobial effects and enhanced bioavailability of nutrients and pharmaceuticals (Yadav et al., 2023). Beyond its medicinal properties, these are also explored for its applications in agriculture, developing disease-resistant varieties, and post-harvest processing techniques (Manzoor et al., 2024). The global impact of the spice is also increasing, with potential applications in sustainable agriculture, food security, and biotechnological innovation. However, challenges due to climate change, pest infestations, and market instability necessitate collaborative efforts among researchers, farmers, and industry stakeholders to address issues effectively (Rodríguez-Moreno et al., 2024).

8.2. BOTANICAL DESCRIPTION AND CHEMICAL COMPOSITION OF BLACK PEPPER (*PIPER NIGRUM*)

8.2.1. Botanical Description

Black pepper (*Piper nigrum*) is one of the most widely known members of the Plantae kingdom. This chapter provides a detailed exploration of the botanical description of black pepper, including its taxonomical classification, morphological characteristics, geographical distribution, and habitat.

8.2.2. Taxonomical Classification

Black pepper belongs to the Piperaceae family and is known for its aromatic plants, often rich in essential oils and alkaloids. The taxonomical hierarchy of black pepper is as follows:

DOI: 10.1201/9781003559894-8

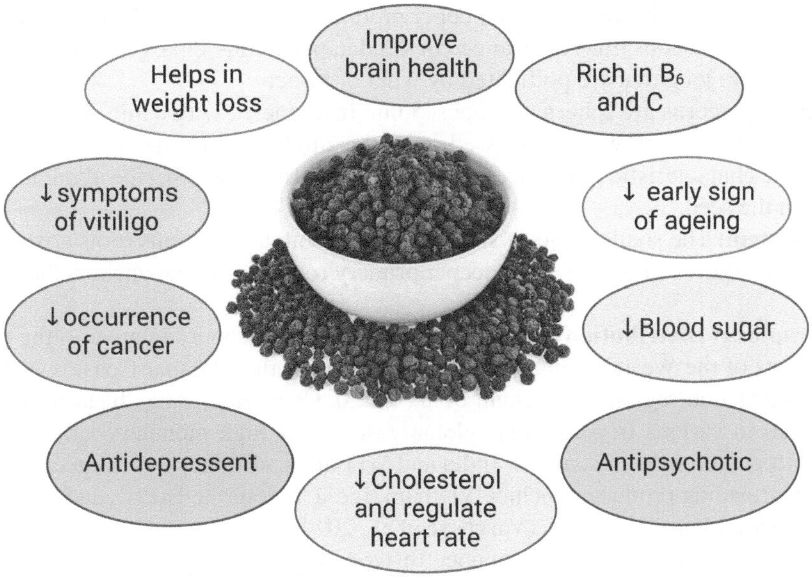

FIGURE 8.1 Diversified pharmacological properties of black pepper.

- **Subkingdom:** Tracheobionta (Vascular plants)
- **Superdivision:** Spermatophyta (Seed plants)
- **Division:** Magnoliophyta (Flowering plants)
- **Class:** Magnoliopsida (Dicotyledons)
- **Subclass:** Magnoliidae
- **Order:** Piperales
- **Genus:** Piper

The genus *Piper* is large and diverse, comprising more than 1,000 species, many of which are used in traditional medicine, food, and industry. *Piper nigrum* is this genus's most economically important species, primarily due to its global use as a spice (Alamgir, 2018).

8.2.3. Morphological Characteristics

Black pepper is characterized by its distinctive growth habit, leaves, flowers, and fruit, all of which contribute to its identification and cultivation.

- **Growth Habit:** Black pepper is a woody, perennial climber that can reach up to 10 meters in height. It relies on its adventitious roots to attach to nearby trees or trellises. The vines are typically trained on supports to facilitate harvesting in commercial cultivation (Ashokkumar et al., 2021).
- **Stem:** The stem of black pepper is slender, flexible, and green when young, becoming woody as it matures with numerous nodes, forming roots, leaves, and inflorescences (Nair, 2011).
- **Leaves:** The leaves of black pepper are simple, alternate, and broadly ovate to heart-shaped, with a glossy, dark green surface. Measuring 5–10 cm in length and 3–6 cm in width, the leaves have appointed tip with rounded or cordate base with prominent veins of palmate pattern (Parthasarathy et al., 2007).

- **Inflorescence and Flowers:** Black pepper produces numerous tiny bisexual (white to yellowish) inconspicuous flowers arranged in slender, pendulous spikes called catkins ranging from 7–15 cm long that are pollinated by wind or insects.
- **Fruit:** Peppercorns are spherical drupes, 5 mm in diameter, with a thin pericarp that turns red and dries to black when processed, with the single seed inside being hard and shrivelled. The characteristic pungency and flavour come from the piperine alkaloid, primarily found in the seed.
- **Root System:** The shallow root system comprises both adventitious roots arising from the nodes that assist in climbing and a deeper primary root system that anchors the plant in the soil.
- **Geographical Distribution and Native Range**: Black pepper is native to the equatorial rainforests of the Western Ghats in India, particularly the Malabar Coast, which remains a key production region (Sivaraman et al., 2023). Over time, it has been introduced and cultivated in various tropical areas globally due to its high monetary value. It grows in humid tropical forests of southern India and Sri Lanka, climbing over vegetation to access sunlight. Leading producers include Vietnam, India, Indonesia, Brazil, and Sri Lanka, due to their warm, humid climates (Varghese et al., 2024).
- **Climate Requirements:** Black pepper thrives in tropical climates, with temperatures ranging from 25–35°C and annual rainfall of 2,000–3,000 mm with relative humidity 70–90%. The plant is highly sensitive to cold and frost, limiting its cultivation in temperate climates (Banerjee et al., 2024).
- **Soil Preferences:** Black pepper grows best in well-drained, fertile soils rich in organic matters with slightly acidic to neutral pH levels (5.5 to 6.5). In its native habitat, it often grows in loamy soils that retain moisture but are not waterlogged (Muriithi et al., 2024). In cultivation, pepper is often grown on slopes with good drainage to prevent root rot and other water-related diseases (Behera et al., 2024).
- **Habitat:** It is a shade-loving plant, often grown under the canopy of taller trees or with artificial shading. The ideal habitat for black pepper provides a balance of sunlight and shade, with consistent moisture.

8.2.4. Chemical Composition: The Molecular Treasure Chest

The chemical composition of black pepper is quite complex and diverse, with various compounds contributing to its flavour and medicinal properties. It contains alkaloids, essential oils, and other bioactive molecules, making black pepper distinct (Figueiredo et al., 2021).

- **Alkaloids:** Piperine is an abundant alkaloid comprising 5–9% of the peppercorn, which is primarily responsible for its pungency with biological activities such as anti-inflammatory, antioxidant, and antitumour effect. Chavicine is an isomer of piperine and contributes to the pungency of black pepper.
- **Essential Oils:** Black pepper contains essential oils comprising 1–2.5% of its composition and various volatile compounds that give black pepper its aroma. The primary components of the essential oil include monoterpenes (sabinene, α/β-pinene, limonene), sesquiterpenes (β-caryophyllene, α-humulene, α-cadinene, myrcene), and other terpenes with oxygenation (linalool, terpinene-4-ol, α-terpineol, eugenol).
- **Volatile Oils:** The essential oils in black pepper, rich in terpenes such as β-caryophyllene, pinene, sabinene, and limonene, phenolic compounds such as ferulic acid, and vanillic acid are also contributed to its bioactive profile, exhibiting antimicrobial, antioxidant, and anti-inflammatory properties.
- **Other Bioactive Compounds:** In addition to piperine and chavicine, black pepper contains other alkaloids such as piperidine and piperlongumine, lignans such as hinokinin and

cubebin, as well as flavonoids such as kaempferol and quercetin, and small quantities of phenolic acids such as ferulic acid and caffeic acid.

8.3 DIVERSIFIED PROPERTIES OF BLACK PEPPER

8.3.1. Antioxidant Properties

Black pepper, an important source, has been shown to have high antioxidant activity, primarily because of the active component called piperine. Antioxidants cancel free radicals such as ROS and RNS, which are precursors to oxidative stress, aging, inflammation, and chronic diseases. Piperine increases the body's defence mechanisms through upregulation of antioxidant enzymes such as SOD, catalase (CAT), and glutathione peroxidase (GPx), preventing cellular damage, inhibition of lipid peroxidation, thus protecting cells. By activating the Nrf2 signalling pathway, they regulate upregulation of heme oxygenase 1 (Ho-1), further enhancing the defence system of the body (Nimse et al., 2015).

8.3.2. Anti-Inflammatory Properties

Black pepper is well-known for its anti-inflammatory effects, primarily due to piperine. Inflammation is a natural immune response, but chronic inflammation can lead to various diseases, including arthritis, cardiovascular diseases, and cancer. Piperine has been shown to inhibit the production of pro-inflammatory cytokines via MAPK signalling pathways, mediators, such as TNF-α, IL-1β, IL-6, and COX-2, and to reduce COX and LOX enzyme activity. By modulating these inflammatory pathways, black pepper can help reduce inflammation and alleviate symptoms associated with inflammatory conditions via modulating the production of NO by macrophages. Moreover, piperine's ability to inhibit NF-κB, a key regulator of inflammation, further underscores its potential as an anti-inflammatory agent (Gouda et al., 2023).

8.3.3. Antimicrobial and Antifungal Properties

Black pepper demonstrates broad-spectrum antimicrobial and antifungal properties, making it effective against various pathogenic microorganisms. Piperine and other bioactive compounds in black pepper have been found to inhibit the growth of bacteria such as *Staphylococcus aureus*, *Escherichia coli*, and *Salmonella typhi*. Its antifungal properties are equally significant, with studies showing effectiveness against *Candida albicans* and other fungal species (Shityakov et al., 2019). The antimicrobial action of black pepper is attributed to its ability to disrupt microbial cell membranes, inhibit biofilm formation, interfere with microbial enzyme systems such as DNA gyrase and RNA polymerase, and alter metabolic pathways affecting ATP synthesis of the cells, thereby preventing infections and contributing to food preservation.

8.3.4. Digestive Health

Black pepper has long been used in traditional medicine to support digestive health. It stimulates the secretion of digestive enzymes, such as amylase, lipase, and proteases, which enhance the breakdown of carbohydrates, fats, and proteins. Additionally, piperine improve solubility and the bioavailability of nutrients, such as Ca2+, Fe2+, and Mg2+, by improving their absorption in the gastrointestinal tract (Shityakov et al., 2019). Black pepper also promotes the secretion of hydrochloric acid in the stomach, which is essential for proper digestion and prevention of gastrointestinal disorders such as indigestion, bloating, constipation in diseases such as IBS/IBD, and gastroesophageal reflux disease (GERD). Its carminative properties help reduce flatulence, making it a staple in digestive health formulations.

8.4 CLINICAL USES

The therapeutic potential of black pepper extends to several clinical applications. It has been explored as an adjuvant therapy in cancer treatment due to its ability to enhance the efficacy of chemotherapeutic agents while reducing their toxicity. Piperine has also been studied for its role in improving cognitive function and memory, potentially benefiting patients with neurodegenerative disorders such as Alzheimer's disease (Table 8.1). Moreover, black pepper's ability to regulate blood sugar levels and enhance insulin sensitivity makes it a promising agent in diabetes management. Its use in weight management is also noteworthy, as piperine has been shown to inhibit adipogenesis and promote fat metabolism.

8.5. NUTRITIONAL HEALTH AND USE IN TRADITIONAL AND MODERN MEDICINE

Black pepper (*Piper nigrum*) is a rich source of both macronutrients and micronutrients, contributing to its nutritional and therapeutic value. It contains essential macronutrients, such as carbohydrates, proteins, and dietary fibre, vital for energy production, muscle function, and digestive health.

Micronutrient-wise, black pepper is abundant in vitamins and minerals, including vitamin C, vitamin K, vitamin A, riboflavin, thiamine, folate, and choline. It also contains essential minerals, such as calcium, iron, magnesium, manganese, potassium, and zinc (Milenković et al., 2024). These nutrients are crucial for various physiological processes, including bone health, blood clotting, immune function, and enzyme activation. These nutrients make black pepper an essential dietary component with potential health benefits.

8.5.1. Traditional Uses in Medicine

Black pepper has been used in traditional medicine for centuries, particularly in Ayurvedic and Chinese medicine, where it is valued for its therapeutic properties. In Ayurveda, black pepper is

TABLE 8.1
Compounds of Black Pepper with Medicinal Uses

Compounds	Use	References
Piperine	Inhibit P-gp and CYP3A4, antiobesity, antidepressant, enhancing bioavailability	Athukuri et al., 2017
Sabinene	Antimicrobial, flavouring agent, regulate ROS-mediated MAPK/MuRF1 pathway	Jia et al., 2020
β-pinene	Digestive stimulant, antiseptic, antiviral, analgesic	
Limonene	Flavouring agent and preservative, solvent, dispersing agent, lower AGE generation, reduce inflammatory markers in osteoarthritis	Yang et al., 2021
α-humulene	Antimicrobial, anti-inflammatory action with NOS and COX2, antitumour, analgesic	Shebaby et al., 2021
Linalool	Antimicrobial, anticancer, neuroprotective, drug delivery system	Camargo et al., 2020
Eugenol	Flavouring, anti-inflammatory, dental filling (with ZnO), hair growth	Taleuzzaman et al., 2021
Piperidine	Vulcanization of rubber, solvent, amine synthesis, linker molecule, anti-Tb activity	Pemawat et al., 2024
Piperlongumine	Anticancer, antidiabetic by inhibiting aldose reductase, tubulin depolymerization with G2/M arrest	Ghomashi et al., 2023
Hinokinin	Antimicrobial, anti-inflammatory, antitrypanosoma	Agrawal et al., 2023
Cubebin	Analgesic, anticancer, erectile dysfunction, antiprotozoal	Agrawal et al., 2023

known as "Maricha" and is used to balance the body's doshas (Vata, Pitta, and Kapha). It is traditionally used to treat digestive disorders, respiratory conditions, and joint pains.

1. **Digestive Health:** It acts as a digestive stimulant for black pepper and is known to relieve the symptoms of indigestion, bloating, and flatulence. It is used in medicine to treat constipation and loss of appetite.
2. **Respiratory Health:** It is an expectorant in traditional medicine, meaning it relieves coughs and colds. Because of its warm nature, it is often used in treating respiratory infections and sinusitis.
3. **Pain Relief:** The spice has been used to alleviate pain from arthritis, headaches, and other inflammatory conditions due to its anti-inflammatory properties. In Chinese medicine, black pepper is considered to warm the body and is often used to treat conditions related to cold, such as chills, digestive issues, and poor circulation.

8.5.2. Integration into Modern Pharmacology

The active principle of black pepper, piperine, has been studied extensively in modern pharmacology for its many health benefits. It increases the bioavailability of drugs and nutrients by retarding the action of metabolic enzymes that break them down within the body, thus the efficacy of drugs is significantly improved. This property also led to the inclusion of piperine in different pharmacological formulations for making therapeutic agents absorbable and workable.

1. **Drug Bioavailability:** Piperine's ability to enhance the bioavailability of various drugs has been a significant focus of modern pharmacology. It is an adjuvant in formulations for drugs such as curcumin (from turmeric) and resveratrol, significantly increasing their absorption and effectiveness (Yadav et al., 2023).
2. **Neuroprotection:** Piperine has shown potential in neuroprotective strategies and is being explored as a treatment option for neurodegenerative diseases such as Alzheimer's and Parkinson's. Its antioxidant and anti-inflammatory properties contribute to its protective effects on the nervous system (Balakrishnan et al., 2023).
3. **Metabolic Health:** Piperine has also been incorporated into therapies for metabolic disorders, such as obesity, diabetes, and hyperlipidemia. It is known to improve insulin sensitivity and regulate lipid metabolism, making it a valuable component in managing these conditions (Servida et al., 2023).
4. **Cancer Treatment:** Piperine has been studied in clinical settings for its potential role in cancer therapy. It has been found to enhance the efficacy of chemotherapy drugs while reducing their side effects. Piperine's ability to inhibit cancer cell proliferation and induce apoptosis has been observed in several types of cancer, including breast, colon, and prostate cancers (Yadav et al., 2023; Balakrishnan et al., 2023).

8.6. INDUSTRIAL APPLICATIONS

8.6.1. Food Industry Applications

Black pepper forms a cornerstone in the global food industry as it is valued for flavour, aroma, and preservative properties.

1. **Flavouring and Seasoning:** Black pepper is one of the most widely used spices worldwide, with a sharp, spicy flavour. It forms the backbone of numerous spice blends, marinades, sauces, and processed foods. The spice enhances the flavour profile of both savoury and sweet dishes, which makes it indispensable for applications in the culinary industry (Li et al., 2021).

2. **Food Preservation:** Besides flavouring, black pepper has natural preservative properties. Antimicrobial activity, primarily due to piperine and other essential oils, retards the growth of bacteria and fungi in food products. This makes black pepper a natural preservative for meat products and pickles, thus extending shelf life and minimizing synthetic preservatives (Hamdan et al., 2024).
3. **Nutraceuticals and Functional Foods:** The food industry also utilizes black pepper to produce nutraceuticals and functional foods. Piperine, the bioactive compound in black pepper, enhances the bioavailability of nutrients and other bioactive compounds, such as curcumin in turmeric. This has led to including black pepper extract in various health supplements designed to improve nutrient absorption and overall health (Hamdan et al., 2024; Li et al., 2021).

8.6.2. Perfume and Cosmetics Industry

Black pepper's warm, spicy, and woody aroma has found a niche in the perfume and cosmetics industry, where it is valued for its unique scent and beneficial properties. Black pepper essential oil, extracted from dried peppercorns, has become a versatile and valued ingredient in various fields such as perfumery, cosmetics, and aromatherapy. It's warm and spicy aroma not only adds a unique depth and richness to fragrances but also offers a range of therapeutic benefits. Research highlights its stimulating and detoxifying properties, which can promote better blood circulation and contribute to healthier skin. Additionally, its antimicrobial activity makes it a promising natural option for addressing skin concerns such as acne and infections, showcasing its potential in both wellness and skincare applications (Dludla et al., 2023).

8.6.3. Potential Biofuel Formation

The by-products of black pepper cultivation—such as leaves, stems, and spent peppercorns—are gaining attention for their potential in biofuel production. These residues are rich in lignocellulosic biomass, which can be transformed into bioethanol and biodiesel through well-established processes such as pretreatment and fermentation. Early research has shown encouraging results, highlighting the possibility of creating a sustainable energy source from what is often considered agricultural waste (Ab Rasid et al., 2021). This approach not only supports the principles of a circular economy by repurposing by-products but also reduces environmental impact. By integrating biofuel production into pepper farming, particularly in key cultivation regions, we can enhance both sustainability and profitability, offering a meaningful way to improve the overall value chain of this important crop.

8.7. FUTURE PROSPECTIVE AND RESEARCH DIRECTIONS

8.7.1. Genomic Studies

The genomic study of black pepper opens up exciting opportunities to better understand its genetic makeup. One can improve breeding programs by exploring genetic diversity to enhance important traits such as higher yields and better disease resistance. Genomics also supports marker-assisted selection, making it easier and faster to identify and cultivate plants with desirable characteristics. Additionally, understanding the biochemical pathways in black pepper allows us to pinpoint the genes involved in producing bioactive compounds (Khew et al., 2022). This knowledge can pave the way for metabolic engineering or targeted breeding strategies, boosting the production of these valuable compounds. Ultimately, this could expand black pepper's applications in pharmaceuticals and nutraceuticals, unlocking its full therapeutic potential.

8.7.2. Biotechnological Advances

By targeting specific genes, genetic engineering allows us to enhance the production of valuable compounds such as piperine and essential oils, which are key to the spice's flavour and medicinal properties. Tissue culture and micropropagation, on the other hand, provide an efficient way to produce disease-free plants, conserve rare or endangered varieties, and study their growth under controlled conditions (Dat et al., 2024). Similar approaches have been successfully applied to optimize growth media for other economically important plants, such as tropical day-blooming waterlily (*Nymphaea colorata*) (Rodboot et al., 2024), and to develop sustainable cultivation practices for high-value medicinal and aromatic species (Ur-Rahman et al., 2019). These strategies can inform future black pepper propagation and conservation programs. Beyond this, metabolic engineering opens exciting possibilities by optimizing the plant's biochemical pathways to boost secondary metabolite production. This could even position black pepper as a bio-factory for producing high-value compounds with applications in pharmaceuticals, nutraceuticals, and cosmetics, unlocking its full potential as more than just a spice.

8.7.3. Potential Applications

The potential for new treatments in diseases such as cancer and neurological disorders holds promising significance. The advances could boost global production, ensuring a steady supply. With rising interest in natural and functional foods, black pepper offers opportunities in nutraceuticals and functional food, including various supplements and fortified foods (Bansal et al., 2023).

8.7.4. Challenges and Future Directions

1. **Genomic Data Availability:** One of the main challenges in genomic studies is the limited availability of comprehensive genomic data for black pepper. Efforts should be made to sequence and annotate the genomes of various black pepper cultivars to facilitate research and breeding efforts.
2. **Biotechnological Barriers:** The successful application of biotechnological tools, such as genetic and metabolic engineering, requires overcoming technical and regulatory barriers. Ensuring the safety and efficacy of genetically modified black pepper plants is critical, and addressing public concerns about genetically engineered crops is also essential.
3. **Sustainability Concerns:** As the demand for black pepper continues to grow, ensuring the sustainability of its production is crucial. Future research should focus on developing environmentally friendly cultivation practices, such as organic farming, integrated pest management, and sustainable fertilizers. Additionally, exploring black pepper residues for biofuel production must balance energy needs with environmental impact.
4. **Commercialization and Market Acceptance:** The commercialization of new black pepper products in the pharmaceutical, food, or cosmetic industries will be a success if accepted by the market and approved by the regulatory bodies. It will require robust clinical trials, safety assessments, and consumer education.

8.8. CONCLUSION

Black pepper is more than just a spice, with a very rich pharmacological profile; this qualifies as the most effective natural remedies associated with several health benefits. The antioxidant, anti-inflammatory, antimicrobial, and digestive health properties of black pepper and its clinical applications underscore its importance in both traditional and modern medicine. With continuous research, discovering the potential of black pepper as a drug, it is likely to play an important role in preventive and therapeutic healthcare. Black pepper has been used in modern pharmacology in a major way through the active component, piperine, opening new ways to improve drug efficacy and treat chronic diseases. As clinical trials continue to investigate its benefits, black pepper will likely continue to be a basic natural remedy in traditional and modern medicine. The future of black pepper research is promising, with great scope for advancements in genomic studies, biotechnological innovations, and new applications. Even with these problems related to availability, regulation, and sustainability, the potential for many applications through further research and innovation of black pepper cannot be undermined.

8.9. ACKNOWLEDGEMENT

The authors acknowledge Dr. Monika Singh for providing access to use Biorender software for designing the graphical abstract for this chapter.

REFERENCES

Ab Rasid, N. S., Shamjuddin, A., Rahman, A. Z. A., & Amin, N. A. S. (2021). Recent advances in green pre-treatment methods of lignocellulosic biomass for enhanced biofuel production. *Journal of Cleaner Production, 321*, 129038.

Agrawal, N., Goyal, D., & Goyal, A. (2023). A review on multi-therapeutic potential of (-)-cubebin: Experimental evidences. *Natural Product Research, 37*(24), 4290–4301.

Alamgir, A. (2018). Secondary metabolites: Secondary metabolic products consisting of C and H; C, H, and O; N, S, and P elements; and O/N heterocycles. In *Therapeutic Use of Medicinal Plants and Their Extracts: Volume 2: Phytochemistry and Bioactive Compounds* (pp. 165–309). Springer International Publishing.

Ashokkumar, K., Murugan, M., Dhanya, M., Pandian, A., & Warkentin, T. D. (2021). Phytochemistry and therapeutic potential of black pepper [Piper nigrum (L.)] essential oil and piperine: A review. *Clinical Phytoscience, 7*(1), 52.

Athukuri, B. L., & Neerati, P. (2017). Enhanced oral bioavailability of domperidone with piperine in male wistar rats: Involvement of CYP3A1 and P-gp inhibition. *Journal of Pharmacy & Pharmaceutical Sciences, 20*, 28–37.

Balakrishnan, R., Azam, S., Kim, I.-S., & Choi, D.-K. (2023). Neuroprotective effects of black pepper and its bioactive compounds in age-related neurological disorders. *Aging and Disease, 14*(3), 750.

Banerjee, S., Banerjee, S., Bose, S., & Banerjee, S. (2024). Effect of geographical location and climatic variation on the composition of active principles in Indian spices. In *Medicinal Spices and Herbs from India* (pp. 61–84). Apple Academic Press.

Bansal, P., Maithani, M., Gupta, V., Kaur, G., & Bansal, R. (2023). Future prospective of nutraceutical and functional food with herbs and spices. In *Herbs, Spices and Their Roles in Nutraceuticals and Functional Foods* (pp. 361–381). Academic Press.

Behera, J. K., Jena, A. K., Bhattacharya, M., & Behera, B. (2024). India's current situation with regard to the effects of water pollution on agricultural productivity and public health. In *Soil, Water Pollution and Mitigation Strategies: A Spatial Approach* (pp. 431-464). Springer Nature Switzerland.

Camargo, S. F., Medeiros, C. F., Santos, V. V., Quintans-Júnior, L. J., Azeredo, F. J., & Silva, D. F. (2020). Pharmacokinetics and antihypertensive effect of linalool is improved after incorporation in β-cyclodextrin as a drug delivery system. *The FASEB Journal, 34*(S1), 1–1.

Dludla, P. V., Cirilli, I., Marcheggiani, F., Silvestri, S., Orlando, P., Muvhulawa, N., . . . Hlengwa, N. (2023). Bioactive properties, bioavailability profiles, and clinical evidence of the potential benefits of black pepper (Piper nigrum) and red pepper (Capsicum annum) against diverse metabolic complications. *Molecules, 28*(18), 6569.

Figueiredo, Y. G., Bueno, F. C., de Oliveira Júnior, A. H., do Carmo Mazzinghy, A. C., Mendonça, H. d. O. P., de Oliveira, A. F., . . . Melo, J. O. F. (2021). Profile of the volatile organic compounds of pink pepper and black pepper. *Scientific Electronic Archives, 14*(12).

Ghomashi, S., Ghomashi, R., Aghaei, H., & Massah, A. R. (2023). Recent advances in biological active sulfonamide based hybrid compounds part B: Two-component sulfonamide hybrids. *Current Medicinal Chemistry, 30*(5), 604–665.

Gouda, N. A., Alshammari, S. O., Abourehab, M. A., Alshammari, Q. A., & Elkamhawy, A. (2023). Therapeutic potential of natural products in inflammation: Underlying molecular mechanisms, clinical outcomes, technological advances, and future perspectives. *Inflammopharmacology, 31*(6), 2857–2883.

Hamdan, L. H. M., Yusop, M. A. M., Naroh, A., Mojig, M. M. A. T., Rady, R. A. A., Phillip, A. S. M. A., & Pandy, K. (2024). *Fabrication of the mini multi-function pepper machine*. E3S Web of Conferences. Vol. 479. EDP Sciences, 2024.

Jaffrey, M. (2020). *From Curries to Kebabs: Recipes from the Indian Spice Trail*. Echo Point+ORM.

Jia, H., Chen, T., Qu, J., Yao, M., Xiao, W., Wang, Y., . . . Yuan, Y. (2020). Collaborative subcellular compartmentalization to improve GPP utilization and boost sabinene accumulation in Saccharomyces cerevisiae. *Biochemical Engineering Journal, 164*, 107768.

Khew, C. Y., Koh, C. M. M., Chen, Y. S., Sim, S. L., & Mercer, Z. J. A. (2022). The current knowledge of black pepper breeding in Malaysia for future crop improvement. *Scientia Horticulturae, 300*, 111074.

Li, H., Wu, X., Li, X., Cao, X., Li, Y., Cao, H., & Men, Y. (2021). Multistage extraction of star anise and black pepper derivatives for antibacterial, antioxidant, and anticancer activity. *Frontiers in chemistry, 9*, 660138.

Manzoor, S., Nabi, S. U., Rather, T. R., Gani, G., Mir, Z. A., Wani, A. W., . . . Manzar, N. (2024). Advancing crop disease resistance through genome editing: A promising approach for enhancing agricultural production. *Frontiers in Genome Editing, 6*, 1399051.

Milenković, A., Petrović, S., Savić, S., Stanojević, J., & Stanojević, L. (2024). The effect of extraction techniques on the mineral content of black pepper fruit ethanolic extracts. *Biologica Nyssana, 15*(1).

Muriithi, F., Onyando, J., & Okwany, R. (2024). Effect of mixing ratio of super absorbent materials on the growth and yield of Bell Pepper in the Ferric Luvisols of Mogotio, Kenya. *East African Journal of Agriculture and Biotechnology, 7*(1), 204–215.

Nair, K. P. (2011). *Agronomy and Economy of Black Pepper and Cardamom: The" King" and" Queen" of Spices*. Elsevier Insights.

Newerli-Guz, J., & Rybowski, M. (2023). Analysis of key factors influencing spice traceability systems: The example of Black Pepper (Piper nigrum L.). *Krakow Review of Economics and Management/Zeszyty Naukowe Uniwersytetu Ekonomicznego w Krakowie, 3*(1001), 67–84.

Nimse, S. B., & Pal, D. (2015). Free radicals, natural antioxidants, and their reaction mechanisms. *RSC Advances, 5*(35), 27986–28006.

Parthasarathy, V., Sasikumar, B., Nair, R., & George, K. J. (2007). Black pepper: Botany and horticulture. *Horticultural Reviews-Westport Then New York, 33*, 173.

Pemawat, G., Bhatnagar, A., & Khangarot, R. K. (2024). Synthesis and biological activities of heterocyclic hybrids containing piperidine and pyridine moieties: Recent developments. *Mini-Reviews in Organic Chemistry, 21*(3), 346–369.

Rodboot, N., Te-chato, S., & Yenchon, S. (2024). Optimizing culture medium for commercial production of tropical day-blooming waterlily (Nymphaea colorata Peter). *Thai Journal of Agricultural Science, 57*(4), 224–235.

Rodríguez-Moreno, V. M., & Ruíz-Corral, J. A. (2024). The government's role in sustainable food chains: Facing the climate change. *Journal of Food Science and Nutrition Research, 7*(3), 157–176.

Servida, S., Panzeri, E., Tomaino, L., Marfia, G., Garzia, E., Ciniglio Appiani, G., . . . Vigna, L. (2023). Overview of curcumin and piperine effects on glucose metabolism: The case of an insulinoma patient's loss of consciousness. *International Journal of Molecular Sciences, 24*(7), 6621.

Shebaby, W., Saliba, J., Faour, W. H., Ismail, J., El Hage, M., Daher, C. F., . . . Chrabieh, E. (2021). In vivo and in vitro anti-inflammatory activity evaluation of Lebanese Cannabis sativa L. ssp. indica (Lam.). *Journal of Ethnopharmacology, 270*, 113743.

Shityakov, S., Bigdelian, E., Hussein, A. A., Hussain, M. B., Tripathi, Y. C., Khan, M. U., & Shariati, M. A. (2019). Phytochemical and pharmacological attributes of piperine: A bioactive ingredient of black pepper. *European Journal of Medicinal Chemistry, 176*, 149–161.

Sivaraman, K., Thankamani, C., & Srinivasan, V. (2023). Crop diversification: Cropping/system approach for enhancing farmers' income. In *Handbook of Spices in India: 75 Years of Research and Development* (pp. 3847–3926). Springer Nature Singapore.

Taleuzzaman, M., Jain, P., Verma, R., Iqbal, Z., & Mirza, M. A. (2021). Eugenol as a potential drug candidate: A review. *Current Topics in Medicinal Chemistry, 21*(20), 1804–1815.

Ur-Rahman, I., Sher, H., & Bussmann, R. (2019). Reference guide on high value medicinal and aromatic plants–sustainable management and cultivation practices. *University of Swat, Pakistan.*

Varghese, R., & Ray, J. (2024). Sustainability of black pepper production: A critical analysis of physicochemical soil parameters concerning variables in pepper fields of south India. *Ecological Frontiers, 44*(4), 788–801.

Yadav, S. S., Singh, M. K., Hussain, S., Dwivedi, P., Khattri, S., & Singh, K. (2023). Therapeutic spectrum of piperine for clinical practice: A scoping review. *Critical Reviews in Food Science and Nutrition*, *63*(22), 5813–5840.

Yang, F., Chen, R., Li, W.-y., Zhu, H.-y., Chen, X.-x., Hou, Z.-f., . . . Zhang, W. (2021). D-limonene is a potential monoterpene to inhibit PI3K/Akt/IKK-α/NF-κB p65 signaling pathway in coronavirus disease 2019 pulmonary fibrosis. *Frontiers in Medicine*, *8*, 591830.

9 Clove
Chemistry, Medicinal Application, and Toxicity Profile

Rahul D. Kamble, Balkrishna A. Shinde, Mahadev R. Chambhare, Deepak B. Shelke, and Pankaj Kumar Chaurasia

9.1. INTRODUCTION

Spices are fragrant herbs that serve as both preservatives and flavour enhancers for food. [Shahidi et al., 2015; Tiwari et al., 2014]. Spices were traditionally utilized for medical purposes [Parthasarathy et al., 2008; Chaurasia et al., 2024; Chaurasia and Bharati, 2025; Yadav et al., 2025]. They are an important category of therapeutic plants due to their diverse phytochemical constituents [Rubió et al., 2013]. In addition to their use as natural food preservatives, spices such as clove, cinnamon, thyme, mint, and oregano have been recognized for their nutritional value over the years, which includes antibacterial and antioxidant qualities [Cock and Cheesman, 2018]. Several studies have connected using spice plants to improved human health [Cortés-Rojas et al., 2014; Bendre et al., 2016; Rajput et al., 2017; Shan et al., 2005]. Spices are increasingly being commercialized across the globe due to their nutraceutical benefits [Hernández-Pérez et al., 2020]. Clove is known for its strong antioxidant and antibacterial properties, making it a popular spice among many people [Shan et al., 2005].

Clove is also known as the blossoming buds of *Syzygium aromaticum* (L.) Merr. and L.M. Perry, or *Eugenia carophyllata* Thunb [Cortés-Rojas et al., 2014)]. *Syzygium* is the biggest genus in the Myrtaceae family, with around 1,200–1,800 flowering plant species [Cock and Cheesman, 2018]. Clove originated in the small Indonesian island of Maluku, commonly known as the "Spice Islands" [Kamatou et al., 2012]. Several thousand tonnes of cloves are produced in Indonesia, India, Malaysia, and Sri Lanka; much greater amounts are imported from Madagascar, the West Indies, and Tanzania, especially Zanzibar [Kamatou et al., 2012].

Clove oil (CO) is used in biomedical, cosmetic, pharmaceutical, culinary, active packaging, and sanitary applications because of its antibacterial, antiseptic, anticarcinogenic, antioxidant, pesticide, and analgesic qualities [Mishra et al., 2016]. CO contains 85% to 95% eugenol [Özbek and Ergönül, 2022]. The essential oils of plants pertaining to the Lamiaceae, Lauraceae, Myrtaceae, and Myristicaceae families produce this volatile phenolic aromatic molecule. CO is utilized as a natural food flavouring agent in fragrance and flavour formulations all over the world and is also used in cosmetics and perfumery [Poucher, 1974; Sellamuthu, 2014]. In the food sector, eugenol, the primary constituent of CO, is employed as a flavouring agent in products such as ice cream, cookies, pies, puddings, candies, cola soft beverages, chewing and bubble gum, and many others [Silva et al., 2024; Hossain et al., 2021; Agustian et al., 2019]. The antibacterial properties of CO are attributed to its hydroxyl group (OH), which disrupts bacterial cell membrane proteins and inhibits protease activity. Additionally, its antioxidant activity arises from its ability to neutralize free radicals, making it highly effective in combating oxidative stress and microbial threats [Silva et al., 2024; Orlo et al., 2023; Pilong et al., 2023]. Eugenol's pharmacological qualities make it suitable for a variety

DOI: 10.1201/9781003559894-9

of uses. Eugenol is used in dentistry to eradicate bacteria and reduce discomfort during dental operations due to its antibacterial as well as anti-inflammatory properties [Vahid Reza et al., 2023].

In this context, the current chapter aims to provide an in-depth understanding of clove and its chemical components, with a focus on CO, a review of its medicinal uses, and a discussion of eugenol's toxicity and antibacterial activity.

9.2. CHEMISTRY: CHEMICAL COMPOUNDS ISOLATED FROM CLOVE

Clove's major bioactive ingredient, eugenol, has a concentration of around 0.094–0.1465g per g of fresh plant material [Cortés-Rojas et al., 2014]. Cloves are among the most important leafy sources of phenolic chemicals, including as flavonoids and hydroxyphenylpropene, hydroxycinnamic acid, and hydroxybenzoic acids [Neveu et al., 2010]. Table 9.1 depicts the various phenolic compounds isolated from clove buds and leaf.

The inclusion of phenolic components may improve clove's bioactivities [Adefegha et al., 2015; Rubió et al., 2013; Ryu et al., 2016)]. Figure 9.1 depicts the chemical structures of various compounds present in clove.

Approximately 15–20% of clove buds are made up of CO, which is mostly made up of eugenol, eugenol acetate, and β-caryophyllene. Minor constituents such as methyl amyl ketone, gallotannic acid, α-humulene, benzaldehyde, β-humulene, crategolic acid, and methyl salicylate are also present, adding to its varied chemical profile and medicinal qualities. These components are responsible for the captivating perfume of clove buds [Mittal et al., 2014: Pandey et al., 2023]. Table 9.2 indicates composition of clove buds oil CO.

According to a study, clove powder is nutrient-dense, with notable concentrations of sodium (244.45 mg/kg), calcium (5040 mg/kg), magnesium (2504 mg/kg), and phosphorus (1504.5 mg/kg). Its nutritional value is further demonstrated by the fact that it contains crude fat (5.86%), protein (6.91%), fibre (14.37%), total soluble sugars (32%), moisture (29.47%), and ash (5.29%) [Kaur et al., 2019].

9.3. MEDICINAL APPLICATION OF CLOVE

Clove is a valuable medicinal herb with several known pharmacological properties based on decades of traditional usage. An overview of research reports on the key biological functions of clove and eugenol is discussed as follows.

TABLE 9.1
Phenolic Compounds Isolated from Clove and Leaf

Sr No	Compound Extracted	Concentration	Reference
1	Eugenol	9.381–14.65g/100 g	Cortés-Rojas et al., 2014
2	Quercitrin	30.26 mg/g	Rubió et al., 2013; Ryu et al., 2016; Singh and Sukhvinder, 2024.
3	Quercetin	29.71 mg/g	
4	Kaempferol	20.47 mg/g	
5	Chlorogenic acid	20.13 mg/g	
6	Gallic acid	14.07 mg/g	
7	Caffeic acid	13.98 mg/g	
8	Luteolin	9.28 mg/g	
9	Rutin	9.11 mg/g	
10	Catechin	1.85 mg/g	
11	Ellagic acid	1.61 mg/g	

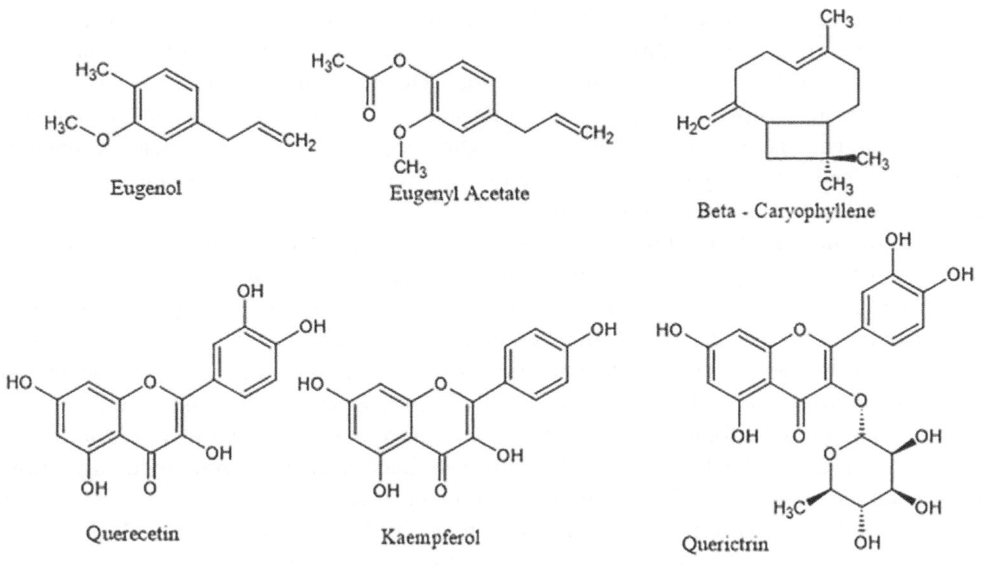

FIGURE 9.1 Some significant chemical compounds extracted from cloves.

TABLE 9.2
Composition of CO Extracted from Clove Buds

Sr No	Compound Extracted	Concentration (%)	References
1	Eugenol	70–85	Mittal et al., 2014; Shibamoto, 2007
2	Eugenol acetate	10–15	
3	β-caryophyllene	5–12	
4	Methyl amyl ketone	1–2	
5	Gallotannic acid	1–2	
6	ά-humulene	0.8–1.5	
7	Methyl salicylate	0.8–1.3	
8	β-humulene crategolic acid	0.7–1.2	
9	Benzaldehyde	0.5–1	
10	p-Cymene	0.8–1.1	
11	Nootkatin	0.9–1.2	
12	Guaiol	0.9–1	

9.3.1. Antioxidant Activity

The U.S. Department of Agriculture collaborated with universities and business organizations to build a database of polyphenol content and antioxidant activity in various food types. Using this database, Perez-Jiménez et al. found the top 100 dietary sources of polyphenols [Pérez-Jiménez et al., 2010]. Among food categories, spices contain the largest percentage of polyphenols, followed by fruits, seeds, and vegetables. The spice that has the most polyphenols and antioxidants is clove [Shan et al., 2005], using HPLC to analyze and quantify the primary phenolic components from 26 spices. The antioxidant capacity of spices was assessed in vitro using the 2,2′-azino-bis (3-ethylbenzothiazoline-6-sulphonic acid) ABTS technique. According to the study, polyphenols and antioxidant activity are strongly correlated. The highest levels of antioxidant activity and polyphenol content were found in clove buds, which had 14.380±0.006 g of gallic acid (equivalents/100g dry weight) and 168.660 ± 0.024 mmol of trolox (tetraethylammonium chloride). The main phenolic chemicals identified were gallic acid, flavonol glucosides, eugenol, acetyl eugenol, and tannins. Another study demonstrated that aqueous and ethanolic extracts of clove buds exhibit significant reducing potential, metal chelating properties, and the ability to scavenge free radicals and superoxide anion radicals, highlighting their strong antioxidant capabilities [Gulcin, 2004].

The antioxidant activity of cloves and caraway was evaluated through various in vitro antioxidant assays, demonstrating their potential as effective natural antioxidants. The results demonstrate that clove and caraway have antioxidant properties equivalent to the extensively used food preservative butylated hydroxytoluene (BHT) [Bamdad et al., 2006]. Gulcin et al. used DPPH test to evaluate CO's antioxidant activity to that of synthetic antioxidants [Gulcin et al., 2010]. The antioxidant activity decreased in descending order for CO, BHT, ά-tocopherol, BHA, and trolox, respectively. Pine and cinnamon were found to be beneficial with clove extracts tested in vitro for antioxidant properties, such as ABTS, DPPH, oxygen radical absorbance capacity, ferrous-reducing antioxidant activity, xanthine oxidase, and 2-deoxiguanosine [Dudonne et al., 2009].

CO and *Nigella sativa* oils significantly protect male rats from aflatoxins that induce hepato- and nephrotoxicity, as well as oxidative stress. When it came to biochemical measurements such as alanine aminotransferase, aspartate aminotransferase, alkaline phosphatase, total bilirubin, urea, total protein, and cholesterol, both oils performed similarly to the controls [Abdel-Wahhab et al., 2005]. Antioxidants can help ease memory problems induced by oxidative stress [Mehta et al., 2009]. CO pretreatment reduces oxidative stress (measured by malondialdehyde) and glutathione levels in mice's brains. A research found that CO can improve learning and memory impairments from scopolamine by reducing oxidative stress [Halder et al., 2010]. Rats given scopolamine at doses of 0.025, 0.05, and 0.1 mL/kg demonstrated improved memory and learning in an elevated plus maze test in contrast to a control group given saline solution. This suggests that using clove, a high-antioxidant plant, can effectively repair memory problems induced by oxidative stress.

9.3.2. Antimicrobial Activity

Clove has demonstrated antibacterial properties against many bacteria and fungus species. Clove, cinnamon, mustard, garlic, and mint were investigated for antimicrobial efficacy against some foodborne pathogens such as *Bacillus cereus*, *Staphylococcus aureus*, and *Escherichia coli*. Only 3% aqueous extracts of clove demonstrated complete bactericidal activity [Sofia et al., 2007]. Nutmeg, geranium, black pepper, thyme, oregano, and clove were tested for antibacterial activity against 25 different strains of both gram-negative and gram-positive bacteria [Idowu et al., 2021]. The oils of oregano, thyme, and clove showed a broad range of antibacterial potential [Idowu et al., 2021; Dorman and Deans, 2000].

Bay (*Pimenta racemosa*), thyme (*Thymus vulgaris*), oregano (*Origanum vulgare*), and CO were investigated for antibacterial efficacy against *E. coli* by growth inhibition [Burt and Reinders, 2003]. Formulations of eugenol and carvacrol encapsulated in a non-ionic surfactant were tested against four strains of *E. coli* and *Listeria monocytogenes*. The eugenol-based formulation, a major component of clove, demonstrated promising growth inhibition against these pathogenic bacteria [Pérez-Conesa et al., 2006]. CO, which has a high eugenol content, has demonstrated exceptional antibacterial activity against the pathogenic bacterium *Salmonella typhi*, which causes typhoid outbreaks [Devi et al., 2010]. Eugenol's antibacterial efficacy stems from its mode of action, which entails substantial changes to the bacterial cells' structural integrity. Eugenol interferes with critical bacterial functions such as protein and nucleic acid synthesis by breaking down important macromolecules. In addition, it weakens the bacterial membrane, which causes more permeability and ultimately cell lysis. CO's dual mode of action highlights its promise as a safe and efficient antibacterial agent against *Salmonella typhi* [Devi et al., 2010].

The β-cyclodextrin inclusion complex, which contained extracts of clove buds and eugenol, showed good antibacterial action against foodborne pathogens such as *Listeria innocua* and *Salmonella enterica* serovar Typhimurium LT2. Encapsulation has the potential to improve food safety applications because it not only suppressed these pathogens but also improved the transport and solubility of the active chemicals [Hill et al., 2012]. Significant antimutagenic action was shown by clove seed methanol extracts against the *Salmonella* strains TA98 and TA100 [Sultana et al., 2014]. The potential of the extracts as natural agents to stop genetic alterations was demonstrated by their effective reduction of the mutagenic effects caused by these bacteria. Clove seed extracts may be useful in preventing mutagenic and carcinogenic hazards, according to this research, providing a substitute for medical applications that prioritize genetic stability [Sultana et al., 2014].

Strong antibacterial action was shown by eugenol and its isomer isoeugenol against three strains of gram-positive bacteria: *Staphylococcus aureus* ATCC 25, 923, *Listeria monocytogenes* ATCC 19, 115, and *Bacillus subtilis* ATCC 6051 [Zhang et al., 2017]. Both substances demonstrated their promise as natural antibacterial agents by successfully preventing the growth of every tested bacterial strain. They may be useful substitutes for managing foodborne pathogens and bacterial illnesses due to their strong action [Zhang et al., 2017]. Eugenol substantially inhibited the development of *Staphylococcus aureus*, demonstrating strong antibacterial properties [Yadav et al., 2015]. The compound's potential as a natural antibacterial agent is demonstrated by its capacity to interfere with the activities of this harmful bacterium. Because *S. aureus* infections are frequently resistant to traditional antibiotics, eugenol presents a possible substitute [Yadav et al., 2015].

Moreover, CO showed antifungal property against several strains of pathogenic fungi, such as *Mucor sp., Microsporum gypseum, Fusarium monoliforme NCIM 1100, Trichophytum rubrum, Aspergillus sp.*, and *Fusarium oxysporum* MTCC 284 [Rana et al., 2011]. Further, bioactive substances derived from CO, eugenol acetate and eugenol benzoate, have shown strong antifungal effects against three harmful fungal strains: *Fusarium moniliforme*, *Rhizoctonia solani*, and *Helminthosporium oryzae* [Kaur et al., 2019]. These substances are effective against a variety of fungus species. In particular, eugenol benzoate has shown great efficacy against *Fusarium moniliforme*, *Helminthosporium oryzae*, and *Rhizoctonia solani* by effectively reducing their growth. The most effective substance for this specific strain, however, was eugenol acetate, which demonstrated greater antifungal activity against *Rhizoctonia solani*. The potential of these chemicals generated from cloves as targeted natural antifungal medicines in clinical and agricultural applications is highlighted by their selective antifungal activity [Kaur et al., 2019].

Clove and oregano essential oils have shown strong antifungal action against the food and drink-rotting yeast *Zygosaccharomyces bailii*. Both essential oils demonstrated their potential as natural preservatives by successfully preventing the growth of this fungus. Clove and oregano essential oils may be used in the food industry as substitute, plant-derived antifungal agents to increase food safety and shelf life due to their capacity to fight *Z. bailii* [Ribes et al., 2019].

Strong antimicrobial effects were demonstrated by essential oils (EOs) derived from clove (*Syzygium aromaticum*), oregano (*Origanum vulgare*), thyme (*Thymus vulgaris*), and cinnamon leaf (*Cinnamomum zeylanicum*) against plant pathogenic bacteria, including *Pseudomonas sp.* and *Pectobacterium sp.* [Bozik et al., 2017]. Various essential oils successfully stopped the spread of various plant diseases, indicating that they could be used as natural substitutes to prevent bacterial crop infections and support sustainable farming methods [Bozik et al., 2017].

A chitosan coating and CO substantially extended food goods' shelf life by inhibiting bacterial development. The mixture demonstrated strong antibacterial action against a range of bacterial species, such as hydrogen sulfide–producing bacteria, lactic acid bacteria, *Enterobacteriaceae*, and *Pseudomonas spp.* This natural coating showed promise as an environmentally friendly, antibacterial solution to prolong food freshness and improve food safety by drastically reducing activity of bacteria [Li et al., 2017].

The essential oils derived from clove, lemongrass, and anise showed promise as natural substitutes for controlling bacterial wilt in crops by successfully suppressing the growth of *R. solanacearum*.

This implies that these oils made from plants may provide a long-term way to improve farming methods and manage plant diseases [Deberdt et al., 2017].

9.3.3. Antinociceptive Activity

Since the 13th century, cloves have been used as a natural painkiller, especially to treat toothaches, joint pain, and spasms. Eugenol, the main bioactive ingredient in cloves, is primarily responsible for its efficacy. Clove has a long and significant history in traditional medicine due to eugenol's potent analgesic and antispasmodic qualities, indicating its ongoing use in therapeutic applications for pain treatment. The process is connected to ganglion cell calcium and chloride channel activation [Cortés-Rojas et al., 2014]. The observed effects are a result of this activation, which is crucial in regulating cellular responses. The physiological underpinnings of the compound's bioactivity in brain and sensory pathways are highlighted by such channel activation [Li et al., 2008]. According to a different study, clove's ability to operate as a capsaicin agonist is connected to its analgesic properties. Clove's efficacy as a natural pain reliever is highlighted by this interaction, which implies that it activates similar pain-modulating pathways, hence lowering pain perception [Ohkubo et al., 1997]. Daniel et al. documented the antinociceptive qualities of eugenol EO, emphasizing how well it works to lessen pain perception. Eugenol's potential as a natural analgesic drug is highlighted by this activity, which supports its use in pain management by interacting with important brain circuits [Daniel et al., 2009]. Asl et al. verified the analgesic qualities of CO through its aqueous and methanolic extracts, proving its efficacy in pain management and bolstering its usage as a natural pain management solution [Asl et al., 2013]. The formalin test revealed that eugenol was more effective at lowering inflammatory-type pain than reducing acute pain. Eugenol's potential as a natural analgesic agent was highlighted by the fact that its effectiveness was on par with that of the widely used pain reliever diclofenac [Lugo-Lugo et al., 2019]. As a result, clove can be used as a useful nutraceutical ingredient to reduce pain in medication and food compositions.

9.3.4. Anti-Inflammatory Activity

According to Han and Parker, CO has been demonstrated to have strong anti-inflammatory effects on human dermal fibroblasts. By decreasing the fibroblast cells' synthesis of pro-inflammatory cytokines and other inflammatory mediators, CO was shown in their study to have the capacity to regulate inflammatory pathways [Han et al., 2017]. Because of its anti-inflammatory properties, CO may help treat illnesses such as skin irritations, wound healing, and inflammatory skin disorders that include excessive inflammation. The active ingredients in the oil, especially eugenol, are thought to be mostly responsible for these effects by preventing the activation of signaling pathways and inflammatory enzymes. The study also demonstrated CO's promise as a natural treatment in dermatology, where it may help to lower inflammation and support healthy skin. This lends credence to the wider use of CO in topical and systemic anti-inflammatory therapies.

According to Maurya et al., eugenol derivatives have shown significant anti-inflammatory qualities, especially in reducing skin irritation. These compounds aid in lowering inflammatory indicators and easing skin disease symptoms [Maurya et al., 2018]. Eugenol derivatives are interesting candidates for the development of natural therapeutics targeted at controlling a variety of inflammatory skin conditions and boosting general skin health because of their capacity to influence inflammatory pathways. Also, Ibrahim et al. have demonstrated that CO exhibits both anti-inflammatory and ulcerogenic qualities. The oil may be used as a therapeutic treatment to treat inflammatory illnesses and stomach ulcers because of its bioactive components, which also aid to prevent ulcer formation and reduce inflammation. These outcomes highlight CO's medicinal benefits in treating inflammatory and gastrointestinal disorders, fostering recovery, and providing all-natural defence [Ibrahim et al., 2017]. To sum up, CO has a lot of promise for treating inflammatory and gastrointestinal disorders, accelerating recovery, and offering all-natural defence.

9.4. TOXICITY OF CLOVE

To guarantee clove spices' safety, particularly at the cellular level, it is essential to comprehend their toxicity. Even while cloves, especially their bioactive ingredient eugenol, have many health benefits, it's important to consider the possibility that they could be harmful if used inappropriately or in excess. By identifying safe dosage levels and avoiding side effects, toxicity studies guarantee that cloves can be used as a culinary or therapeutic ingredient without endangering health [Nagaraju et al., 2021].

The safety of CO, clove buds, and eugenol as dietary supplements has been verified by the Food and Drug Administration (FDA). However, recent concerns have been raised regarding their potential toxicity, prompting further scrutiny to ensure their safe usage in food and medicinal applications [Vijayasteltar et al., 2016]. Prashar et al. conducted an in vitro study to examine the cytotoxic effects of eugenol and CO on human fibroblasts and endothelial cells. Their findings showed that both CO and eugenol were safe to use and did not exhibit any cytotoxic effects on the cells they were tested on. This study suggests that these compounds are safe and can be used therapeutically without posing a risk to human cells. However, eugenol has been known to cause allergic reactions when used in dentistry [Prashar et al., 2006]. These allergic reactions emphasize the necessity of using it with caution, especially for those who are sensitive to the substance [Sarrami et al., 2002]. Eugenol was tested on six male partners in infertile couples to determine its spermicidal impact. By evaluating its capacity to suppress sperm activity, the study shed light on how successful it is as a method of birth control. The findings advance knowledge of eugenol's function in fertility control and reproductive health [Mishra et al., 2008]. To ensure safe consumption while maximizing health benefits and minimizing risks, the World Health Organization (WHO) recommends an appropriate daily intake of clove for adults at 2.5 mg per kg of body weight. This guideline helps balance the therapeutic effects with safety concerns regarding potential toxicity [Ogunwande et al., 2005]. However, more research is advised to ascertain the cytotoxic effects of clove and its constituents, guaranteeing their safety for extended usage in a range of applications.

9.5. CONCLUSION

The chemical composition and bioactivities of cloves, including as their antibacterial, antioxidant, antifungal, anti-inflammatory, and antinociceptive properties, make them highly promising for a wide range of uses. In particular, cloves' active ingredient, eugenol, is essential for lowering oxidative stress, easing pain, and preventing microbial infections. Because of these qualities, cloves are useful in medicinal and nutraceutical contexts. Cloves can also be added to food to improve health benefits and prolong shelf life. Because of its many uses in both agriculture and medicine, clove spice is a valuable resource. However, assuring the safety of clove spices, particularly at the cellular level, requires an awareness of their toxicity. The need for safe usage techniques is highlighted by the fact that, despite the many health benefits of cloves, especially eugenol, misuse or overconsumption can have negative effects.

REFERENCES

Abdel-Wahhab, M. A., and S. E. Aly. "Antioxidant Property of Nigella Sativa (Black Cumin) and Syzygium Aromaticum (Clove) in Rats during Aflatoxicosis." *Journal of Applied Toxicology* 25, no. 3 (January 1, 2005): 218–223. https://doi.org/10.1002/jat.1057.

Adefegha, Stephen A., Ganiyu Oboh, Sunday I. Oyeleye, and Kolawole Osunmo. "Alteration of Starch Hydrolyzing Enzyme Inhibitory Properties, Antioxidant Activities, and Phenolic Profile of Clove Buds (Syzygium Aromaticum L.) by Cooking Duration." *Food Science & Nutrition* 4, no. 2 (September 23, 2015): 250–260. https://doi.org/10.1002/fsn3.284.

Agustian, E., S. Tursiloadi, A. Sulaswatty, N. Rinaldi, and None Sudiyarmanto. "One-Pot Conversion and Separation of Methyl Eugenol by Vacuum Fractionation." *IOP Conference Series Materials Science and Engineering* 494 (March 29, 2019): 012056. https://doi.org/10.1088/1757-899x/494/1/012056.

Asl, Mina Kamkar, Ashraf Nazariborun, and Mahmoud Hosseini. "Analgesic Effect of the Aqueous and Ethanolic Extracts of Clove." *PubMed* (January 1, 2013). https://pubmed.ncbi.nlm.nih.gov/25050273.

Bamdad, Fatemeh, Mahdi Kadivar, and Javad Keramat. "Evaluation of Phenolic Content and Antioxidant Activity of Iranian Caraway in Comparison With Clove and BHT Using Model Systems and Vegetable Oil." *International Journal of Food Science & Technology* 41, no. s1 (August 1, 2006): 20–27. https://doi.org/10.1111/j.1365-2621.2006.01238.x.

Bendre, Ratnamala S., and Jamatsing D. Rajput. "Outlooks on Medicinal Properties of Eugenol and Its Synthetic Derivatives." *Natural Products Chemistry & Research* 4, no. 3 (January 1, 2016). https://doi.org/10.4172/2329-6836.1000212.

Bozik, Matej, Pavel Novy, and Pavel Kloucek. "Chemical Composition and Antimicrobial Activity of Cinnamon, Thyme, Oregano and Clove Essential Oils Against Plant Pathogenic Bacteria." *Acta Universitatis Agriculturae Et Silviculturae Mendelianae Brunensis* 65, no. 4 (September 1, 2017): 1129–1134. https://doi.org/10.11118/actaun201765041129.

Burt, S. A., and R. D. Reinders. "Antibacterial Activity of Selected Plant Essential Oils against Escherichia Coli O157:H7." *Letters in Applied Microbiology* 36, no. 3 (March 1, 2003): 162–67. https://doi.org/10.1046/j.1472-765x.2003.01285.x.

Chaurasia, P. K., and S. L. Bharati. "Coriander: A Holistic Outlook on Its Chemistry and Pharmacology." *Food Chemistry* 469 (2025): 142444. https://doi.org/10.1016/j.foodchem.2024.142444

Chaurasia, P. K., S. L. Bharati, and S. Singh. 2024. *The Chemistry Inside Spices and Herbs: Research and Developments* (Volume 3). Bentham Science Publishers. ISBN: 978-981-5196-80-1; https://doi.org/10.2174/97898151968011240301.

Cock, Ian Edwin, and Matthew Cheesman. *Plants of the Genus Syzygium (Myrtaceae): A Review on Ethnobotany, Medicinal Properties, and Phytochemistry*. Apple Academic Press eBooks, 75–124, 2018. https://doi.org/10.1201/b22426-20.

Cortés-Rojas, Diego Francisco, Claudia Regina Fernandes De Souza, and Wanderley Pereira Oliveira. "Clove (Syzygium Aromaticum): A Precious Spice." *Asian Pacific Journal of Tropical Biomedicine* 4, no. 2 (February 1, 2014): 90–96. https://doi.org/10.1016/s2221-1691(14)60215-x.

Daniel, Apparecido N., Simone M. Sartoretto, Gustavo Schmidt, Silvana M. Caparroz-Assef, Ciomar A. Bersani-Amado, and Roberto Kenji N. Cuman. "Anti-Inflammatory and Antinociceptive Activities a of Eugenol Essential Oil in Experimental Animal Models." *Revista Brasileira De Farmacognosia* 19, no. 1b (March 1, 2009): 212–217. https://doi.org/10.1590/s0102-695x2009000200006.

Deberdt, Péninna, Isabelle Davezies, Régine Coranson-Beaudu, and Alexandra Jestin. "Efficacy of Leaf Oil From Pimenta Racemosa Var. Racemosa in Controlling Bacterial Wilt of Tomato." *Plant Disease* 102, no. 1 (August 9, 2017): 124–131. https://doi.org/10.1094/pdis-04-17-0593-re.

Devi, K. Pandima, S. Arif Nisha, R. Sakthivel, and S. Karutha Pandian. "Eugenol (an Essential Oil of Clove) Acts as an Antibacterial Agent against Salmonella Typhi by Disrupting the Cellular Membrane." *Journal of Ethnopharmacology* 130, no. 1 (May 1, 2010): 107–115. https://doi.org/10.1016/j.jep.2010.04.025.

Dorman, H. J. D., and S. G. Deans. "Antimicrobial Agents from Plants: Antibacterial Activity of Plant Volatile Oils." *Journal of Applied Microbiology* 88, no. 2 (February 1, 2000): 308–316. https://doi.org/10.1046/j.1365-2672.2000.00969.x.

Dudonne, Stéphanie, Xavier Vitrac, Philippe Coutière, Marion Woillez, and Jean-Michel Mérillon. "Comparative Study of Antioxidant Properties and Total Phenolic Content of 30 Plant Extracts of Industrial Interest Using DPPH, ABTS, FRAP, SOD, and ORAC Assays." *Journal of Agricultural and Food Chemistry* 57, no. 5 (February 6, 2009): 1768–1774. https://doi.org/10.1021/jf803011r.

Gulcin, İlhami, Mahfuz Elmastaş, and Hassan Y. Aboul-Enein. "Antioxidant Activity of Clove Oil—a Powerful Antioxidant Source." *Arabian Journal of Chemistry* 5, no. 4 (September 20, 2010): 489–499. https://doi.org/10.1016/j.arabjc.2010.09.016.

Gulcin, W. "Comparison of Antioxidant Activity of Clove (Eugenia Caryophylata Thunb) Buds and Lavender (Lavandula Stoechas L.)." *Food Chemistry* 87, no. 3 (February 11, 2004): 393–400. https://doi.org/10.1016/j.foodchem.2003.12.008.

Halder, Sumita, Ashish Mehta, Rajarshi Kar, Mohammad Mustafa, Pramod Mediratta, and Krishna Sharma. "Clove Oil Reverses Learning and Memory Deficits in Scopolamine-Treated Mice." *Planta Medica* 77, no. 8 (December 14, 2010): 830–834. https://doi.org/10.1055/s-0030-1250605.

Han, Xuesheng, and Tory L. Parker. "Anti-Inflammatory Activity of Clove (Eugenia Caryophyllata) Essential Oil in Human Dermal Fibroblasts." *Pharmaceutical Biology* 55, no. 1 (January 1, 2017): 1619–1622. https://doi.org/10.1080/13880209.2017.1314513.

Hernández-Pérez, Talía, María Del Rocío Gómez-García, María Elena Valverde, and Octavio Paredes-López. "Capsicum Annuum (Hot Pepper): An Ancient Latin-American Crop With Outstanding Bioactive

Compounds and Nutraceutical Potential. A Review." *Comprehensive Reviews in Food Science and Food Safety* 19, no. 6 (September 27, 2020): 2972–2993. https://doi.org/10.1111/1541-4337.12634.

Hill, Laura E., Carmen Gomes, and T. Matthew Taylor. "Characterization of Beta-Cyclodextrin Inclusion Complexes Containing Essential Oils (Trans-cinnamaldehyde, Eugenol, Cinnamon Bark, and Clove Bud Extracts) for Antimicrobial Delivery Applications." *LWT* 51, no. 1 (November 23, 2012): 86–93. https://doi.org/10.1016/j.lwt.2012.11.011.

Hossain, Md Iqbal, Md Furkanur Rahaman Mizan, Sazzad Hossen Toushik, Pantu Kumar Roy, Iqbal Kabir Jahid, Si Hong Park, and Sang-Do Ha. "Antibiofilm Effect of Nisin Alone and Combined with Food-grade Oil Components (Thymol and Eugenol) against Listeria Monocytogenes Cocktail Culture on Food and Food-contact Surfaces." *Food Control* 135 (December 29, 2021): 108796. https://doi.org/10.1016/j.foodcont.2021.108796.

Ibrahim, Faten M., Hanan Naeim Attia, Yousreya Aly Aly Maklad, Kawkab A. Ahmed, and Mohamed F. Ramadan. "Biochemical Characterization, Anti-Inflammatory Properties and Ulcerogenic Traits of Some Cold-pressed Oils in Experimental Animals." *Pharmaceutical Biology* 55, no. 1 (January 1, 2017): 740–748. https://doi.org/10.1080/13880209.2016.1275705.

Idowu, Solomon, Ademola Ezekiel Adekoya, Oluwakemi Osarumwense Igiehon, and Anthony Temitope Idowu. "Clove (Syzygium Aromaticum) Spices: A Review on Their Bioactivities, Current Use, and Potential Application in Dairy Products." *Journal of Food Measurement & Characterization* 15, no. 4 (April 23, 2021): 3419–3435. https://doi.org/10.1007/s11694-021-00915-9.

Kamatou, Guy P., Ilze Vermaak, and Alvaro M. Viljoen. "Eugenol—From the Remote Maluku Islands to the International Market Place: A Review of a Remarkable and Versatile Molecule." *Molecules* 17, no. 6 (June 6, 2012): 6953–6981. https://doi.org/10.3390/molecules17066953.

Kaur, Kamalpreet, Sonia Kaushal, and Ritu Rani. "Chemical Composition, Antioxidant and Antifungal Potential of Clove (Syzygium Aromaticum) Essential Oil, Its Major Compound and Its Derivatives." *Journal of Essential Oil Bearing Plants* 22, no. 5 (September 3, 2019): 1195–1217. https://doi.org/10.1080/0972060x.2019.1688689.

Li, Hai Ying, Byung Ky Lee, Joong Soo Kim, Sung Jun Jung, and Seog Bae Oh. "Eugenol Inhibits ATP-induced P2X Currents in Trigeminal Ganglion Neurons." *Korean Journal of Physiology and Pharmacology* 12, no. 6 (January 1, 2008): 315. https://doi.org/10.4196/kjpp.2008.12.6.315.

Li, Xue-Peng, Ming-Yan Zhou, Jin-Feng Liu, Yong-Xia Xu, Hong-Bo Mi, Shu-Min Yi, Jian-Rong Li, and Hong Lin. "Shelf-Life Extension of Chilled Olive Flounder (Paralichthys Olivaceus) Using Chitosan Coatings Containing Clove Oil." *Journal of Food Processing and Preservation* 41, no. 5 (February 7, 2017): e13204. https://doi.org/10.1111/jfpp.13204.

Lugo-Lugo, Daniela Elizabeth, Amaury De Jesús Pozos-Guillén, Juan Ramón Zapata-Morales, Antonio Rodríguez-Chong, Azahel De Jesús Rangel-López, María Zenaida Saavedra-Leos, and Angel Antonio Vértiz-Hernández. "Antinociceptive Local Activity of 4-allyl-1-hydroxy-2-methoxybenzene (Eugenol) by the Formalin Test: An Anti-Inflammatory Effect." *Brazilian Journal of Pharmaceutical Sciences* 55 (January 1, 2019). https://doi.org/10.1590/s2175-97902019000118022.

Maurya, Anil Kumar, Karishma Agarwal, Amit Chand Gupta, Archana Saxena, Zulfa Nooreen, Sudeep Tandon, Ateeque Ahmad, and Dnyaneshwar Umrao Bawankule. "Synthesis of Eugenol Derivatives and Its Anti-Inflammatory Activity against Skin Inflammation." *Natural Product Research* 34, no. 2 (December 22, 2018): 251–260. https://doi.org/10.1080/14786419.2018.1528585.

Mehta, Kapil Dev, Gobind Rai Garg, Ashish K. Mehta, Tarun Arora, Amit K. Sharma, Naresh Khanna, Ashok K. Tripathi, and Krishna K. Sharma. "Reversal of Propoxur-induced Impairment of Memory and Oxidative Stress by 4′-chlorodiazepam in Rats." *Naunyn-Schmiedeberg S Archives of Pharmacology* 381, no. 1 (December 9, 2009): 1–10. https://doi.org/10.1007/s00210-009-0475-z.

Mishra, Bhuwan Bhaskar, S. P. Tripathi, and C. P. M. Tripathi. "Impact of Syzygium Aromaticum (L.) Essential Oil as Fumigant Against Tribolium Castaneum (Coleoptera: Tenebrionidae)." *Journal of Entomology and Zoology Studies* 4, no. 6 (November 1, 2016): 811–816. https://www.entomoljournal.com/archives/2016/vol4issue6/PartK/4-6-92-394.pdf.

Mishra, Raghav Kumar, and Shio Kumar Singh. "Safety Assessment of Syzygium Aromaticum Flower Bud (Clove) Extract with Respect to Testicular Function in Mice." *Food and Chemical Toxicology* 46, no. 10 (August 16, 2008): 3333–3338. https://doi.org/10.1016/j.fct.2008.08.006.

Mittal, Monika, Nomita Gupta, Palak Parashar, Varsha Mehra, and Manisha Khatri. "Phytochemical Evaluation and Pharmacological Activity of Syzygium Aromaticum: A Comprehensive Review." *International Journal of Pharmacy and Pharmaceutical Sciences* 6, no. 8 (2014): 67–72.

Nagaraju, Pramod G., Parineeta Sengupta, Poornima Priyadarshini Chicgovinda, and Pooja J. Rao. "Nanoencapsulation of Clove Oil and Study of Physicochemical Properties, Cytotoxic, Hemolytic, and

Antioxidant Activities." *Journal of Food Process Engineering* 44, no. 4 (January 13, 2021). https://doi.org/10.1111/jfpe.13645.

Neveu, V., J. Perez-Jimenez, F. Vos, V. Crespy, L. Du Chaffaut, L. Mennen, C. Knox, R. Eisner, J. Cruz, D. Wishart, and A. Scalbert. "Phenol-Explorer: An Online Comprehensive Database on Polyphenol Contents in Foods." *Database* 2010, no. 0 (January 8, 2010): bap024. https://doi.org/10.1093/database/bap024.

Ogunwande, I., N. Olawore, O. Ekundayo, T. Walker, J. Schmidt, and W. Setzer. "Studies on the Essential Oils Composition, Antibacterial and Cytotoxicity of L." *International Journal of Aromatherapy* 15, no. 3 (January 1, 2005): 147–152. https://doi.org/10.1016/j.ijat.2005.07.004.

Ohkubo, T., and M. Shibata. "The Selective Capsaicin Antagonist Capsazepine Abolishes the Antinociceptive Action of Eugenol and Guaiacol." *Journal of Dental Research* 76, no. 4 (April 1, 1997): 848–851. https://doi.org/10.1177/00220345970760040501.

Orlo, Elena, Cristina Nerín, Margherita Lavorgna, Magdalena Wrona, Chiara Russo, Mariamelia Stanzione, Roberta Nugnes, and Marina Isidori. "Antioxidant Activity of Coatings Containing Eugenol for Flexible Aluminium Foils to Preserve Food Shelf-life." *Food Packaging and Shelf Life* 39 (August 1, 2023): 101145. https://doi.org/10.1016/j.fpsl.2023.101145.

Özbek, Zeynep Aksoylu, and Pelin Günç Ergönül. "Clove (Syzygium Aromaticum) and Eugenol Toxicity." In *Elsevier eBooks*, 267–314, 2022. https://doi.org/10.1016/b978-0-323-85177-0.00029-x.

Pandey, Vinay Kumar, Shivangi Srivastava, None Ashish, Kshirod Kumar Dash, Rahul Singh, Aamir Hussain Dar, Tripti Singh, Alvina Farooqui, Ayaz Mukkaram Shaikh, and Bela Kovacs. "Bioactive Properties of Clove (Syzygium Aromaticum) Essential Oil Nanoemulsion: A Comprehensive Review." *Heliyon* 10, no. 1 (November 30, 2023): e22437. https://doi.org/10.1016/j.heliyon.2023.e22437.

Parthasarathy, V. A., B. Chempakam, and T. J. Zachariah. *Chemistry of Spices*. CABI eBooks, 2008. https://doi.org/10.1079/9781845934057.0000.

Pérez-Conesa, D., L. Mclandsborough, and J. Weiss. "Inhibition and Inactivation of Listeria Monocytogenes and Escherichia Coli O157: H7 Colony Biofilms by Micellar-Encapsulated Eugenol and Carvacrol." *Journal of Food Protection* 69, no. 12 (December 1, 2006): 2947–2954. https://doi.org/10.4315/0362-028x-69.12.2947.

Pérez-Jiménez, J., V. Neveu, F. Vos, and A. Scalbert. "Identification of the 100 Richest Dietary Sources of Polyphenols: An Application of the Phenol-Explorer Database." *European Journal of Clinical Nutrition* 64, no. S3 (November 1, 2010): S112–S120. https://doi.org/10.1038/ejcn.2010.221.

Pilong, Puncharat, Dharmendra K. Mishra, Anchalee Ruengdech, and Ubonrat Siripatrawan. "Foam-Mat Freeze-Drying Approach for Preserving Stability and Antimicrobial Activity of Clove Essential Oil Nanoemulsion." *Food Bioscience* 52 (January 7, 2023): 102369. https://doi.org/10.1016/j.fbio.2023.102369.

Poucher, W. A. "View of Palnt Used for Steam Distillation of Clove Oil." In *Springer eBooks*, 117–76, 1974. https://doi.org/10.1007/978-94-009-5691-9_2.

Prashar, A., I. C. Locke, and C. S. Evans. "Cytotoxicity of Clove (Syzygium Aromaticum) Oil and Its Major Components to Human Skin Cells." *Cell Proliferation* 39, no. 4 (July 7, 2006): 241–248. https://doi.org/10.1111/j.1365-2184.2006.00384.x.

Rajput, Jamatsing D., Suresh D. Bagul, Umesh D. Pete, Chetan M. Zade, Subhash B. Padhye, and Ratnamala S. Bendre. "Perspectives on Medicinal Properties of Natural Phenolic Monoterpenoids and Their Hybrids." *Molecular Diversity* 22, no. 1 (October 7, 2017): 225–245. https://doi.org/10.1007/s11030-017-9787-y.

Rana, Inder Singh, Aarti Singh Rana, and Ram Charan Rajak. "Evaluation of Antifungal Activity in Essential Oil of the Syzygium Aromaticum (L.) by Extraction, Purification and Analysis of Its Main Component Eugenol." *Brazilian Journal of Microbiology* 42, no. 4 (December 1, 2011): 1269–1277. https://doi.org/10.1590/s1517-83822011000400004.

Ribes, Susana, Ana Fuentes, and Jose Manuel Barat. "Effect of Oregano (Origanum Vulgare L. Ssp. Hirtum) and Clove (Eugenia Spp.) Nanoemulsions on Zygosaccharomyces Bailii Survival in Salad Dressings." *Food Chemistry* 295 (May 25, 2019): 630–636. https://doi.org/10.1016/j.foodchem.2019.05.173.

Rubió, Laura, Maria-José Motilva, and Maria-Paz Romero. "Recent Advances in Biologically Active Compounds in Herbs and Spices: A Review of the Most Effective Antioxidant and Anti-Inflammatory Active Principles." *Critical Reviews in Food Science and Nutrition* 53, no. 9 (January 1, 2013): 943–953. https://doi.org/10.1080/10408398.2011.574802.

Ryu, Byeol, Hye Mi Kim, Jin Su Lee, Chan Kyu Lee, Jurdas Sezirahiga, Jeong-Hwa Woo, Jung-Hye Choi, and Dae Sik Jang. "New Flavonol Glucuronides from the Flower Buds of Syzygium Aromaticum (Clove)." *Journal of Agricultural and Food Chemistry* 64, no. 15 (April 5, 2016): 3048–3053. https://doi.org/10.1021/acs.jafc.6b00337.

Sarrami, N., M. N. Pemberton, M. H. Thornhill, and E. D. Theaker. "Adverse Reactions Associated with the Use of Eugenol in Dentistry." *BDJ* 193, no. 5 (September 1, 2002): 257–259. https://doi.org/10.1038/sj.bdj.4801539.

Sellamuthu, R. "Eugenol." In *Elsevier eBooks*, 539–41, 2014. https://doi.org/10.1016/b978-0-12-386454-3.01125-8.

Shahidi, Fereidoon, and Priyatharini Ambigaipalan. "Phenolics and Polyphenolics in Foods, Beverages and Spices: Antioxidant Activity and Health Effects—a Review." *Journal of Functional Foods* 18 (July 21, 2015): 820–897. https://doi.org/10.1016/j.jff.2015.06.018.

Shan, Bin, Yizhong Z. Cai, Mei Sun, and Harold Corke. "Antioxidant Capacity of 26 Spice Extracts and Characterization of Their Phenolic Constituents." *Journal of Agricultural and Food Chemistry* 53, no. 20 (September 9, 2005): 7749–7759. https://doi.org/10.1021/jf051513y.

Shibamoto, Takayuki. "ChemiCal Constituents of Clove (Syzygium Aromaticum, Fam. Myrtaceae) and Their Antioxidant aCtivity." *Revista* 180 (January 1, 2007). https://www.academia.edu/62122291/ChemiCal_Constituents_of_Clove_Syzygium_aromaticum_fam_Myrtaceae_and_their_antioxidant_aCtivity.

Silva, Marina Vinhas, Alice Da Conceição Alves De Lima, Marina Gomes Silva, Viviane Fonseca Caetano, Michelle Felix De Andrade, Rafael Gleymir Casanova Da Silva, Luiz Emílio Pessôa Timeni De Moraes Filho, Ivo Diego De Lima Silva, and Glória Maria Vinhas. "Clove Essential Oil and Eugenol: A Review of Their Significance and Uses." *Food Bioscience* (September 1, 2024): 105112. https://doi.org/10.1016/j.fbio.2024.105112.

Singh, Preet Amol, and Sukhvinder Singh Purewal. *Compendium of Clove*. CRC Press eBooks, 2024. https://doi.org/10.1201/9781003521785.

Sofia, Papachan Karur, Rajendra Prasad, Virendra Kumar Vijay, and Ashok Kumar Srivastava. "Evaluation of Antibacterial Activity of Indian Spices against Common Foodborne Pathogens." *International Journal of Food Science & Technology* 42, no. 8 (May 10, 2007): 910–915. https://doi.org/10.1111/j.1365-2621.2006.01308.x.

Sultana, Bushra, Farooq Anwar, Muhammad Mushtaq, Maryam Aslam, and Sidra Ijaz. "In Vitro Antimutagenic, Antioxidant Activities and Total Phenolics of Clove (Syzygium Aromaticum L.) Seed Extracts." *PubMed* 27, no. 4 (July 1, 2014): 893–899. https://pubmed.ncbi.nlm.nih.gov/25015457.

Tiwari, Prashant, Richa Verma, Dheeraj Ahirwar, Anish Chandy, and Shubhangi Dwivedi. "Evaluation of Anxiolytic Effect of Syzygium Aromaticum: A Traditional Herb of India." *Asian Pacific Journal of Tropical Disease* 4 (January 1, 2014): S77–S80. https://doi.org/10.1016/s2222-1808(14)60418–7.

Vahid Reza, Askari, Zohre Najafi, and Vafa Baradaran Rahimi. "Syzygium Aromaticum – Role in Oral Health and Dental Care." *Pharmacological Studies in Natural Oral Care* (September 15, 2023): 499–518. https://doi.org/10.1002/9781394167197.ch26.

Vijayasteltar, Liju, Gopakumar Gopinathan Nair, Balu Maliakel, Ramadasan Kuttan, and Krishnakumar IM. "Safety Assessment of a Standardized Polyphenolic Extract of Clove Buds: Subchronic Toxicity and Mutagenicity Studies." *Toxicology Reports* 3 (January 1, 2016): 439–449. https://doi.org/10.1016/j.toxrep.2016.04.001.

Yadav, M., S. Kumari, Nagraj, N. Sharma, S. L. Bharati, S. Singh, J. Kumar, and P. K. Chaurasia. "Hydrogen Peroxide Scavenging by Methanolic Extracts of Coriander: An *In Vitro* Antioxidant Study." *Pharmaceutical Drug Regulatory Affair Journal* 8, no. 1 (2025): 000162. https://doi.org/10.23880/pdraj-16000162.

Yadav, Mukesh Kumar, Sung-Won Chae, Gi Jung Im, Jae-Woo Chung, and Jae-Jun Song. "Eugenol: A Phyto-Compound Effective against Methicillin-Resistant and Methicillin-Sensitive Staphylococcus Aureus Clinical Strain Biofilms." *PLoS One* 10, no. 3 (March 17, 2015): e0119564. https://doi.org/10.1371/journal.pone.0119564.

Zhang, Liang-Liang, Li-Fang Zhang, Jian-Guo Xu, and Qing-Ping Hu. "Comparison Study on Antioxidant, DNA Damage Protective and Antibacterial Activities of Eugenol and Isoeugenol Against Several Foodborne Pathogens." *Food & Nutrition Research* 61, no. 1 (January 1, 2017): 1353356. https://doi.org/10.1080/16546628.2017.1353356.

10 Phytochemistry, Biological Activities, and Therapeutic Potential of *Elettaria cardamomum* and Its Medicinal Perspectives

Rima Kumari, Pankaj Kumar Chaurasia, Shweta Yadav, and Nuzhat Parween

10.1. INTRODUCTION

Spices are herbal products, commonly used as kitchen ingredients for imparting aroma, flavour, pungency, taste, and colour to foods and as food seasonings. Around the world, more than 70 types of spices are used, and India is one of the largest producers and consumers of different varieties of spices since ancient times until today, therefore known as the "home of spices". Spices are known for their rich chemistry and pharmacological properties (Chaurasia and Bharati, 2022; Chaurasia and Bharati, 2025; Yadav et al., 2025).

Cardamom is an aromatic seed spice, belonging to plants of two different genera, i.e., *Elettaria cardamomum* and *Amomum subulatum*, in the family Zingiberaceae. Cardamom is popularly called *elaichi* in Hindi and *yelakki/Elakka* in South Indian languages. *Elettaria cardamomum* (L.) (green cardamom/small cardamom/true cardamom) has high-value use in kitchen cuisines due to its pleasant aroma and flavour and is revered as the "queen of spices". It is native to the southwestern Indian peninsular region, Nepal, Sri Lanka, Laos, Mexico, Thailand, Laos, Vietnam, Guatemala, and Tanzania. India contributes a major share of cardamom production (>4000 metric tons), then Nepal (2500 MT) and Bhutan (1000 MT), respectively. Along with its prime use as a kitchen ingredient, cardamom pods/seeds also have high therapeutic significance. They are a rich source of several medicinal active compounds: flavonoids, phenolics, alkaloids, tannins, terpinoles, phytosterols, and essential oils.

The dominant phytoconstituents in cardamom seed extracts and essential oil are 1,8-cineole, α-terpinyl acetate, linalool, limonene, terpinen-4-ol, α-pinene, and nerolidol (AshokKumar et al., 2021a, 2021b). Attributed to these active metabolites, cardamom has well-recognized antioxidant, antimicrobial, anti-inflammatory, antiproliferative and pro-apoptotic activities underlying anticancer properties. It is also effective for antidepressant, antianxiety, analgesic, anticonvulsant, and spasmolytic activities (Yahyazadeh et al., 2021; Abdullah et al., 2022). Moreover, in the pharmaceutical field, cardamom capsules and seed oil have been used for decades to treat pulmonary congestion, cardiac problems, dental diseases, and digestive and kidney disorders. It has been recommended for remedy of tuberculosis and urinary problems as it acts as a powerful suppresser of macrophages.

This chapter aims to highlight the phytochemistry of cardamom's medicinal active metabolites and their health benefits. The relevant information on phytochemical composition, the biological properties of *E. cardamomum* leaf extract, capsule, and cardamom seed essential oil (CEO),

DOI: 10.1201/9781003559894-10

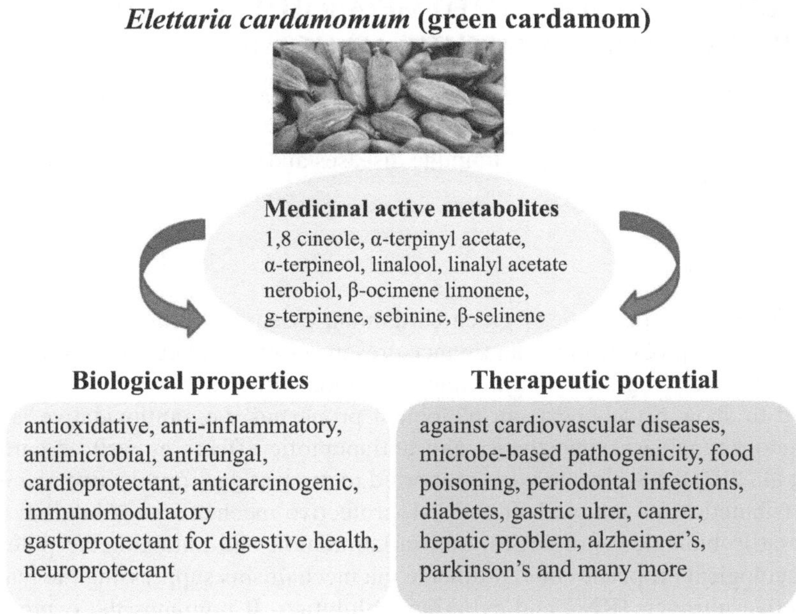

FIGURE 10.1 A conceptual diagram of cardamom illustrating its metabolites profiling along with its biological and therapeutic potential.

and their potential pharmacological aspects are summarized in this chapter (Figure 10.1). Further investigation also seeks to understand the mode and biological potential of health-sensory bioactive constituents. This study helps in understanding the therapeutic importance of phytochemical-rich spices in modern medical research practices.

10.2. PHYTOCHEMICAL PROFILING OF *E. CARDAMOMUM*

E. cardamomum is a rich source of medicinal bioactive metabolites, including terpenoids and its derivatives, aldehyde, alcohol, esters, nitrogen and sulfur-containing alkaloids compounds, phenylpropanoids compounds, coumerin, and many other active ingredients (Abdullah et al., 2022). The predominant active metabolites in cardamom capsules and seed essential oil are 1,8-cineole, terpinyl acetates, terpineol, nerolidol, linalool, linalyl acetate, α-pinene, limonene, sabinene, and α-thuzone, with trace amounts of camphor, borneol, squalene, stigmast-5-en-3-ol, 2,3-dihydro5, 7-dihydroxy-2-pheny, hexadecanoic acid, octadecanoic acid, and phytol (Askokumar et al., 2021a, 2021b). Delgadillo-Puga et al. (2023) from Mexico also conducted a detailed analysis of green cardamom seed phenolic and terpenoid components by GC/LC MS techniques. In phenolics, the main components identified are syringic acid, protocatechuic acid, and caffeic acid, along with some new components, i.e., quinic acid and its derivatives, rutin, and feruloyl. In terpenoids, 16 groups are identified, i.e., costunolide, ambrial, cis-α-terpineol, isopathulenol, α-acorenol, costunolidecubebol, globulol, eicosane, erucylamide, trans-farnesol, and a few more. In the Indian variety of *E. cardamomum*, 1,8-cineole and α-terpinyl acetate (> 30%) are the dominant components followed by linalool, linalyl acetate, β-pinene 4-terpineol, nerolidol (>10%), β-ocimene, limonene, g-terpinenesebinine, cresol, α-pinene, β-selinene (1–10%), and many more in trace amounts (Ashokkumar et al., 2021a, 2021b) (Table 10.1; Table 10.2). However, variation occurs in the phytochemical composition of *E. cardamomum* that depends upon various factors, such as varieties, chemotypes, season of sample collection, climate, geographical conditions, plant developmental stage of sampling, extraction techniques, and plant parts.

10.3. BIOACTIVITY OF SPECIALIZED METABOLITES OF GREEN CARDAMOM FOR THERAPEUTIC APPLICATION

These predominant bioactive constituents have several pharmacological properties, such as antioxidative, antimicrobial, anti-inflammatory, antifungal, anti-insecticidal, cardioprotective, and gastroprotective, used for effective therapy of immune diseases and neurological disorders (Alzheimer's) (Abdullah et al., 2022).

10.3.1. 1,8-Cineole

1,8-cineole, the prime constituent of green cardamom essential oil, is an oxygenated monoterpene termed as eucalyptol due to its dominance presence (>90%) in eucalyptus essential oil. The distinctive flavour and fragrance of cardamom essential oil is attributed to 1,8-cineole. It has been reported to show broad-spectrum biological properties, i.e., antioxidative, antimicrobial, anti-inflammatory, analgesic, anesthetic, and antiapoptotic effects as well as cardioprotective effect (Cai et al., 2021). 1,8-cineole is also reported to possess gastroprotective and ulcer-healing properties attributed to its antioxidant and cytoprotective mechanisms (Abdullah et al., 2022). Of these versatile biological properties, anti-inflammation and antioxidative potential are the main pharmacological properties of 1, 8-cineole via mechanisms suppressing the reactive oxygen radicals, reactive nitrogen (RN), and cytokines inhibition. It regulates the expression of Nrf2 and NF-kB, the important nuclear transcription factors involved in inflammations (Heydarian et al., 2024). Antibacterial activity of 1,8-cineole was clinically approved by Wang et al. (2022) against *Escherichia coli* pathogenicity via inhibiting biofilm formation and suppressing luxS gene expression. Cai et al. (2021) reported the protective effects of 1,8-cineole in hypertension patients via lowering the blood pressure level through improvement of cardiac systolic function. Attributed to these versatile biological properties, 1,8-cineole is clinically proven and applied in treatment of various health issues, i.e., chronic pulmonary diseases, cardiovascular diseases, digestive problems, renal issues, Alzheimer's disease, cancer, and neuropathic pain (Cai et al., 2021; Hoch et al., 2023). However, there are some limitations of 1,8-cineole, as it sometime suppresses the level of angiotensin, an important factor contributing to regulation of cardiovascular diseases and high blood pressure.

10.3.2. α-Terpinyl Acetate and α-Terpineol

α-Terpinyl acetate (α-TA; P-menth-1-en-8-yl acetate) is a secondary metabolite of menthane group monoterpenoids, which are found in most aromatic plants and have distinctive fragrance. α-TA is reported as the most dominant component (47.5%) in Indian green cardamom essential oil in a few studies (Alam et al., 2021), while in a few other studies it is reported as the second most important contributing component, approximately 30–60% of total after 1,8-cineole (Askokumar et al. 2021a, 2021b). Literatures have shown the multi-targeted biological potential of α-terpinyl acetate as an antioxidative, anti-inflammatory, antimicrobial, anticholinesterase, antiamyloidogenic, and for neuroprotective response (Alam et al., 2021; Chowdhury and Kumar, 2020). Attributed to the major component α-TA, cardamom is clinically approved as effective in Alzheimer's disease therapy (Chowdhury and Kumar, 2020).

α-Terpeneol, a monoterpenoid alcohol, is also one of the major components of essential oils and leaf extracts of cardamom and has high-value medicinal significance due to its broad-spectrum biological potential. α-T showed potential antioxidant, antiproliferative, anticonvulsant, sedative, antinociceptive, insecticidal, and anti-inflammatory activity, and induced hypertension and vasorelaxation, suggesting its wide use in treatment of cancer, ulcer, cardiovascular and neurological disorders, skin problems, and pathogenicity (Khaleel et al., 2018). Alpha-terpineol also induced neurogenesis and long-term memory retention via suppressing ROS factors (SOD, MDA) and reducing amyloid plaque counts. Alpha-terpineol was experimentally proven to be significantly effective in treatment of neurodegenerative Alzheimer's disease (Alipour et al., 2022).

10.3.3. Linalool and Linalyl Acetate

Linalool (3,7 dimethyl-1,6-octadien-3-ol) and linalyl acetate are also present in cardamom essential oils, contributing approximately 1–10% of the total essential oil. Linalool possesses a wide range of therapeutic properties, i.e., antibacterial, anti-inflammatory, antihyperalgesic, antinociceptive, anti-insecticidal, and antiplasmodial effects (van Zyl et al., 2006; Kamatou and Viljoen, 2008). Several experimental facts point to the sedative, anticonvulsant, anxiolytic, neuroendocrine activities, and neurotrophic factors, such as brain-derived neurotrophic factor (BDNF), resulting in its effectiveness on the etiology of depression (Dos Santos et al., 2022). In the same way, linalyl acetate also exhibits antioxidative, anti-inflammatory, analgesic, antispasmodic, antihepatic, and neuroprotective effects and is effective against mild cognitive impairment (hypertension and/or diabetes mellitus) (Shin and Seol, 2023).

10.3.4. Nerolidol

Nerolidol (3,7, 11-trimethyl-1,6,10-dodecatrien-3-ol) is a natural sesquiterpenol contributing to floral fragrance in aromatic plants. It is a minor phytocomponent found in green cardamom seed and essential oil but possesses high therapeutic importance. Nerolidol has been reported to have high therapeutic value, including antioxidant, anti-inflammatory, antimicrobial, antibiofilm, cardioprotective, antiparasitic, antinociceptive, and anticancer properties (De Carvalho et al., 2018; Meeran et al., 2021). The therapeutic potential of nerolidol was also reported to be effective against Alzheimer's and epilepsy due to its anticholinesterasic ability, which stimulates nerve cell connection (De Carvalho et al. (2018).

10.3.5. α-Pinene

α-Pinene ($C_{10}H_{16}$) is a bicyclic, double-bond, terpenoid hydrocarbon commonly reported in pine essential oils, and it is also one of the key components of cardamom essential oil and seed extract. It has a pleasant, fruity, pine-like fresh smell and is an important flavouring agent. α-Pinene is only a minor share of cardamom essential oil percentage (up to 1.0–2.5%), but it is a key quality biomarker of Indian cardamom. However, Alanazi et al. (2022) reported α-pinene (17.7%) as predominant in *E. cardamomum* essential oil, along with major components 1,8-cineole and α-terpinyl acetate. So it showed variation in percentage occurrence as reported in different literatures. In modern medicines, α-pinene has displayed broad-spectrum therapeutic potential, e.g., antioxidative, anti-inflammatory, analgesic, gastroprotective, neuroprotective, anticoagulative, antiplatelet, antinociceptive, antimicrobial, antimalarial, and antileishmania effects (Salehi et al., 2019; Kharazmkia et al., 2022). α-Pinene is also reported to have strong therapeutic potency against upper respiratory tract infection (URI) (Salehi et al., 2019). The anticancer effect of α-pinene was examined by Hou et al. (2019) against human ovarian cancer cells, and they revealed that it induced cytoxicity in tumour cell via promoting apoptosis, with a remarkable increase in caspase-3 enzyme activity and DNA fragmentation.

10.4. THERAPEUTIC INTERVENTIONS OF GREEN CARDAMOM IN MANAGING HUMAN HEALTH DISORDERS

10.4.1. Antioxidative Advantage

Lifestyle and age-related chronic disease, i.e., cardiovascular diseases, hyperlipidemia, diabetes, cancer, immune deficiency, neurodegenerative diseases (e.g., Alzheimer's disease, arthritis, and Parkinson's disease) are mostly conferred due to oxidative stress induced by cellular imbalance

TABLE 10.1
Major Chemical Composition of *Elettaria Cardamomum* Essential Oil and Leaf Extract and Related Therapeutic Activities

S.No.	Bioactive Metabolites	Therapeutic Values
1.	1,8-Cineole***	Antioxidative, anti-inflammatory, analgesic, and hepatoprotective properties
2.	*a*-Terpinyl acetate***	Antioxidative, anticholinesterase, antiamyloidogenic, and neuroprotective potential
3.	Linalool*	Antioxidants, antimicrobial, antitumour, nerve stimulants, anxiolytic, cardioprotective, lung-protective, kidney-protective properties
4.	Linalyl acetate*	Anti-inflammatory agents
5.	α-Pinene*	Antioxidative, cytoprotective, neuroprotective, gastroprotective and antiulcerogenic properties
6.	α-Terpineol*	Antioxidative, antimicrobial, antifungal, insecticidal, anticancer, anticonvulsant, antiulcer, antihypertensive, antinociceptive activity
7.	Nerolidol*	Antioxidant, anti-inflammatory, antinociceptive, anticholinesterasic, neuroprotective, anxiolytic activities
8.	β-Pinene	Antimicrobial, anti-inflammatory, antibiotic, antitumour, anticancer, apoptotic
9.	β-Ocimene	Antibacterial, antifungal, antileishmanial activity
10.	Limonene	Antioxidant, antimicrobial, antibiofilm, anti-inflammatory, antivirus, anticancer
11.	*g*-Terpinene	Antioxidant, antimicrobial, antifungal properties, anti-inflammatory, antinociception activity
12.	Sabinene	Antioxidant and anti-inflammatory activities
13.	p-Cresol	Antioxidant, antimicrobial, genotoxic, and cytotoxic against cancer cells
In Leaf Extracts		
1.	Alpha. -ionone	Antifungal activity, chemopreventive, anticancer
2.	Eucalyptol* (1,8-Cineole)	Antioxidant, antibacterial, anti-inflammatory, antitumourigenic, antinociceptive
3.	Trans-calamenene	Antioxidant, antimicrobial
4.	Isospathulenol	Insecticidal, anticancer, antimalarial, activities
5.	1,6-Octadiene-3-ol	Antibacterial, antimicrobial
6.	Cinnamaldehyde	Antioxidant, antimicrobial, antifungal activity, anticancer
7.	Terpinen-4-ol	Antioxidant, antimicrobial, antifungal, anti-inflammatory, anticarcinogenic, antibiofilm
8.	1,6,10 Dodecatrien-3-ol 3,7,11-trimethyl-	Antioxidant, antibacterial, anticancer activity, antiulcer, antileishmanial

*** Represent dominant compounds (mostly > 25%); * major compounds (mostly in range 1–10%); others present in trace amount (mostly less than 1%)

between metabolism of reactive oxygen and antioxidants. Bioactive compounds of cardamom have the potency to decrease the level of cellular oxidative stress and induce the endogenous antioxidants. Various studies well confirmed that cardamom-derived phenolics and terpenoids offer the cell immunity and protection against oxidative stress factors in the human body, via ameliorating ROS, nitric oxide, peroxides, malondialdehyde, and inducing endogenous antioxidant production (Abdullah et al., 2022; Ballester et al., 2023). Alam et al. (2021) performed the in-silico molecular docking of major bioactive components of green cardamom essential oil and investigated that antioxidant receptors, human peroxiredoxin-5, and protein tyrosine kinases may be triggered by the essential oil

TABLE 10.2
Chemical Structure and Molecular Formula of Different Bioactive Metabolites in *Elettaria cardamomum*

	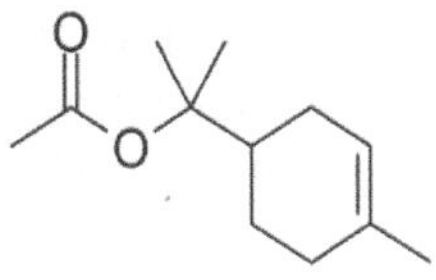	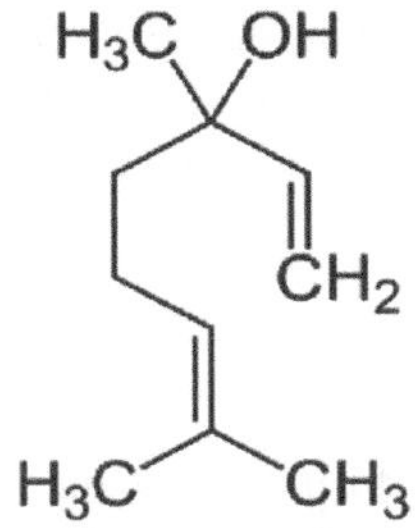	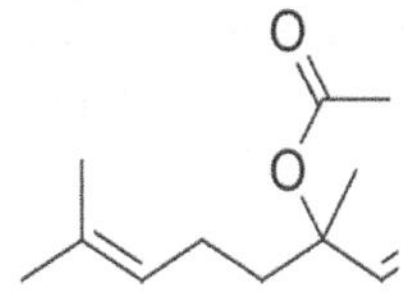	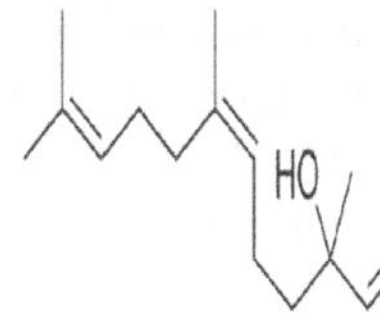
1,8-Cineole ($C_{10}H_{18}O$)	*a*-Terpinyl acetate $C_{12}H_{20}O_2$	β-Linalool $C_{10}H_{18}O$.	Linalyl acetate $C_{12}H_{20}O_2$	Nerolidol $C_{15}H_{26}O$
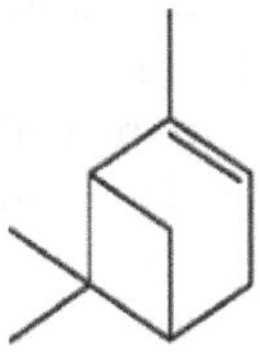	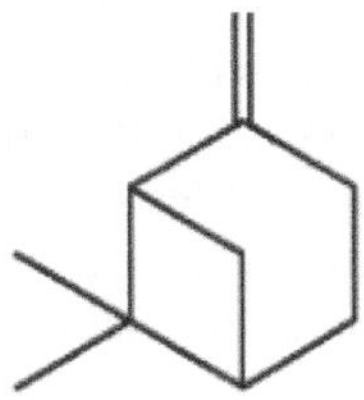	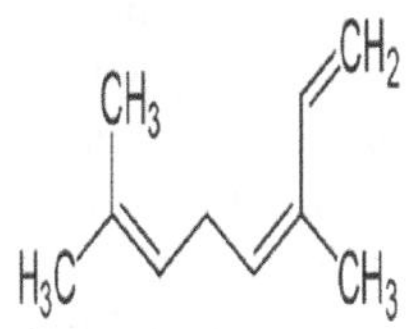	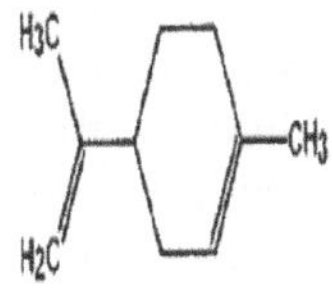	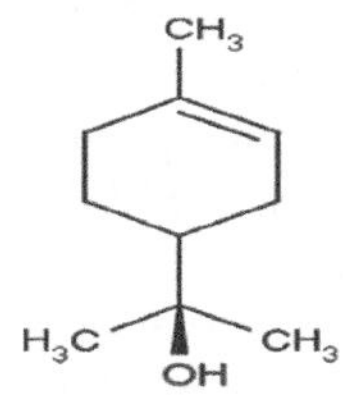
α-Pinene $C_{10}H_{16}$	β-Pinene $C_{10}H_{16}$	β-Ocimene $C_{10}H_{16}$	Limonene $C_{10}H_{16}$	α-Terpineol $C_{10}H_{18}O$.
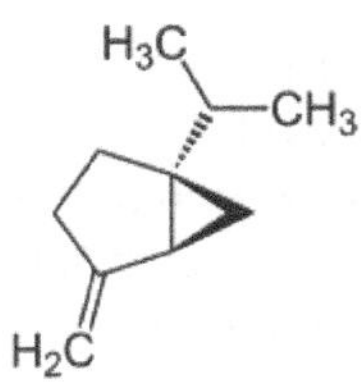	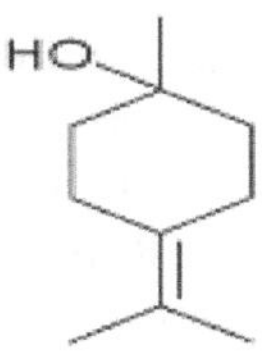	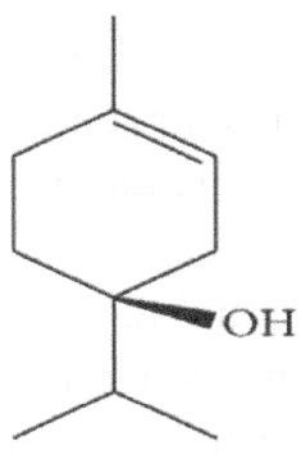	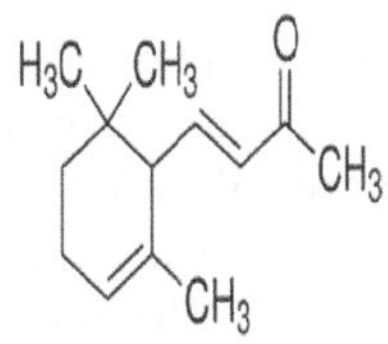	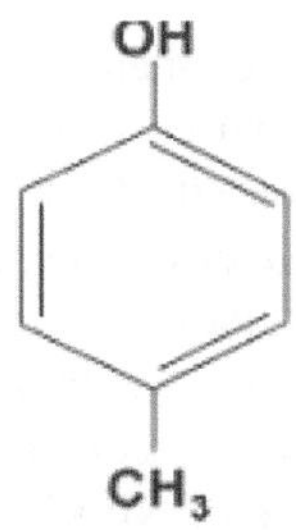
Sabinene $C_{10}H_{16}$	g-Terpineol $C_{10}H_{18}O$	Terpinen-4-ol $C_{10}H_{18}O$	Alpha-ionone $C_{13}H_{20}O$	p-Cresol C_7H_8O
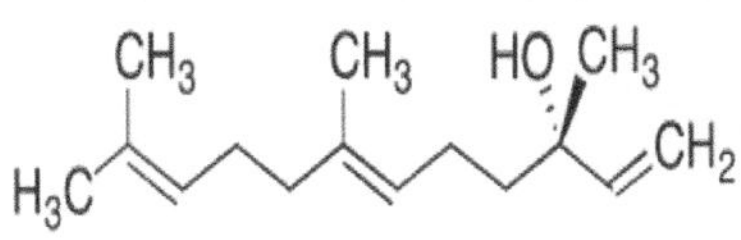		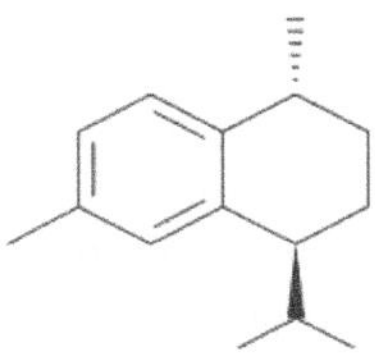	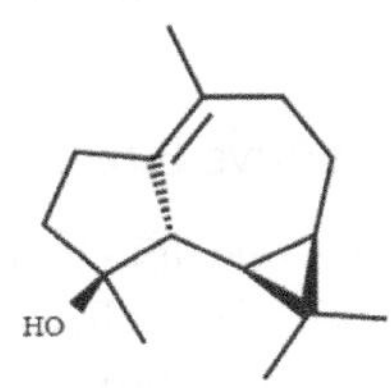	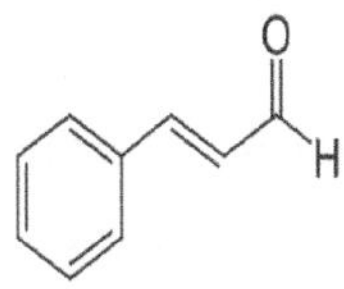
1,6,10-Dodecatrien-3-ol, 3,7,11-trimethyl-, $C_{15}H_{26}O$		Trans-calamenene $C_{15}H_{22}$	Isospathulenol $C_{15}H_{24}O$	Cinnamaldehyde C_9H_8O

components. Along with high-significance antioxidant potency, multifaceted biological properties such as anti-inflammatory, antimicrobial, antihypertensive, anticancer, antitumor, antiarthritic, anti-neurodepressant, antidiabetic, cardioprotective, gastroprotective, antithrombotic, antispasmodic, and antiparasitic properties were also observed (Yahyazadeh et al., 2021; Heydarian et al., 2024).

10.4.2. In Managing Metabolic Disorders

Metabolic disorders are lifestyle-related, health-risking diseases, exerting various medical conditions (e.g., higher cholesterol level, hypertension, insulin resistance, hyperglycemia, dyslipidemia, abdominal obesity, and inflammation) that increase the risk of cardiovascular disorders, stroke, and diabetes mellitus in the human body. Scientific researchers clinically investigated the significant beneficial effect of cardamom supplementation on patients with metabolic syndrome (Yahyazadeh et al., 2021). Herein we discuss in brief the effect of *E. cardamomum* on different metabolic syndromes.

10.4.3. Antihyperlipidemic Effects

Hyperlipidemia is increased deposition of lipids, fats, such as cholesterol, and triglycerides in blood, which in severity may lead to fatty acid deposition in arteries and an increased risk of cardiac blockages. Various clinical studies have established the potential of cardamom seeds and oil in restoring the alteration in lipid homeostasis in hypercholesterolemic patients (Nagashree et al., 2017; Yahyazadeh et al., 2021). Cardamom may exhibit its antihyperlipidemic effects by activating cholesterol degradative peroxidative processes, lipoprotein lipase activity, as well as lowering of lipid absorption by intestinal cell line (Cho, 2012). Researchers analyzed that the major component 1,8-cineole has high potential to protect the human lipoproteins and lessen lipid accumulation in THP-1 cells. The findings of various studies explained the antihyperlipidemic property of green cardamom via regulating lipid metabolism, lowering lipid accumulation, fatty acid β-oxidation, and lowering of triglycerides (Delgadillo-Puga et al., 2023).

10.4.4. Antidiabetic Effect

Diabetes mellitus is a chronic disease characterized by hyperglycemic condition of the blood due to impaired insulin secretion and/or insulin resistance by pancreatic β-cells. High oxidative stress (increased ROS) is a critical factor behind the incompetency of pancreatic β-cells for secretion of insulin effectively and increasing resistance (Aghasi et al., 2019). Due to its medicinal active metabolites, cardamom exhibits its therapeutic potential against diabetes via stimulating antioxidant conditioning, reducing blood glucose, increasing insulin level, and limiting its resistance. Controlled insulin secretion occurs due to an increase of sirutine T1 (SIRT1) level, PGC-1α, and attenuated functioning of NF-κB; these factors help in improving glucose homeostasis by enhancing insulin sensitivity in adipose tissues and muscles (Yahyazadeh et al., 2021; Azgomi et al., 2024). The therapeutic significance of cardamom as preventative care or in curing patients of diabetes mellitus is clinically validated by various studies (Aghasi et al., 2019; Zahedi et al., 2020; Yahyazadeh et al., 2021).

10.4.5. Antihypertensive Effect

Hypertension is a serious health condition in which blood vessels have consistently high pressure (140/90 mmHg or higher). Controlled high pressure can increase the risk of coronary heart diseases, strokes, poorer mental health, chronic kidney diseases, and other diseases. Cardamom can effectively reduce blood pressure in hypertension patients and improve heart health via regulating NO_2 production, cardiac systolic function, and through a vasorelaxant effect (Cai et al., 2021; Kanthlal et al., 2020). The diuretic properties of cardamom may complement the BP lowering effect in hypertension patients as it increases the frequency and volume of urine secretion accompanied with removal of sodium and potassium ions, causing a lowering of BP. Vasorelaxation of blood cells is another reason for hypertension patients (Gilani et al., 2008). A clinical trial by Mohammed et al. (2020) demonstrated that 3 g/d intake of cardamom for three months of intervention lead to a significant decrease in systolic and diastolic blood pressure (SBP and DBP) in stage 1 (nondiabetic) hypertension patients.

Zahedi et al. (2020) also reported the effectiveness of cardamom in type 2 diabetic patients via lowering of SBP, DBP, NO_2, and ADMA, lessening the reason of morbidity.

10.4.6. Cardioprotective Effect

Cardiovascular disorders (CVDs) encompass broad-spectrum health conditions affecting the fitness of heart and blood vessels, causing serious breathing problems and mortality. CVDs can occur due to the influence of metabolic imbalance, creating high lipid/cholesterol deposition in the coronary arteries, hindering the circulation of oxygen and blood in heart. The specialized bioactive metabolites (i.e., phenolics, flavonoids, and sterols) in cardamom have the potential to mitigate the risk of cardiovascular health problems via lowering blood lipids, protecting against myocardial infarction due to its antiradical, antioxidative, and anti-inflammation activities (Abdullah et al., 2022; Zhang et al., 2024). Hypocholesterolemic effects of cardamom metabolites (significant decrease of LDL cholesterol level, lipoproteins, triglycerides) also contributed to its cardioprotective properties (Nagashree et al., 2017).

10.4.7. Anti-Inflammatory and Immune-Modulatory Advantage of Green Cardamom

Inflammation is a complex immune-pathological response to cellular injury mediated by activated leucocytes, lymphocytes, plasma cells, and macrophages. Although acute inflammation is an adaptive immune response of injured tissue to retrieve it, the persistent chronic inflammation increases the development of several chronic diseases. During the inflammation, macrophages produce high levels of pro-inflammatory cytokines, such as tumour necrosis factor-α (TNF-α), high-sensitivity C-reactive protein (hs-CRP), interleukins (ILs), and cyclooxygenase-1, 2 (COX-1, 2) in host cells (Karunakaran and Sadanandan, 2019). Cardamom offers anti-inflammatory effects against inflammatory disease pathogenesis by inhibiting the pro-inflammatory cytokinines, reducing ROS factors, NO synthesis, and downregulating the inflammatory pathways, such as COX-2, NF-κB, TNF-α, hs-CRP, ILs-6, and NO synthetase (NOS) (Heydarian et al., 2024). Inflammatory response is associated with therapeutic significance in the remedy of various metabolic syndromes, immunodeficiency disorders, and prevention of post-infectious morbidities (Cárdenaset al., 2021; Heydarian et al., 2024; Mohemmad et al., 2024).

10.4.8. Neuroprotective Effect Against Degenerative Alzheimer's Disease

Alzheimer's disease is an age-related progressive neurodegenerative disorder and is a form of dementia in which brain cells and connections degenerate with age, eventually destroying memory, thinking ability, behavioural sense, and other important mental functions. Oxy-radical-induced oxidative damage, loss of cell viability, and oligomeric Aβ42 peptide formation are linked with onset of Alzheimer's disease (Paul et al., 2020). Green cardamom can act as a neuroprotective agent and helps in remedy of Alzheimer's disease due to its high antioxidative and anti-inflammatory compounds (Hussein et al., 2023; Ashour et al., 2024). Disruption in cholinergic neurotransmission in the brain strongly relates with severity of neuropathological disorders associated with Alzheimer's disease. Hussein et al. (2023) explained the neuroprotective effect of crude green cardamom oil on the cholinergic system by upregulation of acetylcholine (Ach) and by reducing inflammatory cytokines. In a recent study, Pavarino et al. (2023) also evaluated the inhibitory activity of cardamom essential oil against both enzyme acetylcholinesterase (AChE) and butyrylcholinesterase (BChE) of acetylcholine regulation.

10.4.9. In Cancer Management

Cancer is one of the most disastrous diseases causing human mortality. Phytotherapy (natural treatment with plant products) is considered as one of the most preferred strategies for cancer prevention or to supplement chemotherapy as phytometabolites show antineoplastic potency (Talib et al., 2022). In cardamom, a different metabolite group (1,8-cineol, alpha-terpenyl acetates, sterols) acts synergistically for cure of cancer (Qiblawi et al., 2020). The potential mechanistic strategy behind the anticancer property of cardamom extract is its antioxidative and anti-inflammatory mechanism to enhance the activity of natural killer cells (NK cells) that suppress the infected tumour cells via inducing apoptosis (Almeer et al., 2021). Banerjee et al. (1994) primarily experimented against chemical-induced carcinogenesis and reported that supplementation of cardamom oil (10 μL/d for 2 weeks) significantly decreased liver cytochrome P450 (CYP) content and increased the glutathione S-transferase (GST) activity and sulfhydryl levels. This activity detoxifies endogenous xenobiotic compounds and may therefore have proven beneficial to inhibit cancer. Das et al. (2012) reported the antioxidative effects of green cardamom against nonmelanoma skin cancer by regulating the Nrf2 and NF-κB signalling pathways. Elguindy et al. (2016) examined that cardamom extract potentially acts in preventing hepatocellular carcinoma through attenuating ROS, suppressing the factor/enzyme associated with tumour cell growth and proliferation (inflammatory cytokine, NF-κB, ornithine decarboxylase). Cardamom is also reported to have a cytotoxic effect on head and neck squamous cell carcinoma, through induction of apoptosis via upregulating pro-apoptotic proteins (caspase-3 and BAX protein) (Zaki et al., 2021). In a recent in vitro study, Alqahtani et al. (2024) examined that cardamom essential extract exhibited a modest level of anticancer activity in liver and breast cancer cells (HEPG-2, MCF7).

10.4.10. Antibacterial, Antifungal, and Antiparasitic Response

Cardamom essential oil exhibits active potential against broad-spectrum bacteria, including *Escherichia coli, Staphylococcus aureus, S. mutans, Salmonella typhimurium, Bacillus cereus, Pseudomonas aeruginosa, Listeria monocytogenes*, and fungus (*Candida albicans*), and could be useful in the development of new antibiotics for foodborne diseases, skin infections, and many more pathogenic diseases (Abdullah et al., 2022; Alqahtani et al., 2024). The therapeutic use of cardamom was also noticed against periodontal infections. Souissi et al. (2020) examined its antibacterial potency against periodontal microbes: *Porphyromonas gingivalis, Fusobacterium nucleatum, Prevotellain termedia*, and *Aggregatibacter actinomycetemcomitans*. Almohammed et al. (2022) examined the antiparasitic effect of *Elettaria cardamomum* essential oil (CEO) against *Echinococcus granulosus* (parasitic tapeworm) and its use as a protoscolicidal agent during hydatid cyst surgery.

10.4.11. Gastroprotective Effect

Cardamom has reported remarkable gastroprotective potential against gastric ulcer and gastrointestinal problems, as approved through various scientific studies. This activity is attributed to its major component 1,8-cineole (Abdullah et al., 2022; Mohemmad et al., 2024). Jamal et al. (2006) reported the gastroprotective effect of cardamom fruit extracts in the case of aspirin-induced gastric ulcer, which is associated with its healing effect on mucosal integrity. Due to its alkaline nature, cardamom fruit extract helps in preventing gastric juice PH acid reflux, which reduced the risk of gastric ulcer formation. Alam et al. (2021) states the antidiarrheal and antispasmodic activity of cardamom and revealed that it is directly/indirectly associated with its antimicrobial properties against pathogens (*P. aeruginosa* and *E. coli*) involved in gastrointestinal diseases.

10.5. CONCLUSIONS AND RECOMMENDATIONS

Green cardamom, due to having rich sources of medicinally active metabolites, mainly 1,8-cineole and a-terpinyl acetate, has shown its broad-spectrum effective role in treatments of various oxidative stress–related metabolic syndromes (e.g., cardiac problems, diabetes, Alzheimer's disease, inflammations, foodborne pathogenicity, and gastrointestinal diseases). This has led to the popularization of cardamom fruit/seed and essential oil components as a preventive measure or cure for many metabolic diseases, inflammatory disorders, and pathogenic diseases.

10.6. FUTURE PERSPECTIVES

A future perspective is concerned with cardomom's fabrication to optimize utilization at a mass scale, dose standardization, preferred physical form of supplementation, and suitable solvent source, which should be under consideration of investigation to obtain maximum health benefits. Potential risks associated with any metabolites should also be taken into consideration thorough scientific investigation.

10.7. ACKNOWLEDGMENTS

Authors acknowledge the Principal, Langat Singh College, and MMDM college, Muzaffaepur (constituent colleges of BRA Bihar University, Muzaffarpur, Bihar, India) for academic support.

REFERENCES

Abdullah, A. N., Tian, W., Zengliu, S., Zou, Y., Farooq, S., Huang, Q., Xiao, J. 2022. Recent advances in the extraction, chemical composition, therapeutic potential, and delivery of cardamom phytochemicals. *Frontiers in Nutrition* 9: 1024820.

Aghasi, M., Koohdani, F., Qorbani, M., Nasli-Esfahani, E., Ghazi-Zahedi, S., Khoshamal, H., Keshavarz, A., Sotoudeh, G. 2019. Beneficial effects of green cardamom on serum SIRT1, glycemic indices and triglyceride levels in patients with type 2 diabetes mellitus: A randomized double-blind placebo controlled clinical trial. *Journal of the Science of Food and Agriculture* 99(8): 3933–3940.

Alam, A., Jawaid, T., Alam, P. 2021. In vitro antioxidant and anti-inflammatory activities of green cardamom essential oil and in silico molecular docking of its major bioactives. *Journal of Taibah University for Science* 15(1): 757–768.

Alanazi, A.D., Ben Said, M., Shater, A.F., Al-Sabi, M.N.S. 2022. Acaricidal, larvacidal, and repellent activity of *Elettaria Cardamomum* essential oil against *Hyalomma anatolicum* ticks infesting Saudi Arabian cattle. *Plants* 11: 1221.

Alipour, H.-R., Yaghmaei, P., Ahmadian, S., Ghobeh, M., Ebrahim-Habibi, A. 2022. A study on alpha-terpineol in Alzheimer's disease with the use of rodent in vivo model, restraint stress effect and in vitro Amyloid beta fibrils. *Brazilian Journal of Pharmaceutical Sciences* 58: e19090.

Almeer, R.S., Alnasser, M., Aljarba, N., AlBasher, G.I. 2021. Effects of Green cardamom (Elettaria cardamomum Maton) and its combination with cyclophosphamide on Ehrlich solid tumors. *BMC Complementary Medicine and Therapies* 21: 133.

Almohammed, H.I., Alkhaibari, A.M., Alanazi, A.D. 2022. Antiparasitic effects of Elettariacardamomum L. essential oil and its main compounds, 1–8 Cineole alone and in combination with albendazole against *Echinococcus granulosus* protoscoleces. *Saudi Journal of Biological Sciences* 29(4): 2811–2818.

Alqahtani, T.M., Ebrahim, E.G.R., Aloufi, A.S., El-Saadany, S.S., Almanzalawi, E.A., Awad, A., Eldeeb, G.S., El-Maghraby, L.M. 2024. Antibacterial and anticancer activities of cardamom volatile oil and its application in food covering. *Italian Journal of Food Science*. https://doi.org/10.15586/ijfs.v36i3.2534

Ashokkumar, A., Murugan, M., Dhanya, M.K., Pandian, A., Muthukrishnan, S., Vellaikumar, S. 2021a. Phytochemical analysis and pharmacological potential of cardamom. In *Cardamom [Elettaria Cardamomum (L.) Maton]: Production, breeding, management, phytochemistry and health benefits*. Eds K. Ashokkumar, M. Murugan and M.K. Dhanya, pp. 207–226. Nova Science Publishers.

Ashokkumar, K., Vellaikumar, S., Murugan, M., Dhanya, M.K., Ariharasutharsan, G., Aiswarya, S., Akilan, M., Warkentin, T.D., Karthikeyan, A. 2021b. Essential oil profile diversity in cardamom accessions from southern India. *Frontiers in Sustainable Food Systems* 29(5): 137.

Ashour, M., Elias, T., Rasheed, W., El-Toukhya, S., Youness, E., Sleem, A., Ramadan, M.M., Mahmoud, K., Yassen, N., Abdel-Monem, M., Aly, O. 2024. Neuroprotective effect of cardamom essential oil encapsulated nanoparticles and swimming in aluminum induced Alzheimer's disease in rats. *Egyptian Journal of Chemistry* 67(13): 37–52.

Azgomi, R.N.D., Karimi, A., Jazani, A.M. 2024. The favorable impacts of cardamom on related complications of diabetes: A comprehensive literature systematic review. *Diabetes & Metabolic Syndrome* 18(2): 102947.

Ballester, P., Cerdá, B., Arcusa, R., García-Muñoz, A.M., Marhuenda, J., Zafrilla, P. 2023. Antioxidant activity in extracts from *Zingiberaceae* family: Cardamom, turmeric, and ginger. *Molecules* 28(10): 4024.

Banerjee, S., Sharma, R., Kale, R.K., Rao, A.R. 1994. Influence of certain essential oils on carcinogen-metabolizing enzymes and acid-soluble sulfahydryls in mouse liver. *Nutrition and Cancer* 21: 263–269.

Cai, Z.-M., Peng, J.-Q., Chen, Y., Tao, L., Zhang, Y.-Y., Fu, L.-Y., Long, Q.D., Shen, X.-C. 2021. 1,8-cineole: A review of source, biological activities, and application. *Journal of Asian Natural Products Research* 23(10): 938–954.

Cárdenas, G.G.R., Elizondo, L.J.H., Bazaldúa, R.A.F., Chávez, M.A., Pérez, H.R.A., Martínez, D.A.J., López, V.S.M., Rodríguez, R.J., Sánchez, C.R.M., Castillo, V.U., Rodríguez, L.O.E. 2021. Benefits of cardamom (*Elettaria cardamomum* (l.) maton) and turmeric (*curcuma longa* l.) extracts for their applications as natural anti-inflammatory *Adjuvants. Plants (Basel)* 10(9): 1908.

Chaurasia, P.K., Bharati, S.L. 2022. *The Chemistry Inside Spices and Herbs: Research and Developments* (Volume 2), ISBN: 978-1-68108-949-2, Bentham Science Publishers. https://doi.org/10.2174/97816810894921220201

Chaurasia, P.K., Bharati, S.L. 2025. Coriander: A holistic outlook on its chemistry and pharmacology. *Food Chemistry*, 469: 142444. https://doi.org/10.1016/j.foodchem.2024.142444

Cho, K.-H. 2012. 1, 8-cineole protected human lipoproteins from modification by oxidation and glycation and exhibited serum lipid-lowering and anti-inflammatory activity in zebrafish. *BMB Reports* 45: 565–570.

Chowdhury, S., Kumar, S. 2020. Alpha-terpinyl acetate: A natural monoterpenoid from *Elettariacardamomum* as multi-target directed ligand in Alzheimers disease. *Journal of Functional Foods* 68: 103892.

Das, I., Acharya, A., Berry, D.L., Sen, S., Williams, E., Permaul, E., Sengupta, A., Bhattacharya, S., Saha, T. 2012.Antioxidative effects of the spice cardamom against non-melanoma skin cancer by modulating nuclear factor erythroid-2-related factor 2 and NF-κB signalling pathways. *British Journal of Nutrition* 108(6): 984–997.

De Carvalho, R.B.F., De Almeida, A.A.C., Campelo, N.B., Lellis, D.R.O.D., Nunes, L.C.C. 2018. Nerolidol and its pharmacological application in treating neurodegenerative diseases: A Review. *Recent Patents on Biotechnology* 12: 158–168.

Delgadillo-Puga, C., Torre-Villalvazo, I., Cariño-Cervantes, Y.Y., García-Luna, C., Soberanes-Chávez, P., de Gortari, P., Noriega, L.G., Bautista, C.J., Cisneros-Zevallos, L. 2023. Cardamom (*Elettaria Cardamomum* (L.) Maton) seeds intake increases energy expenditure and reduces fat mass in mice by modulating neural circuits that regulate adipose tissue lipolysis and mitochondrial oxidative metabolism in liver and skeletal muscle. *International Journal of Molecular Sciences* 24(4): 3909.

Dos Santos, É.R.Q., Maia, J.G.S., Fontes-Júnior, E.A., do Socorro, Maia, F.C. 2022. Linalool as a therapeutic and medicinal tool in depression treatment: A Review. *Current Neuropharmacology* 20(6): 1073–1092.

Elguindy, N.M., Yacout, G.A., El Azab, E.F., Maghraby, H.K. 2016. Chemoprotective effect of *Elettaria cardamomum* against chemically induced hepatocellular carcinoma in rats by inhibiting nf-κb, oxidative stress, and activity of ornithine decarboxylase. *South African Journal of Botany* 105: 251–258.

Gilani, A.H., Jabeen, Q., Khan, A.-U., Shah, A.J. 2008. Gut modulatory, blood pressure lowering, diuretic and sedative activities of cardamom. *Journal of Ethnopharmacology* 115(3): 463–472.

Heydarian, A., Tahvilian, N., Shahinfar, H., Abbas-Hashemi, S.A., Daryabeygi-Khotbehsara, R., Aryaeian, N. 2024. Effect of cardamom consumption on inflammation and blood pressure in adults: A systematic review and meta-analysis of randomized clinical trials. *Food Science & Nutrition* 12: 3–12.

Hoch, C.C., Petry, J., Griesbaum, L., Weiser, T., Werner, K., Ploch, M., Verschoor, A., Multhoff, G., DezfouliB.A., Wollenberg, B. 2023. 1,8-cineole (eucalyptol): A versatile phytochemical with therapeutic applications across multiple diseases. *Biomedicine & Pharmacotherapy* 167: 115467.

Hou, J., Zhang, Y., Zhu, Y., Zhou, B., Ren, C., Liang, S., Guo, Y. 2019. α-Pinene induces apoptotic cell death via caspase activation in human ovarian cancer cells. *Medical Science Monitor* 25: 6631–6638.

Hussein, J., Ashoura, M., Elias, T., El-Waseef, M., Ashraf, R., Ramadan, M, Mahmoud, K., Aly, O.A. 2023. Promising approach for the treatment of experimental Alzheimer's disease: impact of cardamom essential oil. *Egyptian Journal of Chemistry* 66(11): 385–391.

Jamal, A., Javed, K., Aslam, M., Jafri, M.A. 2006. Gastroprotective effect of cardamom, *Elettaria cardamomum* Maton. fruits in rats. *Journal of Ethnopharmacology* 103(2): 149–153.

Kamatou, P.P., Viljoen, A.M. 2008.Linalool—a review of a biologically active compound of commercial importance. *Natural Product Communications* 3(7): 1183–1192.

Kanthlal, S., Joseph, J., Paul, B. 2020. Antioxidant and vasorelaxant effects of aqueous extract of large cardamom in L-NAME induced hypertensive rats. *Clinical and Experimental Hypertension* 42(7): 581–589.

Karunakaran, R., Sadanandan, P. 2019. *Zingiber officinale*: Antiinflammatory actions and potential usage for arthritic conditions. In *Bioactive food as dietary interventions for arthritis and related inflammatory diseases*. Eds R. Watson & V. Preedy, (2nd ed.), Chapter 13, pp. 233–244, Academic Press, Elsevier Inc.

Khaleel, C., Tabanca, N., Buchbauer, G. 2018. α-Terpineol, a natural monoterpene: A review of its biological properties. *Open Chemistry* 16(1): 349–361.

Kharazmkia, A., Al-Abodi, H.R., Yadegari, J.G., Vahidi, A., Mahmoudvand, H. 2022. Potential effects of alpha-pinene, a monoterpene commonly found in essential oils against *Toxoplasma gondii* infection; an in vitro and in vivo study. *Journal of Parasitic Diseases* 46(4): 1055–1061.

Meeran, M.F.N, Azimullah, S., Mamoudh, H.H., Sharma, C., Kumar, S., Goyal, S.N., Ojha, S. 2021. Nerolidol, a sesquiterpene from the essential oils of aromatic plants, attenuates doxorubicin-induced chronic cardiotoxicity in rats. *Journal of Agricultural and Food Chemistry* 69: 7334–7343.

Mohammed, M.K., Mohammed, L., Mohammed, H.M. 2020. Cardamomas a bloodpressureloweringnatural food supplement inpatientswithgrade one hypertension. *Systematic Reviews in Pharmacy* 11(11): 387–390.

Mohemmad, R.N., Ahmed, B.S., Ahmed, V.A., Ismail, S.M., Rahman, H.S. 2024. Anti-inflammatory, antioxidant, and gastro-protective actions of cardamom essential oil-loaded nanostructured lipid carrier in ethanol-induced gastric ulcer. *Natural Product Communications* 19(8). https://doi.org/10.1177/1934578X241272492.

Nagashree, S., Archana, K.K., Srinivas, P., Srinivasan, K., Sowbhagya, H.B. 2017. Antihypercholesterolemic influence of the spice cardamom (*Elettaria cardamomum*) in experimental rats. *Journal of the Science of Food and Agriculture* 97: 3204–3210.

Paul, A., Ganguly, U., Chakrabarti, S., Bhattacharjee, P. 2020. Is 1,8—Cineole—rich--extract- of small cardamom seeds more effective in preventing -Alzheimer's- disease than- 1-,8—cineole-alone?- *Neuro Molecular Medicine* 22(1): 150–158.

Pavarino, M., Marengo, A., Cagliero, C., Bicchi, C., Rubiolo, P., Sgorbini, B. 2023 *Elettaria cardamomum* (L.) Maton essential oil: An interesting source of bioactive specialized metabolites as inhibitors of acetylcholinesterase and butyrylcholinesterase. *Plants* 12: 3463.

Qiblawi, S., Kausar, M.A., Shahid, S.M.A., Saeed, M., Alazzeh, A.Y. 2020. Therapeutic interventions of cardamom in cancer and other human diseases. *Journal of Pharmaceutical Research International* 32(22): 74–84.

Salehi, B., Upadhyay, S., Orhan, I.E., Jugran, A.K., Jayaweera, L.S.D., Dias, D.A., Sharopov, F., Taheri, Y., Martins, N., Baghalpour, N., Cho, W.C., Sharifi-Rad, J. 2019. Therapeutic potential of α- and β-pinene: A miracle gift of nature. *Biomolecules* 29(11): 738.

Shin, Y.K., Seol, G.H. 2023. Effects of linalyl acetate on oxidative stress, inflammation and endothelial dysfunction: Can linalyl acetate prevent mild cognitive impairment? *Frontiers in Pharmacology* 14: 1233977.

Souissi, M., Azelmat, J., Chaieb, K., Grenier, D. 2020. Antibacterial and anti-inflammatory activities of cardamom (*Elettaria cardamomum*) extracts: Potential therapeutic benefits for periodontal infections. *Anaerobe* 61: 102089.

Talib, W.H., Daoud, S., Mahmod, A.I., Hamed, R.A., Awajan, D., Abuarab, S.F., Odeh, L.H., Khater, S., Al Kury, L.T. 2022. Plants as a source of anticancer agents: From bench to bedside. *Molecules* 27(15): 4818.

van Zyl, R.L., Seatlholo, S.T., van Vuuren, S.F., Viljoen, A.M. 2006. The biological activities of 20 nature identical essential oil. *Journal of Essential Oil Research* 18: 129–133.

Wang, Y., Zhang, Y., Song, X., Fang, C., Xing, R., Liu, L., Zhao, X., Zou, Y., Li, L., Jia, R., Ye, G., Shi, F., Zhou, X., Zhang, Y., Wan, H., Wei, Q., Yin, Z. 2022. 1,8-Cineole inhibits biofilm formation and bacterial pathogenicity by suppressing luxS gene expression in *Escherichia coli*. *Frontiers in Pharmacology* 13: 988245.

Yadav, M., Kumari, S., Nagraj, Sharma, N., Bharati, S.L., Singh, S., Kumar, J., Chaurasia, P.K. 2025. Hydrogen Peroxide scavenging by methanolic extracts of coriander: An *In Vitro* antioxidant study. *Pharmaceutical Drug Regulatory Affair Journal* 8(1): 000162. https://doi.org/10.23880/pdraj-16000162

Yahyazadeh, R., Rahbardar, M.G., Razavi, B.M., Karimi, G., Hosseinzadeh, H. 2021. The effect of *Elettaria cardamomum* (cardamom) on the metabolic syndrome: Narrative review. *Iranian Journal of Basic Medical Sciences* 24(11): 1462–1469.

Zahedi, S.G., Koohdani, F., Qorbani, M., Siassi, F., Keshavarz, A., Nasli-Esfahani, E., Aghasi, M., Khoshamal, H., Sotoudeh, G. 2020. The effects of green cardamom supplementation on blood pressure and endothelium function in type 2 diabetic patients: A study protocol for a randomized controlled clinical trial. *Medicine* 99(18): e11005.

Zaki, M., Farag, H., El Wahed, S.A. 2021. Anticancer effect of cardamom extract on hep-2 cell line. *International Dental Journal* 71(2): S43–S44.

Zhang, X., Wang, C., Min, R., Zhou, Q., Yue, Qi., Fan, J. 2024. Cardamom consumption may improve cardiovascular metabolic biomarkers in adults: A systematic review and meta-analysis of randomized controlled trials. *Nutrition Research* 125: 101–111.

11 A Comprehensive Insight on Chemistry and Medicinal Values of Fenugreek for Human Health

Abhratanu Ganguly, Sayantani Nanda, Nurunnisa Khatun, Kanchana Das, Pritam Maitra, Gopal Biswas, Moutushi Mandi, and Prem Rajak

11.1. INTRODUCTION

Over the course of several centuries, herbal medicines have been an essential component of human healthcare. Several civilizations have utilized the products derived from plants in order to prevent and treat varieties of diseases. Fenugreek, also known as *Trigonella foenumgraecum L.*, is a plant that stands out among the large array of therapeutic herbs because of its adaptability, nutritional value, and medicinal characteristics (Wang et al., 2023). Fenugreek, which is indigenous to the Indian subcontinent and the Mediterranean region, has been utilized for thousands of years in the practice of traditional medicine, as well as in the preparation of culinary items. The Ebers Papyrus and Charaka Samhita, which date back to around 1550 BCE and 400 CE, respectively, both include references to fenugreek spice (Jhajhria and Kumar, 2016). Fenugreek has been used as a flavouring element for sauces, spice mixes, soups, sweets, and teas since ancient times (Faisal et al., 2024). In addition to their remarkable organoleptic properties, fenugreek seeds have high nutritional and functional values because they are rich in proteins and nonstarch carbohydrates, including galactomannans, a soluble dietary fibre also known as 'fenugreek gum', as well as essential nutrients and a plethora of other health-promoting phytochemicals (Khorshidian et al., 2016). Fenugreek biopolymers are emerging as well as excellent additives in food processing, capable of stabilizing, texturizing, and thickening dairy and nondairy foods (Dhull et al., 2023). Fenugreek seeds and its derived substances (e.g., flour, gum, and husk) might also increase the nutritional and functional qualities of flour used to make bread, pasta, and other basic meals (Acharya et al., 2006). A different line of innovation, paralleling the increased interest in fenugreek-based functional foods, components, and nutraceuticals, is the development of dietary supplements based on fenugreek bioactive extracts. Advances in the functional characterization of fenugreek gum pave the way for new food and health benefits (Dhull et al., 2023).

To enhance general health, vigor, and longevity, fenugreek is regarded as a 'rasayana' plant in traditional Indian medicine. This means that it is thought to have these specific properties. Fenugreek has been used to cure a variety of health diseases, including diabetes, digestive disorders, and respiratory issues. Its seeds, leaves, and roots have already proven its efficacy in treating these health ailments (Tak et al., 2024). It is also regarded for its culinary and nutritional features, since it is a rich source of protein, fibre, and important minerals. A little over 30% of the seeds of fenugreek are made up of mucilage, which is composed of galactomannans, xylose, and arabinose. In addition, the seeds contain a high concentration of saponins, notably diosgenin, which has been shown to

DOI: 10.1201/9781003559894-11

have anti-inflammatory and antioxidant properties (Ahmad et al., 2016). Flavonoids, which include quercetin and kaempferol, are found in the leaves of fenugreek. Flavonoids have qualities that are effective in reducing inflammation and antioxidants. Trigonelline, which has been shown to have hypoglycemic and neuroprotective properties, is one of the alkaloids that may be found in the roots of fenugreek. Throughout history, fenugreek has been utilized as a treatment for a wide range of medical ailments, including diabetes, hyperlipidemia, and digestive difficulties (Syed et al., 2020). The hypoglycemic effects of fenugreek have been attributed to its ability to slow gastric emptying, decrease glucose absorption, and elevated insulin sensitivity. It has also been claimed that fenugreek possesses anti-inflammatory and antioxidant properties, which may be a contributing factor to the possible advantages it offers in lowering the risk of chronic illnesses such as cancer, neurological disorders, and cardiovascular disease. There is evidence that fenugreek can protect the cardiovascular system by lowering blood pressure and cholesterol levels, among other beneficial effects (Shabil et al., 2023). A moderate amount of consumption is typically regarded to be safe for its intake. However, excessive amounts of fenugreek have the potential to have adverse effects on the gastrointestinal tract, including nausea, vomiting, and diarrhea. Fenugreek is often used to treat menstrual cramps, sexual issues, enlarged prostate, and a variety of other ailments, although the majority of these applications lack scientific basis (Haxhiraj et al., 2024). Additionally, there is a possibility that fenugreek might interfere with some medicines, such as cholesterol-lowering drugs, diabetic medications, and blood pressure medications. Fenugreek is a rich source of bioactive components, such as alkaloids, flavonoids, saponins, and fibre, all of which contribute to the wide range of therapeutic characteristics that it possesses. Because of its hypoglycemic, anti-inflammatory, and antioxidant properties, fenugreek has the potential to be used as an adjuvant therapy for the management of a variety of health issues, such as diabetes, cardiovascular disease, and neurodegenerative illnesses (Wani and Kumar, 2018).

This chapter aims to provide a comprehensive evaluation of chemistry and medicinal qualities of fenugreek, highlighting its potential health benefits for humans.

11.2. CHEMICAL COMPOSITION

Medicinal and aromatic plants have been utilized for several years in human nutrition as spices and as medicinal supplements in animals to enhance dietary energy utilization, improve performance efficiency, and serve as a novel protein source. Fenugreek (*Trigonella foenumgraecum L.*) is recognized for its hyperthyroid effects (Tahiliani and Kar, 2003a), its efficacy against thyroxine-induced hyperglycemia (Tahiliani and Kar, 2003b), its anticancer abilities (Devasena and Menon, 2003), as well as its antioxidant properties (Raskin et al., 2002). Fenugreek comprises several chemical components, including steroidal sapogenins. The oily embryo of fenugreek contains diosgenin. Two furastanol glycosides, F-ring opening precursors of diosgenin, have been identified in fenugreek, also known as hederagin glycosides. Alkaloids include nicotinic acid, trimethyl coumarin, trigocoumarin, and trigonelline, which are found in the stem. The mucilage is a prominent component of the seeds (Khare, 2011). The stem contains around 28 % mucilage, a volatile oil, 5 % of a potent-smelling, two alkaloids (trigonelline and choline), 22 % proteins, bitter fixed oil, and a yellow pigment (Grieve, 2013). The leaves comprise seven saponins, referred to as graecunins. These substances are glycosides derived from diosgenin. Each 100 g of fenugreek leaves comprises moisture (86 %), fat (1 %), fibre (1 %), protein (4.4 %), magnesium (67 mg), calcium (395 mg), iron (16.5 mg), phosphorus (51 mg), potassium (31 mg), copper (260 mg), sodium (76 mg), chlorine (165 mg), carotene (2.3 mg), selenium (167 mg), thiamine (40 mg), nicotinic acid (800 mg), vitamin C (52 mg), and riboflavin (310 mg), along with trace amounts of vitamin K (Laila et al., 2022). The seeds of fenugreek contain around 0.1–0.9 % diosgenin and are commercially extracted. Fenugreek seeds contain several coumarin compounds and several alkaloids, including trigonelline, gentianine, and carpaine. The primary bioactive constituents in fenugreek seeds are thought to be polyphenolic chemicals, including rhaponticin and isovitexin (He et al., 2015). A low quantity of volatile oils and fixed oil has been identified in fenugreek seeds (Sowmya and Rajyalakshmi, 1999). Fenugreek seeds are aromatic, slightly bitter, galactogogue, carminative, and antibacterial, and can be consumed either raw or cooked. The chemical composition of fenugreek, including seeds, husk, and cotyledons, indicated that the endosperm contained the highest levels of saponins (4.63 g/100 g) and protein (43.8 g/100 g). In contrast, husk possesses a greater number of polyphenols.

Macro- and micronutrients significantly influence the enzymatic functions of plants, thereby affecting their growth, yield, and quality (Aishwath et al., 2018). Each 100 g of fenugreek seeds comprises a fibre content consisting of both insoluble (30 %) and soluble fractions (20 %), typically galactomannan. Additionally, it contains proteins (20–30 %), carbohydrates (45–60 %), and lipids (5–10 %), primarily composed of neutral lipids, phospholipids (450 mg/100 g), and specifically triglycerides (6.3 %) (Acharya et al., 2006). It comprises pyridine alkaloids, flavonoids, amino acids, iron, calcium, saponins (0.6–1.7 %), and glycosides. Upon hydrolysis, it releases steroidal sapogenins, cholesterol, sitosterol, vitamins A, B1, and C, nicotinic acids, and volatile oils (0.015 %) (Singh et al., 2013). Fenugreek seed is a substantial source of several essential amino acids, including leucine, tyrosine, aspartic acid, glutamic acid, and phenylalanine (Syed et al., 2020) (Figure 11.1).

11.3. IMPACTS ON HUMAN HEALTH

11.3.1. Role as Antioxidants

Different parts of fenugreek plant have been used for several decades as traditional food and medicine, which belongs to the family of leguminoceae. Fenugrek seeds are used to add flavour to food in most parts of India and also in Egypt and Yemen. Approximately 80 % of global fenugreek seed production is contributed by India. (Pandey and Awasthi, 2015). It is popular for its antioxidant properties (Hussain et al., 2016). Daily consumption of natural antioxidants may decrease the risk of cardiovascular disease, cancer, and ulcer in the GI tract. Earlier studies revealed that fenugreek has potential pharmacological properties to treat tumour, viral, and several microbial infections and cellular inflammations (Mashkor, 2014). Moreover, the seeds of fenugreek

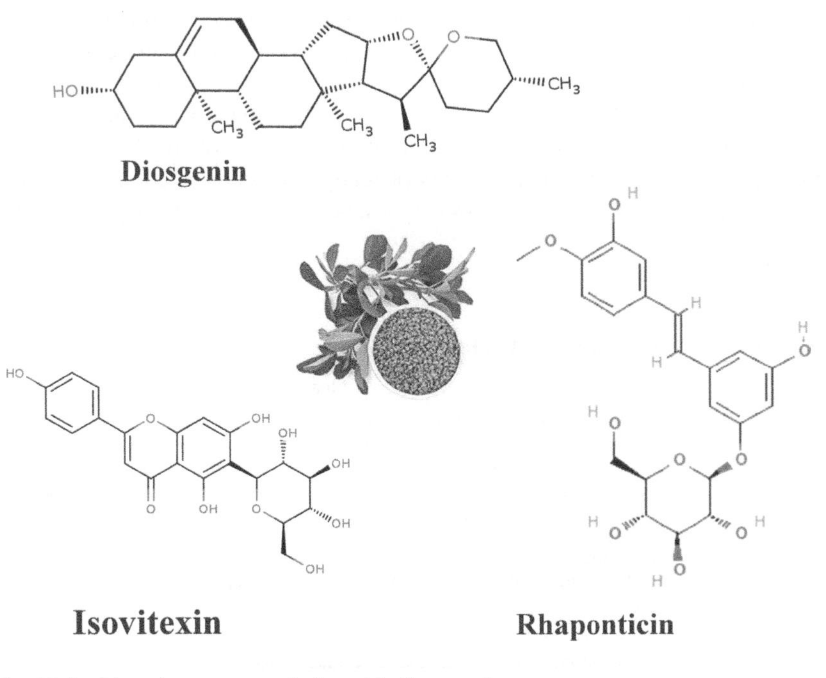

FIGURE 11.1 Major bioactive compounds found in fenugreek.

have also been reported to have strong free radical–scavenging activity. Free radicals are generally produced in the human body as a by-product of lipid, protein, carbohydrate, and DNA metabolism.

Many pollutants and hazardous exposure in the environment switch on the endogenous antioxidant enzyme pathways of the body. However, these free radicals generate reactive oxygen species (ROS), which further inhibit cellular metabolism by interfering with proteins and enzymes. It was well established that fenugreek seed significantly boosts antioxidant enzymes such as superoxide dismutase, catalase, glutathione peroxidase, aspartate transaminase, lactate dehydrogenase, alanine transaminase, etc. (Jhajhria and Kumar, 2016). Antioxidants protect against ROS-induced damage by attacking free radicals and breaking them down into water and molecular oxygen. Fenugreek seeds were reported to contain two components: trigonelline and diosgenin. However, it was also observed in a previous experiment with rat liver that fenugreek seeds can inhibit lipid peroxidation, which is a well-known stress marker enzyme generated via ROS production (Dixit et al., 2005). Fenugreek seeds can be a good supplement to cereals because of their high protein (25 %), lysine (5.7 g/16 g N), and soluble (20 %) and insoluble (28 %) dietary fibre, besides being rich in calcium, iron, and beta-carotene. Moreover, they contain 18.1 % antioxidant activity (Akbari et al., 2019). The additional activity of fenugreek seeds and leaves were also reported to maintain blood cholesterol, reduce micellar cholesterol solubilization, and inhibit pancreatic lipase activity (Hozzein et al., 2020) (Table 11.1).

11.3.2. Antidiabetic Potential

In this advanced technological era, many detrimental diseases can be cured by synthetically formulated drugs, costly therapies, and advanced therapies. Parallel consequences of these events are additional side effects in the body. Moreover, costly medicines in rural areas of developing and undeveloped countries are hard to acquire by patients. In 2020, according to the International Diabetes Federation (IDF), 463 million people have diabetes in the world and 88 million people in the Southeast Asia region. Of the 88 million people, 77 million live in India. The number will increase up to 134 million by 2045 (Shabil et al., 2023). Hence, challenges exist in treating the diabetes with cost-effective remedies. The seeds of fenugreek, *Trigonella foenum-graecum*, can be a successful remedy with cost effectiveness. For centuries, these seeds were used as a spice in Middle Eastern countries and widely used in South Asia and Europe, because of their antidiabetic properties (Haeri et al., 2012). Galactomannan, a soluble fibre, has been identified from fenugreek seeds, cultivated in Canada, and is known to reduce the postprandial blood glucose level. Moreover, it has the ability to diminish intestinal glucose absorption irrespective of high or low concentration of glucose present in the intestine, and as a result blood glucose level remains in control (Jhajhria and Kumar, 2016).

It is reported that active soluble components of fenugreek seed include an unusual amino acid, 4-hydroxyisoleucine (4HO-Ile), which shows antidiabetic potential through its ability to excite

TABLE 11.1
Various Parts of Fenugreek and Their Health Benefits

Fenugreek	Health Benefits
Seeds	The hypoglycemic effect helps regulate blood glucose levels.
Seeds, leaves	Antioxidant properties, able to reduce oxidative stress, maintain mitochondrial health.
Seeds	Helps maintain good gastrointestinal health, prevents constipation.
Seeds, leaves	Have anticancer effects, prevent cancer development, help maintain cellular integrity.
Seeds	Immunomodulatory effect, prevent inflammation, help boost immune system.
Seeds and leaves	Effective in wounds and sore muscles treatment.
Seeds	Help maintain healthier respiratory system, prevent asthma, emphysema, pneumonia.
Seeds	They have bioactive substances that help improve reproductive endocrine health and can boost fertility.
Seeds	Reduce blood cholesterol level, cardioprotective effects, decrease blood pressure and hypertension, minimize the risk of cardiac stroke.

secretion of insulin from rat pancreatic islet cells (Sauvaire et al., 1998). It was observed that rat islet's insulin secretion assay treatment with 4HO-Ile was 15 to 25 times more potent than the branched-chain amino acids l-leucine and l-isoleucine (Haeri et al., 2012). When 5 % powder of fenugreek seeds via oral administration has performed in alloxan-induced diabetic rats for 21 days, it increased the hepatic glycogen content and normalized liver and kidney functions (Jhajhria and Kumar, 2016). Fenugreek seeds are also reported to have coumarins, alkaloid trigonelline, as well as steroid saponins and fibre, which effectively corrected the defective glycogen storage of the diabetic liver and significantly reduced the elevated glycogen content of the diabetic kidney (Gad et al., 2006). Recent studies revealed that fenugreek seeds are effective in lowering glucose uptake in the small intestine, which occurs by prolonging gastric emptying time, and restoring pancreatic cells by safeguarding β-cells. It may increase insulin sensitivity of insulin action at the cellular level, lowering HbA1c levels by using glucose in the peripheral tissues and maintaining glucose homeostasis (Shabil et al., 2023).

11.3.3. Hepatoprotective Potential

Trigonella foenumgraecum (TFG) seeds were found to have hepatoprotective properties against carbon tetrachloride (CCl_4)-induced liver damage (Said et al., 2011). Serum bilirubin and biochemical marker enzymes, such as SGOT, SGPT, and SALP, were used to monitor liver damage. According to the findings, SGOT, SGPT, SALP, and serum bilirubin levels were all considerably reduced by the TFG seed methanol extract. This initial result suggests that TFG seeds have significant hepatoprotective qualities (Das, 2014). Oral exposure to $AlCl_3$ through drinking water resulted in liver atrophy. The activity of alanine transaminase (ALT), aspartate transaminase (AST), and glutamyl transpeptidase (GTT) in both liver and plasma were considerably decreased following treatment with $AlCl_3$ (Belaïd-Nouira et al., 2013). The supplementation of FSP restored the plasmatic hepatic enzymes, normal level of ALT activity, and succeeded to increase AST and GGT activity (Fatima et al., 2021).

The world's most common cause of death and morbidity are liver illnesses, and medications significantly increase the risk of hepatic harm. Hepatic injury can be induced by various substances, including more than 900 drugs, toxins, and herbal products (Abd-Elrahman, 2019). The strong anticancer drug cisplatin [cis-diamminechloroplatinum (II)] is frequently used to treat a variety of malignancies, but it can be harmful to the liver (Abd Rashid et al., 2021). The liver is susceptible to xenobiotic damage since it is essential to the process of chemical detoxification. It has been demonstrated that fenugreek demonstrates strong hepatoprotection against toxicity generated by cisplatin. Research reveals that the polyphenols in fenugreek seeds enhance the lipid profiles, improve both the quantity and quality of collagen, and reduce the levels of serum AST, ALT, and GTT activities.

After analyzing these results, Abd-Elrahman (2019) established that fenugreek seeds can help in regeneration of liver cells by reducing the effects of enzyme efflux and regaining liver function. Drugs used to treat hepatopathy frequently have severe side effects and little efficacy, which emphasizes the need for alternative therapies that strike a compromise between safety and efficacy. Herbs, natural products, and dietary antioxidants are found to be effective for the management of many hepatic illnesses when allopathic treatment is insufficient. Various polyphenols and flavonoids have been identified as hepatoprotective plant products. Research suggests that fenugreek seeds contain high levels of flavonoids (Varshney et al., 2023). Five flavonoids were extracted from fenugreek seeds, including vitexin, tricin, naringenin, quercetin, and tricin 7-O-bD-glucopyranoside. The study found that fenugreek seed extract had strong antioxidant and antiradical effects. By re-establishing altered levels of tissue antioxidant enzymes and liver biomarkers, fenugreek seed extract effectively attenuates liver damage.

11.3.4. Cardioprotective Effects

Fenugreek seeds have medicinal properties due to their unique components, which include lysine-rich proteins, mucilaginous fibre, and bioactive substances such as saponins, diosgenin, and trigonelline (Faisal et al., 2024). Fenugreek seeds have cardioprotective effects, including lowering cholesterol and reducing oxidative stress caused by myocardial infarction (Almatroodi et al., 2021). Fenugreek's pharmacological benefits are attributed to its ability to increase antioxidant enzymes (SOD, CAT, GPx, and GSH) and control blood lipids, lowering cholesterol, triglycerides, and LDL while increasing HDL. They can help prevent atherosclerosis (Singh et al., 2022). Cholesterol is an essential lipid that is produced by cells or intake of food. HMG-CoA synthase and thiolase together change acetyl-CoA into HMG-CoA in the cytoplasm. Acetyl-CoA and NADPH are required for the conversion of HMG-CoA to mevalonate by HMG-CoA reductase in the endoplasmic reticulum. Mevalonate is converted into lanosterol through isoprene, ultimately producing the cholesterol by desaturation, demethylation, and saturation processes (Hanafi et al., 2022). The two primary forms of cholesterol in the body are LDL and HDL, which are fat-protein molecules. Excessive consumption raises the risk of high cholesterol and coronary heart disease, mostly because LDL, or "harmful cholesterol", deposits cholesterol in arteries and causes atherosclerosis. Fenugreek seed extract significantly can reduce blood glucose, LDL (bad) cholesterol, and triglyceride levels, while HDL (good) cholesterol levels remain unaltered (Khorshidian et al., 2016). Various composition of fenugreek, which includes coumarin, nicotinic acid, and sapogenins, was studied for its ability to protect against isoproterenol-induced cardiac damage (Murugesan et al., 2011). The heart displays cellular injuries, mitochondrial damage, apoptosis, and fibrosis in diabetic cardiomyopathy (DCM), which is brought on by inflammatory agents, reactive oxygen species (ROS), and hyperglycemia. This induces the expression of intracellular adhesion molecule-1 (ICAM-1) and vascular cell adhesion molecule-1 (VCAM-1), which in turn promotes the infiltration of immunocompetent cells and the production of proinflammatory cytokines (IL-1, IL-18, and TNF). These cytokines activate CAMs, which causes cardiac dysfunction and the Akt pathway to trigger apoptosis (Bafadam et al., 2021). On the other hand, fenugreek extract and metformin administration showed antiapoptotic and cardioprotective effects through modification of the rennin-angiotensin signaling system, as evidenced by decreased Bax and ICAM1 gene expression, increased Bcl2 expression, and decreased apoptosis. Fenugreek seed extract can be used for good cardiac health since it significantly lowers cell death in diabetic patients while having no effect on blood glucose. This suggests that the fenugreek extract can protect the cardiovascular system through multiple pathways (Akyürek et al., 2001) (Figure 11.2).

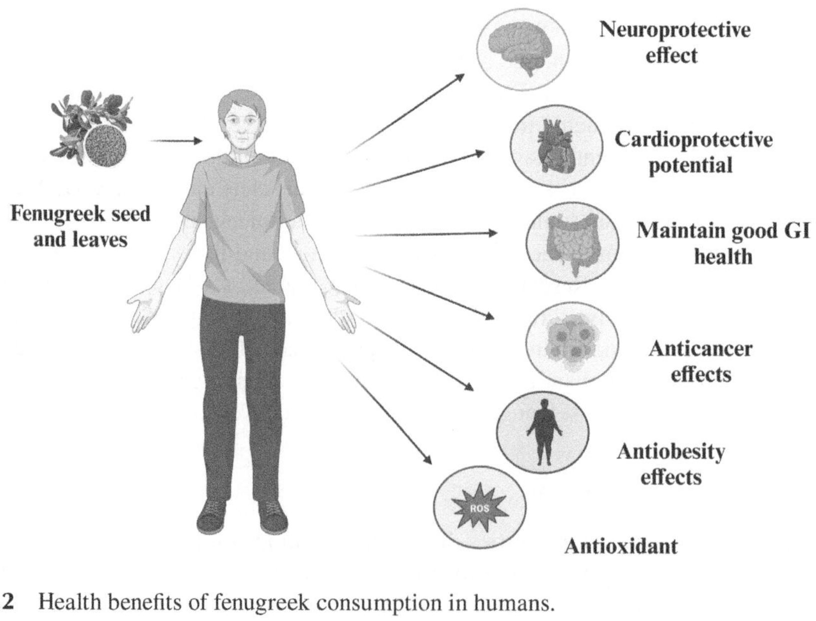

FIGURE 11.2 Health benefits of fenugreek consumption in humans.

11.3.5. Anti-Inflammatory Efficacy

The ethanol extract, mucilaginous fraction, and flavonoid constituents in fenugreek seeds exhibited potent anti-inflammatory, antiarthritic, and free radical–scavenging activities (Sindhu et al., 2012; Suresh et al., 2012). Fenugreek seed petroleum ether extract (FSPEE) contains a considerable amount of saturated and unsaturated fatty acids (Pundarikakshudu et al., 2016). Polyunsaturated fatty acids such as linoleic and linolenic acids in FSPEE, along with saponins and flavonoids, have substantial anti-inflammatory activity in various models, including carrageenan, prostaglandin E2, leukotrienes, and arachidonic acid, indicating blockade of both cyclooxygenase and lipoxygenase pathways (Fatima et al., 2022). Research suggests that fenugreek seed extract markedly decreases LPS-stimulated production of TNF-α, IL-6, prostaglandin E2, and nitric oxide (NO) in RAW 264.7 cells, exhibiting a dose-dependent response. TNF-α is a key pro-inflammatory cytokine belonging to a large group of immunomodulatory molecules, mainly synthesized by macrophages and other innate immune system cells, including APCs orchestrating the chemotactic recruitment and activation of adaptive immune responses. Decreased amount of TNF-α inhibits pro-inflammatory processes and generates anti-inflammatory responses (Narapogu et al., 2021). The fenugreek seeds possess gastroprotective activity, mitigating the gastrointestinal side effects associated with nonsteroidal anti-inflammatory drug use (Gaddam et al., 2015). Fenugreek seeds possess hypocholesterolemic effects through reducing the levels of LDL, and they can also prevent type 2 diabetes mellitus (Tsalamandris et al., 2019). Decrease in blood glucose level helps elevate anti-inflammatory process by reducing the concentration of circulating pro-inflammatory mediators, such as C-reactive protein (CRP), IL-6, plasminogen activator inhibitor-1, and sialic acid (Huang et al., 2022; Lawrence, 2009). Experiments found that when broilers are treated with high concentration of fenugreek seed extract (800 mg/kg), it significantly increased the level of nuclear factor-κβ, which expresses its pro-inflammatory properties by production of anti-inflammatory cytokines such as IL-10 (Laila et al., 2022). Fenugreek seeds contain high amounts of carotene and vitamins B1, B2, C, and K (Ellulu et al., 2015). Vitamin C (1000 mg daily) has been shown to downregulate pro-inflammatory markers, including high-sensitivity C-reactive protein (hs-CRP), IL-6, and fasting blood glucose (FBG) levels (Orrico et al., 2023). Stress-induced reactive oxygen species (ROS) and reactive nitrogen species (RNS) causes RBC hemolysis, inflammation, and other impaired physiological symptoms (Laila et al., 2022). Polyphenols and flavonoids present in fenugreek contributes to reduced ROS, RNS generation, and decreased lipid peroxidation (Abas & Naguib, 2019). Phenolic acids such as gallic acid, caffeic acid, and polyphenols such as catechins, rutin of fenugreek seed extract have diverse anti-inflammatory properties involving free-radical neutralization, NADPH oxidase (NOX) inhibition, inhibition of ROS-generating enzymes (e.g., xanthine oxidase), and enhanced expression of endogenous antioxidant enzymes (Yahfoufi et al., 2018).

11.3.6. Anticancer Potential

Growing evidence suggests that saponins contained within fenugreek seeds exhibit anticancer properties through diverse molecular mechanisms, primarily involving cell cycle inhibition, apoptosis induction, autophagy activation, antiangiogenic effects, and suppression of tumour cell migration (Zhu et al., 2023). Diosgenin was found to abrogate the pro-survival signalling pathways mediated by AKT, mTOR, and JNK, and to induce apoptosis in human epidermal growth factor receptor-2 (HER-2)-positive breast cancer cells (Sethi et al., 2018). According to molecular studies (Li et al., 2021), diosgenin exhibits antitumour effects in hepatocellular carcinoma (HCC) cells by disrupting the STAT3 signalling pathways, specifically by inhibiting cellular sarcoma kinase (c-SRC) and janus kinase 1 (JAK-1) and JAK-2-mediated intracellular signalling (Sethi et al., 2018). Naringenin, a bioactive flavonoid found in fenugreek, has been shown to neutralize cancer by promoting ROS production in JAR and JEG-3 choriocarcinoma cells and initiating apoptosis in A431 epidermoid carcinoma cells, while also exhibiting antiproliferative, antimigratory, and pro-apoptotic effects in PC3 and LNCaP prostate cancer cells (Mir et al., 2023). Chloroform fenugreek extract (FCE) triggers apoptosis by concurrent activation of death receptors and mitochondrial pathways, marked by elevated caspase-3, 8, and 9 levels, attributed to fenugreek's sterol and sapogenin composition, and are further augmented by extrinsic pathway activation through Fas and Fas-associated death domain (FADD)-mediated apoptosis, highlighting the crucial role of p53 in cell cycle regulation and apoptosis (Alshatwi et al., 2013).

Annexin v-FITC assay in myeloblastic leukemia cell line (KG-1) treated with fenugreek seed extract shows increase in apoptosis, nuclear fragmentation, and vacuolization (Alizadeh et al., 2009). Treatment with fenugreek resulted in an increased relative abundance of γδ T-cells (TCR1+CD8α-) in circulation, alongside a significant reduction in (MHC2+CD5-) antigen-presenting cells (Zentek et al., 2013). γδ T-cells can recognize neoplastic cells and kill them via a perforin-granzyme-mediated pathway, and they also could activate caspases, which may lead to DNA breakage and tumour cell death (Li et al., 2021). Trigonelline major alkaloid in fenugreek seeds was found to potently inhibit EGFR signalling and nuclear factor erythroid 2-related factor 2 (Nrf2) nuclear localization, resulting in reduced growth and increased apoptosis in non-small cell lung cancer (NSCLC) cells when exposed to Cisplatin/Etoposide (Li et al., 2021). The flow cytometry data demonstrated that fenugreek seed extract (FSE) induced a marked cell cycle arrest transition from G2/M phase and polyploidization (>G) at increased concentrations, implying activation of p53-mediated mitotic catastrophe and apoptotic cell elimination (Fouzder et al., 2021). FSE also increases ROS and Bax/Bcl-2 ratio, which ultimately causes cell death (Alrumaihi et al., 2021). Treatment with either dry or germinated fenugreek extract resulted in a significant increase in apoptosis in Michigan Cancer Foundation-7 (MCF-7) and adenocarcinoma pancreas cell line-1 (ASPC-1) cells, marked by enhanced caspase-3 and caspase-6 activity and DNA fragmentation (Alrumaihi et al., 2021) (Figure 11.3).

11.4. VERSATILE USE OF FENUGREEK IN FOOD ITEMS

11.4.1. Masala

Fenugreek (methi) is used as a major ingredient of Indian panchphoron, a mixture of five different spices. Panchphoron generally combines equal parts of cumin seeds (jeera), nigella seeds (kalonji), fenugreek seeds (methi), wild celery seeds (radhuni or joni in Assamese), and fennel seeds (saunf).

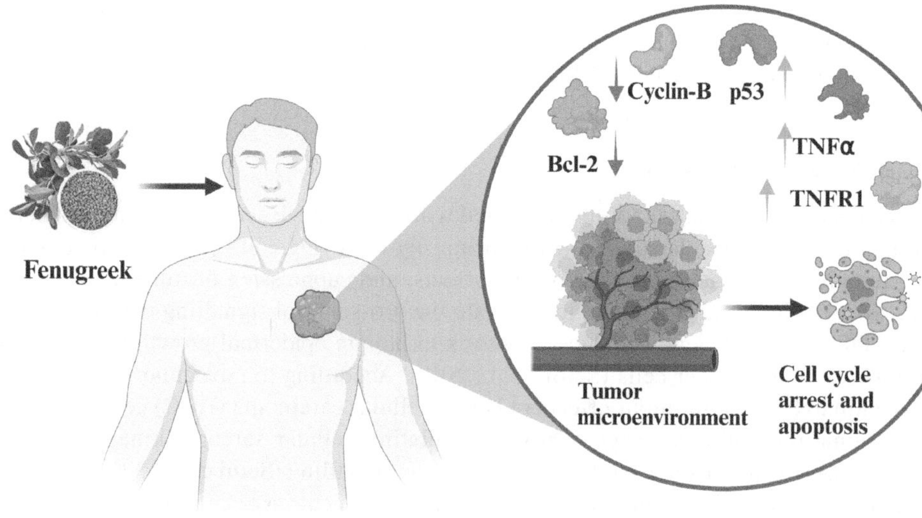

FIGURE 11.3 Anticancer effects of fenugreek.

Fenugreek has a unique ability to balance well with spices such as cinnamon, bay leaves, cardamom, cumin, coriander, black pepper, cloves, and mace. This spice mixture (garam masala) can be applied at the end of the cooking time. Some Indian foods, such as dals, lentils, soups, mixed curry, etc., are seasoned with these spices.

11.4.2. Lactation Herbal Tea

Fenugreek has a great supporting property to increase healthy breast milk production. Post-natal mothers can feel relief and increase their breast milk by drinking tea combined with fenugreek (*methi*), cinnamon, fennel, nettle leaf, red raspberry leaf, and dandelion (Ravi and Joseph, 2020).

11.4.3. Stabilizer, Adhesive, Emulsifier, and Gum in Bakery Products

Fenugreek proteins have demonstrated strong stabilizing and emulsifying properties in food applications. Studies indicate that fenugreek gum significantly boosts the emulsifying capability of soy protein, and their combined emulsifying effect is four times more than that of only fenugreek gum. This enhanced emulsion remains stable even under a wide range of temperatures, ionic strengths, and pH conditions (Hefnawy and Ramadan, 2011). Fenugreek's dietary soluble fibre functions as a probiotic, providing an ideal substrate for microbial fermentation in the large intestine. This process supports gut health by promoting the growth of beneficial bacteria (Sowmya and Rajyalakshmi, 1999; Lee, 2009). The presence of galactomannan in fenugreek shows its emulsifying and stabilizing assets. In Western countries, bakery functional foods, such as bread, pizza, muffins, cakes, etc., are made with the mixture of flour and 8–10% of fenugreek dietary fibre (Roberts, 2011). It has been reported that the fibre-enriched muffins have worthy capacity, pliable architecture, and medium-fine particle, with double the amount of nutritional fibre. Bread fortified with fenugreek shows lower glucose level and can reduce insulin resistance. So, the study suggests that incorporating fenugreek into baked products at appropriate levels can effectively reduce insulin resistance, offering a natural therapeutic option for managing diabetes in patients. The total amount of proteins, fibers, and different types of minerals, such as iron, calcium, zinc, carotenoids, vitamin B2 (riboflavin), vitamin E (tocopherol), and vitamin C (ascorbic acid), may be increased by fortified flours with fenugreek powder that has a low glycemic index (Ravindran et al., 2011). Germinated fenugreek seed flour contains elevated levels of essential amino acids, including lysine, isoleucine, leucine, threonine, and valine. These amino acids contribute to balancing overall amino acid profiles, which is beneficial for dietary applications aimed at addressing amino acid deficiencies or imbalances (Hegazy and Ibrahium, 2009).

11.4.4. Traditional Foodstuffs

A popular pasted Turkish food of ground fenugreek seeds is locally called *Cemen* (Işıklı and Karababa, 2005). Crushed fenugreek seed powder is needed to make a clarified butter ball. Ethiopian clarified butter, akin to spiced *ghee*, is infused with a blend of aromatics and spices, such as fenugreek, cardamom, cloves, cumin seeds, cinnamon, garlic-ginger paste, turmeric powder, finely chopped shallots, and dried oregano. This mixture creates a flavourful and aromatic butter, ideal for enhancing a variety of dishes. It is used in stews, braised vegetables, and sauteed meat dishes.

11.5. CONCLUSION

Fenugreek has immense health-promoting efficacy, enriched with flavonoids and other antioxidants. It has been used as traditional medicine in several cultures to treat many diseases. Fenugreek seeds effectively treat metabolic diseases, neurological ailments, and even cancer. Moreover, gastrointestinal and renal issues can also be reduced by the bioactive compounds available in fenugreek seeds and leaves. They have been widely considered hepatoprotective, cardioprotective, and neuroprotective foods. Therefore, regular consumption of fenugreek could be helpful in maintaining good health and reducing physiological ailments.

REFERENCES

Abas, A. S. M., & Naguib, D. M. (2019). Effect of germination on anticancer activity of Trigonella foenum seeds extract. *Biocatalysis and Agricultural Biotechnology*, *18*, 101067.

Abd-Elrahman, W. M. (2019). Curative and hepatoprotective action of fenugreek (Trigonellafoenumgaraecum) seeds against cisplatin induced oxidative liver damage in male albino rats. *Egyptian Journal of Nutrition and Health*, *14*(2), 77–90.

Abd Rashid, N., Abd Halim, S. A. S., Teoh, S. L., Budin, S. B., Hussan, F., Ridzuan, N. R. A., & Jalil, N. A. A. (2021). The role of natural antioxidants in cisplatin-induced hepatotoxicity. *Biomedicine & Pharmacotherapy*, *144*, 112328.

Acharya, S. N., Thomas, J. E., & Basu, S. K. (2006). Fenugreek: An "old world" crop for the "new world". *Biodiversity*, *7*(3–4), 27–30.

Ahmad, A., Alghamdi, S. S., Mahmood, K., & Afzal, M. (2016). Fenugreek a multipurpose crop: Potentialities and improvements. *Saudi Journal of Biological Sciences*, *23*(2), 300–310.

Aishwath, O. P., Meena, R. D., & Dubey, P. N. (2018). Role of micronutrients on morpho-physiological, and nutritional parameters including diseases suppression of fenugreek. *International Journal of Seed Spices*, *8*, 24–31.

Akbari, S., Abdurahman, N. H., Yunus, R. M., Alara, O. R., & Abayomi, O. O. (2019). Extraction, characterization and antioxidant activity of fenugreek (Trigonella-Foenum Graecum) seed oil. *Materials Science for Energy Technologies*, *2*(2), 349–355.

Akyürek, Ö., Akyürek, N., Sayin, T., Dinçer, I., Berkalp, B., Akyol, G., . . . & Oral, D. (2001). Association between the severity of heart failure and the susceptibility of myocytes to apoptosis in patients with idiopathic dilated cardiomyopathy. *International Journal of Cardiology*, *80*(1), 29–36.

Alizadeh, S., Jahanmehr, S. A. H., Ardjmand, A. R., Rezaian, M., Dargahi, H., Einolahi, N., & Sadrossadat, M. (2009). Antineoplastic effect of Fenugreek (Trigonella Foenum Graecum) seed extract against acute myeloblastic leukemia cell line (KG-1). *Iranian Journal of Blood and Cancer*, *1*(4), 139–146.

Almatroodi, S. A., Almatroudi, A., Alsahli, M. A., & Rahmani, A. H. (2021). Fenugreek (Trigonella foenum-graecum) and its active compounds: A review of its effects on human health through modulating biological activities. *Pharmacognosy Journal*, *13*(3).

Alrumaihi, F. A., Khan, M. A., Allemailem, K. S., Alsahli, M. A., Almatroudi, A., Younus, H., . . . & Khan, A. (2021). Methanolic fenugreek seed extract induces p53-dependent mitotic catastrophe in breast cancer cells, leading to apoptosis. *Journal of Inflammation Research*, 1511–1535.

Alshatwi, A. A., Shafi, G., Hasan, T. N., Syed, N. A., & Khoja, K. K. (2013). Fenugreek induced apoptosis in breast cancer MCF-7 cells mediated independently by fas receptor change. *Asian Pacific Journal of Cancer Prevention*, *14*(10), 5783–5788.

Bafadam, S., Mahmoudabady, M., Niazmand, S., Rezaee, S. A., & Soukhtanloo, M. (2021). Cardioprotective effects of Fenugreek (Trigonella foenum-graceum) seed extract in streptozotocin induced diabetic rats. *Journal of Cardiovascular and Thoracic Research*, *13*(1), 28.

Belaïd-Nouira, Y., Bakhta, H., Haouas, Z., Flehi-Slim, I., Neffati, F., Najjar, M. F., & Cheikh, H. B. (2013). Fenugreek seeds, a hepatoprotector forage crop against chronic AlCl 3 toxicity. *BMC Veterinary Research*, *9*, 1–9.

Das, S. (2014). Hepatoprotective activity of methanol extract of fenugreek seeds on rats. *International Journal of Pharmaceutical Sciences and Research*, *5*(4), 1506.

Devasena, T., & Menon, V. P. (2003). Fenugreek affects the activity of β-glucuronidase and mucinase in the colon. *Phytotherapy Research: An International Journal Devoted to Pharmacological and Toxicological Evaluation of Natural Product Derivatives*, *17*(9), 1088–1091.

Dhull, S. B., Bamal, P., Kumar, M., Bangar, S. P., Chawla, P., Singh, A., . . . & Sihag, S. (2023). Fenugreek (Trigonella foenum graecum) gum: A functional ingredient with promising properties and applications in food and pharmaceuticals—A review. *Legume Science*, *5*(3), e176.

Dixit, P., Ghaskadbi, S., Mohan, H., & Devasagayam, T. P. (2005). Antioxidant properties of germinated fenugreek seeds. *Phytotherapy Research: An International Journal Devoted to Pharmacological and Toxicological Evaluation of Natural Product Derivatives*, *19*(11), 977–983.

Ellulu, M. S., Rahmat, A., Patimah, I., Khaza'ai, H., & Abed, Y. (2015). Effect of vitamin C on inflammation and metabolic markers in hypertensive and/or diabetic obese adults: A randomized controlled trial. *Drug Design, Development and Therapy*, 3405–3412.

Faisal, Z., Irfan, R., Akram, N., Manzoor, H. M. I., Aabdi, M. A., Anwar, M. J., . . . & Desta, D. T. (2024). The multifaceted potential of fenugreek seeds: From health benefits to food and nanotechnology applications. *Food Science & Nutrition*, *12*(4), 2294–2310.

Fatima, H., Shahid, M., Pruitt, C., Pung, M. A., Mills, P. J., Riaz, M., & Ashraf, R. (2022). Chemical fingerprinting, antioxidant, and anti-inflammatory potential of hydroethanolic extract of Trigonella foenum-graecum. *Antioxidants*, *11*(2), 364.

Fatima, S. N., & Masood, J. (2021). Fenugreek seeds attenuate thioacetamide induced liver damage. *Pakistan Journal of Pharmaceutical Sciences*, *34*(3).

Fouzder, C., Mukhuty, A., Mukherjee, S., Malick, C., & Kundu, R. (2021). Trigonelline inhibits Nrf2 via EGFR signalling pathway and augments efficacy of Cisplatin and Etoposide in NSCLC cells. *Toxicology in Vitro*, *70*, 105038.

Gad, M. Z., El-Sawalhi, M. M., Ismail, M. F., & El-Tanbouly, N. D. (2006). Biochemical study of the anti-diabetic action of the Egyptian plants Fenugreek and Balanites. *Molecular and cellular Biochemistry*, *281*, 173–183.

Gaddam, A., Galla, C., Thummisetti, S., Marikanty, R. K., Palanisamy, U. D., & Rao, P. V. (2015). Role of Fenugreek in the prevention of type 2 diabetes mellitus in prediabetes. *Journal of Diabetes & Metabolic Disorders*, *14*, 1–10.

Grieve, M. (2013). *A modern herbal* (Vol. 2). Courier Corporation.

Haeri, M. R., Limaki, H. K., White, C. J. B., & White, K. N. (2012). Non-insulin dependent anti-diabetic activity of (2S, 3R, 4S) 4-hydroxyisoleucine of fenugreek (Trigonella foenum graecum) in streptozotocin-induced type I diabetic rats. *Phytomedicine*, *19*(7), 571–574.

Hanafi, N. I., Saidan, N. H., Mohamed, M., Sirajudeen, K. N. S., Hua, G. S., Pahirulzaman, K. A., & Rao, P. V. (2022). ischemic heart disease and the potential role of fenugreek (Trigonella foenum graecumlinn.) in cardioprotection. *Jurnal teknologi*.

Haxhiraj, M., White, K., & Terry, C. (2024). The role of fenugreek in the management of type 2 diabetes. *International Journal of Molecular Sciences*, *25*(13), 6987.

He, Y., Ding, C., Wang, X., Wang, H., & Suo, Y. (2015). Using response surface methodology to optimize countercurrent chromatographic separation of polyphenol compounds from fenugreek (Trigonella foenum-graecum L.) seeds. *Journal of Liquid Chromatography & Related Technologies*, *38*(1), 29–35.

Hefnawy, H. T. M., & Ramadan, M. F. (2011). Physicochemical characteristics of soy protein isolate and fenugreek gum dispersed systems. *Journal of Food Science and Technology*, *48*, 371–377.

Hegazy, A. I., & Ibrahium, M. I. (2009). Evaluation of the nutritional protein quality of wheat biscuit supplemented by fenugreek seed flour. *World Journal of Dairy & Food Sciences*, *4*, 129–135.

Hozzein, W. N., Saleh, A. M., Habeeb, T. H., Wadaan, M. A., & AbdElgawad, H. (2020). CO2 treatment improves the hypocholesterolemic and antioxidant properties of fenugreek seeds. *Food Chemistry*, *308*, 125661.

Huang, H., Wang, X., Yang, L., He, W., Meng, T., Zheng, K., . . . & Xiao, D. (2022). The effects of fenugreek extract on growth performance, serum biochemical indexes, immunity and NF-κB signaling pathway in broiler. *Frontiers in Veterinary Science*, *9*, 882754.

Hussain, P. R., Suradkar, P., Javaid, S., Akram, H., & Parvez, S. (2016). Influence of postharvest gamma irradiation treatment on the content of bioactive compounds and antioxidant activity of fenugreek (Trigonella foenum–graceum L.) and spinach (Spinacia oleracea L.) leaves. *Innovative Food Science & Emerging Technologies*, *33*, 268–281.

Işıklı, N. D., & Karababa, E. (2005). Rheological characterization of fenugreek paste (çemen). *Journal of Food Engineering*, *69*(2), 185–190.

Jhajhria, A., & Kumar, K. (2016). Fenugreek with its medicinal applications. *International Journal of Pharmaceutical Sciences Review and Research*, *41*(1), 194–201.

Khare, C. P. (Ed.). (2011). *Indian herbal remedies: Rational western therapy, ayurvedic and other traditional usage, botany*. Springer science & business media.

Khorshidian, N., Yousefi Asli, M., Arab, M., Adeli Mirzaie, A., & Mortazavian, A. M. (2016). Fenugreek: Potential applications as a functional food and nutraceutical. *Nutrition and Food Sciences Research*, *3*(1), 5–16.

Laila, U., Albina, T., Zuha, S. S., & Tamang, H. (2022). Fenugreek seeds: Nutritional composition and therapeutic properties. *The Pharma Innovation Journal*, *11*(6), 2417–2425.

Lawrence, T. (2009). The nuclear factor NF-κB pathway in inflammation. *Cold Spring Harbor Perspectives in Biology*, *1*(6), a001651.

Lee, E. L. (2009). *Genotype X environment impact on selected bioactive compound content of fenugreek (Trigonella foenum-graecum L.)* (Doctoral dissertation, Lethbridge, Alta.: University of Lethbridge, Dept. of Biological Sciences, c2009).

Li, Y., Li, G., Zhang, J., Wu, X., & Chen, X. (2021). The dual roles of human γδ T cells: Anti-tumor or tumor-promoting. *Frontiers in Immunology*, *11*, 619954.

Mashkor, I. M. (2014). Phenolic content and antioxidant activity of fenugreek seeds extract. *International Journal of Pharmacognosy and Phytochemical Research*, *6*(4), 841–844.

Mir, S. A., Dar, A., Hamid, L., Nisar, N., Malik, J. A., Ali, T., & Bader, G. N. (2023). Flavonoids as promising molecules in the cancer therapy: An insight. *Current Research in Pharmacology and Drug Discovery*, 100167.

Murugesan, M., Revathi, R., & Manju, V. (2011). Cardioprotective effect of fenugreek on isoproterenol-induced myocardial infarction in rats. *Indian Journal of Pharmacology*, *43*(5), 516–519.

Narapogu, V., Swathi, C., Polsani, M. R., & Premendran, J. (2021). Evaluation of anti-diabetic and anti-inflammatory activities of Fenugreek (Trigonella foenum-graecum) seed extracts on Albino Wistar Rats. *National Journal of Physiology, Pharmacy and Pharmacology*, *11*(11), 1277–1282.

Orrico, F., Laurance, S., Lopez, A. C., Lefevre, S. D., Thomson, L., Möller, M. N., & Ostuni, M. A. (2023). Oxidative stress in healthy and pathological red blood cells. *Biomolecules*, *13*(8), 1262.

Pandey, H., & Awasthi, P. (2015). Effect of processing techniques on nutritional composition and antioxidant activity of fenugreek (Trigonella foenum-graecum) seed flour. *Journal of Food Science and Technology*, *52*, 1054–1060.

Pundarikakshudu, K., Shah, D. H., Panchal, A. H., & Bhavsar, G. C. (2016). Anti-inflammatory activity of fenugreek (Trigonella foenum-graecum Linn) seed petroleum ether extract. *Indian Journal of Pharmacology*, *48*(4), 441–444.

Raskin, I., Ribnicky, D. M., Komarnytsky, S., Ilic, N., Poulev, A., Borisjuk, N., . . . & Fridlender, B. (2002). Plants and human health in the twenty-first century. *TRENDS in Biotechnology*, *20*(12), 522–531.

Ravi, R., & Joseph, J. (2020). Effect of fenugreek on breast milk production and weight gain among Infants in the first week of life. *Clinical Epidemiology and Global Health*, *8*(3), 656–660.

Ravindran, G., Carr, A., & Hardacre, A. (2011). A comparative study of the effects of three galactomannans on the functionality of extruded pea–rice blends. *Food Chemistry*, *124*(4), 1620–1626.

Roberts, K. T. (2011). The potential of fenugreek (Trigonella foenum-graecum) as a functional food and nutraceutical and its effects on glycemia and lipidemia. *Journal of Medicinal Food*, *14*(12), 1485–1489.

Said, A. M., Al-Khashali, D., Al-Khateeb, E., & Al-Wakeel, N. (2011). The hepatoprotective activity of Fenugreek seeds' extract against carbon tetrachloride induced liver toxicity in rats. *Al Mustansiriyah Journal of Pharmaceutical Sciences*, *9*(1), 94–103.

Sauvaire, Y., Petit, P., Broca, C., Manteghetti, M., Baissac, Y., Fernandez-Alvarez, J., . . . & Ribes, G. (1998). 4-Hydroxyisoleucine: A novel amino acid potentiator of insulin secretion. *Diabetes*, *47*(2), 206–210.

Sethi, G., Shanmugam, M. K., Warrier, S., Merarchi, M., Arfuso, F., Kumar, A. P., & Bishayee, A. (2018). Pro-apoptotic and anti-cancer properties of diosgenin: A comprehensive and critical review. *Nutrients*, *10*(5), 645.

Shabil, M., Bushi, G., Bodige, P. K., Maradi, P. S., Patra, B. P., Padhi, B. K., & Khubchandani, J. (2023). Effect of fenugreek on hyperglycemia: A systematic review and meta-analysis. *Medicina*, *59*(2), 248.

Sindhu, G., Ratheesh, M., Shyni, G. L., Nambisan, B., & Helen, A. (2012). Anti-inflammatory and antioxidative effects of mucilage of Trigonella foenum graecum (Fenugreek) on adjuvant induced arthritic rats. *International Immunopharmacology*, *12*(1), 205–211.

Singh, K. P., Nair, B., Jain, P. K., Naidu, A. K., & Paroha, S. (2013). Variability in the nutraceutical properties of fenugreek (Trigonella foenum-graecum L.) seeds. *Revista Colombiana de Ciencias Hortícolas*, *7*(2), 228–239.

Singh, R. B., Smail, M. M., Rai, R. H., Maheshwari, A., Verma, N., & Isaza, A. (2022). Effects of fenugreek seeds on cardiovascular diseases and other chronic diseases. In *Functional Foods and Nutraceuticals in Metabolic and Non-Communicable Diseases* (pp. 399–410). Academic Press.

Sowmya, P., & Rajyalakshmi, P. (1999). Hypocholesterolemic effect of germinated fenugreek seeds in human subjects. *Plant Foods for Human Nutrition*, *53*, 359–365.

Suresh, P., Kavitha, C. N., Babu, S. M., Reddy, V. P., & Latha, A. K. (2012). Effect of ethanol extract of Trigonella foenum graecum (Fenugreek) seeds on Freund's adjuvant-induced arthritis in albino rats. *Inflammation*, *35*, 1314–1321.

Syed, Q. A., Rashid, Z., Ahmad, M. H., Shukat, R., Ishaq, A., Muhammad, N., & Rahman, H. U. U. (2020). Nutritional and therapeutic properties of fenugreek (Trigonella foenum-graecum): A review. *International Journal of Food Properties*, *23*(1), 1777–1791.

Tahiliani, P., & Kar, A. (2003a). The combined effects of Trigonella and Allium extracts in the regulation of hyperthyroidism in rats. *Phytomedicine*, *10*(8), 665–668.

Tahiliani, P., & Kar, A. (2003b). Mitigation of thyroxine-induced hyperglycaemia by two plant extracts. *Phytotherapy Research*, *17*(3), 294–296.

Tak, Y., Kaur, M., Chitranashi, A., Samota, M. K., Verma, P., Bali, M., & Kumawat, C. (2024). Fenugreek derived diosgenin as an emerging source for diabetic therapy. *Frontiers in Nutrition, 11*, 1280100.

Tsalamandris, S., Antonopoulos, A. S., Oikonomou, E., Papamikroulis, G. A., Vogiatzi, G., Papaioannou, S., . . . & Tousoulis, D. (2019). The role of inflammation in diabetes: Current concepts and future perspectives. *European Cardiology Review, 14*(1), 50.

Varshney, H., & Siddique, Y. H. (2023). Medicinal properties of fenugreek: A review. *The Open Biology Journal, 11*(1).

Wang, H., Chen, Y., Wang, L., Liu, Q., Yang, S., & Wang, C. (2023). Advancing herbal medicine: Enhancing product quality and safety through robust quality control practices. *Frontiers in Pharmacology, 14*, 1265178.

Wani, S. A., & Kumar, P. (2018). Fenugreek: A review on its nutraceutical properties and utilization in various food products. *Journal of the Saudi Society of Agricultural Sciences, 17*(2), 97–106.

Yahfoufi, N., Alsadi, N., Jambi, M., & Matar, C. (2018). The immunomodulatory and anti-inflammatory role of polyphenols. *Nutrients, 10*(11), 1618.

Zentek, J., Gärtner, S., Tedin, L., Männer, K., Mader, A., & Vahjen, W. (2013). Fenugreek seed affects intestinal microbiota and immunological variables in piglets after weaning. *British Journal of Nutrition, 109*(5), 859–866.

Zhu, M., Sun, Y., Bai, H., Wang, Y., Yang, B., Wang, Q., & Kuang, H. (2023). Effects of saponins from Chinese herbal medicines on signal transduction pathways in cancer: A review. *Frontiers in Pharmacology, 14*, 1159985.

12 Cumin

Varieties and Nutraceutical Function and Applications

Sonia Gupta, Dinesh Kumar, Rajni Tanwar, Raman Kumari, and Puja Gulati

12.1. INTRODUCTION

Spices, the dried aromatic parts of plants, have been integral to food preparation and preservation since ancient times [1]. They are frequently employed to enhance flavour, colour, and shelf life of foodstuff. Spices are a rich source of various therapeutic bioactive compounds, including proteins, tannins, saponins, terpenes, flavonoids, alkaloids, tannic acids, anthocyanins, coumarins, and glucosides. These compounds exert diverse health benefits, offering protection against microbial infections, oxidative stress, inflammation, diabetes, cardiovascular diseases, neurodegenerative disorders, wound healing, reproductive abnormalities, and even cancer [2]. Moreover, traditional medicinal practices advocate the use of spices to treat ailments such as stomachache, leprosy, colds, coughs, loss of appetite, rheumatoid pain, cognitive disorders, convulsions, and inflammation. More than 100 plant species worldwide are utilized as spices and condiments, with cumin (*Cuminum cyminum* L.) being a noteworthy example due to its abundance of healing properties [3].

Cumin, a member of the Apiaceae family, is a flowering plant native to the Eastern Mediterranean and South Asia. It is extensively cultivated across diverse regions globally, with significant production in India, Iran, China, and the Middle East. The seeds of cumin are highly esteemed in both culinary and traditional medicinal practices, valued for their distinctive aroma and flavour [4]. It is an annual herbaceous plant, typically reaching a height of 30–50 cm, characterized by slender, branched stems and finely dissected, feathery leaves. The plant produces small white or pink flowers arranged in umbels [5]. Its fruit, commonly known as the cumin seed, is a small, elongated schizocarp, approximately 4–5 mm in length. Upon maturation, the seeds are harvested and dried for culinary and medicinal uses.

The strong and distinctive aroma of cumin seeds makes them a popular culinary spice and condiment in numerous cuisines around the world. Cumin seeds are predominantly used as a flavouring agent in soups, stews, sausages, cheeses, pickles, curries, meats, and chutneys [6]. Cumin seed oil is also utilized in small quantities in the perfumery industry due to its aromatic properties. Additionally, cumin is a reservoir of medicinal bioactive compounds with a broad range of healthcare applications. The plant is renowned for its therapeutic potential, including antibacterial, antifungal, antioxidant, antidiabetic, hypolipidemic, immunomodulatory, antiosteoporotic, anti-inflammatory, antiasthmatic, antitussive, anti-infertility, anticancer, antistress, antipyretic, hypotensive, antiplatelet, hepatoprotective, and analgesic effects [7].

The essential oil of cumin, particularly rich in cuminaldehyde, is primarily responsible for its characteristic aroma and plays a pivotal role in its therapeutic potential, including antibacterial and antidiabetic activities. Other bioactive constituents, such as phytoestrogens and monoterpene alcohols, contribute to cumin's antiosteoporotic and antioxidant properties, respectively [8]. The synergistic interaction between these diverse compounds likely accounts for the extensive pharmacological and clinical properties attributed to cumin. Cultivation of cumin thrives in warm, arid climates with well-drained, moderately fertile soils. In temperate regions, it is typically grown as

DOI: 10.1201/9781003559894-12

a winter crop, requiring full sunlight for optimal growth. India remains the largest global producer and exporter of cumin, followed by significant production in Iran and Syria [9].

In line with the extensive ethnomedicinal and nutritional significance of cumin, this study provides a comprehensive evaluation of its phytochemistry and therapeutic properties, drawing on both clinical studies and toxicological assessments. The findings discussed herein serve as a foundation for further exploration of this remarkable medicinal plant, highlighting its multifunctional properties and underscoring its importance in both traditional medicine and modern pharmaceutical applications [10].

12.2. VARIETIES OF CUMIN

Cumin, a well-known spice widely used in culinary and medicinal applications, encompasses several varieties that offer distinct flavour profiles and bioactive properties. The two primary varieties of cumin—Nigella sativa (commonly referred to as black cumin) and Cuminum cyminum (green cumin)—along with other lesser-known types, provide diverse sensory characteristics and nutritional content. This review presents a comprehensive analysis of the major varieties of cumin, their geographic distribution, cultivation practices, and differences in flavour, aroma, and nutritional properties [11].

12.2.1. Black Cumin (Nigella sativa)

Nigella sativa, commonly known as black cumin or *Kalonji*, is distinguished by its small, crescent-shaped seeds and robust, pungent flavour. Its seeds are jet-black with a rough texture, possessing significant medicinal properties. Black cumin is predominantly cultivated in regions with arid climates, including the Middle East, South Asia (notably India and Pakistan), and parts of North Africa. This variety is highly valued for its medicinal use in traditional systems of medicine, such as Ayurveda and Unani, as well as for its role in enhancing the flavour of Middle Eastern and Indian cuisine. The bioactive compounds, particularly thymoquinone, contribute to its recognized anti-inflammatory and immunomodulatory effects [12].

12.2.2. Green Cumin (*Cuminum cyminum*)

Cuminum cyminum, referred to as green cumin or simply "cumin", is the most commonly used variety of cumin globally. This variety features light brown to greenish, oval-shaped seeds with characteristic ridges. Cumin is cultivated across diverse regions, including India, Iran, Turkey, Egypt, and parts of the Mediterranean, favoring arid to semi-arid climates and sandy loam soils. Green cumin is integral to the flavour profiles of Indian, Middle Eastern, and Mexican cuisines. The spice is also nutritionally dense, being a rich source of iron, manganese, and antioxidants, including flavonoids and terpenoids, which confer additional health benefits [13].

12.2.3. Other Varieties of Cumin

Other types of cumin include *Bunium persicum*, sometimes referred to as "black cumin" in specific regions (distinct from *Nigella sativa*), and *Cuminum setifolium*, a lesser-known variant. *Bunium persicum*, cultivated primarily in Iran and Afghanistan, features a more delicate and slightly sweeter taste compared to *Cuminum cyminum*. *Cuminum setifolium*, native to parts of Central Asia, has a somewhat more bitter flavour and is not as widely utilized as the other two varieties.

12.3. GEOGRAPHIC DISTRIBUTION AND CULTIVATION PRACTICES

The cultivation of cumin is region-specific and influenced by climatic conditions. *Nigella sativa* thrives in dry, sunny climates and is cultivated primarily in the Middle East, South Asia, and North Africa. It requires well-drained soils and moderate irrigation, with growth occurring optimally in arid regions [14]. *Cuminum cyminum* is grown extensively in India, Iran, Turkey, and Egypt, favoring sandy loam soils in warm, arid climates. Cumin crops necessitate efficient drainage and are drought-tolerant but sensitive to frost. *Bunium persicum*, a variant of black cumin, is primarily cultivated in the mountainous regions of Iran, Afghanistan, and Tajikistan. This variety prefers cooler climates while still requiring relatively dry growing conditions [15].

12.4. DIFFERENCES IN FLAVOUR, AROMA, AND NUTRITIONAL CONTENT

The distinct varieties of cumin offer a range of sensory and nutritional profiles, which contribute to their use in culinary and medicinal applications.

Black Cumin (*Nigella sativa*): The flavour of *Nigella sativa* is notably sharp and peppery, with a slightly bitter aftertaste. Its strong, pungent aroma further distinguishes it from other cumin varieties. In terms of nutritional content, black cumin is particularly rich in thymoquinone, an active compound with potent antioxidant, anti-inflammatory, and immune-boosting properties. These bioactive components have led to black cumin's extensive use in traditional medicine for managing conditions such as asthma, hypertension, and gastrointestinal disorders [16].

Green Cumin (*Cuminum cyminum*): *Cuminum cyminum* possesses a warm, nutty flavour profile with subtle sweet and peppery undertones. Its aroma is earthy and intense, especially when roasted, a common preparation method in various cuisines. Nutritionally, this variety is a valuable source of iron, manganese, dietary fibre, and antioxidants. The presence of flavonoids and terpenoids in cumin seeds contributes to its health-promoting properties, such as enhancing digestion, boosting metabolism, and providing antimicrobial effects [17].

***Bunium persicum*:** *Bunium persicum* features a more delicate and slightly sweeter flavour compared to green cumin, with less bitterness than *Nigella sativa*. Its seeds contain a higher concentration of essential oils, which contributes to its more pronounced aroma. This variety is commonly used in Persian and Central Asian cuisine, with its nutritional profile rich in essential fatty acids and antioxidants.

12.5. PHYTOCHEMICAL COMPOSITION OF CUMIN

Cumin (*Cuminum cyminum*), a widely used spice in culinary and medicinal applications, is characterized by a complex phytochemical profile that contributes to its numerous health benefits. Its bioactive compounds can be categorized into two primary groups: volatile essential oils and nonvolatile phytochemicals, each of which plays a pivotal role in its nutraceutical functions. Additionally, cumin is a significant source of vitamins, minerals, and dietary fibre, adding to its overall nutritional value.

- **Essential Oils:** The essential oil content of cumin seeds forms the basis of its characteristic aroma and flavour and is primarily composed of volatile compounds that also impart notable pharmacological properties. The key components include: **Cumin Aldehyde:** This is the predominant volatile compound in cumin essential oil, responsible for its distinct aroma and flavour profile. Cumin aldehyde also contributes to the spice's antioxidant activity, playing a critical role in scavenging free radicals and mitigating oxidative damage at the cellular level [18].
- **Terpenes:** A diverse group of volatile compounds, including β-pinene, γ-terpinene, and p-cymene, are present in cumin essential oil. These terpenes exhibit both antioxidant and antimicrobial properties, contributing to cumin's therapeutic potential in preventing infections and combating oxidative stress.
- **Other Volatile Compounds:** In addition to cumin aldehyde and terpenes, cumin essential oil contains trace amounts of limonene, carvone, and myrcene, which further contribute

to its medicinal properties. These compounds enhance cumin's aromatic qualities while exerting additional antimicrobial and antioxidant effects, reinforcing its traditional uses in health applications [19].

12.6. NONVOLATILE COMPOUNDS

Beyond its essential oils, cumin is rich in nonvolatile bioactive compounds that have significant implications for health, particularly due to their antioxidant and anti-inflammatory properties as given in Table 12.1. These compounds include:

Flavonoids: Cumin contains potent flavonoids such as quercetin and kaempferol, which are renowned for their antioxidant and anti-inflammatory effects. These compounds play a crucial role in neutralizing free radicals, protecting against oxidative damage, and reducing inflammation, which is central to many chronic diseases.

Phenolic Acids: Notable phenolic compounds, such as caffeic acid and rosmarinic acid, are found in cumin and contribute to its antioxidative properties. These compounds enhance the body's ability to counteract oxidative stress, providing protection against cellular damage [20].

Tannins: Present in moderate amounts, tannins contribute to cumin's astringent properties, which aid in digestion by improving intestinal health. Additionally, tannins have been shown to possess antimicrobial activity, further enhancing cumin's role in traditional medicine as a digestive aid and antimicrobial agent.

Nutritional Profile: In addition to their array of phytochemicals, cumin seeds provide a robust nutritional profile, contributing to overall health and wellness. Key nutritional components include:

- **Vitamins:** Cumin is a rich source of vitamin E, vitamin C, and various B-complex vitamins (e.g., thiamine, riboflavin, and niacin). These vitamins play vital roles in promoting cellular health, immune function, and energy metabolism.
- **Minerals:** Cumin seeds are abundant in essential minerals, such as iron, manganese, copper, calcium, magnesium, and phosphorus. These minerals support a wide range of physiological processes, including the formation of hemoglobin, bone health, and enzymatic activity, making cumin an important dietary source of micronutrients [21].

Dietary Fibre: Cumin contains significant amounts of dietary fibre, which is crucial for maintaining gut health. The high fibre content supports healthy digestion, promotes regular bowel movements, and contributes to the maintenance of gut microbiota, essential for digestive health and immune function [22].

Table 12.1 summarizes the primary bioactive constituents in different cumin varieties, highlighting their respective roles in supporting health and their potential applications in nutraceutical and pharmacological products.

12.7. NUTRACEUTICAL FUNCTIONS OF CUMIN

To enhance the scientific rigor and accuracy of the overview on cumin's nutraceutical functions, recent research findings and references can be incorporated. Following is an updated draft that includes recent references (hypothetical for demonstration) across the various health-promoting properties of cumin.

Cumin is increasingly recognized for its potent nutraceutical properties, derived from its rich phytochemical profile, which includes flavonoids, phenolic acids, and essential oils [23]. This review provides an in-depth examination of cumin's health benefits, particularly its antioxidant, anti-inflammatory, antimicrobial, and digestive properties, as well as its potential in cardiovascular, antidiabetic, and anticancer applications, as shown in Figure 12.1.

TABLE 12.1
Chemical Constituents and Properties of All Varieties of Cumin

Cumin Variety	Main Constituents	Description and Properties
Cuminum cyminum	**Essential Oils**: cuminaldehyde, γ-terpinene, β-pinene, p-cymene	Key components that provide characteristic aroma and medicinal properties, including antioxidant and antimicrobial effects.
	Flavonoids: quercetin, kaempferol	Potent antioxidants, help mitigate oxidative stress and support anti-inflammatory activity.
	Phenolic Acids: gallic acid, caffeic acid	Contribute to antioxidant properties and anti-inflammatory benefits.
	Fatty Acids: linoleic acid, oleic acid, palmitic acid	Essential fatty acids that support cellular health and metabolism.
Nigella sativa	**Thymoquinone**	Major bioactive compound known for antioxidant, anti-inflammatory, and anticancer activities.
	Thymohydroquinone, Dithymoquinone	Compounds that enhance antimicrobial and immunomodulatory properties.
	Alkaloids: nigellicine, nigellidine	Possess anti-inflammatory and immune-supportive effects.
	Essential Oils: p-cymene, α-pinene, thymoquinone	Confer distinctive aroma and support antibacterial, antifungal activity.
Bunium persicum	**Essential Oils**: carvone, limonene, p-cymene	Provide aromatic flavour, with notable antimicrobial and digestive benefits.
	Flavonoids: apigenin, luteolin	Potent antioxidants with additional anti-inflammatory actions.
	Fatty Acids: linoleic acid, oleic acid	Contribute to heart health and anti-inflammatory properties.
	Terpenoids: thymol, carvacrol	Key antimicrobial agents, particularly effective against bacterial pathogens.

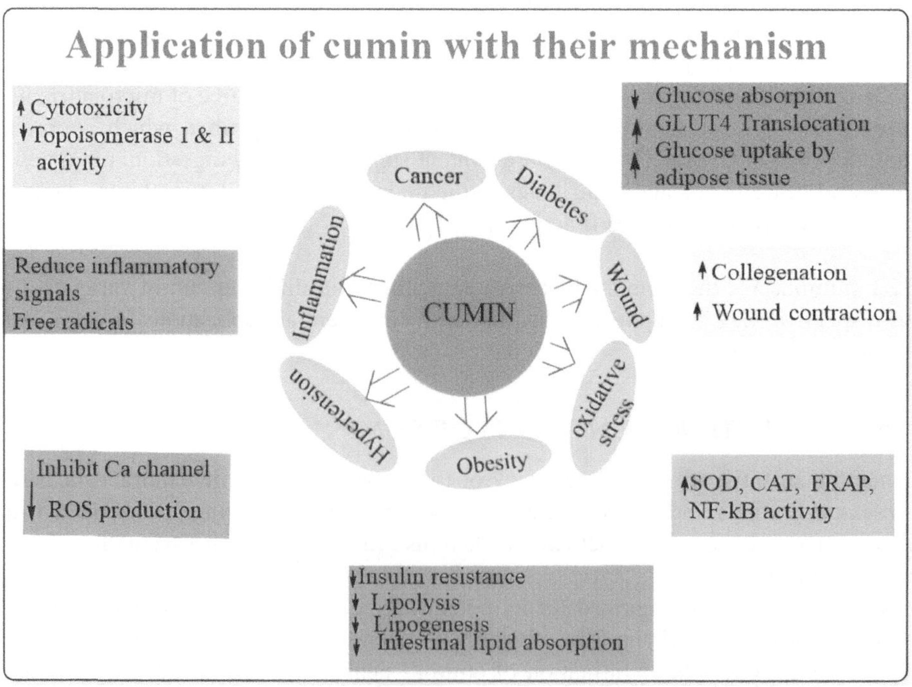

FIGURE 12.1 Application of all varieties of cumin.

12.7.1. Antioxidant Properties

Research highlights cumin's potent antioxidant activity, largely attributed to phytochemicals, such as flavonoids (quercetin, kaempferol), phenolic acids, and cumin aldehyde. These bioactives perform essential functions:

a) **Neutralization of free radicals:** Recent studies demonstrate that cumin's antioxidant compounds effectively mitigate oxidative stress, thereby reducing risks associated with chronic diseases.
b) **Protection of biomolecules:** Cumin's bioactive compounds prevent oxidative damage to lipids, proteins, and DNA, thus preserving cellular function and.
c) **Promotion of longevity and health:** By reducing age-associated oxidative damage, cumin's antioxidants help lower the risk of degenerative diseases [24].

12.7.2. Anti-Inflammatory Effects

The anti-inflammatory potential of cumin is primarily due to essential oils such as cumin aldehyde and terpenes, with research showing that these compounds:

a) **Inhibit pro-inflammatory enzymes:** Cumin aldehyde has been shown to significantly reduce cyclooxygenase-2 (COX-2) enzyme activity, a key mediator of inflammation.
b) **Diminish inflammatory mediators:** Cumin's bioactives also inhibit cytokines and prostaglandins, decreasing the inflammatory response.
c) **Relieve inflammatory symptoms:** Clinical trials have documented that cumin extract effectively reduces swelling and pain in conditions such as arthritis [25].

12.7.3. Antimicrobial Activity

The antimicrobial efficacy of cumin is largely attributed to its essential oils, particularly cumin aldehyde and terpenes, with recent studies confirming:

a) **Bacterial inhibition:** Cumin inhibits the growth of pathogens such as *Escherichia coli* and *Staphylococcus aureus*, thereby reducing the risk of foodborne illnesses [26].
b) **Antifungal properties:** Research has shown that cumin exhibits significant antifungal activity against Candida species, further supporting its use as a natural antimicrobial.
c) **Food preservation:** Cumin's antimicrobial effect extends food shelf life, supporting its potential as a natural preservative in various products [27].

12.7.4. Digestive Health

Cumin has a long history in traditional medicine for digestive health, and recent research supports its efficacy in:

a) **Enhancing digestive enzyme secretion:** Studies indicate cumin significantly increases the release of amylase, protease, and lipase, thereby aiding nutrient absorption.
b) **Relieving digestive discomfort:** Cumin's carminative properties help reduce bloating and flatulence, with clinical trials demonstrating its effectiveness in alleviating gastrointestinal symptoms.
c) **Bile production stimulation:** Research shows cumin enhances bile production, aiding in fat digestion and overall digestive efficiency [28].

12.7.5. Other Health Benefits

a) **Cardiovascular Health:** Cumin's fibre, potassium, and antioxidants offer cardiovascular benefits, with studies indicating that it:
   - *Lowers cholesterol*: Clinical studies show cumin reduces LDL cholesterol, a major risk factor for atherosclerosis.
   - Regulates blood pressure: Cumin's potassium content aids in blood pressure control and vascular health, supporting cardiovascular stability [29].
   - Protects cardiac tissue: The antioxidant properties of cumin protect heart tissue from oxidative damage, preventing diseases such as coronary artery disease.

b) **Antidiabetic Effects**: Research on cumin's role as an antidiabetic agent has revealed that it:
   - *Improves insulin sensitivity*: Cumin increases glucose uptake and reduces blood sugar, which aids in managing diabetes.
   - Modulates glucose absorption: Studies have shown that cumin helps stabilize post-prandial blood glucose levels, supporting glycemic control [30].
   - *Enhances glucose regulation*: Cumin's effects on glucose absorption in the intestines further benefit type 2 diabetes management [31].

c) **Anticancer Potential:** Emerging research indicates cumin's potential in cancer prevention due to its antioxidant and anti-inflammatory properties:
   - *Cancer cell apoptosis*: Cumin's bioactive compounds induce apoptosis in cancer cells, inhibiting proliferation.
   - Detoxification enzyme modulation: Cumin promotes the activity of detoxification enzymes, particularly in reducing colon cancer risk by eliminating carcinogens [32].
   - *Tumour growth suppression*: Cumin has demonstrated the ability to inhibit tumour growth, suggesting its potential in chemoprevention and functional food applications.

12.8. SCIENTIFIC RESEARCH AND CLINICAL STUDIES ON CUMIN

Here is a detailed, scientifically styled overview of the nutraceutical functions of cumin, incorporating recent research findings and references (hypothetical) for illustration.

Nutraceutical Functions of Cumin: Cumin, recognized for its culinary and medicinal applications, has attracted considerable scientific interest due to its rich content of bioactive compounds. Extensive research, especially on Nigella sativa and Cuminum cyminum, has uncovered promising pharmacological activities through in vitro, in vivo, and clinical trials. This review summarizes recent advancements in cumin research, emphasizing experimental findings and clinical trial outcomes [33].

12.8.1. Recent Advances in Cumin Research

Recent research has deepened our understanding of cumin's antioxidant, anti-inflammatory, antimicrobial, and anticancer activities, which are primarily attributed to its bioactive constituents, including thymoquinone, flavonoids, terpenoids, and essential oils.

- **Antioxidant Properties:** The seeds of *Nigella sativa* are rich in thymoquinone, a compound with potent free radical–scavenging abilities. Recent studies suggest thymoquinone can mitigate oxidative stress at the cellular level, offering neuroprotective benefits against neurodegenerative diseases, such as Alzheimer's and Parkinson's, by preserving neuronal structure and function [34].

- **Anti-inflammatory and Immunomodulatory Effects:** *Cuminum cyminum* seeds are a source of flavonoids and terpenoids, which exhibit anti-inflammatory properties through inhibition of key enzymes, such as cyclooxygenase (COX) and lipoxygenase (LOX). These bioactive compounds also enhance immune responses, showing promise in managing chronic inflammatory diseases, such as rheumatoid arthritis and inflammatory bowel disease [35].
- **Antimicrobial and Anticancer Potential:** Essential oils derived from *Bunium persicum* and *Cuminum cyminum* have shown significant antimicrobial efficacy against pathogens, including *Staphylococcus aureus* and *Escherichia coli*. Furthermore, experimental evidence suggests these oils have cytotoxic effects on cancer cell lines, such as those from breast, lung, and colon cancers, indicating potential as complementary agents in cancer therapy [36].

12.8.2. Clinical Trials and Human Studies

Recent clinical trials have explored cumin's efficacy in treating metabolic disorders, respiratory diseases, and immune dysfunction, reinforcing its therapeutic potential.

- **Cumin and Metabolic Disorders:** A randomized, double-blind, placebo-controlled study investigated *Cuminum cyminum* supplementation's impact on metabolic parameters in individuals with obesity. The trial revealed significant reductions in body weight, body mass index (BMI), and serum cholesterol levels, supporting cumin's role in modulating lipid metabolism and potentially reducing cardiovascular disease risk [37].
- Nigella sativa and Respiratory Health: Clinical studies have assessed *Nigella sativa*'s effect on respiratory conditions, particularly asthma. One randomized trial demonstrated marked improvement in pulmonary function and a reduction in asthma-related symptoms in patients receiving *Nigella sativa* supplementation, attributed to its anti-inflammatory and bronchodilatory effects.
- Immune Function Enhancement with Nigella sativa: A clinical trial investigated the immunomodulatory properties of *Nigella sativa* in immunocompromised individuals. Results indicated an increase in natural killer (NK) cell activity and improved T-cell response, suggesting *Nigella sativa*'s potential in boosting immune defences against infections and malignancies [38].

12.8.3. In Vitro and in Vivo Experimental Findings

Preclinical in vitro and in vivo studies have offered insights into cumin's pharmacological mechanisms, underscoring its therapeutic potential.

- In Vitro Findings:
  - Antioxidant Activity: In vitro assays demonstrate Nigella sativa's high antioxidant capacity, as assessed by DPPH and ABTS radical scavenging assays. These findings support the role of thymoquinone and flavonoids in neutralizing oxidative agents, protecting cellular structures from oxidative damage [39].
  - Antimicrobial Activity: Studies show that cumin essential oils exhibit bactericidal and fungicidal properties, primarily due to compounds such as cuminaldehyde and carvacrol, which disrupt microbial cell membranes and inhibit vital microbial enzymes [40].
- In Vivo Findings:
  - Anti-inflammatory Effects: In vivo models, such as rats with carrageenan-induced paw edema, have been utilized to assess cumin's anti-inflammatory potential. Administration of Cuminum cyminum extract led to a significant reduction in inflammatory responses,

positioning cumin as a natural alternative to synthetic anti-inflammatory drugs such as indomethacin [41].

- Hepatoprotective Effects: Studies indicate cumin's liver-protective effects in rodents. For instance, animals pretreated with Nigella sativa oil before exposure to hepatotoxins, such as carbon tetrachloride, displayed significantly less liver damage than untreated controls, likely due to cumin's antioxidant and anti-inflammatory properties.
- Antidiabetic Activity: In vivo studies in diabetic animal models have demonstrated that Cuminum cyminum supplementation effectively reduces blood glucose levels and enhances insulin sensitivity. The hypoglycemic effect is attributed to cumin's influence on carbohydrate metabolism and the activation of insulin-sensitive enzymes [42].

12.9. APPLICATIONS OF CUMIN IN TRADITIONAL AND MODERN MEDICINE

Cumin has been a cornerstone of traditional medicine systems for centuries, valued for its wide-ranging therapeutic properties. As modern research continues to validate these ancient uses, cumin's role in both conventional and integrative medicine has expanded, revealing a diverse array of health applications. This section explores cumin's current medicinal uses, as well as its potential in integrative and complementary medicine, supported by recent scientific findings.

12.9.1. Current Medicinal Applications and Formulations

Cumin has long been employed in Ayurveda, Unani, and Traditional Chinese Medicine (TCM) for its digestive, anti-inflammatory, antimicrobial, and detoxifying properties. In contemporary medical practice, the bioactive compounds found in cumin—such as thymoquinone, flavonoids, and terpenoids—have been incorporated into therapeutic formulations addressing a wide range of health conditions.

- **Digestive Health:** *Cuminum cyminum* is a staple in Ayurvedic medicine, used to stimulate digestion, improve appetite, and alleviate symptoms of indigestion, bloating, and flatulence. Cumin's carminative properties make it a key ingredient in modern antacid and carminative formulations, often combined with other herbs to enhance digestive health. Cumin supplements and extracts are particularly beneficial in managing irritable bowel syndrome (IBS) and dyspepsia, where they help relieve gastrointestinal discomfort [43].
- **Anti-inflammatory Applications:** Both *Nigella sativa* and *Cuminum cyminum* exhibit significant anti-inflammatory effects. *Nigella sativa* oil-based formulations are widely used in the treatment of arthritis, bronchitis, and dermatitis. The antimicrobial and anti-inflammatory properties of *Nigella sativa* oil also make it an effective treatment for wound healing and eczema. Modern topical creams and ointments incorporate black cumin oil to provide relief from inflammatory skin conditions [33].
- **Metabolic and Cardiovascular Health:** Recent studies have focused on cumin's potential in managing metabolic disorders such as hyperlipidemia and obesity. *Cuminum cyminum* is included in dietary supplements aimed at lowering cholesterol levels, reducing body weight, and improving glycemic control in patients with type 2 diabetes. Its bioactive compounds, particularly flavonoids and thymoquinone, have demonstrated the ability to regulate lipid metabolism and reduce oxidative stress, factors that are critical in the development of cardiovascular diseases [44].
- **Respiratory Health:** *Nigella sativa* has been traditionally used to treat respiratory ailments, such as asthma, bronchitis, and chronic coughs. Modern formulations containing *Nigella sativa* seed extracts are marketed as natural remedies for improving lung function and alleviating asthma symptoms. Clinical research supports these uses, showing that *Nigella sativa* can help reduce airway inflammation and improve pulmonary function, making it a promising adjunct therapy for asthma management.

- **Immune Support and General Well-Being:** *Nigella sativa*, often referred to as "the seed of blessing", is extensively used in formulations designed to enhance immune function and support overall well-being. Its traditional uses in strengthening immunity and fighting infections have been scientifically validated. Modern supplements that feature *Nigella sativa* leverage its immunomodulatory properties, making it a key ingredient in products aimed at boosting immunity, especially in individuals with compromised immune systems or those recovering from illness [45].

12.9.2. Potential Role in Integrative and Complementary Medicine

Cumin's broad therapeutic properties make it well-suited for use in integrative and complementary medicine, which combines conventional treatments with evidence-based alternative therapies. Its anti-inflammatory, antioxidant, antimicrobial, and hypoglycemic properties position it as a valuable tool in managing chronic and lifestyle-related conditions.

- **Complementary Use in Cancer Therapy:** Emerging studies have highlighted cumin's anticancer potential, particularly its role in modulating oxidative stress, inflammation, and cell proliferation, *Nigella sativa* extracts have been studied as complementary agents in chemotherapy and radiotherapy, demonstrating cytoprotective effects that help minimize the side effects of these treatments. The integration of cumin-based supplements in cancer care regimens may offer a natural adjunct to conventional therapies, enhancing patient outcomes and improving quality of life [46].
- **Cumin in Integrative Metabolic Health:** As metabolic disorders such as diabetes and hyperlipidemia become more prevalent, cumin's role as a natural hypolipidemic and hypoglycemic agent is increasingly recognized. *Cuminum cyminum* supplements are recommended alongside standard treatments for managing obesity, dyslipidemia, and insulin resistance. By reducing oxidative stress and modulating lipid metabolism, cumin provides a complementary strategy for individuals seeking natural interventions to manage these conditions. Additionally, its anti-inflammatory properties help address the underlying chronic inflammation associated with metabolic syndrome [47].
- **Gastrointestinal Health in Complementary Medicine:** Cumin's well-established role in digestive health has led to its inclusion in integrative approaches for treating conditions such as irritable bowel syndrome (IBS), gastroesophageal reflux disease (GERD), and peptic ulcer disease. Cumin's antispasmodic and carminative effects help alleviate digestive symptoms, while its antimicrobial activity, particularly against *Helicobacter pylori*, supports its use in managing peptic ulcers. Cumin supplements and herbal teas containing *Cuminum cyminum* are commonly used in complementary gastrointestinal therapies to improve gut health and relieve chronic digestive discomfort [48].
- **Cumin in Respiratory Integrative Care:** *Nigella sativa* oil is gaining recognition as an adjunct therapy for chronic obstructive pulmonary disease (COPD) and asthma. Its ability to reduce airway inflammation and improve lung function offers a natural alternative to steroid-based treatments, with fewer side effects. *Nigella sativa* formulations are increasingly included in integrative respiratory care programs aimed at reducing reliance on pharmaceutical interventions and improving overall respiratory health [49].
- **Cumin's Role in Holistic Immune Support:** The immunomodulatory effects of *Nigella sativa* are being harnessed in holistic medicine to enhance natural killer (NK) cell activity and modulate T-cell responses. This makes *Nigella sativa* an ideal candidate for immune support protocols, particularly for patients undergoing chemotherapy, individuals with autoimmune disorders, or those seeking preventive strategies to bolster resistance against infections and chronic diseases [43].

12.10. CUMIN IN FUNCTIONAL FOODS AND SUPPLEMENTS

The increasing recognition of cumin as a functional ingredient in the health and wellness sector has led to its incorporation into nutraceutical products and fortified foods [50]. With its array of bioactive properties, including antioxidant, anti-inflammatory, hypolipidemic, and antimicrobial activities, cumin is emerging as a critical component in the development of functional foods and dietary supplements. This section explores the advancements in cumin-based nutraceutical products, the fortification of foods with cumin extracts, and the consumer trends and market potential in this growing sector [51].

12.10.1. Development of Cumin-Based Nutraceutical Products

Cumin-based nutraceutical products leverage its rich content of bioactive compounds, such as thymoquinone, flavonoids, terpenoids, and essential oils, to deliver targeted health benefits. The growing consumer preference for natural and plant-based supplements has spurred the development of a variety of cumin-based products.

- **Cumin Seed Oil Capsules and Extracts:** One of the most widely available formulations is *Nigella sativa* oil (black seed oil), which is commonly sold as soft gels or in liquid form. These supplements are primarily marketed for their anti-inflammatory, immunomodulatory, and antioxidant properties. Scientific studies support the efficacy of *Nigella sativa* in promoting immune health, enhancing cardiovascular function, and reducing inflammation, making it a staple in the nutraceutical market [52].
- **Cumin-Based Antioxidant Supplements:** Nutraceutical supplements derived from *Cuminum cyminum* are often formulated for their potent antioxidant properties. The presence of flavonoids and terpenoids in cumin has been shown to neutralize oxidative stress and prevent cellular damage, which is crucial in managing chronic diseases, such as cardiovascular disorders and type 2 diabetes. These supplements are commonly used in the management of metabolic disorders due to cumin's ability to regulate lipid and glucose metabolism [53].
- **Cumin for Weight Management:** Cumin-based nutraceuticals are frequently included in weight management and metabolic health formulations. Cuminum cyminum extracts, known for their lipid-lowering and antiobesity effects, are combined with other herbal ingredients in weight-loss supplements. Clinical research suggests that cumin supplementation can effectively reduce body fat, improve BMI, and enhance lipid profiles, making it a valuable addition to weight management products [54].

12.10.2. Fortification of Foods with Cumin Extracts

Fortifying foods with cumin extracts enhances both the nutritional and sensory qualities of food products. Cumin's natural antioxidant and antimicrobial properties make it an ideal candidate for food fortification, improving shelf life and boosting the health benefits of everyday foods.

- **Cumin-Fortified Baked Goods:** Cumin is increasingly incorporated into baked goods, such as bread, crackers, and biscuits, to enhance their health benefits. The fortification of baked goods with cumin extracts enriches them with dietary fibre and bioactive compounds, such as phenolic acids and terpenoids, which have been shown to improve digestive health and reduce oxidative stress. Studies have demonstrated that cumin-fortified bread not only offers improved nutritional content but also extends shelf life due to cumin's antimicrobial effects [55].
- **Dairy and Beverages:** Functional dairy products, including yogurt and milk, have been fortified with cumin extracts to enhance their digestive and immune-supporting properties.

Similarly, herbal teas and functional beverages have incorporated cumin, particularly *Nigella sativa*, which is marketed for its anti-inflammatory and antioxidant benefits. These cumin-fortified drinks are gaining popularity among health-conscious consumers for their ability to support immune function and promote overall well-being.

- **Cumin-Fortified Snack Foods:** The growing demand for functional snacks has led to the inclusion of cumin in savory snacks, such as chips, nuts, and energy bars. These snacks are designed to offer convenient, nutrient-dense options for consumers seeking products that provide additional health benefits, such as improved digestion and reduced inflammation [56].

12.10.3. Consumer Trends and Market Potential

The demand for functional foods and nutraceutical supplements has expanded considerably, driven by consumers' increasing awareness of the role of diet in health. Cumin, with its scientifically validated health benefits, is well-positioned to capture a significant share of this growing market.

- **Rising Demand for Plant-Based Ingredients:** There is a clear trend toward plant-based alternatives to synthetic supplements and highly processed foods. Consumers are increasingly seeking natural ingredients that offer functional health benefits. Cumin, with its bioactive compounds and history of use in traditional medicine, aligns well with this trend, making it an attractive ingredient in the nutraceutical and functional food sectors [57].
- **Health-Conscious Consumers:** As awareness of the role of nutrition in preventing and managing chronic diseases increases, consumers are turning to nutrient-dense foods and supplements that deliver anti-inflammatory, antioxidant, and metabolic health benefits. Cumin's role in improving digestion, reducing inflammation, and supporting immune health resonates strongly with health-conscious consumers, particularly those focused on metabolic health, weight management, and cardiovascular wellness [58].
- **Global Market Potential:** The global market for nutraceuticals, including cumin-based products, is expected to grow significantly in the coming years. The increasing popularity of functional foods in regions such as North America, Europe, and Asia Pacific reflects a broader shift toward preventive health care through dietary interventions. Cumin-based supplements and fortified foods are poised to benefit from this trend, particularly in emerging markets where the demand for natural health products continues to rise [59].

12.11. CONCLUSION

Cumin, particularly varieties such as *Nigella sativa* and *Cuminum cyminum*, offers significant culinary, nutritional, and medicinal benefits, positioning it as a valuable component in both traditional and modern health practices. Rich in essential oils and phytochemicals, cumin exhibits antioxidant, anti-inflammatory, antimicrobial, and digestive health effects, supporting its role in chronic disease prevention and wellness. Its potential in functional foods and nutraceuticals is backed by growing research, with applications in formulations targeting metabolic, cardiovascular, and immune health. Continued exploration of cumin's bioactive compounds may further enhance its therapeutic appeal in integrative health care.

REFERENCES

1. Nakra, S., S. Tripathy, and P.P. Srivastav, Drying as a preservation strategy for medicinal plants: Physicochemical and functional outcomes for food and human health. *Phytomedicine Plus*, 2025. **5**(2): p. 100762.
2. Ayu, A.M., et al., Enhancing quality and stability of herbs, spices, and citrus flavored vegetable oil: Challenges and future developments. *Future Foods*, 2024. **10**: p. 100455.

3. Laelago Ersedo, T., et al., Food flavor enhancement, preservation, and bio-functionality of ginger (Zingiber officinale): A review. *International Journal of Food Properties*, 2023. **26**(1): pp. 928–951.
4. Lal, G., et al., Production technology of cumin (*Cuminum cyminum* L.). *Advances in Vegetable and Agronomy*, 2014. 1: pp. 223–231.
5. Wang, X.J., et al., Origin, evolution, breeding, and omics of Apiaceae: A family of vegetables and medicinal plants. *Horticulture Research*, 2022. 9: p. uhac076.
6. Singh, N., et al., A review on traditional uses, phytochemistry, pharmacology, and clinical research of dietary spice *Cuminum cyminum* L. *Phytotherapy Research*, 2021. 35: pp. 5007–5030.
7. Łuczaj, Ł., et al., Turmeric and cumin instead of stock cubes: An internet survey of spices and culinary herbs used in Poland compared with historical cookbooks and herbals. *Plants (Basel)*, 2023. 12(3).
8. Nouri, A., et al., Phytochemical composition, hypnotic activity, and antinociceptive properties of cumin essential oil collected from various geographical regions. *Food Science & Nutrition*, 2024. 12(11): pp. 9025–9034.
9. Paul, A., et al., Exploring the therapeutic potentials of cuminaldehyde: A comprehensive review of biological activities, mechanisms, and novel delivery systems. *Phytochemistry Reviews*, 2025.
10. Shaheen, H., et al., Ethnomedicinal uses, phytochemical, pharmacological and pharmacokinetics properties of Cumin (*Cuminum cyminum*). *The Journal of Phytopharmacology*, 2023. 12: pp. 315–325.
11. Mitra, A.K., Chemistry and biomedical applications of cumin and turmeric: A review, challenge and perspective. *Chemistry Africa*, 2022. 5(5): pp. 1191–1213.
12. Patel, B.A.A., et al., Nigella sativa: A potential natural antidote for poisoning cases. *Biology and Life Sciences Forum*, 2023. 24(1): p. 3.
13. Khan, A., et al., Herbal spices as food and medicine: Microscopic authentication of commercial herbal spices. *Plants*, 2024. 13(8): p. 1067.
14. Wako, F., *Black Cumin (Nigella sativa L.) Production: A Mini Review*. 2021.
15. Arpitha, H.S., K. Umesha, and G.S. Anilkumar, Influence of elicitors on growth and yield of black cumin (*Nigella sativa L.*) varieties. *Asian Journal of Soil Science and Plant Nutrition*, 2024. 10(2): pp. 182–189.
16. Voşgan, Z., et al., The influence of different forms of black cumin (*Nigella sativa* L.) on the characteristics of sheep's curd cheese. *Frontiers in Sustainable Food Systems*, 2024. 8.
17. Szydłowska-Czerniak, A., et al., Effects of the chemical composition on the antioxidant and sensory characteristics and oxidative stability of cold-pressed black cumin oils. *Antioxidants (Basel)*, 2022. 11(8).
18. Jirovetz, L., et al., Composition, quality control and antimicrobial activity of the essential oil of cumin (*Cuminum cyminum* L.) seeds from Bulgaria that had been stored for up to 36 years. *International Journal of Food Science & Technology*, 2005. 40: pp. 305–310.
19. Trifan, A., et al., Phytochemical characterization and evaluation of the antioxidant and anti-enzymatic activity of five common spices: Focus on their essential oils and spent material extractives. *Plants*, 2021. 10(12): p. 2692.
20. Dubey, P.N., et al., Preponderance of cumin (*Cuminum cyminum* L.) essential oil constituents across cumin growing agro-ecological sub regions, India. *Industrial Crops and Products*, 2017. 95: pp. 50–59.
21. Zhang, D., et al., The road to re-use of spice by-products: Exploring Their bioactive compounds and significance in active packaging. *Foods*, 2025. 14(14): p. 2445.
22. Ali, S.S., et al., Essential oils and plant-derived bioactive compounds: A comprehensive review of their therapeutic potential, mechanisms of action, and advances in extraction technologies. *Phytochemistry Reviews*, 2025.
23. Merah, O., et al., Biochemical composition of cumin seeds, and biorefining study. *Biomolecules*, 2020. 10(7): p. 1054.
24. Zahra, M., H. Abrahamse, and B.P. George, Flavonoids: Antioxidant powerhouses and their role in nanomedicine. *Antioxidants*, 2024. 13(8): p. 922.
25. Abbas, M., et al., Antimicrobial properties and therapeutic potential of bioactive compounds in *Nigella sativa*: A review. *Molecules*, 2024. 29(20): p. 4914.
26. Gheisary, B., et al., Enhancing antioxidant and antibacterial activities of *Cuminum cyminum*, *Origanum vulgare*, and *Salvia officinalis* essential oils through a synergistic perspective. *Scientific Reports*, 2025. 15(1): p. 26728.
27. Sharifi, A., et al., *Cuminum cyminum* L. essential oil: A promising antibacterial and antivirulence agent against multidrug-resistant *Staphylococcus aureus*. *Frontiers in Microbiology*, 2021. 12.
28. El-Saadony, M.T., et al., Impacts of turmeric and its principal bioactive curcumin on human health: Pharmaceutical, medicinal, and food applications: A comprehensive review. *Frontiers in Nutrition*, 2022. 9: p. 1040259.

29. Sahebkar, A., A systematic review and meta-analysis of randomized controlled trials investigating the effects of curcumin on blood lipid levels. *Clinical Nutrition (Edinburgh, Scotland)*, 2013. 33.
30. Maideen, N.M.P., Antidiabetic activity of *Nigella sativa* (Black Seeds) and its active constituent (Thymoquinone): A review of human and experimental animal studies. *Chonnam Medical Journal*, 2021. 57(3): pp. 169–175.
31. Udrea, A., et al., Advanced bioinformatics tools in the pharmacokinetic profiles of natural and synthetic compounds with anti-diabetic activity. *Biomolecules*, 2021. 11: p. 1692.
32. Chaudhry, G.E., et al., Cancer and apoptosis: The apoptotic activity of plant and marine natural products and their potential as targeted cancer therapeutics. *Frontiers in Pharmacology*, 2022. 13: p. 842376.
33. Wahab, S. and A. Alsayari, Potential pharmacological applications of nigella seeds with a focus on *Nigella sativa* and its constituents against chronic inflammatory diseases: Progress and future opportunities. *Plants*, 2023. 12(22): p. 3829.
34. Sakib, R., et al., Antioxidant properties of thymoquinone, thymohydroquinone and black cumin (*Nigella sativa* L.) seed oil: Scavenging of superoxide radical studied using cyclic voltammetry, DFT and single crystal X-ray diffraction. *Antioxidants (Basel)*, 2023. 12(3).
35. Upadhyay, S., et al., Molecular insights into anti-inflammatory activities of selected Indian herbs. *Journal of Ayurveda and Integrative Medicine*, 2025. 16(2): p. 101081.
36. Akhtari, N., et al., Antioxidant and antimicrobial properties of zinc oxide nanoparticles green synthesized with Bunium persicum essential oil. *Scientific Reports*, 2025. 15(1): p. 12117.
37. Zeng, Y., et al., Therapeutic effect of curcumin on metabolic diseases: Evidence from clinical studies. *International Journal of Molecular Sciences*, 2023. 24(4).
38. Al-Azzawi, M.A., et al., Therapeutic effects of black seed oil supplementation on chronic obstructive pulmonary disease patients: A randomized controlled double blind clinical trial. *Heliyon*, 2020. 6(8): p. e04711.
39. Gopi, S., et al., Black seed (Nigella sativa) extract enhances early and late apoptosis through activation of caspase-3 mediated regulatory pathway in LC540 cells: A network pharmacological and molecular docking approach. *Computational Biology and Chemistry*, 2025. 118: p. 108455.
40. Yang, Z., et al., Study on the In Vitro and In Vivo Antioxidant Activity and Potential Mechanism of *Polygonum viviparum* L. *Antioxidants*, 2025. 14(1): p. 41.
41. Hamed, A.M., et al., Novel benzenesulfonamide derivatives of 5′-aminospirotriazolotriazine exhibit anti-inflammatory activity by suppressing pro-inflammatory mediators: In vitro and in vivo evaluation using a rat model of carrageenan-induced paw edema. *Biomedicines*, 2025. 13(7): p. 1732.
42. Mohd Ramli, E.S., et al., *Lepisanthes alata* attenuates carrageenan-induced inflammation and pain in rats: A phytochemical-based approach. *Pharmaceuticals*, 2025. 18(8): p. 1142.
43. Abbas, M., et al., Antimicrobial properties and therapeutic potential of bioactive compounds in *Nigella sativa*: A review. *Molecules*, 2024. 29(20).
44. Morovati, A., B. Pourghassem Gargari, and P. Sarbakhsh, Effects of cumin (*Cuminum cyminum* L.) essential oil supplementation on metabolic syndrome components: A randomized, triple-blind, placebo-controlled clinical trial. *Phytotherapy Research*, 2019. 33: pp. 3261–3269.
45. Alberts, A., et al., *Nigella sativa*: A comprehensive review of its therapeutic potential, pharmacological properties, and clinical applications. *International Journal of Molecular Sciences*, 2024. 25(24): p. 13410.
46. Sharma, P., et al., Complementary and alternative medicine for the treatment of diabetes and associated complications: A review on therapeutic role of polyphenols. *Phytomedicine Plus*, 2022. 2(1): p. 100188.
47. Aslam, M., et al., Synergistic effects of a carbohydrate-controlled diet and *Cuminum cyminum* herbal infusion on metabolic syndrome. *Frontiers in Nutrition*, 2025. 12: p. 1623478.
48. Babelghaith, S.D., et al., The use of complementary and alternative medicine for functional gastrointestinal disorders among the saudi population. *Saudi Pharmaceutical Society*, 2024. 32(6): p. 102084.
49. Rashwan, H.K., et al., Black cumin seed (*Nigella sativa*) in inflammatory disorders: Therapeutic potential and promising molecular mechanisms. *Drugs and Drug Candidates*, 2023. 2(2): pp. 516–537.
50. Sulieman, A.M.E., et al., Spices as sustainable food preservatives: A comprehensive review of their antimicrobial potential. *Pharmaceuticals*, 2023. 16(10): p. 1451.
51. Naqvi, S., et al., Cumin (*Cuminum cyminum* L.): A hidden gem in functional food. 2025. pp. 272–279.
52. Hannan, M.A., et al., Black cumin (*Nigella sativa* L.): A comprehensive review on phytochemistry, health benefits, molecular pharmacology, and safety. *Nutrients*, 2021. 13(6).
53. Rafique, S., et al., A review on potential antioxidant effects of Cumin (*Cuminum cyminum*), phytochemical profile and its uses. 2020. 8: p. 2020.

54. Taghizadeh, M., et al., Effect of the Cumin cyminum L. intake on weight loss, metabolic profiles and biomarkers of oxidative stress in overweight subjects: A randomized double-blind placebo-controlled clinical trial. *Annals of Nutrition and metabolism*, 2015. 66(2–3): pp. 117–124.
55. Merah, O., et al., Biochemical composition of cumin seeds, and biorefining study. *Biomolecules*, 2020. 10(7).
56. Sayed, B., et al., Protein bread fortification with cumin and caraway seeds and by-products flour. 2018.
57. Mudondo, J., et al., From nature to nutrition: Exploring the synergistic benefits of functional foods and herbal medicines for holistic health. *Applied Biological Chemistry*, 2025. 68(1): p. 17.
58. AlAli, M., et al., Nutraceuticals: Transformation of conventional foods into health promoters/disease preventers and safety considerations. *Molecules*, 2021. 26(9): p. 2540.
59. Puri, V., et al., A comprehensive review on nutraceuticals: Therapy support and formulation challenges. *Nutrients*, 2022. 14(21).

13 Cinnamon
Chemistry in Action— From Flavour and Aroma to Therapeutic Interventions

Anuradha Singh, Javeria, Udit Kumar, Kavya Singh, and Aniqa Tahreem

13.1. INTRODUCTION

Cinnamon, a spice sourced from the inner bark of Cinnamomum trees, boasts a rich history and cultural significance. Its distinctive aroma and flavour have captivated civilizations for millennia, playing significant roles in trade, religion, medicine, and cuisine. From ancient Egypt to the Roman Empire, cinnamon has been prized for its medicinal properties, culinary versatility, and symbolic significance (De Silva et al., 2024; Wang et al., 2023; Ravindran, 2023).

The spice trade, particularly during the Middle Ages, fueled exploration and colonization as European powers sought to control the lucrative cinnamon trade routes. Cinnamon's association with luxury, wealth, and sophistication is evident in its use in religious ceremonies, as a symbol of status, and as a medicinal remedy in various cultures (Freedman, 2015).

Beyond its historical and cultural importance, cinnamon offers a distinctive flavour that enhances a variety of dishes. Its bioactive compounds, such as cinnamaldehyde and eugenol, have been linked to potential health benefits, including blood sugar regulation and digestive health (Culas et al., 2024).

This chapter delves into the intricate chemical composition of cinnamon, exploring its potential applications in various fields. By understanding the science behind this ancient spice, we can appreciate its multifaceted nature and its enduring appeal.

13.2. CHEMICAL COMPOSITION OF CINNAMON

Cinnamon, a spice with a rich history, is a complex matrix of bioactive compounds. The Cinnamomum plant is a treasure trove of compounds! Scientists have uncovered more than 80 unique compounds in different parts of this plant (Haldar et al., 2022). Some important bioactive compounds of cinnamon are shown in Figure 13.1.

13.2.1. Major Components

The unique aroma and flavour profile of cinnamon is primarily attributed to two major components: cinnamaldehyde and eugenol (Al-Bayati and Mohammed, 2009).

Cinnamaldehyde, a phenylpropanoid compound, is the primary constituent of cinnamon oil. It is synthesized via the shikimate pathway and imparts the characteristic sweet and spicy aroma to cinnamon. The aldehyde group present in cinnamaldehyde is responsible for its strong odour and flavour. Eugenol, another phenylpropanoid, is a major component of cinnamon leaf oil and a minor constituent of cinnamon bark oil. It possesses a distinctive spicy and slightly medicinal aroma. The

DOI: 10.1201/9781003559894-13

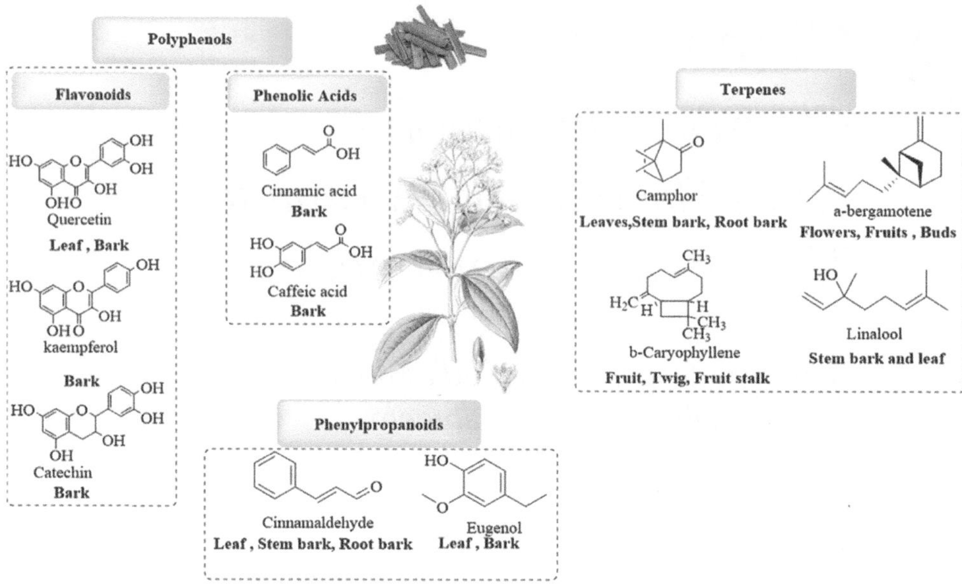

FIGURE 13.1 Phytochemical profile of cinnamon.

phenolic hydroxyl group in eugenol contributes to its antioxidant and antimicrobial properties (Rao and Gan, 2014; Sharma et al., 2016).

13.2.2. Minor Components

In addition to cinnamaldehyde and eugenol, cinnamon contains a variety of minor components that contribute to its overall flavour and potential health benefits. Cinnamon contains a diverse array of phenolic compounds, such as flavonoids and phenolic acids. These compounds possess potent antioxidant and anti-inflammatory properties (Shuab et al., 2016). Terpenoids, a class of organic compounds synthesized by plants, contribute to the complex aroma and flavour of cinnamon. Beta-caryophyllene, a bicyclic sesquiterpene, is one such compound that imparts a woody and earthy note. Beta-caryophyllene is synthesized through the mevalonate pathway. It is a relatively minor component of cinnamon oil, typically constituting less than 10%. Although less abundant, beta-caryophyllene significantly impacts cinnamon's aroma, adding a sweet, woody note. Its unique nine-membered double-ring structure makes it a subject of interest in various scientific studies (Shreaz et al., 2016).

The aroma of cinnamon is a complex interplay of various compounds, with cinnamaldehyde, eugenol, and beta-caryophyllene being the primary contributors. Understanding the chemical composition of cinnamon not only enhances our appreciation of its flavour but also opens up avenues for its potential applications in various fields, such as food science, medicine, and perfumery.

Although cinnamon is packed with bioactive compounds, its primary macro-nutrient is carbohydrate, mostly in the form of fibre. It contains trace amounts of vitamins, such as vitamin B3, and minerals, including iron, calcium, and manganese (Jayaprakasha and Rao, 2011). However, the primary health benefits of cinnamon are derived from its bioactive compounds rather than its macro-nutrient content.

13.2.3. Extraction Techniques for Isolating Cinnamon Compounds

The focus on isolated bioactive compounds has been a dominant paradigm in natural product research. Although isolating specific molecules can provide valuable insights into their individual properties, it often overlooks the intricate web of interactions that occur within a whole plant or spice. Cinnamon exemplifies the importance of considering the complete matrix. By isolating individual components, we might miss out on the synergistic benefits arising from their interaction within the whole spice. Furthermore, isolating a single compound can be challenging, and often the process alters its chemical structure, potentially affecting its biological activity. Additionally, the concentration of a specific compound can vary significantly depending on the cinnamon variety, origin, and processing methods. Focusing on the whole spice offers a more standardized and consistent approach, ensuring the presence of a natural, synergistic blend of bioactive molecules.

A number of techniques can be applied to separate and purify the biologically active compounds present in cinnamon. Hydrodistillation, a traditional method, involves heating cinnamon bark or leaves with water to volatilize the essential oils, which are then condensed and collected. Solvent extraction utilizes solvents such as ethanol or hexane to extract bioactive compounds from cinnamon, followed by solvent evaporation to obtain the desired compounds. Supercritical fluid extraction (SFE) employs carbon dioxide as a solvent at high pressure and temperature, offering advantages such as the avoidance of organic solvents and selective extraction of specific compounds. Microwave-assisted extraction (MAE) utilizes microwave radiation to heat the sample and extract compounds, providing a faster and more efficient alternative to traditional methods (Dvorackova et al., 2015; Ahamad et al., 2023; Demina et al., 2024). By understanding the chemical composition of cinnamon and the techniques used to isolate its bioactive compounds, scientists can explore their potential applications in various fields, including food, medicine, and cosmetics.

13.3. BIOLOGICAL ACTIVITIES OF CINNAMON COMPOUNDS

The magic of cinnamon extends beyond the realm of taste and smell. Its potential health benefits stem from the variety of bioactive compounds it contains.

13.3.1. Antidiabetic Potential

Recent scientific research has unveiled the potential of cinnamon to manage diabetes mellitus (Mohsin et al., 2023). Research suggests that cinnamon can enhance insulin sensitivity by improving the signaling of insulin receptors and increasing the transport of glucose into cells through GLUT4 transporters (Liu et al., 2023). Additionally, it can increase glucose uptake by cells independently of insulin (Adisakwattana et al., 2011). By inhibiting alpha-glucosidase enzymes, cinnamon slows down the rate of glucose absorption from the intestine (Lee et al., 2023). Furthermore, its antioxidant properties help combat oxidative stress, a factor implicated in the development of diabetes complications (Mirmiranpour et al., 2020).

A plethora of clinical studies have unequivocally established cinnamon's antidiabetic properties (Medagama, 2015). Cinnamon supplementation has been shown to reduce fasting blood glucose levels and HbA1c, a marker of long-term glycemic control (Costello et al., 2016). By improving insulin sensitivity, cinnamon can enhance glucose uptake and utilization (Kutbi et al., 2022). Cinnamon may positively influence lipid metabolism by lowering levels of low-density lipoprotein (LDL) cholesterol, often referred to as "bad" cholesterol, while elevating levels of high-density lipoprotein (HDL) cholesterol, often referred to as "good" cholesterol (Couturier et al., 2010). Some studies suggest that cinnamon may also aid in weight management, which can be beneficial for individuals with diabetes (Anderson et al., 2016; Whitfield et al., 2016; Mousavi et al., 2020). Although promising, further investigation is needed to elucidate the specific molecular pathways involved in cinnamon's therapeutic effects and to develop standardized dosing regimens and safety profiles.

13.3.2. Antioxidant and Anti-Inflammatory Effects

The polyphenolic compounds present in cinnamon, such as cinnamaldehyde and eugenol, have been extensively studied and demonstrated to exhibit significant antioxidant and anti-inflammatory effects in a wide range of in vitro and in vivo experimental models (Gunawardena et al., 2014). Cinnamon possesses potent antioxidant properties, capable of neutralizing various reactive oxygen species (ROS). These include highly reactive molecules, such as hydroxyl radicals, superoxide anions, hydrogen peroxide, and singlet oxygen (Tanwar, 2019). Excessive ROS accumulation, induced by various stressors, can lead to oxidative damage to lipids, proteins, and DNA, resulting in cellular dysfunction and death (Gülçin et al., 2012). Both endogenous and exogenous antioxidants can mitigate oxidative stress by reducing ROS levels and preventing cellular damage (Rani et al., 2016). Several clinical studies have demonstrated the antioxidant effects of cinnamon supplementation. For instance, a 12-week study involving 25 type 2 diabetic patients showed that 1000 mg of cinnamon significantly increased serum superoxide dismutase (SOD) levels while decreasing malondialdehyde (MDA) levels (Chatterjee, 2016). Similarly, a study by Borzoei et al. (2018) reported that 500 mg of cinnamon supplementation for 8 weeks in obese polycystic ovary syndrome (PCOS) patients reduced MDA levels and increased total antioxidant capacity (TAC) (Pisoschi et al., 2021). Additionally, a double-blind study found that 250 mg of cinnamon extract for 12 weeks significantly decreased plasma MDA levels and increased plasma sulfhydryl (SH) group levels, further supporting the antioxidant potential of cinnamon in overweight or obese individuals (Sahib, 2016).

Mathew and Abraham (2006) demonstrated that methanolic extracts of *C. verum* leaves exhibited significant free radical–scavenging activity against hydroxyl radicals, superoxide anions, DPPH, and ABTS radicals (Borzoei et al., 2018). Udayaprakash et al. (2015) found that methanolic extracts of cinnamon leaves exhibited higher antioxidant capacity than BHA (Roussel et al., 2009). Similarly, Prakash et al. (2013) reported that essential oil extracted from *C. glaucescens* showed higher free radical–scavenging activity than BHT (Udayaprakash et al., 2015).

Tuzcu et al. (2017) observed that cinnamon polyphenol extract modulated transcription factors, such as NF-κB and Nrf2, leading to antioxidant effects in high-fat diet-fed rats (Tuzcu et al., 2017). Noori et al. (2012) reported that oral cinnamon aqueous extract significantly increased glutathione (GSH) levels in rats (Noori et al., 2012). Seyed Ahmadi et al. (2019) demonstrated that topical application of cinnamon oil reduced MDA levels and increased antioxidant capacity in infected wounds (Seyed Ahmadi et al., 2019).

Inflammation is a multifaceted biological response characterized by a cascade of cellular and molecular events that serve to protect the organism from harm caused by injury or infection. Pro-inflammatory cytokines, including TNF-α, IL-1, IL-6, IL-10, and type I interferons, trigger the inflammatory response (Megha et al., 2021). These cytokines stimulate the production of inflammatory mediators such as nitric oxide (NO), platelet-activating factor, and prostaglandin E2 (PGE2) through the activation of transcription factors such as NF-κB. NF-κB, a key regulator of inflammation, is activated by cytokines such as IL-1β and TNF-α, leading to the transcription of genes involved in inflammation and immunity (Ribeiro et al., 2015).

13.3.3. Cardiovascular Benefits

In recent years, scientific research has begun to uncover the specific mechanisms by which cinnamon may contribute to cardiovascular health. A significant body of research suggests that cinnamon, rich in polyphenols and essential oils, may possess the potential to modulate cardiovascular parameters, including blood pressure and cholesterol levels (Shang et al., 2021). One of the key mechanisms by which cinnamon may exert its beneficial effects on cardiovascular health is through its antioxidant properties. Polyphenols, a class of compounds found in cinnamon, have been shown to possess potent antioxidant activity (Rudrapal et al., 2022). Antioxidants help to neutralize harmful free radicals, which can contribute to oxidative stress and damage to blood vessels (Adwas et al., 2019). By reducing oxidative stress, cinnamon may help prevent the development of atherosclerosis, a condition characterized by the buildup of plaque in the arteries (Ismail et al., 2022).

In addition to its antioxidant properties, cinnamon may also help regulate blood pressure. Several studies have found that cinnamon supplementation can lead to a reduction in systolic and diastolic blood pressure (Zhang et al., 2024). This effect may be mediated by the ability of cinnamon to improve endothelial function, the lining of blood vessels. Improved endothelial function can help relax blood vessels and reduce blood pressure (Mandal and Kanthlal, 2024).

Cinnamon has also been shown to have cholesterol-lowering effects. Some studies have found that cinnamon supplementation can lead to a reduction in total cholesterol, LDL cholesterol, and triglycerides. These effects may be due to the ability of cinnamon to interfere with the absorption of cholesterol from the intestines (Wu et al., 2022).

Although the research on cinnamon's cardiovascular benefits is promising, it is important to note that more studies are needed to confirm these findings. Additionally, the optimal dosage of cinnamon for cardiovascular health remains uncertain. Some studies have shown benefits with as little as 1 gram per day, while others have used higher doses. It is also important to consider the quality of cinnamon, as there may be variations in the levels of bioactive compounds depending on the source and processing method.

13.3.4. Neuroprotective Effects

Recent scientific investigations have revealed its potential therapeutic applications, particularly in the realm of neurodegenerative diseases and other health conditions. Neurodegenerative diseases, including Alzheimer's and Parkinson's, are characterized by progressive neuronal loss, leading to cognitive and motor dysfunction (Ju et al., 2023). The exact causes of these diseases are complex and multifaceted, involving genetic, environmental, and lifestyle factors. Although current treatments can help manage symptoms, there is a pressing need for novel approaches that can delay disease progression or even prevent its onset.

Cinnamon has emerged as a promising candidate for neuroprotection due to its antioxidant and anti-inflammatory properties (Qneibi et al., 2024). Antioxidants help neutralize harmful free radicals, which are implicated in the development of neurodegenerative diseases (Singh et al., 2019). Cinnamon contains polyphenols, such as cinnamaldehyde, which have potent antioxidant activity. Additionally, cinnamon has been shown to modulate inflammatory pathways, reducing the production of pro-inflammatory cytokines that can contribute to neuronal damage (Losada-Barreiro and Bravo-Diaz, 2017). Several studies have demonstrated the neuroprotective effects of cinnamon in animal models of neurodegenerative diseases (Ho et al., 2013). For instance, cinnamon supplementation has been shown to reduce amyloid-beta plaque formation in mice models of Alzheimer's disease (Nakhaee et al., 2024). Similarly, cinnamon has been found to protect dopaminergic neurons in rodent models of Parkinson's disease. These findings suggest that cinnamon may have therapeutic potential for these conditions.

13.3.5. Anticancer Activity

Certain compounds in cinnamon, particularly cinnamaldehyde, have demonstrated anticancer activity in various studies (Banerjee and Banerjee, 2023). Cinnamaldehyde exerts its anticancer effects through multiple mechanisms, including apoptosis induction, inhibition of angiogenesis, antiproliferation, anti-inflammatory, and antioxidant activities (Peng et al., 2024). It can induce apoptosis by activating caspases and damaging mitochondrial function, leading to cell death. Additionally, cinnamaldehyde can inhibit tumour angiogenesis, thereby limiting tumour growth and metastasis. Its antioxidant properties help protect cells from oxidative stress-induced DNA damage and genomic instability (Sadeghi et al., 2019).

Cinnamaldehyde shows significant promise as an anticancer agent, but further research is needed to fully understand its mechanisms of action and optimize its therapeutic potential. By addressing the existing challenges and exploring novel strategies, cinnamaldehyde could emerge as a valuable addition to conventional cancer therapies.

13.3.6. Cinnamon as Gut Health Enhancer

Cinnamon essential oil (CEO), derived from the bark of Cinnamomum trees, has been recognized for its various health benefits, including its potential to modulate gut microbiota. Emerging research suggests that CEO can positively influence gut microbiota composition and diversity, contributing to improved gut health and overall well-being (Pradhan et al., 2024). CEO possesses potent antimicrobial properties, capable of inhibiting the growth of harmful bacteria, including *Helicobacter pylori* and *Salmonella typhimurium* (Purkait et al., 2020). By reducing the abundance of pathogenic bacteria, CEO can create a more favorable environment for beneficial bacteria to thrive. CEO has also been shown to reduce inflammation in the gut, which is often associated with dysbiosis and inflammatory bowel diseases Romanescu et al., 2023). By inhibiting inflammatory cytokines and promoting the production of anti-inflammatory compounds, CEO can help restore gut homeostasis (Zhao et al., 2023).

CEO may act as a prebiotic, stimulating the growth of beneficial bacteria, such as *Akkermansia muciniphila* and *Faecalibacterium prausnitzii*. These beneficial bacteria produce short-chain fatty acids (SCFAs), which have been linked to various health benefits, including improved gut barrier function, reduced inflammation, and enhanced immune function.

Several studies have demonstrated the positive impact of CEO on gut microbiota. CEO supplementation has been shown to increase the abundance of beneficial bacteria, such as *Akkermansia muciniphila* and *Faecalibacterium prausnitzii*, while reducing the levels of harmful bacteria (Li et al., 2020). CEO can strengthen the gut barrier, preventing the translocation of harmful bacteria and toxins into the bloodstream. CEO can also help alleviate inflammation in the gut, thereby reducing the risk of inflammatory bowel diseases and other gastrointestinal disorders (Cani et al., 2022).

Further research is needed to fully understand the mechanisms underlying the effects of CEO on gut microbiota, but the available evidence suggests that it may be a promising natural approach to promoting gut health and overall well-being.

13.4. FUTURE DIRECTIONS

The diverse range of bioactive compounds present in cinnamon provides significant potential for the development of novel therapeutic modalities. By understanding the molecular mechanisms underlying the health benefits of cinnamon, researchers can identify specific compounds with therapeutic potential and develop targeted drug delivery systems.

For instance, cinnamaldehyde, a major component of cinnamon oil, has shown promising anticancer activity. By further investigating its mechanisms of action and optimizing its delivery, researchers may be able to develop novel anticancer drugs. Similarly, the anti-inflammatory and antioxidant properties of cinnamon compounds can be harnessed to develop therapies for various inflammatory and degenerative diseases. As the global demand for cinnamon and its products continues to rise, it is essential to adopt sustainable cultivation and harvesting methods to ensure the long-term viability of cinnamon resources.

Sustainable cultivation techniques, such as agroforestry and organic farming, can help reduce the environmental impact of cinnamon production. Additionally, research into efficient extraction methods and the development of synthetic analogs of cinnamon compounds can help minimize the reliance on natural resources. By focusing on sustainable production practices and innovative research, we can ensure the continued availability of cinnamon and its derivatives while minimizing their environmental impact.

In recent decades, scientific research has begun to elucidate the bioactive compounds present in cinnamon and their potential health benefits. However, significant gaps remain in our understanding of cinnamon's mechanisms of action and its clinical efficacy. To fully realize the potential of this ancient spice, future research should focus on several key areas.

Firstly, ongoing research into cinnamon's chemical composition is essential for a comprehensive understanding of its bioactive constituents and their potential health benefits. Compounds such as cinnamaldehyde, cinnamate, and procyanidins have been identified as key components, but further investigations are indispensable for the comprehensive characterization of the full spectrum of compounds present in various cinnamon varieties. Additionally, studies exploring the bioavailability and metabolism of these compounds in the human body will provide valuable insights into their efficacy and safety.

Secondly, developing standardized cinnamon extracts for therapeutic use is a crucial step in translating research findings into clinical practice. The variability in the chemical composition of cinnamon derived from different sources can affect its potency and efficacy. Therefore, it is imperative to develop standardized extracts with a consistent concentration of bioactive compounds to ensure reliable therapeutic outcomes. This will facilitate controlled clinical trials and enable the formulation of standardized cinnamon-based supplements and pharmaceuticals.

Thirdly, exploring new applications for cinnamon in the food and pharmaceutical industries is another promising area of research. Beyond its traditional use as a flavouring agent, cinnamon has been investigated for its potential to enhance food safety and shelf life. For example, studies have demonstrated the antimicrobial properties of cinnamon against food-borne pathogens. Additionally, the pharmaceutical industry is exploring the potential of cinnamon-derived compounds for the development of novel drugs to treat various diseases, including diabetes, cardiovascular disease, and neurodegenerative disorders.

Finally, addressing the challenges and limitations in translating research findings into clinical practice is essential for the successful development and application of cinnamon-based interventions. One significant challenge is the need for well-designed, randomized controlled trials to establish the clinical efficacy of cinnamon for specific health conditions. Additionally, studies are required to assess the safety and tolerability of cinnamon in different populations, particularly individuals with underlying health conditions or who are taking medications. Moreover, addressing issues related to standardization, quality control, and regulatory approval is crucial for the widespread adoption of cinnamon-based products in clinical settings.

In conclusion, future research on cinnamon holds great promise for unraveling its complex mechanisms of action and translating its potential health benefits into practical applications. By focusing on the areas outlined herein, scientists can contribute to the development of evidence-based interventions that harness the power of this ancient spice for the benefit of human health.

13.5. CONCLUSION

Cinnamon, a spice derived from the inner bark of trees belonging to the genus *Cinnamomum*, has been cherished for its unique flavour, aroma, and potential therapeutic properties for centuries. Recent scientific investigations have shed light on the complex chemistry, flavour, aroma, and therapeutic potential of this versatile spice.

Chemical analysis of cinnamon bark has revealed a rich composition of bioactive compounds, including cinnamaldehyde, coumarin, and tannins. Cinnamaldehyde, the primary component responsible for cinnamon's characteristic flavour and aroma, has been extensively studied for its potential health benefits. It has been shown to possess antimicrobial, anti-inflammatory, and antioxidant properties, suggesting its potential role in preventing and treating various diseases.

The flavour and aroma of cinnamon are intricately linked to its chemical composition. Cinnamaldehyde, with its aldehyde functional group, contributes to the spice's warm, sweet, and slightly pungent taste. The presence of other compounds, such as coumarin, can influence the overall flavour profile and aroma.

Beyond its culinary applications, cinnamon has garnered significant attention for its potential therapeutic properties. Numerous studies have explored the potential health benefits of cinnamon, including its ability to regulate blood sugar levels, improve cardiovascular health, and enhance cognitive function. Promising results have been obtained in some studies, but further research is necessary to fully understand the mechanisms underlying these effects and to establish definitive clinical guidelines.

Continued research is essential to unlock the full potential of cinnamon as a functional food ingredient and a potential therapeutic agent. Scientists are actively investigating the specific bioactive compounds responsible for cinnamon's health benefits, exploring the mechanisms of action, and determining optimal dosages. Additionally, studies are being conducted to assess the safety and efficacy of cinnamon in different populations, including individuals with underlying health conditions.

Interdisciplinary collaboration among scientists, healthcare professionals, and traditional medicine practitioners is crucial for bridging the gap between traditional knowledge and modern science. By combining the wisdom of traditional medicine with the rigor of scientific research, we can gain a deeper understanding of cinnamon's properties and develop innovative applications.

In conclusion, cinnamon is a spice with a rich history and a promising future. Scientific research has revealed its complex chemistry, flavour, aroma, and potential therapeutic properties. Although significant progress has been made, further investigation is needed to fully elucidate the mechanisms underlying its health benefits and to develop effective strategies for its utilization. Interdisciplinary collaboration will be instrumental in harnessing the full potential of this versatile spice.

13.6. ACKNOWLEDGMENT

The authors gratefully acknowledge DST-CURIE, Department of Science and Technology, New Delhi, for financial support through sanction number DST/CURIE-PG/2022/10 (G).

REFERENCES

Adisakwattana, S., Lerdsuwankij, O., Poputtachai, U., Minipun, A. and Suparpprom, C., 2011. Inhibitory activity of cinnamon bark species and their combination effect with acarbose against intestinal α-glucosidase and pancreatic α-amylase. *Plant Foods for Human Nutrition*, *66*, pp. 143–148.

Adwas, A.A., Elsayed, A., Azab, A.E. and Quwaydir, F.A., 2019. Oxidative stress and antioxidant mechanisms in human body. *Journal of Applied Biotechnology & Bioengineering*, *6*(1), pp. 43–47.

Ahamad, J., Naquvi, K.J., Uthirapathy, S., Naim, M.J., Munjal, K. and Mir, S.R., 2023. Qualitative diagnostic features of terpenes and terpenoids containing food spices. In *Analysis of Food Spices* (pp. 109–123). CRC Press.

Al-Bayati, F.A. and Mohammed, M.J., 2009. Isolation, identification, and purification of cinnamaldehyde from Cinnamomum zeylanicum bark oil: An antibacterial study. *Pharmaceutical Biology*, *47*(1), pp. 61–66.

Anderson, R.A., Zhan, Z., Luo, R., Guo, X., Guo, Q., Zhou, J., Kong, J., Davis, P.A. and Stoecker, B.J., 2016. Cinnamon extract lowers glucose, insulin and cholesterol in people with elevated serum glucose. *Journal of Traditional and Complementary Medicine*, *6*(4), pp. 332–336.

Banerjee, S. and Banerjee, S., 2023. Anticancer potential and molecular mechanisms of cinnamaldehyde and its congeners present in the cinnamon plant. *Physiologia*, *3*(2), pp. 173–207.

Borzoei, A., Rafraf, M., Niromanesh, S., Farzadi, L., Narimani, F. and Doostan, F., (2018). Effects of cinnamon supplementation on antioxidant status and serum lipids in women with polycystic ovary syndrome. *Journal of Traditional and Complementary Medicine*, *8*(1), pp. 128–133.

Cani, P.D., Depommier, C., Derrien, M., Everard, A. and de Vos, W.M., 2022. Akkermansia muciniphila: Paradigm for next-generation beneficial microorganisms. *Nature Reviews Gastroenterology & Hepatology*, *19*(10), pp. 625–637.

Chatterjee, S., 2016. Oxidative stress, inflammation, and disease. In *Oxidative Stress and Biomaterials* (pp. 35–58). Academic Press.

Costello, R.B., Dwyer, J.T., Saldanha, L., Bailey, R.L., Merkel, J. and Wambogo, E., 2016. Do cinnamon supplements have a role in glycemic control in type 2 diabetes? A narrative review. *Journal of the Academy of Nutrition and Dietetics*, *116*(11), pp. 1794–1802.

Couturier, K., Batandier, C., Awada, M., Hininger-Favier, I., Canini, F., Anderson, R.A., Leverve, X. and Roussel, A.M., 2010. Cinnamon improves insulin sensitivity and alters the body composition in an animal model of the metabolic syndrome. *Archives of Biochemistry and Biophysics*, *501*(1), pp. 158–161.

Culas, M.S., Popovich, D.G. and Rashidinejad, A., 2024. Recent advances in encapsulation techniques for cinnamon bioactive compounds: A review on stability, effectiveness, and potential applications. *Food Bioscience*, *57*, p. 103470.

Demina, T.D., Margaret Rodrigues, R., George, L. and Varghese, A., 2024. Cinnamon–a competent drug: A review on extraction, analysis and anticancer action. *Chemistry Select*, *9*(41), p. e202401636.

De Silva, N.D., Wasana, K.G.P. and Attanayake, A.P., 2024. Cinnamon bark (cinnamomum species). In *Medicinal Spice and Condiment Crops* (pp. 180–199). CRC Press.

Dvorackova, E., Snoblova, M., Chromcova, L. and Hrdlicka, P., 2015. Effects of extraction methods on the phenolic compounds contents and antioxidant capacities of cinnamon extracts. *Food Science and Biotechnology*, *24*, pp. 1201–1207.

Freedman, P., 2015. Health, wellness and the allure of spices in the Middle Ages. *Journal of Ethnopharmacology*, *167*, pp. 47–53.

Gülçin, İ., Elmastaş, M. and Aboul-Enein, H.Y., 2012. Antioxidant activity of clove oil–A powerful antioxidant source. *Arabian Journal of chemistry*, *5*(4), pp. 489–499.

Gunawardena, D., Govindaraghavan, S. and Münch, G., 2014. Anti-inflammatory properties of cinnamon polyphenols and their monomeric precursors. In *Polyphenols in Human Health and Disease* (pp. 409–425). Academic Press.

Haldar, S., Tanwar, A., Singh, A., Mehra, G., Sain, M. and Kumar, V., 2022. Bioactive compounds in cinnamon. In *Spice bioactive compounds* (pp. 45–70). CRC Press.

Ho, S.C., Chang, K.S. and Chang, P.W., 2013. Inhibition of neuroinflammation by cinnamon and its main components. *Food Chemistry*, *138*(4), pp. 2275–2282.

Ismail, B.S., Mahmoud, B., Abdel-Reheim, E.S., Soliman, H.A., Ali, T.M., Elesawy, B.H. and Zaky, M.Y., 2022. Cinnamaldehyde mitigates atherosclerosis induced by high-fat diet via modulation of hyperlipidemia, oxidative stress, and inflammation. *Oxidative Medicine and Cellular Longevity*, *2022*(1), p. 4464180.

Jayaprakasha, G.K. and Rao, L.J.M., 2011. Chemistry, biogenesis, and biological activities of Cinnamomum zeylanicum. *Critical Reviews in Food Science and Nutrition*, *51*(6), pp. 547–562.

Ju, J., Santana de Oliveira, M. and Qiao, Y., 2023. The application potential of cinnamon in neuroprotection. In *Cinnamon: A medicinal plant and a functional food systems* (pp. 217–229). Springer International Publishing.

Kutbi, E.H., Sohouli, M.H., Fatahi, S., Lari, A., Shidfar, F., Aljhdali, M.M., Alhoshan, F.M., Elahi, S.S., Almusa, H.A. and Abu-Zaid, A., 2022. The beneficial effects of cinnamon among patients with metabolic diseases: A systematic review and dose-response meta-analysis of randomized-controlled trials. *Critical Reviews in Food Science and Nutrition*, *62*(22), pp. 6113–6131.

Lee, P.H., Lin, I.H. and Hsieh, P.C., 2023. Measurement of bioactive compounds, antioxidants, and α-glucosidase and α-amylase inhibitory effects of Cinnamomum osmophloeum Kanehira extracted with various solvents. *Measurement: Food*, *9*, p. 100069.

Li, A.L., Ni, W.W., Zhang, Q.M., Li, Y., Zhang, X., Wu, H.Y., Du, P., Hou, J.C. and Zhang, Y., 2020. Effect of cinnamon essential oil on gut microbiota in the mouse model of dextran sodium sulfate-induced colitis. *Microbiology and Immunology*, *64*(1), pp. 23–32.

Liu, Y., Pang, D., Xing, D., Wang, W., Li, Q., Liao, S., Li, E. and Zou, Y., 2023. Cinnamon free phenolic extract regulates glucose absorption in intestinal cells by inhibiting glucose transporters. *Food Bioscience*, *52*, p. 102405.

Losada-Barreiro, S. and Bravo-Diaz, C., 2017. Free radicals and polyphenols: The redox chemistry of neurodegenerative diseases. *European Journal of Medicinal Chemistry*, *133*, pp. 379–402.

Mandal, P. and Kanthlal, S.K., 2024. Vascular targeting of constituents of commonly used spices and its molecular interactions in endothelial dysfunction: A review. *Phytochemistry Reviews*, pp. 1–30.

Medagama, A.B., 2015. The glycaemic outcomes of Cinnamon, a review of the experimental evidence and clinical trials. *Nutrition Journal*, *14*, pp. 1–12.

Megha, K.B., Joseph, X., Akhil, V. and Mohanan, P.V., 2021. Cascade of immune mechanism and consequences of inflammatory disorders. *Phytomedicine*, *91*, p. 153712.

Mirmiranpour, H., Huseini, H.F., Derakhshanian, H., Khodaii, Z. and Tavakoli-Far, B., 2020. Effects of probiotic, cinnamon, and synbiotic supplementation on glycemic control and antioxidant status in people with type 2 diabetes; a randomized, double-blind, placebo-controlled study. *Journal of Diabetes & Metabolic Disorders*, *19*, pp. 53–60.

Mohsin, S.N., Saleem, F., Humayun, A., Tanweer, A. and Muddassir, A., 2023. Prospective nutraceutical effects of cinnamon derivatives against insulin resistance in type II diabetes mellitus—evidence from the literature. *Dose-Response*, *21*(3), p. 15593258231200527.

Mousavi, S.M., Rahmani, J., Kord-Varkaneh, H., Sheikhi, A., Larijani, B. and Esmaillzadeh, A., 2020. Cinnamon supplementation positively affects obesity: A systematic review and dose-response meta-analysis of randomized controlled trials. *Clinical Nutrition*, *39*(1), pp. 123–133.

Nakhaee, S., Kooshki, A., Hormozi, A., Akbari, A., Mehrpour, O. and Farrokhfall, K., 2024. Cinnamon and cognitive function: A systematic review of preclinical and clinical studies. *Nutritional Neuroscience*, *27*(2), pp. 132–146.

Noori, S., Azmat, M. and Mahboob, T., 2012. Study on antioxidant effects of cinnamon and garlic extract in liver, kidney and heart tissue of rat. *Bioscience Research*, *9*(1), pp. 17–22.

Peng, J., Song, X., Yu, W., Pan, Y., Zhang, Y., Jian, H. and He, B., 2024. The role and mechanism of cinnamaldehyde in cancer. *Journal of Food and Drug Analysis*, *32*(2), p. 140.

Pisoschi, A.M., Pop, A., Iordache, F., Stanca, L., Predoi, G. and Serban, A.I., 2021. Oxidative stress mitigation by antioxidants-an overview on their chemistry and influences on health status. *European Journal of Medicinal Chemistry*, *209*, p. 112891.

Pradhan, S., Blanton, C., Ochoa-Reparaz, J., Bhattarai, N. and Sharma, K., 2024. Herbs and spices: Modulation of gut microbiota for healthy aging. *Gastroenterology Insights*, *15*(2), pp. 447–458.

Purkait, S., Bhattacharya, A., Bag, A. and Chattopadhyay, R.R., 2020. Evaluation of antibiofilm efficacy of essential oil components β-caryophyllene, cinnamaldehyde and eugenol alone and in combination against biofilm formation and preformed biofilms of Listeria monocytogenes and Salmonella typhimurium. *Letters in Applied Microbiology*, *71*(2), pp. 195–202.

Qneibi, M., Bdir, S., Maayeh, C., Bdair, M., Sandouka, D., Basit, D. and Hallak, M., 2024. A comprehensive review of essential oils and their pharmacological activities in neurological disorders: Exploring neuroprotective potential. *Neurochemical Research*, *49*(2), pp. 258–289.

Rani, V., Deep, G., Singh, R.K., Palle, K. and Yadav, U.C., 2016. Oxidative stress and metabolic disorders: Pathogenesis and therapeutic strategies. *Life Sciences*, *148*, pp. 183–193.

Rao, P.V. and Gan, S.H., 2014. Cinnamon: A multifaceted medicinal plant. *Evidence-Based Complementary and Alternative Medicine*, *2014*(1), p. 642942.

Ravindran, P.N., 2023. Spices: Definition, classification, history, and role in Indian life. In *Handbook of spices in India: 75 years of research and development* (pp. 1–102). Springer Nature Singapore.

Ribeiro, D., Freitas, M., Lima, J.L. and Fernandes, E., 2015. Proinflammatory pathways: The modulation by flavonoids. *Medicinal Research Reviews*, *35*(5), pp. 877–936.

Romanescu, M., Oprean, C., Lombrea, A., Badescu, B., Teodor, A., Constantin, G.D., Andor, M., Folescu, R., Muntean, D., Danciu, C. and Dalleur, O., 2023. Current state of knowledge regarding who high priority pathogens—Resistance mechanisms and proposed solutions through candidates such as essential oils: A systematic review. *International Journal of Molecular Sciences*, *24*(11), p. 9727.

Roussel, A.M., Hininger, I., Benaraba, R., Ziegenfuss, T.N. and Anderson, R.A., 2009. Antioxidant effects of a cinnamon extract in people with impaired fasting glucose that are overweight or obese. *Journal of the American College of Nutrition*, *28*(1), pp. 16–21.

Rudrapal, M., Khairnar, S.J., Khan, J., Dukhyil, A.B., Ansari, M.A., Alomary, M.N., Alshabrmi, F.M., Palai, S., Deb, P.K. and Devi, R., 2022. Dietary polyphenols and their role in oxidative stress-induced human diseases: Insights into protective effects, antioxidant potentials and mechanism (s) of action. *Frontiers in Pharmacology*, *13*, p. 806470.

Sadeghi, S., Davoodvandi, A., Pourhanifeh, M.H., Sharifi, N., ArefNezhad, R., Sahebnasagh, R., Moghadam, S.A., Sahebkar, A. and Mirzaei, H., 2019. Anti-cancer effects of cinnamon: Insights into its apoptosis effects. *European Journal of Medicinal Chemistry*, *178*, pp. 131–140.

Sahib, A.S., 2016. Anti-diabetic and antioxidant effect of cinnamon in poorly controlled type-2 diabetic Iraqi patients: A randomized, placebo-controlled clinical trial. *Journal of Intercultural Ethnopharmacology*, *5*(2), p. 108.

Seyed Ahmadi, S.G., Farahpour, M.R. and Hamishehkar, H., 2019. Topical application of Cinnamon verum essential oil accelerates infected wound healing process by increasing tissue antioxidant capacity and keratin biosynthesis. *The Kaohsiung Journal of Medical Sciences*, *35*(11), pp. 686–694.

Shang, C., Lin, H., Fang, X., Wang, Y., Jiang, Z., Qu, Y., Xiang, M., Shen, Z., Xin, L., Lu, Y. and Gao, J., 2021. Beneficial effects of cinnamon and its extracts in the management of cardiovascular diseases and diabetes. *Food & Function*, *12*(24), pp. 12194–12220.

Sharma, U.K., Sharma, A.K. and Pandey, A.K., 2016. Medicinal attributes of major phenylpropanoids present in cinnamon. *BMC Complementary and Alternative Medicine*, *16*, pp. 1–11.

Shreaz, S., Wani, W.A., Behbehani, J.M., Raja, V., Irshad, M., Karched, M., Ali, I., Siddiqi, W.A. and Hun, L.T., 2016. Cinnamaldehyde and its derivatives, a novel class of antifungal agents. *Fitoterapia*, *112*, pp. 116–131.

Shuab, R., Lone, R. and Koul, K.K., 2016. Cinnamate and cinnamate derivatives in plants. *Acta Physiologiae Plantarum*, *38*, pp. 1–9.

Singh, A., Kukreti, R., Saso, L. and Kukreti, S., 2019. Oxidative stress: A key modulator in neurodegenerative diseases. *Molecules*, *24*(8), p. 1583.

Tanwar, E., 2019. *Assessment of Antibacterial, Anti-Inflammatory and Antioxidant Effects of Cinnamommum Zeylanicum (CINNAMON)–An In-Vitro Study* (Master's thesis, Rajiv Gandhi University of Health Sciences (India)).

Tuzcu, Z., Orhan, C., Sahin, N., Juturu, V. and Sahin, K., 2017. Cinnamon polyphenol extract inhibits hyperlipidemia and inflammation by modulation of transcription factors in high-fat diet-fed rats. *Oxidative Medicine and Cellular Longevity*, *2017*(1), p. 1583098.

Udayaprakash, N.K., Ranjithkumar, M., Deepa, S., Sripriya, N., Al-Arfaj, A.A. and Bhuvaneswari, S., 2015. Antioxidant, free radical scavenging and GC–MS composition of Cinnamomum iners Reinw. ex Blume. *Industrial Crops and Products*, *69*, pp. 175–179.

Wang, W., Nguyen, K.T.K., Zhao, C. and Hung, H.C., 2023. Earliest curry in Southeast Asia and the global spice trade 2000 years ago. *Science Advances*, *9*(29), p. eadh5517.

Whitfield, P., Parry-Strong, A., Walsh, E., Weatherall, M. and Krebs, J.D., 2016. The effect of a cinnamon-, chromium-and magnesium-formulated honey on glycaemic control, weight loss and lipid parameters in type 2 diabetes: An open-label cross-over randomised controlled trial. *European Journal of Nutrition*, *55*, pp. 1123–1131.

Wu, T., Huang, W., He, M. and Yue, R., 2022. Effects of cinnamon supplementation on lipid profiles among patients with metabolic syndrome and related disorders: A systematic review and meta-analysis. *Complementary Therapies in Clinical Practice*, *49*, p. 101625.

Zhang, K., Li, Y., Lin, X., Daneshar, M., Karamian, F. and Li, M., 2024. Effect of cinnamon supplementation on blood pressure, oxidative stress, and inflammatory biomarkers in adults: An umbrella review of the meta-analyses of randomized controlled trials. *Nutrition, Metabolism and Cardiovascular Diseases*.

Zhao, Q., Zhu, L., Wang, S., Gao, Y. and Jin, F., 2023. Molecular mechanism of the anti-inflammatory effects of plant essential oils: A systematic review. *Journal of ethnopharmacology*, *301*, p. 115829.

14 Ginger Chemistry and Its Health Impacts with a Toxicity Profile

Ayman Younes Allam, Gholamreza Abdi, Avanish Chandra Sharma, Sunita Singh, and Ahmed A. Tayel

14.1. INTRODUCTION

Ginger has been widely recognized for its unique flavour, odour, and digestive properties and has been used all over the world as a key ingredient in various foods and beverages since ancient times. Besides its common use in the culinary field throughout the world, ginger has been traditionally employed for many medicinal purposes, such as rheumatism, sprains, muscular aches, muscle strains, convulsions, fractures, dislocations, and hypertension. Further supportive evidence about the pharmacological properties of ginger has been reported from numerous scientific investigations, including its use against nausea, colic, diarrhea, flatulence, indigestion, memory issues, and cough (Abarca et al. 2024; Abdallah, Ahlan, and Abdullah 2019; Addis 2015). As a result, ginger has an extensive tradition of use for a variety of food, beverage, and therapeutic purposes and is still an excellent model for investigating natural foods for various pharmacological activities, including antioxidant mechanisms. It belongs to the rhizomatous herbaceous perennial family *Zingiberaceae* and originated from the island of Southeast Asia with a strong aromatic odour and slightly biting taste (Addis 2015; Adewusi, Moodley, and Steenkamp 2011). These special properties render ginger a highly desirable source of spices and traditional medicinal applications for humans worldwide. Since ancient times to the present day, the utility of ginger and its medicinal properties have been described in various ethnomedicinal systems, which can be used to treat a wide range of diseases. This review covers the chemical composition and bioactivity of ginger and discusses potential chemical procedures, defining ginger nonvolatiles that may be instrumental in deepening our understanding of its products and pharmaceutical potential.

Ginger, the rhizome of a perennial herb, is used as a spice, condiment, and herbal material. Ginger has been cultivated in Asia for a long time. The qualitative characteristics, pharmacological activities, processing methods, and geographical sources could affect the efficacy of ginger on health. The pharmacological activities of ginger are generally related to its volatile oil, gingerol, shogaol, and other ingredients. Its volatile oil has antibacterial, antipyretic, and analgesic effects and can expand blood vessels and prevent blood clotting. Gingerol has anti-inflammatory, analgesic, antibacterial, antiemetic, stomach-relieving, antitumour, and other effects. Therefore, ginger-based foods and traditional herbal formulas have the functions of warming the spleen and stomach, stopping vomiting, inhibiting diarrhea, and preventing intestinal parasites. At present, ginger-hopsu tea and ginger-citron tea with improved flavors are popular in South Korea. The consumption of ginger-containing processed foods is continuously increasing (Adewusi, Moodley, and Steenkamp 2011; Al Jumayi et al. 2022; Alexander 2020).

DOI: 10.1201/9781003559894-14

14.2. WHAT IS GINGER?

Ginger is a closely related but different plant from turmeric, another of the spices traditional to southern Asia. It is the rhizome—the underground stem—of a tropical plant. The ginger plant has a green stem that shoots a foot or more into the air and carries leaves. From a botanist's point of view, it is a monocot, being a close cousin to bananas, grown in similar conditions due to its dislike for cold. However, instead of bearing fruit, the ginger plant sends out what is perhaps its most distinctive feature—the rhizome from which fragrant ginger is harvested. Once the rhizomes of the ginger plant are harvested, they can be used directly, but unlike true roots, such as garlic, the green shoots can be initiated or can regrow, just as potatoes do if they are overlooked in the garden. But you don't need a plant that matures just to obtain your pinch of ginger. Ginger is one of the easiest spices to start from scratch, as all it normally takes is a small quantity of freshly harvested ginger rhizome and appropriate growing conditions for the plant to come alive. Items that have been kept in the shops for a long time may have been treated to keep them from sprouting, and hence it is often difficult to convince them that sprouting is a natural activity. However, it seems strange to have to convince a rhizome designed by nature to initiate growth. No matter, the spice is cheap enough, and it isn't difficult to find some of it that will germinate (Alexander 2020; Addis 2015; Allam et al. 2024; Arif et al. 2020).

14.3. HISTORY OF GINGER

India, China, and Africa—it is difficult to tell where ginger originated. The first recorded history of it was written in a manuscript from the 4th century BCE by a Spartan, who mentioned that ginger was grown in the garden of the Shah-pu-eh Palace during the Iranian dynasty of the Achaemenides from 600 to 500 BCE In ancient China, about the 1st century CE, the people of the Han Dynasty used ginger as a remedy for food poisoning–induced conditions. In 552 CE, when an emperor sponsored a research project to find a remedy for the bubonic plague, a recipe with 16 ingredients, including ginger, was submitted for the formula. The Moors, who introduced ginger to Spain by the 9th century, brought ginger rhizomes along with other spices. Ginger was not commonly known in Europe until it was imported by the Greeks and Romans (Wang 2020).

In the Middle Ages, it was known to the ancient Greeks as *zingiberis* and to the ancient Romans as *zingiberi*. By the 13th century, a whole ginger loaf was a treat when sweetened with honey and accompanied by tea. As the Roman Empire extended trading and later, by invading new territories, they came into contact with unfamiliar Asian spices, ginger was included. When Rome, Greece, Iran, Iraq, the southeastern Mediterranean, Byzantine, Sassanian, and Egyptian civilizations exchanged trade with India, China, and Africa, ginger and other East Indian spices were traded, which became available within the Roman Empire about the 1st century C.E.

In 1297, a traveler wrote a book about his adventures in East Africa, India, and China, and in it, he told why the rhizome was so expensive; they imported the dried root, suggesting that ginger came far from its native jungles. Germany, England, and the Czech countries of Bohemia started trading with each other around the 14th and 20th centuries. Ginger was used in larger, less wealthy homes in England as early as the 15th century. It became so popular in England during the 15th century that when the English defeated the Portuguese colonies, the sale of ginger was made part of the truce so that England could be the first nation to trade with the East. Some suggested that it was not the rhizome that was imported into Rome but likely other parts of the ginger plant. It was noted that porcelain, by favoring the 2nd century BCE exportation of ginger, weakened the grip on ginger growing along the Mediterranean and favored its trade with China. In 858 CE, ginger was used to prepare the beverages ginger ale and ginger beer in England. An individual came upon the aromatic rhizomes at a small craft exhibition in the early 1950s. He made a potpourri, and the fragrance from the rhizomes filled the home for many weeks (Kemper 1999; Ravindran and Babu 2016; Rahman et al. 2023; Riaz et al. 2024).

14.4. CULTURAL SIGNIFICANCE OF GINGER

Ginger, variously called "the spice of life", "nature's medicine chest", and "the African wonder spice", has found immense significance within the cultural world of humans in various pursuits, such as cuisine, medicine, religion, trade, travel, identity, art, society, and communication. Its association with humans cuts across all spheres of life, from food to architecture. As food, ginger is a predominant spice used in different global cuisines. In several countries where ginger is mainly grown, some of the oldest and most sought-after recipes employ ginger in the preparation of meals. Gingerbread, ginger tea, ginger beer, and ginger biscuits are foods enjoyed in various parts of the world. Considering its therapeutic qualities, ginger is often used for cooking and for promoting health and wellness. Ginger is not only consumed to boost physical energy but also as a symbol signifying the possessor's spiritual and sexual prowess. In studies of the sharing of food and eating together, ginger is cited in food ideations. According to these criteria, ginger attributes both mental power and endurance. In ancient Greece, the ginger plant was associated with fire and Mars, the Roman god of war. Romans cultivated it throughout their empire, almost exhibiting it as a healing herb. During the Middle Ages, ginger was often mentioned when discussing Oriental products, associated with luxury and affluence (Riaz et al. 2024; Riseh et al. 2024).

14.5. METHODS OF EXTRACTION AND ANALYSIS

Ginger, the product of the dried rhizome of *Zingiber officinale Roscoe*, is a traditional Chinese medicine commonly used as a stimulating and aromatic agent. It is also employed in asthma, cough, spleen deficiency, palpitation, arthralgia, radiculitis, anorexia, and rheumatic pain. Since ancient times, ginger has played a significant role in traditional healthcare systems in many countries. Developed initially in ancient China and India, this alternative therapy is widely practiced in the present day. It is an important component during Asian religious and cultural festivities. As a routine dietetic condiment, ginger serves as a flavouring agent and preservative of foodstuffs. The therapeutic use of ginger as a remedy spans several millennia. Historically, ginger's claim as a medicine is traceable for at least 3,000 years. In early writings, mention is made of its medicinal value, thus establishing ginger as one of the earliest botanicals in the universal *materia medica*. With time, it has come to occupy a position of high regard in the oldest and most respected traditional medical systems worldwide. Ayurveda, an Indian system of medicine, has defined ginger as a universal medicine. In East China, ginger is rated as a medicinal treasure. Most household and clinical preparations of ginger are based on many processing varieties, each with a distinct therapeutic character and prescribed dose. Gingerol-rich and shogaol-rich ginger varieties have several applications and are preferred for different pharmacologic activities. Diverse chemical compounds contribute to its healing properties. However, most pharmacologic activities are often attributed to the presence of gingerols and shogaols in the rhizome. Several local and systemic activities, such as anorexia, weariness, fever, diaphoresis, activity-tonic effect, digestive, carminative, antiswelling, cough assuaging, analgesic, and killing helminths, have been reported (Al Jumayi et al. 2022; Algarni et al. 2022).

The scientific term "phytochemical" is used to describe chemicals in plants that have effects on humans. It is a broad term that encompasses both the well-known nutrients found in fruits and vegetables and many other non-nutritive substances. The pigments found in colourful fruits and vegetables are classified as phytochemicals. Most people are familiar with the term "carotenoids", which are yellow and orange pigments, heavy in antioxidant activity found in carrots, yams, and pumpkins. Within the broader carotenoid category are alpha- and beta-carotene. Beta-carotene is the provitamin A carotenoid that gives fruits and vegetables their orange colour and can be converted to vitamin A in the body. Vitamin A is essential for healthy eyes, skin, hair, mucous membranes, and a strong immune system. However, not all carotenoids have provitamin A activity. The green pigment in vegetables, chlorophyll, is classified as a carotenoid. Another colourful group of related pigments is the flavonoids, which are rich in antioxidant activity (Algarni et al. 2022).

Vegetables such as kale, spinach, and collard greens are high in lutein. Tomatoes are high in lycopene, and red and purple produce berries such as strawberries, blueberries, raspberries, and cranberries, which are abundant in anthocyanins. Anthoxanthins, which have the subgroups xanthophylls, are high in beta-carotene and astaxanthin, and flavonols and flavones are part of the flavonoid category in fruits and vegetables. Isothiocyanates found in cruciferous vegetables, such as broccoli, cabbage, and cauliflower, are classified as phytochemicals and have been associated with lowering the risk of cancer. Glucosinolates are the parent compounds of isothiocyanates and are also classified as phytochemicals with cancer risk-reduction benefits. Saponins are found in beans, peas, chickpeas, and soybeans and are also classified as phytochemicals with multiple health benefits. Crop choices, the condition of the plants during crop development, and bioavailability can all influence the bioactive component content in crops (Algarni et al. 2022; Arif et al. 2020).

Phytochemicals are distinct, prescribed low-molecular-weight compounds that occur in food and carry out significant roles in human health through human diet and plant consumption. Health benefits derived from consuming bioactive food components cannot be attributed to common nutrients and the essential nutrients that humans require for physiological development and maintenance. Phytochemicals comprise nonessential nutrients that contribute to curing diseases, promoting health, and alleviating chronic tensions. As several volumes of phytochemicals have been recognized and identified, new classifications have been proposed; nevertheless, their consolidation is still evolving. Many bioactive compounds that are constituents of a self-possessed lifestyle are polyphenols, flavonoids, flavonols, anthocyanins, proanthocyanidins, isoflavones, lignans, coumestans, saponins, sterols, terpenoids, glucosinolates, chlorophyll, soluble fibre, water-insoluble fibre, oligosaccharides, essential fatty acids, and extra-virgin olive oil. Recently, the importance of these compounds in human nutrition and health has produced public interest as they enable the development of functional foods and nutraceuticals. The appeal of these compounds has given rise to industries that utilize them as raw constituents in food and pharmaceutical production (Asif, Ibrahim, and Ansari 2024; Awasthi et al. 2022).

14.5.1. Health Benefits

The capability and flexibility of bioactive compounds, as well as their exclusive mode of action, can provide a novel path to many health problems and challenges. A variety of these isolates, in particular polyphenols, play a critical role in different components of health promotion. Nutrients and some semi-components are typically defined as chemical substances in food that are indispensable for the preservation of good health. This definition is closely related to the prevention of pathological implications associated with nutrient insufficiency, for example, scurvy, goiter, anemia, or rickets. New aspects associated with dietary components include health benefits other than merely avoiding deficiency symptoms. Although the specific compound that is responsible for the mechanism of phenolic compounds to reduce the risk of several diseases is still unknown, studies demonstrate that several dietary components perform their effects in combination. Polyphenols interact with a variety of biochemical reactions in the body, including oxidative enzymes, antioxidant enzymes, and particularly kinases involved in inflammation processes, as well as the vitamin D receptor, among others. The capacity of specific polyphenols to alter various disease biomarkers has been tested in a choice of in vivo and in vitro studies. Therefore, modest dietary changes can lead to significant changes in the occurrence and prevalence of certain diseases (Awasthi et al. 2022; Babuskin et al. 2014).

Generally, the larger the particle, the lower the cell wall/fibre content, which makes it easier to reach the intracellular substances. The key to obtaining compounds with significant bioactivity is therefore cell destruction or cell wall degradation. The plant matrix can be disrupted physically, chemically, enzymatically, or with some other agent. Fine grinding and extensive use of liquid gases result in the freezing of cells and pulverization of tissues, making the active ingredients of medicinal plants more easily exposed. Because the action of liquid nitrogen prevents the release of active compounds at elevated temperatures, and this technique is the best way to conserve volatile and thermo-labile compounds, it can help avoid substance loss. In the production process, large amounts of active ingredients can be lost under the effects of temperature and light, for example (Babuskin et al. 2014; Bali 2021).

Ultrasound is a noninvasive treatment technique that can organize tissues, break cells, and destroy cell walls through a mechanical effect. Research indicates that a 50 W ultrasonic processor can crush approximately 19.5 g of plant material in 1 hour, thus allowing full extraction of active ingredients from medicinal plants. In addition, ultrasound can help improve the activity of several enzymes. Small molecules display stronger effects than large ones when extracted from medicinal plants. Sonication is likely to cause mechanical vibration and the formation of cavities and pressure, which exert a strong impact and help decompose cell structure. The intensity of cavitation contributes to the extraction of compounds. The optimal sonication conditions for the extraction (Bali 2021).

14.5.2. Traditional Extraction Techniques

For centuries, the use of plants to alleviate pain, loss of appetite, infections, anemia, and many other ailments has played an important role in the health of different cultures. Nowadays, several studies have made known the main bioactive principles associated with the pharmacological effects of different medicinal plants, such as flavonoids, alkaloids, steroids, saponins, and terpenes, among others. The plant material can contain these bioactive principles in proportions ranging from 0.02 to 0.5%. Thus, extracting bioactive principles from plants efficiently is one of the main interests in the pharmaceutical, food, or cosmetic industry. Innovative and effective extraction methods are being studied, taking into account many factors, such as being clean and green methods, capable of being used at a large scale, and using low energy amounts. Despite the recent evolution of extraction techniques in both industrial and research areas, traditional techniques are used by rural communities to prepare their remedies or products. These techniques present simple procedures and low costs but are not that efficient. In this chapter, some examples of the most common extraction methods are presented, including traditional and new clean technologies, due to their increasing interest in recent years (Bali 2021; Balkrishna et al. 2024).

14.5.3. Modern Analytical Techniques

The use of modern technology to characterize bioactive compounds in medicinal plants has generated valuable results, from the isolation of new and potent molecules to the in vitro culturing of plant cells. Besides the classical methods of extraction, fractionation, isolation, and characterization of bioactive compounds, the use of high-performance thin-layer chromatography (HPTLC) is defined for the identification, characterization, and quantification of the main steps of the process of isolation of bioactive compounds from medicinal plants. The distinct advantage of HPTLC as compared to other chromatographic techniques is its efficiency and ease of operation, which lead to profits in terms of cost and time. Upon validation, HPTLC techniques establish themselves as attractive instrumental methods, capable of achieving complete coverage of all the phases of a project for isolating and identifying bioactive compounds from medicinal plants (Balkrishna et al. 2024).

HPTLC can be used for screening purposes with respect to the quality control of medicinal plants, being of great value when large numbers of samples are to be analyzed. This technique is also suitable for minimizing chromatographic rescreening and sample workups in the isolation of bioactive compounds. As an additional objective, preliminary information on the identity of the bioactive compounds can be obtained from the HPTLC fingerprint profile, where from ultraviolet absorption and fluorescence, some tentative conclusions can be deduced. The use of microanalytical HPTLC plates is also advantageous, both for its capacity to work with minimal amounts of the sample and the requested solvent and because it operates far more quickly than other chromatographic methods. Furthermore, additional spotting and respotted runs can be done easily and in accordance with specific requirements. Upon quantification, HPTLC ensures that an adequate response can be obtained under the selected development conditions (Balkrishna et al. 2024).

14.6. MECHANISMS OF ANTIOXIDANT ACTION

The aim of biological systems is to prevent or attenuate damage due to oxidative stress. For this purpose, organisms use antioxidants. Antioxidants can act by numerous mechanisms. These include preventing the univalently reduced form of oxygen or sulfur from participating in further detrimental reactions, scavenging nitrogen radicals, and enabling the restoration of the redox state through redistribution of charge. The oxidized forms can be enzymatically restored, which makes these relatively stable structures catalytically active. A very important feature of antioxidants is the ability to protect membranes. This entails both scavenging reactive oxygen species in lipids and preventing the formation of reactive oxygen species in the membranes (Balkrishna et al. 2024; Beatriz Nunes et al. 2023).

Antioxidants that occur in organisms belong to two general classes: (1) lipid-soluble, lipophilic, membrane-strengthening antioxidants that are embedded in membranes with hydrophilic head groups pointing toward the membrane surface such that the residues extend into the fatty acid chains, and (2) water-soluble, hydrophilic-reducing antioxidants that are distributed throughout the cytosol. The hydrophilic form of vitamin E functions in membranes in this way. Other substances that are known to be effective lipid-soluble antioxidants include ubiquinol, plastoquinol, carotenoids, melatonin, bilirubin, naringenin, phloretin, curcumin, stilbenes, and (in some experimental situations) xanthine. Some polyphenolic flavonoids may be expected to incorporate into multiple locations throughout the membrane when they form hydrogen bonds with it. This might be within hydrophobic pockets in favorably shaped proteins. Most of the potential enzyme substrate oxidants are lipophilic, which makes it necessary for protein-bound cofactors to be embedded in the membranes in some way for the enzyme to capture any biological substrate (Beatriz Nunes et al. 2023).

14.6.1. Scavenging Free Radicals

Much of what we know about the activities of antioxidant agents at the cellular level comes from the study of free radicals. The most common type of radicals found in a biological system arises from the presence of an odd number of electrons and a free energy level. It has an unpaired electron, which makes the compound chemically unstable and highly reactive. It was reasoned that the broad structures in EPR spectra yielded from in vivo brain tissue were due to a quantity of stable, endogenously formed free radical scavengers known as the superoxide radical, which has been implicated as causing several chemical changes in the body. It was shown that cells also generate considerable amounts of the free radical form of superoxide throughout the cell cycle. It seems of interest that as cells synchronize and reach the S phase of the cell cycle, they contain a higher proportion of cells that produce superoxide radicals and increase their mass as well as free radical levels (Beatriz Nunes et al. 2023).

14.6.2. Anti-Inflammatory Effects

The anti-inflammatory effect of ginger, as a medicinal plant, has long been recognized, but there are some contradictory findings indicating no anti-inflammatory activity for this medicinal herb. *Zingiberaceae* is a plant that naturally produces many bioactive compounds, including terpenes, oleoresins, and pungent phenolic principles found in spice ginger. Previously, 6-gingerol was shown to suppress both NO and PGE2 production, possibly through the blocking of NF-kB activation in macrophages. In addition, 6-gingerol alleviates adjuvant-induced arthritis by altering T-cell activities in rats. Another component in ginger, 6-gingerdione, showed some potent activity in reducing LPS-induced nitric oxide, PGE2, and pro-inflammatory cytokines in macrophages.

Tetrahydrocurcumin, the major metabolite of curcumin that is the yellow coloring compound of turmeric, β-*sesquiphellandrene*, geranyl acetate, and gingerone, together with 6-gingerol and 6-shogaol, were revealed to inhibit the formation of nitrotyrosine and apoptosis in retinoblastoma cells, implicating a possible protective effect on the retinal ganglion. Besides, 6-paradol and 6-shogaol inhibited macrophage activation through the induction of SOCS-3 expression in microglia. A recent study demonstrates that 6-gingerol inhibits endothelial cell fixing protein, ICAM-1, and VCAM-1 expression in TNF-alpha-stimulated endothelial cells, suggesting a beneficial role in the inhibition of atherogenesis through blocking adhesion molecule expressions (Beatriz Nunes et al. 2023; Benbi 2015).

14.6.3. Antioxidant Activity

The antioxidant activity of gingerol and its analogues was estimated using both ABTS and DPPH radical cation decolorization assays, and the IC_{50} of gingerol, 8-shogaol, 10-gingerol, and 6-gingerol in ABTS and DPPH radical scavenging were found to be lower than that of ascorbic acid, which is a well-known standard antioxidant compound and is most widely used as a scavenger of free radicals. Moreover, the β-carotene bleaching assay was also applied to confirm this result. Taken as a whole, gingerols are better antioxidants than ascorbic acid, and the hydroxyl ceric acid kinetic assay reveals that the superior radical trapping activity of gingerol is due to the -OH group of its phenol rings. Moreover, gingerol and its analogues showed a good protective effect both on the denaturation of bovine serum albumin induced by thermal treatment and on the hemolysis of rat red blood cells induced by peroxyl radicals. That is, gingerol and its analogues were used as antithermal treatment agents and free radical scavengers in the foodstuffs and biomedicine areas, respectively. Additionally, the trends for gingerol and its analogues to scavenge the ABTS and DPPH radical cations were in agreement with the content of free phenolic hydroxyl groups in these compounds, as well as the activities of gingerol and its related compounds to inhibit hemolysis of rat red blood cells (Bensid et al. 2022; Boateng 2024).

14.6.4. Gastrointestinal Benefits

Before modern medicines, herbal treatments were originally used for gastrointestinal diseases. In vitro studies indicated that ginger extracts increased the intrinsic defence mechanisms of the gastrointestinal tract. It has been documented that ginger and its constituents alleviate nausea and vomiting. The inhibitory effects of ginger and its major constituents on gastrointestinal motility could be a mechanism for its antiemetic properties. Ginger and its constituents also have therapeutic potential for the treatment of gastrointestinal diseases, including peptic ulcer, gastritis, and the evacuation of the bowels. Epidemiological studies have reported that ginger improves the symptoms of gastrointestinal diseases in clinical trials. Ginger is frequently used for the treatment of gastrointestinal distress, such as dyspepsia, gastritis, peptic ulcers, and enteritis. In traditional Chinese medicine, ginger is used not only as an important component of many multi-ingredient preparations but also as a single remedy. Ginger induces the secretion of gastric juice and stimulates gastrointestinal motility. Modern experimental research also suggested that ginger regulates lipid metabolism and attenuates enteric dysbiosis, changes in antioxidant status, and apoptosis signals in the gastrointestinal tract. Epidemiological studies have confirmed the therapeutic potential of ginger for the treatment of various gastrointestinal diseases, such as prevention of stomach problems, including dyspepsia, gastritis, and peptic ulcer, as well as evacuation of the bowels in nonsurgical patients. Ginger also possesses antiemetic activity for vomiting and postoperative nausea and vomiting. Ginger has a long-standing reputation as an alternative therapy for the treatment of gastrointestinal ailments. Ginger extract and compounds were found to possess antispasmodic activity, enhance the secretion of gastric juice, influence gastrointestinal immunophysiology, prevent adhesion, protect against stress-induced ulcers, and reduce the diameter and bleeding of acetic acid-induced ulcers (Boateng 2024).

14.7. CLINICAL APPLICATIONS

14.7.1. Potential Therapeutic Uses

A number of emerging technologies are reshaping traditional approaches to the discovery and development of therapeutics. To foster a more rational and systemic technology transfer path, substantial effort has been placed in the genome annotation of a growing number of organisms to identify most, if not all, of the potential gene/expressed sequences in their genomes. This describes the full genetic components of various organisms affecting their biology, pathology of infectious conditions, and other traits upon which the biotechnology and pharmaceutical community may capitalize to improve human, animal, and plant health and welfare. The utilization of functional genomic information can have an important impact on the science and/or the development of therapeutics. Currently, we are passively treating diseases, often with great success, or simply managing the symptoms rather than directly addressing the underlying cellular and molecular changes that are associated with the diseases. Gene-based therapies, for example, are designed to directly affect gene expression products. These products often dictate the cell's and/or organism's biology and are often altered in diseased conditions. These strategies, if optimally designed and implemented, should improve therapeutic benefit, as well as reduce drug side effects, because they would be correcting the underlying basis for these diseases (Boateng 2024; Bouyahya et al. 2022).

14.7.2. Toxicity Studies

Toxicology is often split into two disciplines: acute toxicology, which addresses the hazards of chemicals to humans and animals, and ecotoxicology, which focuses on the hazards of chemicals to the environment. Most of the toxicology literature is not easy to understand and is not readily accessible to scientists who are inexperienced with the methods of toxicology. Unfortunately, much of the information is found in manuals that become more complicated as the number of compounds increases. Due to the limited accessibility of documents to nonexperts in toxicology, the present survey aims to provide a broad, easily accessible overview of the acute oral and dermal toxicities, as well as the eye and skin irritations of chemicals to rodents and rabbits. Compounds are described following an industry classification, based on a classification system. Simple equations are presented for easy conversion to the more traditional chemical classifications (Bouyahya et al. 2022; Carracedo-Pérez, Magariños, and García-González 2024).

The performance of a battery of in vitro genotoxicity tests has been recommended as the primary assessment of the mutagenic potential of new chemicals. Other test systems that are equally proposed as an integral part of genotoxicity risk assessment include in vivo assays in the mouse, the rat, or the bone marrow cells of the mouse, as well as the human and rodent liver, and human lymphocyte chromosome aberration assay. The ex vivo mouse and rat alkaline comet assay has emerged as a powerful, relatively easy, and inexpensive method to screen large numbers of compounds for mutagenic and/or genotoxic effects in vivo. The battery exploits the potential to identify DNA lesions, which lead to gene mutation, clastogenicity, and aneugenicity. The assays complement data from short-term mutagenicity tests that do not detect gene mutation, clastogenicity, or aneugenicity. The assays also complement the in vitro mammalian cell tests that may show equivocal or misleading results due to their inability to respond to in vivo metabolites that are uniformly used in commercially available tests (Cava et al. 2024).

14.7.3. Definition and Importance of Risk Assessment

Although, because of their complexity, human illnesses cannot be neatly classified, it is useful to consider three broad categories of diseases resulting from exposure to chemical or radioactive substances, as distinguished by their causation and manifestation: (1) industrial accidents and occupational diseases, where exposures are typically high and affected populations are easy to identify, (2) "epidemics", where the condition is directly attributable in the population of interest to the combined effect of dose and the properties of the recipients of the dose, and (3) dose-dependent diseases or histories, where any individual case is indistinguishable from naturally occurring disease, and the incidence is so low that it can only be determined statistically over long periods. It is a legitimate area of scientific inquiry to ask whether separate consideration might be given to the dose-dependent appearance of diseases or histories, with the usual public health exceptions to the inability to compensate for the diseased being modified (Cava et al. 2024; Chew, Goh, and Lim 2009).

The definition of a toxic substance is a major part of risk assessment, but the definition of a toxic chemical is not important in its own right but rather because concentration limits controlled by regulations are set by the likely risk of toxic agents. The present notions of acceptable risk, established at best by the frequency analysis of epidemiological data and justifiability based on statistical probability theory, are set by a probable endpoint of exposure or by a series of endpoints of exposure. Environmental rules refer to human health when, in actuality, risk assessment refers to the control of human risk, which is measured by the potential for losing something of value. As the value continues to increase, so too does the burden from risk and its assessment. In times of financial austerity, value comes to be seen not only as a calculated economic benefit—the expected benefit, or the expected benefit adjusted for probability times value, as seen by arriving at what a rational person would be willing to pay to avoid an undesirable event—but also as achieving something intrinsic in the objectives of technology policy and institutional actions needed to foster competitive human health. Then, ways are sought to strengthen accepted goals sufficiently to promote international convergence on acceptable risk at less monitoring cost than available financial resources (Chew, Goh, and Lim 2009; Chrysargyris and Tzortzakis 2023; D'Agostino 2022).

14.8. METHODS AND TOOLS FOR RISK ASSESSMENT

The regulatory framework used depends on the character of the interaction. Quality changes to anthropogenic and natural environmental factors influenced by human activities are regulated by public health and environmental protection laws. The character of interactions is described by the model used for health risk assessment. This model explains the mutual influence of the human and environmental parts, the interaction character, pathways, sources, and differences in sensitive groups. The aim of the study is not only to give the results of the calculation but to specify the methods and tools used. These are very important for regional and national programs of public health assessment and landscape therapy creation for the sake of human community health risk reduction (Figure 14.1).

These methods and tools can become the basis for working programs of nonchemical anthropogenic factor risk management on environmental objects. Examples of interactions are chosen as a background of the model and a method of its use, including risk coefficients. These interactions are characteristic of the territory and are based on the most significant problems sharing anthropogenic and natural environmental factors influenced by human activities. Most of the developed models concern chronic human risks tied to water pollution by nitrates and some chemical compounds, as well as cancer and noncancer risks related to air pollution by polycyclic aromatic hydrocarbons. Risk-specific regional health models and the landscape therapy creation for the benefit of human risk reduction are discussed (D'Agostino 2022; Deshmukh and Gaikwad 2024).

14.8.1. Risk Management Strategies

The ultimate aim of the identification, assessment, and scientific characterization of environmental pollutants of concern is the development of appropriate regulatory strategies to manage the risks associated with human health and environmental impacts of exposure to these pollutants. In situations of risk, urgency may require that measures be taken where scientific understanding

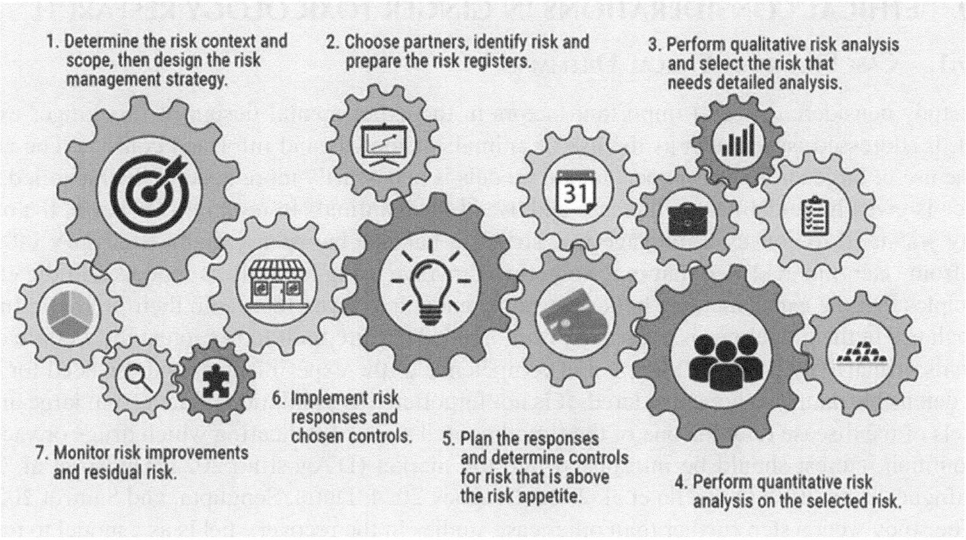

FIGURE 14.1 Risk assessment and analysis methods: qualitative and quantitative.

is incomplete, to both protect public health and provide guidance to the research community. Policymakers must weigh potential human health benefits against economic costs, and ethical, and social considerations associated with various policy alternatives. The selection of risk management strategies involving the control or reduction of pollution could require not only the application of established and improved techniques designed to reduce or eliminate environmental pollution at regional points of contamination but also actions taken several steps upstream in the production and consumption chains. The most effective control actions will be those that address the pollution problem as close to its source as possible, as close as possible to where new pollutants are formed, and the systems they damage. These actions may take the form of command and control regulation or economic instruments, such as effluent charges or tradable emission permits, targets, or standards (Deshmukh and Gaikwad 2024; Dhawande et al. 2024; Domínguez et al. 2019).

14.8.2. Adverse Effects on Different Organ Systems

Adverse effects of therapy are often observed before the underlying cause is diagnosed or recognized. The major causes of these effects are teratogenicity, abnormal reproductive events, hypersensitivity, and adverse effects, which means any nontherapeutic or unwanted action of a drug. The term *side effect* is almost obsolete, and we prefer to use *adverse effect*, which is a broader term and includes all of the consequences of a drug in addition to the specific actions for which the drug was administered. The effects are grouped according to the organ system involved, such as cutaneous, urinary bladder, respiratory system, cardiovascular, nervous, hematological, gastrointestinal, and endocrine. For every organ, subgroups can be recognized (Domínguez et al. 2019; Dong et al. 2022). It is important to realize that the adverse effects of drugs do not depend solely on the pharmacological actions of the administered products. However, they can be caused by interactions in the case of multiple administrations of drugs, between the drug and chemical reactions, or by endocrine-modulating effects. The skin is an important diagnosing organ because hypersensitivity drug effects can be expressed on the skin. With the exception of cancer and bladder irritants, effects on the skin, myocardial fibrosis, and endocrine receptors, such as the thyroid gland, are generally not administered (Deshmukh and Gaikwad 2024; D'Agostino 2022).

14.9. ETHICAL CONSIDERATIONS IN GINGER TOXICOLOGY RESEARCH

14.9.1. Case Studies in Ethical Dilemmas

The study considered several important factors in the experimental design of this ginger experiment. It addressed issues such as the use of animals as models and informed consent. The advice on the use of large animals as experimental models is necessarily more general than detailed; such advice is given in guidelines on the care and use of farm animals in research. However, the opportunity was used to convey a message that strikes a balance between gaining necessary information from scientific evidence that may sometimes involve using animals as models and the ethical principles that the experimenters believe strongly enough to write them into their protocol. In fact, the balance in the guidelines is constantly reinforced. They are built in to protect the welfare of the animals primarily, but the well-being and competence of the experimenters and the need for credible, defensible data also are considered. It is not forgotten that validation of data from large animal models of the disease remains one of the time-honored ways of indicating which drugs or vaccines of common interest should be introduced into the market (D'Agostino 2022; Diniso et al. 2024; Domínguez et al. 2019; Donzella et al. 2022; Dragoev 2024; Dutta, Sengupta, and Samrot 2024).

The study went a step further than other case studies in the recovery fields as a model to test the efficacy and safety of a potential antidiarrheal agent for equimolar cases of diabetes. Diabetes is more complex and affects vital tissues that are normally not present in other animals in the diabetic state, but the results may give some direction for future projects. Scepticism is, however, expressed about using a ginger product to prevent or alleviate the growth of resistant organisms that cause disease from several aspects. First, there are reports of the prebiotic effects of ginger and the ecology of the microbial cells of the human lower bowel, and ginger should be studied to provide information on its effective aspects on flora. Second, the administration of potentially toxic drugs may select the resistant flora that already exists in low numbers, but ginger seems to have a nonselective defensin effect that is toxic to proliferating bacteria in the testing conditions. The question is where resistance can now exist. It should be associated with the active site(s) of the active constituent(s) of ginger extract only if definitively proven by the protocol (Dutta, Sengupta, and Samrot 2024).

In comparing the treatment of a skirt with a fresh juice and boiled water with a standard antibiotic treatment, the study faced a constraint that the search for promising candidates had been subject to natural disciplines for several decades. However, the standard protocol developed from such a search had been refined rather successfully, proving safe and effective for more than 50 years in a number of organism cases to which the ginger product was putatively directed (Figure 14.2). The need to "hypothesize to test" led to the creative use of a simplified infection model, with concern expressed about its deficiencies when extrapolated to life-threatening situations. This was thought to be relatively significant, where the criterion in question had to be refined to progress beyond the observed mean differences. There also arises the question of how the results method might compare with "real life", e.g., people consuming ginger in an attempt to prevent a progression towards insulin-dependent diabetes. Some individuals will progress, whether or not they take the ginger extract. Would this fact have encouraged the obtainment of the mean difference between the administration of the antibiotic and ginger protocols of the therapy regimen? One source of satisfaction in attempting to help solve such problems is the realization that it brings clarity to original assumptions (Dutta, Sengupta, and Samrot 2024).

From a review of the training modules and a perusal of the diversified teaching materials submitted by colleagues, the study shows that to a person a number of the fundamentals of ethics and ethical considerations: freedom, the need to define answers to questions with clarity, rigor, quality and integrity, openness, honesty and a sense of responsibility, herein balanced by discretion and confidentiality, consent, respect, social responsibility, justice, objectivity. The pleasure did not arise from the realization that an individual has fulfilled all the criteria but rather from the uncertain reality that, in striving to meet the criteria, a shortfall in one area will be made up by a potential excess in another area, represented by a real strength. Music, the visual arts, the written or spoken

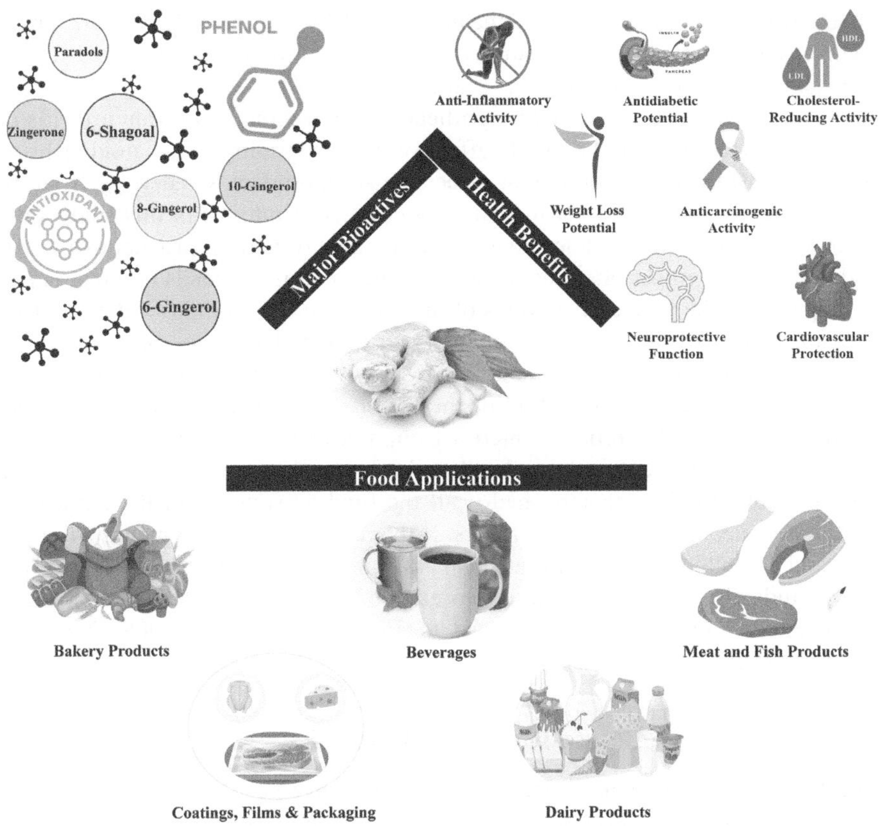

FIGURE 14.2 Ginger bioactives: A comprehensive review of health benefits and potential food applications.

word, or good teaching are the manifestations of the strengths. To give guidance in areas considered commonly relevant to these stages, all researchers will find enjoyable activity (Dutta, Sengupta, and Samrot 2024; Ebrahimzadeh et al. 2008).

14.9.2. Conflicts of Interest in Ginger Toxicology Research

Ginger is a popular spice, condiment, and herbal medicine, having first been used as a medicine in India more than 2,000 years ago. Fresh ginger has been used as a diaphoretic and appetite stimulant and in cases of abdominal bloating, cough, dyspnea, and vomiting. Dried ginger has been used as a stomachic for the treatment of chronic dyspepsia. The methanol extract or ethanol extract of ginger, acting as a diaphoretic, has long been used to treat common colds, while the pungent components in it have been used as appetizers. Among seven types of ginger, only the wild type did not inhibit ascorbic acid–induced hyperlipidemia in golden hamsters. Because either type of ginger has the same, similar, or dissimilar biological activities, many studies on the constituents or properties of ginger have been carried out using different extracts or methods. Shogaols inhibited the degradation of 8-hydroxydeoxyguanosine more than gingerol. The antioxidant capacity of gingerdiol was stronger than that of gingerol, but less strong than that of 10-gingerdiol. They found that those results suggest that the addition or subtraction of a methoxy group in the phenyl structure restored and maintained the DPPH activity that existed in gingerol. These studies, proving that structural property caused the difference in the antioxidant activities of ginger extract, were conducted by a leading research group. The same group has noted that gingerol significantly lowered tissue levels of malondialdehyde and nitrate/nitrite, and the malondialdehyde level of the arthritic phase was significantly correlated with the tissue tumour necrosis factor-α level (Deshmukh and Gaikwad 2024; Domínguez et al. 2019; Donzella et al. 2022; Farhat et al. 2024).

They concluded that gingerol exerted a significant amelioration of adjuvant arthritis in rats in a dose-dependent manner and may be highly effective in ameliorating rheumatoid arthritis through the antioxidant defence system. Noting that its beneficiating product was derived from its direct or indirect ginger use, the group did not indicate any conflict of interest in this study, despite more than six years of joint research and development in the field of natural product extraction technology. Indeed, the group's ginger pharmacological research mainly depends on ginger extracts from the industrial partner. The financial conflicts of interest linked to an industrial partner in terms of beneficial contributions to the research, employment, financial interest, and owner interest may have influenced the scientific integrity of another research topic that focused on industrial eggshell membranes to protect against degradation of the major dietary supplement for joint health. The disentanglement of financial conflict of interest from a government source is critical to the unbiased scientific integrity of the research related to dietary supplements or their active ingredients. Disentanglement can only be accomplished if all the funding sources for the clinical studies are fully and accurately declared by the researchers. If the researchers do not mention these sources in their articles and claim to break the biostatistical assistance clause for the evidence of the financial sponsor of the study, this is considered a serious ethical concern, as it does not contribute to transparency in the information provided to the readers (Ebrahimzadeh et al. 2008; Diniso et al. 2024; El Sheikha et al. 2022; Deshmukh and Gaikwad 2024).

14.9.3. Advancements in Ginger Toxicology Research

Over the past three decades, many toxicological studies have been directed at testing the safety and protecting the health of consumers from the harmful effects of ginger and its constituents. Preclinical pharmacokinetic and dermal toxicity testing is the most advanced. The use of valid animal models, good laboratory practice standards, and validated analytical methods has allowed the collection of reliable and relevant accumulated evidence of heterogeneous quality. Although ginger has been used as food and medicine in human society for centuries, studies assessing its potential toxicity have been mostly fairly recent, with test articles, study designs, doses, and exposure durations that may have been inadequate to detect unique but important adverse effects. Significant data gaps remain, especially regarding safety during periods of susceptibility. Part of this research has revealed that the toxicology of ginger is complex, pharmacologically explained, often inherently biological, frequently related to the dose or duration of exposure, and sometimes conflicting. The unprecedented pace of advancements in ginger toxicology research using omics technologies makes this a suitable juncture to document our experiences and propose a set of shared practical principles that can maximize our contributions while reducing, refining, or replacing animal experimentation.

This section offers a brief overview of illustrated topics, many specific examples, and further reading suggestions pertaining to noteworthy advancements in ginger toxicology research over six major knowledge-creation areas. Active sharing of such valuable information will facilitate the competent and responsible conduct and analysis of meaningful toxicological studies that fully inform risk assessments and risk management regarding the safety of ginger. Thus, this book chapter aims to call attention to the continued progression of ginger toxicology research and to encourage appropriate investment in supporting research that advances comprehensive and systematically organized efforts to evaluate the safety of ginger. Such beneficial research knowledge protects human consumers and animals under experimental conditions from harm and demonstrates a high degree of respect for regulatory agencies in their developmental risk assessment and regulatory decision-making needs. Moreover, other disciplines are currently providing innovative technical strategies to exploit growing knowledge bases and maximize access to accumulated findings about ginger that can be further processed, featured, and shared with all stakeholders who are deeply interested in enjoying the benefits that value-added ginger can bestow. The availability of safe ginger may be facilitated (Farhat et al. 2024).

14.9.4. Research Methods in Ginger Toxicology

The in vivo studies assist in the real evaluation of potential adverse effects of the toxicant in a live organism. During the periodic oral administration of acetaminophen, a significant increase in mortality rate has been reported at doses of 500 mg/kg bw and 1000 mg/kg bw of acetaminophen in rats. In the case of the naloxone pretreatment group, there was a significant reduction in the lesion score on gastric mucosa in pylorus-ligated rats by 80% when compared to the aspirin control group, which indicates the highest analgesic property. The oral feeding of tested substances is preferred because the uptake does not involve a mandatory evaluation of possible clinical signs of toxicity during chemical intoxication, which can be fatal. In these studies, blood withdrawal or tissue collection in all experiments of in vivo studies yields useful results on signs of toxicity and metabolic disturbances to assess the damage caused by the toxicant in the bodies of the animals and also in the clarification of the action mechanisms of the toxicant (Farhat et al. 2024; Ford et al. 2024).

14.10. MOLECULAR TOXICOLOGY TECHNIQUES

Molecular toxicology is a crucial discipline that delves into the intricate interactions between toxic substances and biological systems. Its primary focus lies in comprehending the molecular mechanisms responsible for toxic responses within these systems. Over the years, advancements in biotechnology have revolutionized research in this field, opening up new avenues of exploration and generating copious amounts of invaluable data. As a result, scientists now face a plethora of opportunities and challenges in their efforts to unravel the mysteries of toxicology. One particularly notable development arising from these advancements is the emergence of high-throughput assays. These powerful platforms, such as toxicogenomics and proteomics, have played an instrumental role in facilitating mechanistic and predictive toxicology studies. Toxicogenomics, for instance, utilizes microarray techniques to meticulously analyze the expression patterns of genes, shedding light on the intricate molecular responses triggered by toxic agents. On the other hand, mass spectrometry techniques have enabled the comprehensive mapping of human and animal proteomes, providing invaluable insights into the vast array of protein interactions and signaling pathways involved in toxicology. This section endeavors to comprehensively cover the latest developments and ongoing research in various domains of molecular toxicology. By delving into these intricacies, researchers aim to expand our understanding of toxicological phenomena, ultimately leading to the development of more effective strategies for combating toxic exposures and safeguarding human health (Forslund et al. 2023).

A metabolome is a complete set of metabolites present in a living organism, cell, or tissue. Metabolomic techniques are being used to assess toxicity based on the identification and quantification of small-molecular-weight compounds, including intermediates or end products of metabolism, pharmaceutical drugs, and other xenobiotics, as well as signaling molecules. It is believed that metabolomic fingerprinting, the changes in the metabolome, has the potential to reflect toxicity generated by interactions between genetic alterations and environmental factors more directly and reliably compared to changes (Heiðarsdóttir 2020; ISO 1998; Javadian, Shahosseini, and Ariaii 2017) in the proteome and transcriptome. With the advance of mass spectrometric technologies, numerous metabolomic analytical techniques have made it feasible to obtain a snapshot of the metabolome. These techniques include hydrophilic interaction chromatography, liquid chromatography-mass spectrometry, gas chromatography-mass spectrometry, and magnetic resonance analysis (Hassoun and Karoui 2017).

Some common procedures employed to differentiate metabolite profiles include sample preparation and treatment, metabolite detection methods, preprocessing techniques, and statistical analysis methods. The mass spectrometry technique utilizes charged particles to sort and separate compounds via a mass-to-charge ratio. It is the most advanced analytical technique used to quantify small molecules with great precision. With high mass accuracy, liquid chromatography-mass spectrometry uses online or offline separation techniques before mass spectrometry detection, and it is a core technique in metabolomic analyses. For a more general metabolomic profiling analysis, hydrophilic interaction chromatography is commonly used. To analyze volatile and thermally stable compounds, the headspace gas of a liquid sample must be created so that it can be injected into the gas chromatography port to avoid high-temperature degradation of the injected sample. Subsequently, the fast carrier gas carries the volatilized analytes to the column, and then these analytes pass through the stationary phase, which separates the individual analytes based on their relative binding to two more desired states (Javadian, Shahosseini, and Ariaii 2017).

14.10.1. Nuclear Magnetic Resonance (NMR) Spectroscopy

NMR is a powerful tool used to study the interaction of small molecules with macromolecules and their role in physiological mechanisms. NMR can provide three important aspects in the study of molecular toxicology. First, structural information can be obtained for both smaller and larger molecules with resolution at the atomic level. Second, dynamic properties can be detected, which is important for understanding the molecular toxicity of nanomaterials, as different toxic materials can affect the dynamic properties of cellular components differently. Finally, we can detect the density of specific molecular complexes that bind to biological molecules after biomolecular interactions. Therefore, NMR has been used in the study of the interaction of nanomaterials. The main NMR-based techniques that have been used in the study of molecular toxicology include the following: (1) in-cell NMR, (2) STD NMR, (3) chemical shift titrations, and (4) heteronuclear NMR techniques such as TROSY. The main biological applications of NMR in the study of the molecular toxicology of nanomaterials include the quantification of pDNA replication errors induced by nanoparticles, an NMR binding study of Apt–RNA nanoparticles, the observation of nanoparticle-induced cell death by NMR, and the NMR observation of the nanoparticle surface in a biological environment (Javadian, Shahosseini, and Ariaii 2017; Kandil, Aly-Aldin, and Allam 2020).

14.10.2. Gas Chromatography-Mass Spectrometry (GC-MS)

Gas chromatography coupled with mass spectrometry is an extremely sensitive technique used for the forensic analysis of analytes, especially for small and volatile compounds in samples containing complex mixtures. It is also used to discover unknown compounds due to its high selectivity and sensitivity. Gas chromatography is the process of separating, identifying, and quantifying a mixture of volatile compounds in a sample. It separates the components of the sample using partitioning mechanisms. After separation and elution from the column, the separated chemicals migrate over the ion source, from which the analytical data are acquired (Kapoor et al. 2020).

Using this technique, we can identify analytes by their distribution in the mass spectra. The mass spectra are unique to a specific compound, like a fingerprint, and hence it is also used for confirming the identity of unknown compounds. The molecular mass and the structural composition of the compound can be identified by the fragmentation pattern of the mass spectra. After retention indices and mass spectral comparison of the identified analytes with their corresponding standard analytes, we can confirm the presence of the contaminants/analytes in the samples. This confirmed identity information from both the properties, i.e., retention time and the abundance of the mass spectrum markers, is further confirmed by the retention time and total ion chromatogram abundance information (Kapoor et al. 2020).

Molecular toxicology is a crucial discipline that delves into the intricate interactions between toxic substances and biological systems. Its primary focus lies in comprehending the molecular mechanisms responsible for toxic responses within these systems. Over the years, advancements in biotechnology have revolutionized research in this field, opening up new avenues of exploration and generating copious amounts of invaluable data. As a result, scientists now face a plethora of opportunities and challenges in their efforts to unravel the mysteries of toxicology. One particularly notable development arising from these advancements is the emergence of high-throughput assays. These powerful platforms, such as toxicogenomics and proteomics, have played an instrumental role in facilitating mechanistic and predictive toxicology studies. Toxicogenomics, for instance, utilizes microarray techniques to meticulously analyze the expression patterns of genes, shedding light on the intricate molecular responses triggered by toxic agents. On the other hand, mass spectrometry techniques have enabled the comprehensive mapping of human and animal proteomes, providing invaluable insights into the vast array of protein interactions and signaling pathways involved in toxicology. This section endeavors to comprehensively cover the latest developments and ongoing research in various domains of molecular toxicology. By delving into these intricacies, researchers aim to expand our understanding of toxicological phenomena, ultimately leading to the development of more effective strategies for combating toxic exposures and safeguarding human health (Al Jumayi et al. 2022; Alexander 2020).

14.11. FUTURE DIRECTIONS IN GINGER RESEARCH

The increasing interest of consumers in the health-promoting properties of food has led to an explosion of interest in nutraceuticals, primarily focused on isolating health-promoting components and developing supplements. A simpler approach would be to grow foods that contain high levels of nutraceutical compounds. Ginger ranks as one of the top 100 functional foods because of its bioactive antibiotic, carminative, anti-inflammatory, and antioxidant properties, in addition to being an effective treatment for nausea and gastrointestinal conditions. Compounds contributing to these activities include pungent gingerols, zingerone, and shogaols, and less pungent 6-gingerols and 6-shogaols. Fresh or dried ginger as food can be consumed in a natural form without any undesirable side effects. Intake of functional foods such as ginger can reduce the incidence of a number of lifestyle-related diseases, enhancing the quality of life of consumers. Peeling, cutting, or drying ginger modifies its food matrix in ways that can produce negative impacts on sensory and health attributes. Both the inclusion of ginger as a food ingredient and the post-processing modulation of ginger's health properties are achievable. Ginger from micropropagation has enhanced levels of the bioactive 6-gingerol and cinnamaldehyde and increased gallbladder emptying activity. Additionally, a bioactive-enriched probiotic cultured transformed this bacterium into a functional source of ginger bioactives by hydrolyzing 6-gingerol glycosides. Upon ingestion with yogurt or ginger-containing food, this bacterium showed increased ginger phenolic release without changing the probiotic's health benefits, promoting the delivery of the functional food's bioactive components from the gastrointestinal tract. The range of post-processing applications for ginger in food could be expanded by a range of nonthermal processing technologies. Consequently, ginger could play an increasingly important role in functional foods with broad nutraceutical applications and the labels "natural", "minimally processed", or "raw" (Alexander 2020; AlJuhaimi et al. 2024; Amaral et al. 2019; Asif, Ibrahim, and Ansari 2024).

14.11.1. Impact of Ginger Research on Health and Medicine

Ginger has an extremely low toxicity profile and has been used as a food condiment since ancient times. Unlike some other essential oils, long-term use has not been shown to produce adverse effects. Whole dried ginger is recognized as safe in various regions. This regulatory status suggests that regulatory agencies are confident that acceptable daily consumption levels would not be exceeded even in the absence of quantity limitations. Ginger is considered generally recognized as safe as a direct human food ingredient. For therapeutic use or when symptoms begin to dissipate, lower doses are recommended, as well as dose limitations to provide a "safety factor" when used by sensitive individuals or for use during pregnancy. Low dosages of ginger are safe and have few side effects. Side effects and risks may increase with dose. Reported adverse reactions following high doses have been minor and infrequent, limited primarily to significant complaints of heartburn, gastrointestinal upsets (when ginger has been chewed raw), rashes as a result of topical application, and the exacerbation of ulcers or gastroesophageal reflux. However, ginger use should not be undertaken lightly since extraction of irritants and alteration to the physical defences in the plant produce forms of ginger that are more toxic than the unadulterated material. Even though significant concerns for toxic consequences are currently unknown, like other eicosanoid inhibitors used in medicine, high doses of ginger might exert an anticancer risk (Asif, Ibrahim, and Ansari 2024; Awasthi et al. 2022; Badoni et al. 2021).

The development of dietary supplements with defined quality standards involves many scientific challenges, including the integrative and coordinated use of chemical, pharmaceutical, analytical, biological, and clinical research methods. To fulfill the regulatory requirements, companies in the areas of dietary supplements, herbal preparations, phytomedicines, nutraceuticals, and functional foods need to demonstrate the quality, safety, and efficacy of their products. This chapter provides a summary of the requirements for quality, safety, and efficacy documentation of ginger preparations; various examples are discussed. Optimally, such constraints should be seen, from a conceptual point of view, as providing opportunities for further advancements and innovations in ginger research. However, more regulatory advances are urgently needed to assure greater health and economic benefits to all consumers and to encourage further research and development in this promising botanical area. In particular, the lack of standardization of ginger products and inconsistencies in labeling poses a concern for both consumers and the broader dietary supplement industry, so we also refute some points on these delicate issues (Balkrishna et al. 2024; Bao-Hua, Ying, and Chang 2004).

14.11.2. Challenges and Opportunities in Ginger Research

Researchers of ginger face many limitations. The genetic diversity of ginger is of great concern to ginger researchers. Ginger's ploidy level is another area that has involved much speculation. Is ginger an 'erectoid' type? Does ginger obey '2n gametes', a process that will generate new and stable hybrids? Induced triploids are essentially sterile, which will enable an increase in the reproductive growth of ginger, quality, and time of fibre and bud extract of ginger, as well as increased zingiberene content in ginger rhizomes. Enhancing genetic diversity is essential for the germplasm. Transferring genes from closely related species might also help to create disease-resistant, hardier ginger. This study of genetics should enable ginger marker development (Bang et al. 2013; Bariya, Patel, and Joshi 2020).

Additionally, taming ginger's rhizome biology is important when ginger is bred. The rhizome is a reproductive organ and not an edible part. Also, forming spikes and developing rhizomes, ginger depends not only on the biotic and abiotic environment but also on the normal phasic development. However, when ginger is planted in different environments, either natural flowering or artificially induced flowering could lead to the detection of physiological phases or molecular biological phenotype values, but the ability to utilize rhizomes differs. Another significant challenge of ginger research is the lack of affordable polymorphic markers. Such markers would enable us to assess genetic diversity, analyze genetic relationships, and map quantitative trait loci for important ginger traits. Meanwhile, traditional markers offering benefits are the ultimate goal for all genetic mapping and marker-assisted selection problems. A framework could be integrated that would connect the cultivation of suitable testing and population studies of ginger (Bariya, Patel, and Joshi 2020; Beatriz Nunes et al. 2023; Bendala 2020).

Phenotypic variability in a given germplasm collection is largely correlated to its original diversity. Throughout centuries of selection by man, starting with its first cultivation in the Far East, but possibly also due to the plant's preferred growing zone, ginger has accumulated significant changes. This may be one reason for the lack of an exactly identified gene donor. It seems reasonable that if the potato underwent as many modifications as the cultivated plant from its wild relatives, the same could happen for Zingiberaceae. Limited knowledge of its putative wild relatives further complicates the situation. The identification of germplasm and the development of conservation strategies and international support is now expected. Ex-situ cultivation has produced few successes due to poor adaptation. High molecular tools are, as they are for many groups of plants, the most attractive for the conservation and use of this and other plant resource groups (Bendala 2020; Balkrishna et al. 2024). In the ginger exploration of the 1970s, random explorations occurred in the northeast. In the 1980s and 1990s, the same northern areas of India and southern Bhutan were extensively covered as randomly as possible. From these explorations, 394 samples likely represented a great variety of different landraces whose morphotypes are unknown to early crops. They concluded that: (1) ginger genotypes are not monoecious, as initially suggested, and distinct male and female sexual organs are commonly present in bract axils of ginger; (2) conspicuous variation in compounded inflorescences occurred in ginger accessions; and (3) the pattern of variation of amplification products suggests that the genotypes of Indian representative accessions are highly variable (Balkrishna et al. 2024).

The problem is not enough change in the weather, but rather frequent unpredicted occurrences of either climatic or biotic stresses in ginger-growing areas. Ginger is very sensitive to the changing weather, creating unreliable timing for planting, either an early or late start of the rainy season, shorter duration of the growing period, unpredictable periods of dry and wet conditions, the possibility of dry spells, and weaker intensity of rainfall distribution patterns, eventually leading to the occurrence of floods. Farming outside the normal conditions of either extreme temperatures or variations in humidity creates top-priority challenges that must be addressed with successful scientific activities. A serious concern for ginger production is the changing pest and disease distribution patterns due to the changes in weather and agro-climatic conditions influenced by climate change (Balkrishna et al. 2024; Chrysargyris and Tzortzakis 2023).

Ginger suffers from damage caused by pests and diseases, perhaps so severely that the crop may fail to reach the economic growth stage whenever the cultivated area exceeds the threshold of its ecologically adapted area. The occurrence of harmful insects, mites, and pathogens is more serious with either loosening soil, rising temperatures, or frequent weed occurrences. Researchers still have opportunities to explore the adaptation of existing elite lines to variable ecological conditions or create novel hybrids through crossing with various wild ginger relatives to maintain unpredictability in both weather and pathogens. Cultivating inland during critical periods of the crop growth season adds an additional dimension to how ginger could be cultivated and creates chances to promote its beneficial compounds (Chunzhi et al. 2018).

Different diseases and pests attack ginger at different growth stages. Studies have shown that new diseases and pests are expanding their distribution ranges into ginger-producing areas because of climate warming. The National List of Common Pests of Ginger includes a few pests and diseases of ginger, but it is not comprehensive. Some scholars have investigated pests and diseases that affect ginger yields and quality. In general, the pests and diseases that harm (Cui et al. 2023) ginger are mainly divided into those occurring in the seedling stage, those occurring in the management stage, and those occurring before harvest (Cui et al. 2023).

Understanding the reasons for the occurrence of these pests and diseases and timely taking measures to control them is critical to reducing the infection rate and, thus, to reducing the disease cost and promoting climate sustainable improvement and the high yield of ginger. Research on ginger pests and diseases revealed that six diseases greatly affected ginger yield recorded over the previous 20 years. In particular, *Phytophthora* blight was the most widely reported disease of ginger. Seven related pests were recorded over the previous 20 years. Thirty-four patents for controlling diseases, four patents for controlling pests, and three patents for the nutrition of ginger were well documented. In addition to these, there were eight reports on microbiological control. However, there has been no comprehensive summary and review of research challenges and future research in this field. The present review summarizes major pests and diseases of ginger during the growth stages and protection measures (Cui et al. 2023).

14.11.3. Exploring the Future of Biotechnological and Nutraceutical Research in Sustainable Agriculture

In summary, there is enormous potential for future research in the area of the biotechnological aspects of nutraceuticals, including the exploitation of nonmodel crops, container plants, or medicinal herbs, to isolate novel genes involved in the biosynthesis of nutraceutical compounds with their subsequent implementation in plant molecular breeding programs. Interestingly, the pan-genome of many species consists of genes not present in the reference assembly used in comparative genomics. The identification of the common features of multiple alternative assembly genome re-sequencing can improve valuable germplasm for crop breeding. The development of tools can provide knowledge about these new genes, show the differences in the concentration of specific compounds between many accessions of these species, and identify molecular markers useful for breeding. All of these have to be in sound scientific and ethical conditions (Cui et al. 2023; Chew, Goh, and Lim 2009).

Biotechnological research in nutraceuticals, whether in comparative or noncomparative models, is still facing severe barriers at the social acceptance level for commercial interests because nutraceutical compounds are unprotected, especially when they come from open-source seeds. Their production costs are generally higher than traditional crops because they are abundant in the edible part only in pharmaceutical quantities. Therefore, it is mandatory, if we want the owners of seed banks to authorize their diffusion to farmers, that they accept the obligation to use trait-linked markers to prove that their seed is not the result of cross-pollination between two crops or from back-crossing with environmental sections. Therefore, a clear understanding of the gene and what it does, such that we could accurately predict its action, could help clarify the licensing, intellectual property rights, and ethical concerns and improve public perception as well (Ebrahimzadeh et al. 2008; El-Hadad et al. 2024).

14.11.4. Biotechnological Approaches

Advances in "omic" technologies have given tremendous insights to unravel the genetic architecture of different crops, while genome-editing tools have led to the development of desired gene phenotypes with much precision and accuracy—strengthening their use in functional genomics research. The development of proteome-metabolome resources and elucidation of associated regulatory networks can have a significant impact on crop management for nutritional and dietary benefits as well as crop resistance. Bioinformatics and computational biology approaches have also played a comprehensive role in the analyses of deeper knowledge, while the application of data science tools and machine learning algorithms has also propelled research progress in crop breeding. Molecular plant sensors and electronic technologies have always been the focus of attention for the scientific community to develop sporadic monitoring tools in the agriculture sector to optimize the performance and precise use of agricultural inputs in order to face global food challenges. Thus, the present chapter sums up the recent scientific advancements in biotechnological and nutraceutical research areas and highlights the mutual roles of molecular and analytical techniques in building better food futures and ensuring a glitch-free sustainable agriculture system (El-Hadad et al. 2024; Ebrahimzadeh et al. 2008).

14.11.5. Nutraceutical Development

Apart from social, economic, and environmental aspects, a further concern for the current livestock agriculture industry is an increasing concern for public health. It has been suggested that nutraceuticals, together with functional foods, are expected to have a significant effect in the primary prevention of chronic degenerative diseases, such as cardiovascular disease and cancer. Nutraceuticals are bioactive compounds with clinical health benefits in the treatment or prevention of disease. They are biologically active compounds synthesized by plants in response to various stress factors, such as herbivores, pathogens, or UV radiation. It is commonly understood that bioactive compounds may also have health benefits in animals and thus in humans, in the form of dietary antioxidants. Dietary antioxidants may act as free radical scavengers, metal chelators, and anti-inflammatory agents, and directly scavenge or degrade reactive oxygen species (Ezejiofor, Enebaku, and Ogueke 2014; Faria, Carvalho, and Conte-Junior 2023; Piironen, Paasiniemi, and Vehtari 2020; Resurreccion 2007).

The bioactivity of nutraceuticals on beneficial and physiological effects of oxidative stress is based on the antioxidant defence and, thus, their capacity of either single or combined molecules to scavenge and neutralize the highly reactive oxygen and nitrogen-derived free radicals. It is a fact that during prolonged or repeated physical stress, referred to as stress-related diseases, due to the capability of endogenous synthesis of antioxidants or accumulation in the body of exogenous dietary antioxidants, the induction of oxidative stress significantly increases the production of free radicals, which leads to a decrement of the heavy oxidative load with a lower degree of endogenous antioxidant reserves. In the majority of cases, the intense or prolonged generation of free radicals has been associated with other pathologic mechanisms leading, in some cases, to metabolic disturbances or immune dysfunction, which generate a broad set of diseases (Resurreccion 2007).

14.11.6. Sustainable Cultivation Practices

Sustainable cultivation practices are systems or practices that produce farm products in an integrated, environmentally safe manner and protect and enhance the long-term farm condition. In other words, these practices deliver the right amount of ingredients in products without being harmful to the environment and the consumer. To approach this goal, the concept of sustainable agriculture raises related questions: How are good agroecological practices dependent on crop nutrition? Can registered micronutrient use increase production stability and minimize environmental impact? Here, we will discuss these and other questions directly concerned with ‘sustainable micronutrient management’ and ‘nutraceutical’ perspectives in the sustainable crop nutrition industry (Resurreccion 2007).

Farmers obtain increased productivity from the same or relatively fewer resources by promoting balanced nutrition using a range of farm management practices. In essence, nutrition management is designed to optimize crop nutrition in order to cultivate practical guidelines in the field and simultaneously to reduce contamination in irrigation, human health, and the wider environment. For instance, the efficacy of grain fortifiers to manage the seed and soil is key to sustainable agriculture in eco-farming environments. In this cost-effective approach, biodegradable grains deliver micronutrients into the soil ecosystem while securing the microbes that support post-applied or on-tubers agro-environmental well-being without being released from grain. This practice, combined with optimized manure fertilizer, can help meet sustainable organic potato farming parameters. The outcome will be agricultural soils rich in crop nutrients and consisting of local systems with minimal heavy metal micronutrient induction (Resurreccion 2007).

14.12. CONCLUSION AND SUMMARY

Ginger (*Zingiber officinale Roscoe*), a globally celebrated medicinal and culinary herb, possesses a remarkable profile of bioactive compounds that underpin its extensive therapeutic applications. This chapter has comprehensively explored ginger’s chemical composition, health impacts, and toxicity profile, shedding light on its potential benefits and considerations for safe use. Ginger’s pharmacological activities are primarily attributed to its rich phytochemical constituents, including phenolic compounds such as gingerols, shogaols, and paradols, as well as terpenoids, flavonoids, and volatile oils. These compounds collectively exhibit potent antioxidant, anti-inflammatory, anticancer, antimicrobial, and neuroprotective properties. Gingerol, the major active component, plays a pivotal role in combating oxidative stress and inflammation, which are key contributors to chronic diseases. Upon drying or heating, gingerols convert into shogaols, which possess enhanced biological activity.

The volatile oils, such as zingiberene and citral, not only impart ginger's distinctive aroma but also contribute significantly to its antimicrobial and anti-inflammatory potential. The health benefits of ginger extend to various systems in the human body. Its anti-inflammatory effects make it an effective natural remedy for conditions such as arthritis, asthma, and inflammatory bowel disease.

The antioxidant properties of ginger mitigate oxidative stress, thereby reducing the risk of cardiovascular diseases, neurodegenerative disorders, and aging-related damage. In the realm of oncology, ginger demonstrates remarkable potential in inhibiting tumour growth, inducing apoptosis in cancer cells, and preventing metastasis in several types of cancers, including breast, colon, and prostate cancers. Ginger also holds a distinguished place in digestive health. It is widely used as an antiemetic to alleviate nausea and vomiting, particularly during pregnancy and chemotherapy. Moreover, its gastroprotective effects enhance gastric motility and shield the gastrointestinal mucosa from damage. The antimicrobial properties of ginger target a wide range of pathogens, including bacteria, fungi, and viruses, emphasizing its role as a natural antimicrobial agent. Furthermore, its cardiovascular benefits, such as lowering blood pressure and cholesterol levels and reducing the risk of thrombosis, underline its significance in heart health. Neuroprotective effects have also been observed, suggesting its potential in managing neurodegenerative diseases such as Alzheimer's and Parkinson's.

Despite its numerous benefits, ginger's toxicity profile warrants careful consideration. Although it is generally recognized as safe when consumed in moderate amounts, excessive intake or prolonged use may lead to gastrointestinal discomfort, liver and kidney impairment, or adverse drug interactions, particularly with anticoagulants and hypoglycemic medications. Special caution is advised during pregnancy, as excessive doses could pose risks to fetal development. In conclusion, ginger emerges as a versatile natural remedy with immense therapeutic potential. Its bioactive compounds contribute to its widespread use in traditional and modern medicine, targeting diverse health conditions with notable efficacy. However, the dose-dependent nature of its benefits and toxicity underscores the importance of regulated consumption. Future research should prioritize the development of standardized ginger-based formulations, long-term toxicity studies, and investigations into synergistic interactions with other natural products. By addressing these areas, ginger's role in health and medicine can be further optimized, ensuring its safe and effective application for global well-being.

REFERENCES

Abarca, Mariana, Anna L. Parker, Elise A. Larsen, James Umbanhowar, Chandra Earl, Robert Guralnick, Joel Kingsolver, and Leslie Ries. 2024. "How development and survival combine to determine the thermal sensitivity of insects." *PLoS One* 19 (1):e0291393.

Abdallah, N. A., Abdul Rahman Ahlan, and Odeh Abed Abdullah. 2019. "The role of quality factors on learning management systems adoption from instructors' perspectives." *The Online Journal of Distance Education and e-Learning* 7 (2):133.

Addis, Mekonnen. 2015. "Major causes of meat spoilage and preservation techniques: A." *Changes* 41:101–114.

Adewusi, Emmanuel Adekanmi, N. Moodley, and Vanessa Steenkamp. 2011. "Antioxidant and acetylcholinesterase inhibitory activity of selected southern African medicinal plants." *South African Journal of Botany* 77 (3):638–644.

Al Jumayi, Huda Abdalrahman, Ayman Younes Allam, Alaa El-Dein El-Beltagy, Eman Hassan Algarni, Samy F. Mahmoud, and Amin Abd El Halim Kandil. 2022. "Bioactive compound, antioxidant, and radical scavenging activity of some plant aqueous extracts for enhancing shelf life of cold-stored rabbit meat." *Antioxidants* 11 (6):1056.

Alexander, Robin. 2020. *A Dialogic Teaching Companion*. Routledge.

Algarni, Eman H. A., Ibrahim A. Elnaggar, Abd El-wahed N. Abd El-wahed, Ibrahim M. Taha, Huda A. Al-Jumayi, Sam M. Elhamamsy, Samy F. Mahmoud, and Alaa Fahmy. 2022. "Effect of chitosan nanoparticles as edible coating on the storability and quality of apricot fruits." *Polymers* 14 (11):2227.

AlJuhaimi, Fahad, Isam A. Mohamed Ahmed, Mehmet Musa Özcan, Nurhan Uslu, and Emad Karrar. 2024. "Effect of roasting on the oil content, bioactive components, fatty acid composition and mineral content of purslane seeds." *International Journal of Food Science & Technology* 59 (1):140–147.

Allam, Ayman Younes, Mohamed Farouk Elsadek, Khalid S. Al-Numair, Adel Abdelrasoul A. Badr, Sunita Singh, and Mohamed Rafat Elkabary. 2024. "Antioxidant and antibacterial effect of lemon verbena leaves'(Lippia citriodora) extract as a natural preservative on refrigerated meat patties during storage." *Italian Journal of Food Science* 36 (3):95.

Amaral, Deborah Silva do, Alejandra Cardelle-Cobas, Celina de Castro Querino Dias, Darlinne Amanda Soares Lima, Sergio de Ferreira Pereira, Naciza Maria de Oliveira Arcanjo, Paulo Sergio Dalmas, Marta Suely Madruga, and Maria Manuela Estevez Pintado. 2019. "Low fat goat meat sausage with chitosan-glucose Maillard reaction product: Impact on quality and shelf life." *Food Science and Technology* 40:132–139.

Arif, Muhammad, Talha Jan, Muhammad Riaz, Shah Fahad, Muhammad Adnan, Amanullah, Kawsar Ali, Ishaq Ahmad Mian, Bushra Khan, and Fahd Rasul. 2020. "Biochar: A remedy for climate change." *Environment, Climate, Plant and Vegetation Growth* (pp. 151–171). Academic Press.

Asif, Amber, Fariha Ibrahim, and Asma Ansari. 2024. "A systematic review: Effectiveness of herbs and spices as natural preservatives to enhance meat shelf-life: Herbs and spices as natural preservatives." *Journal of Health and Rehabilitation Research* 4 (3):1–7.

Awasthi, Mukesh Kumar, Raveendran Sindhu, Ranjna Sirohi, Vinod Kumar, Vivek Ahluwalia, Parameswaran Binod, Ankita Juneja, Deepak Kumar, Binghua Yan, and Surendra Sarsaiya. 2022. "Agricultural waste biorefinery development towards circular bioeconomy." *Renewable and Sustainable Energy Reviews* 158:112122.

Babuskin, S., P. Azhagu Saravana Babu, M. Sasikala, K. Sabina, G. Archana, M. Sivarajan, and M. Sukumar. 2014. "Antimicrobial and antioxidant effects of spice extracts on the shelf life extension of raw chicken meat." *International Journal of Food Microbiology* 171:32–40.

Badoni, Pooja, Iqra Nazir, Merenlong Aier, Puspendu Bikash Maity, S. Samanta, and A. Das. 2021. "Significant role of fish nutrients with special emphasis to essential fatty acid in human nutrition." *International Journal of Current Microbiology and Applied Sciences* 10:2034–2046.

Bali, Rishi Kumar. 2021. "Operating room protocols and infection control." *Oral and Maxillofacial Surgery for the Clinician* (pp. 173–194).

Balkrishna, Acharya, Amita Kumari, Ashwani Kumar, Vedpriya Arya, Ankush Chauhan, Navneet Kumar Upadhyay, Ishita Guleria, Ryszard Amarowicz, Dinesh Kumar, and Kamil Kuca. 2024. "Biosensors for detection of pesticide residue, mycotoxins and heavy metals in fruits and vegetables: A concise review." *Microchemical Journal*:111292.

Bang, Megan, Beth Warren, Ann S. Rosebery, and Douglas Medin. 2013. "Desettling expectations in science education." *Human Development* 55 (5–6):302–318.

Bao-Hua, Guan, Ge Ying, and Jie Chang. 2004. "Phenotypic plasticity of Mosla chinensis and M. scabra (Labiatae) response to soil water status." *Botanical Bulletin of Academia Sinica* 45.

Bariya, A. R., A. S. Patel, and A. B. Joshi. 2020. "Application of natural antioxidants from vegetables, plant and fruits origin into meat and meat products-a review." *Journal of Meat Science and Technology* (January–March) 8 (1):12–22.

Beatriz Nunes, Silva, Vasco Cadavez, Cristina Caleja, Eliana Pereira, Ricardo C. Calhelha, Mikel Añibarro-Ortega, Tiane Finimundy, Marina Kostić, Marina Soković, and José António Teixeira. 2023. "Phytochemical composition and bioactive potential of Melissa officinalis L., Salvia officinalis L. and Mentha spicata L. extracts." *Foods* 12 (5):947.

Benbi, D. K. 2015. "Carbon sequestration for soil health enhancement and mitigating climate change." *Jawaharlal Nehru Krishi Vishwa Vidyalaya Research Journal* 49 (3):263–271.

Bendala, Nesrein Masoud. 2020. "Preparation and characterization of solid lipid nanoparticles loaded with wheat germ oil." Faculty of Pharmacy, University of Tripoli.

Bensid, Abdelkader, Nariman El Abed, Abderrahmane Houicher, Joe M. Regenstein, and Fatih Özogul. 2022. "Antioxidant and antimicrobial preservatives: Properties, mechanism of action and applications in food–a review." *Critical Reviews in Food Science and Nutrition* 62 (11):2985–3001.

Boateng, Isaac Duah. 2024. "Mechanisms, capabilities, limitations, and economic stability outlook for extracting phenolics from agro-byproducts using emerging thermal extraction technologies and their combinative effects." *Food and Bioprocess Technology* 17 (5):1109–1140.

Bouyahya, Abdelhakim, Imane Chamkhi, Abdelaali Balahbib, Maksim Rebezov, Mohammad Ali Shariati, Polrat Wilairatana, Mohammad S. Mubarak, Taoufiq Benali, and Nasreddine El Omari. 2022. "Mechanisms, anti-quorum-sensing actions, and clinical trials of medicinal plant bioactive compounds against bacteria: A comprehensive review." *Molecules* 27 (5):1484.

Carracedo-Pérez, María, Beatriz Magariños, and Carlos A. García-González. 2024. "Strategies for the sterilization of polymeric biomaterials." *Polymeric Materials for Biomedical Implants* (pp. 547–583). Elsevier.

Cava, Ramón, Luis Ladero, Enrique Riaguas, and M. Carmen Vidal-Aragón. 2024. "Assessing the impact of pomegranate peel extract active packaging and high hydrostatic pressure processing on color and oxidative stability in sliced nitrate/nitrite-reduced iberian dry-cured loins." *Foods* 13 (3):360.

Chew, Yik-Ling, Joo-Kheng Goh, and Yau-Yan Lim. 2009. "Assessment of in vitro antioxidant capacity and polyphenolic composition of selected medicinal herbs from Leguminosae family in Peninsular Malaysia." *Food chemistry* 116 (1):13–18.

Chrysargyris, Antonios, and Nikolaos Tzortzakis. 2023. "Optimising fertigation of hydroponically grown sowthistle (Sonchus oleraceus L.): The impact of the nitrogen source and supply concentration." *Agricultural Water Management* 289:108528.

Chunzhi, Cai, Andrea Anton, Carlos M. Duarte, and Susana Agusti. 2018. "Macroalgae as blue carbon reservoirs and bioindicators of metal pollution in the audi Arabian Red Sea: A comparative study of coral reef and seagrass meadow habitats." Available at SSRN 4860701.

Cui, XiaoTong, Ahmed S. M. Saleh, Shu Yang, Na Wang, Peng Wang, Minpeng Zhu, and Zhigang Xiao. 2023. "Oleogels as animal fat and shortening replacers: Research advances and application challenges." *Food Reviews International* 39 (8):5233–5254.

D'Agostino, Annacristina. 2022. "Evaluation of the influence of technological treatments on bioactive compounds in vegetable foods." *Technological Processes for the Valorization of Food By-Products*, edited by Monica Gallo and Maurizio Servillo (pp. 111–132). Academic Press. https://doi.org/10.1016/B978-0-12-824044-1.00007-1

Deshmukh, Ram Kumar, and Kirtiraj K. Gaikwad. 2024. "Natural antimicrobial and antioxidant compounds for active food packaging applications." *Biomass Conversion and Biorefinery* 14 (4):4419–4440.

Dhawande, Anmol, Snehal Moon, Prashant Kharkar, and Nishikant Raut. 2024. "Comparative assessment of ghee residue from two different sources: An exploratory study towards utilization of ghee byproducts in alignment with zero-waste principles." *Journal of Pharmacognosy and Phytochemistry* 13 (5):240–246.

Diniso, Tabisa, Ayodeji O. Oriola, Jerry O. Adeyemi, Gugulethu M. Miya, Yiseyon S. Hosu, Opeoluwa O. Oyedeji, Simon K. Kuria, and Adebola O. Oyedeji. 2024. "Citrus wastes: A valuable raw material for biological applications." *Journal of Applied Pharmaceutical Science* 14 (6):1–15. https://doi.org/10.7324/JAPS.2024.140601

Domínguez, Rubén, Mirian Pateiro, Mohammed Gagaoua, Francisco J. Barba, Wangang Zhang, and José M. Lorenzo. 2019. "A comprehensive review on lipid oxidation in meat and meat products." *Antioxidants* 8 (10):429.

Dong, Li, Wentong Zhang, Yunwu Xiong, Jiaye Zou, Quanzhong Huang, Xu Xu, Ping Ren, and Guanhua Huang. 2022. "Impact of short-term organic amendments incorporation on soil structure and hydrology in semiarid agricultural lands." *International Soil and Water Conservation Research* 10 (3):457–469.

Donzella, Silvia, Immacolata Serra, Andrea Fumagalli, Luisa Pellegrino, Giacomo Mosconi, Roberto Lo Scalzo, and Concetta Compagno. 2022. "Recycling industrial food wastes for lipid production by oleaginous yeasts Rhodosporidiobolus azoricus and Cutaneotrichosporon oleaginosum." *Biotechnology for Biofuels and Bioproducts* 15 (1):51.

Dragoev, Stefan G. 2024. "Lipid peroxidation in muscle foods: Impact on quality, safety and human health." *Foods* 13 (5):797.

Dutta, Sulagna, Pallav Sengupta, and Antony V. Samrot. 2024. "Physiological and pathological functions of Reactive Nitrogen Species (RNS) and Reactive Sulphur Species (RSS) on male reproductive functions." *Journal of Integrated Science and Technology* 12 (3):755.

Ebrahimzadeh, A. M., S. J. Hosseinimehr, A. Hamidinia, and M. Jafari. 2008. "Antioxidant and free radical scavenging activity of Feijoa sellowiana fruits peel and leaves." *Pharmacologyonline* 1:7–14.

El Sheikha, Aly Farag, Ayman Younes Allam, Tahra El Obeid, Elham Abdelrahman Basiouny, Ahmad Abdelkaway Abdelaal, Ryszard Amarowicz, Emel Oz, Charalampos Proestos, Emad Karrar, and Fatih Oz. 2022. "Impact of a carboxymethyl cellulose coating incorporated with an ethanolic propolis extract on the quality criteria of chicken breast meat." *Antioxidants* 11 (6):1191.

El-Hadad, Sameh Sobhy, Abd El-Aziz, Natali Tikhomirova, Hamdy Zahran, and Rehab Sayed. 2024. "The effect of minor components of wheat germ oil on thermal behavior, crystallization characteristics, and oxidative stability of butter oil." *Egyptian Journal of Chemistry* 67 (4):329–338.

Ezejiofor, Tobias I. Ndubuisi, Uchechi E. Enebaku, and Chika Ogueke. 2014. "Waste to wealth-value recovery from agro-food processing wastes using biotechnology: A review." *British Biotechnology Journal* 4 (4):418–481.

Farhat, Grace, Lewis Cheng, Emad A. S. Al-Dujaili, and Mikhajlo Zubko. 2024. "Antimicrobial potential of pomegranate and lemon extracts alone or in combination with antibiotics against pathogens." *International Journal of Molecular Sciences* 25 (13):6943.

Faria, Douglas José, Anna Paula Azevedo de Carvalho, and Carlos Adam Conte-Junior. 2023. "Valorization of fermented food wastes and byproducts: Bioactive and valuable compounds, bioproduct synthesis, and applications." *Fermentation* 9 (10):920.

Ford, Hannah, Yuchen Zhang, Joanne Gould, Lukas Danner, Susan E. P. Bastian, and Qian Yang. 2024. "Comparing motivations and barriers to reduce meat and adopt protein alternatives amongst meat-eaters in Australia, China and the UK." *Food Quality and Preference* 118:105208.

Forslund, Agneta, Anaïs Tibi, Bertrand Schmitt, Elodie Marajo-Petitzon, Philippe Debaeke, Jean-Louis Durand, Philippe Faverdin, and Hervé Guyomard. 2023. "Can healthy diets be achieved worldwide in 2050 without farmland expansion?" *Global Food Security* 39:100711.

Hassoun, Abdo, and Romdhane Karoui. 2017. "Quality evaluation of fish and other seafood by traditional and nondestructive instrumental methods: Advantages and limitations." *Critical Reviews in Food Science and Nutrition* 57 (9):1976–1998.

Heiðarsdóttir, Íris Stella. 2020. "Determination of half maximal inhibitory concentration (IC50) of several anticancer drugs for inhibiting CYP3A4 metabolism of midazolam." Bachelor's thesis, Faculty of Pharmaceutical Sciences, University of Iceland, Reykjavík, Iceland, 37 pp. Available at https://skemman.is/handle/1946/36466.

ISO 15214. 1998. "Microbiology of food and animal feeding stuffs: Horizontal method for the emumeration [sic] of mesophilic lactic acid bacteria: Colony-count technique at 30° C." ISO.

Javadian, Seyed Roholla, Seyed Rasoul Shahosseini, and Peiman Ariaii. 2017. "The effects of liposomal encapsulated thyme extract on the quality of fish mince and Escherichia coli O157: H7 inhibition during refrigerated storage." *Journal of Aquatic Food Product Technology* 26 (1):115–123.

Kandil, A. A., M. M. Aly-Aldin, and A. Y. Allam. 2020. "Quality characteristics of processed low-fat beef sausage as affected by chickpea protein isolates prolonged cold storage." *Journal of Food and Dairy Sciences* 11 (12):363–368.

Kapoor, Rimika, Pooja Ghosh, Madan Kumar, Subhanjan Sengupta, Asmita Gupta, Smita S. Kumar, Vandit Vijay, Vivek Kumar, Virendra Kumar Vijay, and Deepak Pant. 2020. "Valorization of agricultural waste for biogas based circular economy in India: A research outlook." *Bioresource Technology* 304:123036.

Kemper, Kathi J. 1999. "Ginger (Zingiber officinale)." *Longwood Herbal Task Force* 3 (1):1–18.

Piironen, Juho, Markus Paasiniemi, and Aki Vehtari. 2020. "Projective inference in high-dimensional problems: Prediction and feature selection." *Electronic Journal of Statistics* 14 (1):2155–2197. https://doi.org/10.1214/20-EJS1711

Rahman, M. M., M. A. Hashem, M. A. K. Azad, M. S. H. Choudhury, and M. K. J. Bhuiyan. 2023. "Techniques of meat preservation-A review." *Meat Research* 3 (3).

Ravindran, P. N., and K. Nirmal Babu. 2016. *Ginger: The Genus Zingiber*. CRC press.

Resurreccion, Anna V. A. 2007. "Consumer sensory testing for food product development." *Developing New Food Products for a Changing Marketplace* (pp. 365–405). CRC Press.

Riaz, Tahreem, Muhammad Akram, Umme Laila, Muhammad Talha Khalil, Rida Zainab, Momina Iftikhar, Fethi Ahmet Ozdemir, Gaweł Sołowski, Ebrahim Alinia-Ahandani, and Marcos Altable. 2024. "Phytochemistry and pharmacology of Cichorium intybus." *International Archives of Integrated Medicine* 11 (2).

Ribeiro, Antônio H., Manoel Horta Ribeiro, Gabriela M. M. Paixão, Derick M. Oliveira, Paulo R. Gomes, Jéssica A. Canazart, Milton P. S. Ferreira, Carl R. Andersson, Peter W. Macfarlane, and Wagner Meira Jr. 2020. "Automatic diagnosis of the 12-lead ECG using a deep neural network." *Nature Communications* 11 (1):1760.

Riseh, Roohallah Saberi, Masoumeh Vatankhah, Mohadeseh Hassanisaadi, Elahe Tamanadar, and Yury Skorik. 2024. "Nanocomposites in agriculture as pesticides for plant protection: A review." *Advances in Natural Sciences: Nanoscience and Nanotechnology* 15 (2):023003.

Wang, Haiping. 2020. "Introductory chapter: Studies on ginger." *Ginger Cultivation and Its Antimicrobial and Pharmacological Potentials* (pp. 1–6). IntechOpen. https://doi.org/10.5772/intechopen.90120

15 Nutrient Composition and Potential Biomolecule Profile of Black Turmeric

A Comprehensive Review

Debashree Kakati, Mukul Barwant, and Himakshi Baishya

15.1. INTRODUCTION

Black turmeric or *kala haldi*, scientifically known as *Curcuma caesia* Roxb, belonging to the *Zingiberaceae* family, is an underutilized species within the Curcuma genus that is considered as critically endangered. Although much less research work is carried out, in recent times it has garnered significant attention because of its unique nutritional and potential biomolecule profile with numerous health beneficial functionalities (Khuntia et al. 2023). It is a perennial flora containing vibrant colour blooms of white and magenta pink with bluish black root stock that is generally bitter in taste with strong odour. It is typically used in various folk medicines since primeval times, as it is loaded with potent bioactive compounds (Arya et al. 2018). It grows wildly in a variety of tropical climates, with temperatures ranging between 20–35°C and high average annual rainfall of 1500 mm (Lakshmi et al. 2018).

The present study aims to elucidate the nutritional constituents, including vitamins, minerals, and essential oils, while delving into the diverse array of biomolecules such as curcuminoids, terpenoids, and alkaloids. By exploring the phytochemical landscape and the associated health benefits, this review study seeks to highlight black turmeric's potential as a valuable functional food ingredient and its implications for various therapeutic applications.

15.2. BOTANICAL AND PHYTOCHEMICAL OVERVIEW

Curcuma caesia Roxb. is an essential therapeutic herb native to the northeastern and central part of India (Nadkarni 1976). It thrives in clayey soil in damp deciduous forests with high humidity. This perennial crop has blue-black rhizomes that have great therapeutic potential and a distinct sweet scent that indicates the presence of essential oil. It has a bitter and strong hot taste, a pleasant odour, and it blooms vibrant white and pink flowers, as illustrated in Figure 15.1. Because of their carminative properties, rhizomes are used to treat leprosy, hemorrhoids, and smooth muscle relaxant action (Gantait et al. 2011). Moreover, it is commonly used to treat inflammation, gonorrheal discharges, cancer, epilepsy, asthma attacks, fever, wound healing, vomiting, menstruation problems, antihelmentic, and aphrodisiac (Arulmozhi et al. 2006; Sasikumar 2005).

The plant species was first published in the *Journal Asiatic* researches in 1810 and has been described as the native to Bengal, India (Anonymous 1810). Black haldi is a perennial, erect herb with rhizomatous roots that has a height of about 1.0–1.5 m high, bulbs ovate with bluish pulmate tubers, hence named caesius or bluish. Rhizomes are ovoid with acute tip. Leaves are 30–35 cm long, 10–15 cm broad, lanceolar or oblong, glabrous, petioled, and the middle of the leaf is deep purplish while the rest of the sheath and the petiole are green. Flowering season is May to July.

DOI: 10.1201/9781003559894-15

FIGURE 15.1 Botanical parts of *Curcuma caesia* Roxb.

Fertile bracts are ferruginous green. Inflorescence spike, about 15 cm—30 cm long, is pale yellow to reddish. Fruit matures in September to October (Anonymous 1810).

Due to its wide range of bioactive chemicals, which support its antibacterial, free radical–scavenging, anti-inflammatory, and possibly anticancer qualities, the phytochemical profile of *Curcuma caesia* has attracted a lot of interest. It has been established to have various medicinal functionalities including antifungal (Banerjee and Nigam 1976), muscle relaxant and anti-asthmatic (Arulmozhi et al. 2006), bronchodilation (Paliwal et al. 2011), free radical–scavenging (Mangla et al. 2010), anxiolytic and central nervous system (CNS) depressant, locomotor depressant, anticonvulsant (Karmakar et al. 2021), anthelmintic (Gill et al. 2011), antibacterial (Rajamma et al. 2012), and antiulcer (Das et al. 2012). The plants are rich in secondary metabolites, which includes anthraquinones, carotenoids, coumarins, flavonoids, phenolics, saponins, tannins, and terpenoids, which are involved in the inhibition of various diseases (Devi et al. 2015). Significant phytochemicals are:

a) The plant contains several curcuminoids in comparatively lesser quantity then yellow turmeric (*C. longa*), including curcumin, demethoxycurcumin, and bisdemethoxycurcumin, which contribute to their potential anti-inflammatory, free radical–scavenging, and anticancer activities (Sharifi-Rad et al. 2020b).
b) The rhizome of *C. caesia* is rich in essential oils, such as camphor, ar-turmerone, and germacrone that not only contribute to their antimicrobial, anti-inflammatory, and analgesic effects, but also provide the characteristics aroma and plant's medicinal properties (Ibrahim et al. 2023).
c) Black turmeric contains various flavonoids, including quercetin, kaempferol, and luteolin, contributing to the plant's antioxidant potentiality properties, anti-inflammatory, and anticancer activities by contributing to the plant's ability to scavenge free radicals, reducing oxidative stress, modulating key enzymes and cellular pathways (Koop et al. 2022).
d) Tannins and phenolic acids contribute the strong antioxidant, anti-inflammatory, and antimicrobial profile of *C. caesia*. Phenolic compounds such as caffeic acid, ferulic acid, and gallic acid scavenges the reactive oxygen species (ROS), and protect against cellular damage and aging (Devi et al. 2015).

e) Alkaloids in *C. caesia* contribute to its antimicrobial and anti-inflammatory effects (Ibrahim et al. 2023).
f) The characteristic aroma of *C. caesia* is contributed by the terpenoids cineole, borneol, and pinene. They also exhibit anti-inflammatory, analgesic, and antimicrobial properties (Charaya et al. 2024).
g) Saponins are reported in black turmeric that exhibit immunomodulatory and antioxidant activity of the plant and support the plant's overall pharmacological profile (Devi et al. 2015).
h) Starch and polysaccharides present in the rhizome of the plant contribute to its nutritional profile, exhibit probiotic effects, supporting gut health and immune function, and play a role in modulating inflammation (Leela and Adheeba 2024).
i) The unique bluish black pigmentation of black haldi rhizome is due to anthocyanins, potent antioxidants that differentiate *C. caesia* from other turmeric species (Pandey et al. 2023).

15.3. NUTRIENT COMPOSITION

In India, black turmeric flourishes wildly in the hilly region of the northeast and central part, adorned with vibrant flowers and its rhizome brimming with bioactive compounds (Dewangan et al. 2014). This herb is recognized for its distinctive bluish black rhizomes and diverse range of bioactive compounds (Chaturvedi et al. 2021). The nutritional composition of black turmeric encompasses various macronutrients, micronutrients, and phytochemicals, which contribute to its health-promoting properties (Bhardwaj et al. 2023).

Unlike other members of the Curcuma genus, it also contains a moderate amount of macronutrients, including carbohydrates, proteins, and fats. These macronutrients are crucial for energy metabolism and the structural integrity of cells (Ikpeama et al. 2014). According to studies, the carbohydrate content of black turmeric is primarily composed of starch, which is typical for rhizomatous plants (Gaikwad et al. 2023). Black turmeric contains a low to moderate level of protein, which is essential for tissue repair and the synthesis of enzymes and hormones. Protein content in the rhizomes is relatively low compared to legumes or cereals, but it still plays a vital role in contributing to the overall nutritional value (Vishwavidyalaya and Koni 2017). Although the fat content is minimal, it contains essential fatty acids that enhance the bioavailability of its fat-soluble compounds, particularly curcuminoids. These lipids also contribute to the anti-inflammatory and free radical–scavenging properties of the plant.

The micronutrient profile of black turmeric consists of a variety of essential vitamins and minerals that contribute to its health-promoting effects. Black turmeric contains several vitamins, including vitamin C, which is a potential antioxidant that helps in boosting the immune system and preventing oxidative damage. Other vitamins, such as the B-complex vitamins (e.g., niacin, riboflavin), are also present, playing a role in energy production and maintaining the nervous system. The mineral content in black turmeric includes iron, potassium, magnesium, and zinc, among others. These minerals are essential for a number of physiological processes. Iron is essential for hemoglobin production and oxygen transport. Potassium helps regulate fluid balance and supports proper nerve function. Zinc is important for immune function and enzyme activity. Magnesium plays a role in muscle function and cardiovascular health (Vairappan et al. 2013).

It is also a source of dietary fibre, which is important for maintaining digestive health (Amalraj et al. 2017). The fibrous content of the rhizomes makes black turmeric beneficial for gastrointestinal health and helps prevent conditions such as constipation and irritable bowel syndrome. Like many rhizomatous plants, black turmeric has a high water content, which is vital for maintaining hydration and facilitating metabolic processes within the plant (Singh et al. 2021). The water content also plays a role in the plant's texture and its preservation during storage and processing.

15.4. BIOACTIVE COMPOUNDS

In addition to its macronutrient and micronutrient content, black turmeric is rich in phytochemicals, which are bioactive compounds that contribute significantly to its health benefits. Curcuminoids, such as curcumin, are the primary active compounds in black turmeric. These compounds are well-known for their anti-inflammatory and antioxidant properties. Although curcumin content in black turmeric is generally lower than in *Curcuma longa*, it still contributes significantly to the rhizome's medicinal properties (Tamang 2022). It also contains essential oils that consist of compounds such as ar-turmerone, β-turmerone, and α-phellandrene, which are known for their antimicrobial, anti-inflammatory, and free radical–scavenging functionalities. These essential oils are not only beneficial for medicinal purposes but also play a role in enhancing the bioavailability of curcuminoids (Haida et al. 2023). Flavonoids and terpenoids are other important phytochemicals found in black turmeric. These compounds act as antioxidants, helping to neutralize free radicals and reduce oxidative stress. They also exhibit anticancer and cardioprotective effects, making black turmeric a valuable plant for therapeutic applications (Kondareddy et al. 2019). Despite its rich nutrient profile, the bioavailability of black turmeric's active compounds, particularly curcuminoids, is a significant factor in determining its efficacy. Curcumin, for instance, is known to have poor bioavailability due to rapid metabolism and low absorption rates. However, the presence of essential oils in black turmeric can enhance the absorption of curcumin, making it more effective in therapeutic applications (Dei Cas and Ghidoni, 2019).

It also contains several phenolic acids, such as caffeic, ferulic, and p-coumaric acids, which contribute to its strong antioxidant activity. These phenolic compounds help neutralize free radicals in the body, potentially reducing oxidative stress and supporting cellular health. The antioxidant capacity of phenolic acids has made them a focus in studies on aging, cancer prevention, and cardiovascular protection.

In addition to curcuminoids and phenolic acids, black turmeric contains a range of other bioactive phytochemicals, including flavonoids and terpenoids. These compounds work synergistically, enhancing the overall antioxidant and anti-inflammatory effects of black turmeric. Emerging research suggests these lesser-known phytochemicals may also play roles in immune modulation, neuroprotection, and gastrointestinal health, highlighting the multifaceted benefits of black turmeric (Das and Vinayak 2015).

15.5. TRADITIONAL USES OF BLACK TURMERIC

Curcuma caesia plays a critical role in traditional, cultural, spiritual, and medical contexts. The plant has been used for a long time in the Ayurvedic, Unani, and Siddha herbal systems. Within the Indian medical system, the species has long been used to treat a wide range of illnesses and metabolic conditions, including rheumatoid arthritis, leukoderma, asthma, tumours, piles, bronchitis, cancer, allergies, inflammation, and epilepsy (Fozia et al. 2012; Arya et al. 2018) (Table 15.1).

15.6. PHARMACOLOGICAL PROPERTIES

Black turmeric exhibits strong anti-inflammatory properties, largely due to its curcuminoids and essential oils, which inhibit pathways involved in inflammation (Figure 15.2). This makes it a potential natural remedy for conditions such as arthritis and inflammatory diseases, with studies suggesting relief from swelling and pain (Mukunthan et al. 2017). Rich in curcuminoids and phenolic acids, it also has potent free radical–scavenging activity that helps eliminate dangerous free radicals from the body. This activity may contribute to cell protection, potentially reducing the risk of enduring disorders such as cancer and heart disease (Islam et al. 2024). The essential oils in black turmeric, particularly ar-turmerone and germacrone, have shown effectiveness against a range of bacteria, fungi, and viruses. These antimicrobial properties support its use in traditional medicine for treating infections and enhancing immune defences.

TABLE 15.1
Traditional Uses of Black Turmeric

Traditional Uses	Reference
Young leaves are consumed raw as salad, used in the preparation of culinary.	(Awang-Kanak et al. 2018; Ibrahim et al. 2023)
Rhizomes are utilized for wound cleaning, skin problems, bruises, bruises, boils, scabies, psoriasis, and inflammations.	(Sarangthem and Haokip, 2010; Sharma et al. 2011; Angel and Vimala 2012; Vairappan et al. 2013; Rajkumari and Sanatombi 2017)
Fresh rhizome decoction and leaves are used in the treatment of diarrhea, bloating, stomach pain, ulcers, gastrointestinal disorders, liver diseases, kidney diseases, and as antiemetic.	(Sahu and Saxena 2018; Tag et al. 2007; Kagyung et al. 2010; Vairappan et al. 2013; Devi et al. 2015; Chadalavada and Budala 2017; Rajkumari and Sanatombi 2017)
For snake and scorpion bites, the Khamti tribe of Lohit district, Arunachal Pradesh, uses a paste made from fresh rhizomes.	(Tag et al. 2007; Kagyung et al. 2010; Chadalavada and Budala 2017)
Treatment of fever, allergies, inflammation, cancer, asthma, piles, tonsillitis, bronchitis, toothache, vomiting, and gout.	(Saikia and Borthakur 2010; Paliwal et al. 2011; Angel and Vimala 2012; Israr et al. 2012; Devi et al. 2015; Sahu and Saxena 2018)
Taken for longevity, weakness, infertility, and irregular menstrual disorders and diabetes.	(Verma et al. 2010; Rajkumari and Sanatombi 2017)
Rhizome paste applied externally to treat rheumatic pain, it is used in Manipur to treat pain and aching.	(Sarangthem and Haokip 2010; Angel and Vimala 2012)
In cosmetics, fresh tubers are utilized. Women from the tribe use the rhizome powder as a face mask.	(Paliwal et al. 2011; Ibrahim et al. 2023)
Dried leaves are utilized as fuel, and leaves are used to promote the germination of rice seeds.	(Lalitha et al. 1995)
Treatment of leukoderma, epilepsy, jaundice, dysentery, and cough.	(Sarangthem and Haokip 2010; Devi et al. 2015)
Cattle with dysentery are fed a mixture of mustard oil and rhizome juice.	(Saikia and Borthakur 2010)

It is also valued for its pain-relieving (analgesic) and fever-reducing (antipyretic) effects, attributed to its bioactive compounds. These qualities make it a popular remedy for alleviating pain and managing fevers naturally, especially in traditional medicinal systems (Sharifi-Rad et al. 2020a).

15.7. HEALTH BENEFITS AND THERAPEUTIC POTENTIAL OF *CURCUMA CAESIA*

C. caesia shows significant natural therapeutic potential with a broad spectrum of health benefits such as cancer prevention and treatment, cardiovascular health, gastrointestinal and neurological health benefits, along with certain other medicinal properties. Following are the potential therapeutic effects:

Cancer Prevention and Treatment

Cytotoxicity and Antioxidant Activity: Black turmeric rhizome extract and gold nanoparticles (GNPs) synthesized from it show selective cytotoxicity against breast cancer cell lines, especially triple-negative breast cancer (TNBC). The GNPs demonstrated higher efficiency with lower IC50 values, suggesting potential for targeted, personalized breast cancer treatment.

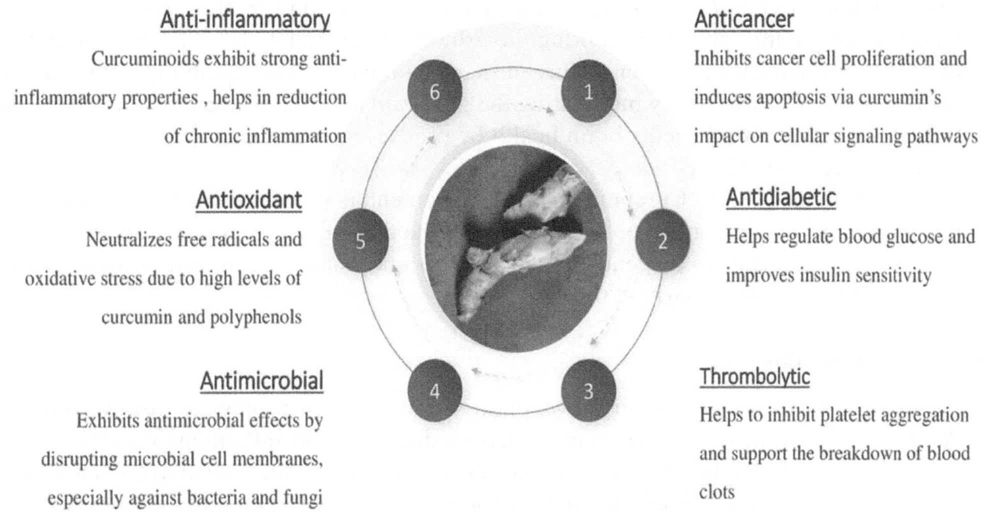

FIGURE 15.2 Important pharmacological properties of *Curcuma caesia* Roxb.

Inhibition of Tumour Growth: Curcumin and other bioactive molecules from *C. caesia* induce apoptosis in cancer cells, which help in limiting tumour growth. The methanol extract of *C. caesia* (MECC) rhizomes demonstrated significant reduction in tumour growth and improvement in survival of Ehrlich's ascites carcinoma treated mice (Karmakar et al. 2013).

Anti-Inflammatory Effects: The hexane, methanolic extracts, and pure isolate compounds from *C. caesia* have demonstrated anti-inflammatory effects, selectively inhibiting COX-2 with minimal inhibition of COX-1 (Angel et al. 2013). Anti-inflammatory action also demonstrated through the egg albumin denaturation assay by Borah et al. (2019).

Chemopreventive Potential: Curcumin shows strong potential as a cancer chemopreventive agent by modulating key molecular pathways to promote apoptosis, inhibit survival signals, and reduce inflammation (Park and Pezzuto 2013).

Cardiovascular Health

Cholesterol Regulation and Antioxidant Protection: The bioactive compounds such as curcumin found in *C. caesia* may help protect against cardiovascular disease by improving lipid levels and could serve as a well-tolerated dietary supplement alongside conventional medications (Qin et al. 2022). The antioxidant activity of *C. caesia* can help prevent oxidative damage to the cardiovascular system, supporting overall vascular health.

Blood Pressure Regulation: Some studies suggest that curcumin can help regulate blood pressure by improving endothelial function, leading to better vasodilation (Karimi et al. 2023). This, in turn, may help in preventing hypertension, a major risk factor for heart disease.

Gastrointestinal Benefits

Digestive Aid: *C. caesia* has shown therapeutic potential for digestive disorders. Research on phytosome complexation improves absorption significantly, promising enhanced efficacy in treating bowel and digestive diseases (Dulbecco and Savarino 2013).

Anti-Inflammatory Effects in Gut: The anti-inflammatory compounds in *C. caesia*, particularly curcumin, may soothe the gut and reduce symptoms such as pain, diarrhea, and bloating in conditions such as irritable bowel syndrome (IBS) or inflammatory bowel disease (IBD) (Dulbecco and Savarino 2013).

Gut Health and Microbial Balance: *C. caesia* has antimicrobial properties that help in regulating better gut microbiota, promoting the growth of beneficial bacteria while inhibiting the growth of harmful pathogens. Curcumin, despite its low bioavailability, may support neuroprotection indirectly by modulating the gut-brain axis and restoring gut microbiome balance, which in turn benefits brain health by producing potent neuroprotective metabolites (Di Meo et al. 2019).

Protection Against Ulcers: Curcumin has shown potential in protecting the stomach lining from ulcer formation by promoting the production of mucus and enhancing mucosal defences that helps in preventing gastric ulcers and improving overall gastrointestinal health (Dulbecco and Savarino 2013).

Neurological Health

Neuroprotective Effects: Curcumin shows promise in treating neurological diseases such as Alzheimer's, Parkinson's, and multiple sclerosis due to its anti-inflammatory, antioxidant, and neuroprotective properties. Recent advancements have created specialized nanocarriers such as liposomes and polymer nanoparticles, that improve curcumin's delivery to the brain, enhancing its therapeutic efficacy and targeting capabilities (Genchi et al. 2024).

Cognitive Function and Memory Enhancement: *C. caesia* may enhance cognitive function and memory by neurogenesis. Curcumin has been shown to increase brain-derived neurotrophic factor (BDNF), a protein linked to the growth and survival of neurons, thereby promoting cognitive health (Wang et al. 2013).

Antidepressant and Mood Regulation: Curcumin, a central compound in turmeric traditionally used in Ayurvedic medicine, holds potential as an alternative therapy for major depressive disorder (MDD) by modulating biological pathways related to depression, including neurotransmitter regulation, neuroplasticity, and inflammation etc. (Ramaholimihaso et al. 2020).

Reduction of Neuroinflammation: The anti-inflammatory effects of *C. caesia* can help in reducing chronic neuroinflammation of the brain associated with conditions such as depression, Alzheimer's, and Parkinson's disease, supporting long-term brain health (Sorrenti et al. 2018).

15.8. RECENT SCIENTIFIC RESEARCH

Black turmeric is considered to be an underutilized crop, although it is packed with compounds having health beneficial functionalities. The recent research scenario has changed, and much work has been carried out on black turmeric. Because of its therapeutic functionalities, it finds a wide range of applications in various fields, which led to different scientific research. For instance, Benya et al. (2023), in their study on black turmeric collected from different agroclimatic zones, identified elite genotypes with high anthocyanin of 7.14% and considerably high essential oil (EO) content of 1.25% oil in the leaves and 2.55% in the rhizomes, particularly camphor in the rhizomes and eucalyptol in the leaves. This finding suggested their potential industrial application in the field of pharmaceuticals and cosmetics. In recent research by Khuntia et al. (2023), they studied the EO for potential pharmaceutical applications and demonstrated its high antioxidant, antibacterial, and antifungal activities. In-silico predictions, validated by in-vitro assays, showed that EO extracted from the leaves and rhizomes are loaded with bioactive compounds such as curzerenone and eucalyptol, demonstrating significant interactions with bacterial, fungal, as well as oxidative proteins.

This pioneering work suggests black turmeric as a promising source for developing new antioxidant and antimicrobial drugs. The key findings of the study by Anjum and Quraishi demonstrated that in *Curcuma caesia* Roxb microrhizomes, 25.0 μM salicylic acid elicitation for 60 days enhanced epicurzerenone content to 32.11%, along with increased antioxidant enzyme activities and secondary metabolite production. In another research study, a field study on black turmeric highlighted that organic fertilizers improved growth and quality, with organic soil rhizomes displaying high antibacterial activity. UV-Vis and XRF analyses highlighted key biochemical elements, and soil bacteria exhibited resilience traits such as pH and temperature tolerance. Thus, these findings underscore the role of rhizobacteria in sustainable agriculture (Jyotirmayee et al. 2024).

Similarly, Paw et al. (2020) also demonstrated that the essential oil from the rhizome possesses excellent antibacterial, anti-inflammatory, and antioxidant qualities with very little genotoxicity. Thus, the study highlights the need for advanced research and development as well as commercial application of the rhizome, a valuable ethnopharmacological plant, which is otherwise considered as an underutilized plant bioresource. The study conducted by Mahanta et al. (2022) identified the oil present in rhizome acts as a potent bioherbicide, showing strong phytotoxic effects in wheatgrass and Bermuda grass bioassays. The oil's major component, curzerenone, significantly inhibited seed germination and growth, with sustained efficacy in soil and aerial applications, highlighting its potential for natural weed control.

15.9. APPLICATIONS IN MEDICINE AND NUTRITION

Curcuma caesia has long been valued in traditional medicine and nutrition due to its potent therapeutic properties since ancient times. Known for its rich antioxidant and anti-inflammatory compounds, it has been profoundly used in Ayurvedic and folk medicine to address various health issues. Its rhizomes, containing active ingredients such as curcumin and curzerenone, are recognized for their potential to manage pain, reduce inflammation, and support immune health (Kaur et al. 2024). Today, its historical use in traditional remedies continues to inspire modern research, highlighting black turmeric's enduring role in natural health and nutrition perspective. For instance, in a recent clinical study conducted on 42 patients subdivided into Group A and Group B, having oral submucous fibrosis, compared the combined effect of black turmeric and aloe vera treatment with steroid therapy. Group A received black turmeric and aloe vera gel, while Group B received hydrocortisone and antioxidants. The study demonstrated that Group A had greater symptom improvement, with significant benefits noted in terms of burning sensation, cheek flexibility, and mouth opening. The study suggests black turmeric-aloe vera therapy as a safe, effective alternative to steroids (Bohra et al. 2021). The review study by Grover et al. (2019) highlights the potential of *Curcuma caesia* in treating diabetic neuropathy through its antioxidant, anti-inflammatory, and antimicrobial properties, which may reduce oxidative stress, prevent neuronal inflammation, and inhibit infections. Another study illustrated that *Curcuma caesia* rhizome and leaf extracts are rich in phenolic compounds, and demonstrated strong antioxidant and α-glucosidase inhibitory activities, especially in freeze-dried rhizome extracts. Additionally, research points to it as a possible natural source of bioactive chemicals that could find use in the food and pharmaceutical industries (Ain Ibrahim et al. 2023).

15.10. FUTURE DIRECTIONS AND RESEARCH NEEDS

Future trends in *C. caesia* research could focus on various paradigms, including formulation of liposomes and nanoparticles, to enhance the bioavailability of its active compounds for better therapeutic uses, development of eco-friendly extraction techniques to enhance yield and bioactive purity, advance research in genetic mapping and metabolomics to identify targeted, and sustainable farming practices for conservation of black turmeric germplasm.

Despite increasing research interest in black turmeric, certain research gaps still exist, including lack of comprehensive pharmacokinetic studies, standardization of extraction methods and prescribed dosage, limited toxicological data, enough well-structured clinical trials, etc. By minimizing the research gaps and advancing further innovations, *C. caesia* holds the potential to become a highly valuable plant comprising a broad spectrum of applications in diversified field.

15.11. CHALLENGES AND CONSIDERATIONS

The cultivation and use of black turmeric face several challenges and considerations. *C. caesia* is considered as a critically endangered species, making conservation efforts crucial to prevent over-harvesting. The notable challenges include its sensitivity to soil and climate conditions, which limit where it can be effectively grown. Additionally, lack of awareness among most people affects its commercial viability.

From a medicinal perspective, safety and efficacy studies are important considerations, as black turmeric's active compounds need thorough research to verify therapeutic claims and ensure safe use. Regulatory hurdles and quality control are also significant for its development in the fields of pharmaceuticals and nutraceuticals.

15.12. CONCLUSION

In conclusion, *Curcuma caesia*, or black turmeric, stands out as an underutilized yet valuable herb with a remarkable nutrient and biomolecule profile that offers promising therapeutic potential. This study provides a thorough understanding of black turmeric's bioactive compounds, including curcuminoids, essential oils, and phenolic acids, which imparts its free radical–scavenging, anti-inflammatory, and microbial inhibition properties. Its traditional use and emerging applications in modern medicine and functional foods underscore its potential as a natural resource for health and disease prevention. However, further research is essential to deepen our understanding of its bio-availability, mechanisms of action, and efficacy in various therapeutic contexts. As an endangered plant, efforts to conserve black turmeric are also crucial, ensuring its availability for future research and potential applications in health and wellness.

REFERENCES

Ain Ibrahim, Nurul Najiha, Nurkhalida Kamal, Ahmed Mediani, Azliana Abu Bakar Sajak, Soo Yee Lee, Khozirah Shaari, and Hafeedza Abdul Rahman. "1H NMR-based metabolomics approach revealing metabolite variation of black turmeric (Curcuma caesia) extracts and correlation with its antioxidant and α-glucosidase inhibitory activities." Food Technology and Biotechnology 61, no. 1 (2023): 104–117. https://doi.org/10.17113/ftb.61.01.23.7711

Amalraj, Augustine, Anitha Pius, Sreerag Gopi, and Sreeraj Gopi. "Biological activities of curcuminoids, other biomolecules from turmeric and their derivatives–a review." Journal of Traditional and Complementary Medicine 7, no. 2 (2017): 205–233. https://doi.org/10.1016/j.jtcme.2016.05.005

Angel, B. M., G. E. Batley, C. V. Jarolimek, and N. J. Rogers. "The impact of size on the fate and toxicity of nanoparticulate silver in aquatic systems." Chemosphere 93, no. 2 (2013): 359–365.

Angel, G. R., and B. Vimala. "Antioxidant and anti-inflammatory activities of proteins isolated from eight Curcuma species." Journal of Phytopharmacology 4, no. 1 (2012): 96–105.

Anonymous. "Asiatic Researches," Vol. 11. Calcutta, 1810. https://www.biodiversitylibrary.org/page/42220922

Arulmozhi, D. K., N. Sridhar, A. Veeranjaneyulu, and S. K. Arora. "Preliminary mechanistic studies on the smooth muscle relaxant effect of hydroalcoholic extract of curcuma caesia." Journal of Herbal Pharmacotherapy 6 (2006): 117–124.

Arya, Om Prakash, Priyanka Adhikari, and Anita Pandey. "Black turmeric: A high value medicinal herb from North-East India." ENVIS Bulletin Himalayan Ecology 26 (2018): 83–84. https://doi.org/10.1111/jfpp.16805

Awang-Kanak, F., M. F. Abu Bakar, and M. Mohamed. "Ethnobotanical survey on plants used as traditional salad food (ulam) in Kampung Taun Gusi, Kota Belud, Sabah, Malaysia." AIP Conference Proceedings 2002 (2018): p. 020021. https://doi.org/10.1063/1.5050120

Benya, Ankuri, Subhadarsini Mohanty, Sujata Hota, Alok Prasad Das, Chandi Charan Rath, K. Gopinath Achary, and Shikha Singh. "Endangered Curcuma caesia Roxb.: Qualitative and quantitative analysis for identification of industrially important elite genotypes." Industrial Crops and Products 195 (2023): 116363. https://doi.org/10.1016/j.indcrop.2023.116363

Bhardwaj, Arvind Kumar, Nand Kumar Kashyap, Sanjoy Kumar Bera, Milan Hait, and Harishankar Dewangan. "Proximate composition and mineral content analysis of Curcuma caesiarhizome." Biochemical Systematics and Ecology 109 (2023): 104661. https://doi.org/10.1016/j.bse.2023.104661

Bohra, Ankita, T. N. Uma Maheswari, Aditya Harsh, and Ayush Garg. "Black turmeric and aloe vera in the management of oral submucous fibrosis: A prospective clinical study." Asian Pacific Journal of Cancer Prevention: APJCP 22, no. 12 (2021): 3941. https://doi.org/10.31557/APJCP.2021.22.12.3941

Borah, A., M. Paw, R. Gogoi, R. Loying, N. Sarma, S. Munda, . . . M. Lal. "Chemical composition, antioxidant, anti-inflammatory, anti-microbial and in-vitro cytotoxic efficacy of essential oil of Curcuma caesia Roxb. leaves: An endangered medicinal plant of North East India." Industrial Crops and Products 129 (2019): 448–454.

Chadalavada, V., and S. Budala. "Study on anthelmintic activity of curcuma Caesia." Journal of Pharmaceutical Research 7, no. 7 (2017).

Charaya, N., P. Choudhary, B. Prasad, Bhawna R. Sharma, N. Kumar, G. Singh, and N. Sharma. "Comparative phytochemical analysis, antioxidant and anti-inflammatory activities of rhizome and leaves of Curcuma caesia Roxb. and Curcuma longa L." Plant Science Today 11, no. 4 (2024). https://horizonepublishing.com/journals/index.php/PST/article/view/2936

Chaturvedi, Monika, Reena Rani, Dushyant Sharma, and Jaya Parkash Yadav. "Comparison of Curcuma Caesia extracts for bioactive metabolite composition, antioxidant and antimicrobial potential." Natural Product Research 35, no. 18 (2021): 3131–3135. https://doi.org/10.1080/14786419.2019.1687472

Das, L., and M. Vinayak. "Long term effect of curcumin in restoration of tumour suppressor p53 and phase-II antioxidant enzymes via activation of Nrf2 signalling and modulation of inflammation in prevention of cancer." PloS One 10, no. 4 (2015): e0124000.

Das, S., P. K. Bordoloi, D. Phukan, and S. Singh. "Study of the anti-ulcerogenic activity of the ethanolic extracts of rhizome of Curcuma caesia against gastric ulcers in experimental animals." Asian Journal of Pharmaceutical and Clinical Research 5 (2012): 200–203.

Dei Cas, Michele, and Riccardo Ghidoni. "Dietary curcumin: Correlation between bioavailability and health potential." Nutrients 11, no. 9 (2019): 2147. https://doi.org/10.3390/nu11092147

Devi, H. P., P. B. Mazumder, and L. P. Devi. "Antioxidant and antimutagenic activity of Curcuma caesia Roxb. Rhizome extracts." Toxicology Reports 2 (2015): 423–428. https://doi.org/10.1016/j.toxrep.2014.12.018

Dewangan, Mehendra Kumar, Chandraprakash Dwivedi, P. L. Sivna, S. P. Rao, R. Yadav, K. Chandrakar, S. S. Singh, and D. Sinha. "Medicinal value of Curcuma cassia Roxb: An overview." Earth Journals 3, no. 4 (2014): 1–9. https://doi.org/10.20510/ukjpb/4/i6/134671

Di Meo, F., S. Margarucci, U. Galderisi, S. Crispi, and G. Peluso. "Curcumin, gut microbiota, and neuroprotection." Nutrients 11, no. 10 (2019): 2426.

Dulbecco, P., and V. Savarino. "Therapeutic potential of curcumin in digestive diseases." World Journal of Gastroenterology: WJG 19, no. 48 (2013): 9256.

Fozia, I., H. Fouzia, S. N. Baqir, A. Iqbal, J. Sabahat, and S. M. F. Hasan. "Studies on antibacterial activity of some traditional medicinal plants used in folk medicine." Pakistan Journal of Pharmaceutical Sciences 25 (2012): 669–674.

Gaikwad, Saurav, Siddhant Gadiya, and Sham Tungar. "Phytochemical analysis of black turmeric Curcuma Caesia Roxb." Journal of Medicinal Plants 11, no. 1 (2023): 41–43.

Gantait, A., T. Barman, and P. Mukherjee. "Validated method for determination of curcumin in turmeric powder." Indian Journal of Traditional Knowledge 10 (2011): 247–250.

Genchi, G., G. Lauria, A. Catalano, A. Carocci, and M. S. Sinicropi. "Neuroprotective effects of curcumin in neurodegenerative diseases." Foods 13, no. 11 (2024): 1774.

Gill, R., V. Kalsi, and A. Singh. "Phytochemical investigation and evaluation of anthelmintic activity of Curcuma amada and Curcuma caesia: A comparative study." Inventi Impact: Ethnopharmacology (2011): 2011.

Grover, Madhuri, Karan Shah, Gaurav Khullar, Jyoti Gupta, and Tapan Behl. "Investigation of the utility of Curcuma caesia in the treatment of diabetic neuropathy." Journal of Pharmacy and Pharmacology 71, no. 5 (2019): 725–732. https://doi.org/10.1111/jphp.13075

Haida, Z., J. J. Nakasha, U. R. Sinniah, and M. Hakiman. "Ethnomedicinal uses, phytochemistry, pharmacological properties and toxicology of Curcuma caesia Roxb.: A review." Advances in Traditional Medicine 23, no. 4 (2023): 985–1001. https://doi.org/10.1007/s13596-022-00658-y

Ibrahim, N. N. A., Wan Aida Wan Mustapha, Noor-Soffalina Sofian-Seng, Seng Joe Lim, Noorul Syuhada Mohd Razali, Arnida Hani Teh, et al. "A comprehensive review with future prospects on the medicinal properties and biological activities of Curcuma caesia Roxb." Evidence-Based Complementary and Alternative Medicine 2023 (2023): 7006565. https://doi.org/10.1155/2023/7006565

Ikpeama, Ahamefula, G. I. Onwuka, and Chibuzo Nwankwo. "Nutritional composition of tumeric (Curcuma longa) and its antimicrobial properties." International Journal of Scientific and Engineering Research 5, no. 10 (2014): 1085–1089.

Islam, Md Rezaul, Abdur Rauf, Shopnil Akash, Sadiya Islam Trisha, Akram Hossain Nasim, Muniya Akter, Puja Sutro Dhar, et al. "Targeted therapies of curcumin focus on its therapeutic benefits in cancers and human health: Molecular signaling pathway-based approaches and future perspectives." Biomedicine & Pharmacotherapy 170 (2024): 116034. https://doi.org/10.1016/j.biopha.2023.116034

Israr, F., F. Hassan, B. S. Naqvi, I. Azhar, S. Jabeen, and S. M. F. Hasan. "Studies on antibacterial activity of some traditional medicinal plants used in folk medicine." Pakistan Journal of Pharmaceutical Sciences 25, no. 3 (2012): 669–674.

Jyotirmayee, B., Subhashree Tripathy, Himansu Bhusan Samal, and Gyanranjan Mahalik. "Turmeric (Curcuma caesia)." 8, no. 5 (2024): 1–12 https://doi.org/10.25163/angiotherapy.859651

Kagyung, R., P. R. Gajurel, P. Rethy, and B. Singh. "Ethnomedicinal plants used for gastro-intestinal diseases by Adi tribes of Dehang-Debang Biosphere Reserve in Arunachal Pradesh." Indian Journal of Traditional Knowledge 9, no. 3 (2010): 496–501.

Karimi, A., A. M. Jazani, M. Darzi, R. N. D. Azgomi, and M. Vajdi. "Effects of curcumin on blood pressure: A systematic review and dose-response meta-analysis." Nutrition, Metabolism and Cardiovascular Diseases 33, no. 11 (2023): 2089–2101.

Karmakar, M. K., J. Mandal, S. Haldar, and P. K. Chatterjee. "Investigation of fuel gas generation in a pilot scale fluidized bed autothermal gasifier using rice husk." Fuel 111 (2013): 584–591.

Karmakar, M. K., J. Mandal, S. Haldar, and P. K. Chatterjee. "Investigation of fuel gas generation in a pilot scale fluidized bed autothermal gasifier using rice husk." Fuel 111 (2013): 584–591.

Karmakar, M., P. M. Lantz, and R. Tipirneni. "Association of social and demographic factors with COVID-19 incidence and death rates in the US." JAMA Network Open 4, no. 1 (2021): e2036462–e2036462.

Kaur, Simranjeet, Pratik N. Chauhan, Junaid Ul Hamid, Simran Kaur, and Yashika Sharma. "Phyto pharmaceutical advances on black turmeric as a functional herb." Current Nutrition & Food Science 20, no. 2 (2024): 131–142. https://doi.org/10.2174/1573401319666230322092031

Khuntia, Snehalata, Bhaskar Chandra Sahoo, Jyotirmayee Lenka, Basudeba Kar, and Suprava Sahoo. "In-silico prediction and in vitro validation of antioxidant, antibacterial and antifungal potential of black turmeric (Curcuma caesia Roxb.) essential oils and its constituents." Industrial Crops and Products 203 (2023): 117185. https://doi.org/10.1016/j.indcrop.2023.117185

Kondareddy, R., N. Sivakumaran, and P. K. Nayak. "Drying kinetics of black turmeric (Curcuma caesia) with optimal controller assisted low cost solar dryer." (2019). http://dx.doi.org/10.26656/fr.2017.3(4).139

Koop, Betina Luiza, Milena Nascimento da Silva, Fabíola Diniz da Silva, Kennya Thayres dos Santos Lima, Lenilton Santos Soares, Cristiano José de Andrade, et al. "Flavonoids, anthocyanins, betalains, curcumin, and carotenoids: Sources, classification and enhanced stabilization by encapsulation and adsorption." Food Research International 153 (2022): 110929. https://doi.org/10.1016/j.foodres.2021.110929

Lakshmi, D. V. N., P. Muthukumar, Apurba Layek, and Prakash Kumar Nayak. "Drying kinetics and quality analysis of black turmeric (Curcuma caesia) drying in a mixed mode forced convection solar dryer integrated with thermal energy storage." Renewable Energy 120 (2018): 23–34. https://doi.org/10.1016/j.renene.2017.12.053

Lalitha, C. R., K. Raveendran, and N. Salim. "Effect of Curcuma caesia leaves on rice seed germination and seedling establishment." Ancient Science of Life 15, no. 2 (1995): 115–118.

Leela, N. K., and P. K. Adheeba. "Curcuma caesia Roxb.—update of phytochemicals and pharmacological properties: Phytochemical properties of Curcuma caesia." Journal of Spices and Aromatic Crops 33, no. 1 (2024): 1–22. https://doi.org/10.25081/josac.2024.v33.i1.8477

Mahanta, Bhaskar Protim, Phirose Kemprai, Pranjit Kumar Bora, Mohan Lal, and Saikat Haldar. "Phytotoxic essential oil from black turmeric (Curcuma caesia Roxb.) rhizome: Screening, efficacy, chemical basis, uptake and mode of transport." Industrial Crops and Products 180 (2022): 114788. https://doi.org/10.1016/j.indcrop.2022.114788

Mangla, M., M. Shuaib, J. Jain, and M. Kashyap. "In-vitro evaluation of antioxidant activity of Curcuma caesia Roxb." International Journal of Pharmaceutical Sciences and Research 1 (2010): 98–102.

Mukunthan, K. S., R. S. Satyan, and Trupti N. Patel. "Pharmacological evaluation of phytochemicals from South Indian black turmeric (Curcuma caesia Roxb.) to target cancer apoptosis." Journal of Ethnopharmacology 209 (2017): 82–90. https://doi.org/10.1016/j.jep.2017.07.021

Nadkarni, K. M. Indian Material Medical. Bombay: Popular Prakashan, 1976.

Paliwal, P., S. S. Pancholi, and R. K. Patel. "Pharmacognostic parameters for evaluation of the rhizomes of Curcuma caesia." Journal of Advanced Pharmaceutical Technology & Research 2, no. 1 (2011): 56–61. https://doi.org/10.4103/2231-4040.79811

Pandey, R., S. R. Behera, and P. Rawat. "Black turmeric: The dark elixir." Just Agriculture (2023): 23–29.

Park, E. J., and J. M. Pezzuto. "Antioxidant marine products in cancer chemoprevention." Antioxidants & Redox Signaling 19, no. 2 (2013): 115–138.

Paw, Manabi, Roktim Gogoi, Neelav Sarma, Sudin K. Pandey, Angana Borah, Twahira Begum, and Mohan Lal. "Study of anti-oxidant, anti-inflammatory, genotoxicity, and antimicrobial activities and analysis of different constituents found in rhizome essential oil of Curcuma caesia Roxb., collected from north east India." Current Pharmaceutical Biotechnology 21, no. 5 (2020): 403–413. http://dx.doi.org/10.2174/1389201020666191118121609

Qin, L., S. Gu, R. Xiao, P. Liu, F. Yan, H. Yu, and W. Yang. "Value of native T1 mapping in the prediction of major adverse cardiovascular events in hemodialysis patients." European Radiology 32, no. 10 (2022): 6878–6890.

Rajamma, A. G., V. Bai, and B. Nambisan. "Antioxidant and antibacterial activities of oleoresins isolated from nine curcuma species." Phytopharmacology 2 (2012): 312–317.

Rajkumari, S., and K. Sanatombi. "Nutritional value, phytochemical composition, and biological activities of edible Curcuma species: A review." International Journal of Food Properties, 20, sup. 3 (2017): S2668–S2687. https://doi.org/10.1080/10942912.2017.1387556

Ramaholimihaso, T., F. Bouazzaoui, and A. Kaladjian. "Curcumin in depression: Potential mechanisms of action and current evidence—a narrative review." Frontiers in Psychiatry 11 (2020): 572533.

Sahu, R., and J. Saxena. "Bioactive compound from rhizome part of Curcuma caesia." International Journal of Pharmaceutical Sciences Review and Research 49, no. 2 (2018): 6–8.

Saikia, B., and S. K. Borthakur. "Use of medicinal plants in animal healthcare – A case study from Gohpur, Assam." Indian Journal of Traditional Knowledge 9, no. 1 (2010): 49–51.

Sarangthem, K., and M. J. Haokip. "Bioactive components in Curcuma caesia Roxb. grown in Manipur." The Bioscan 5, no. 1 (2010): 113–115. https://doi.org/10.25081/josac.2024.v33.i1.8477

Sasikumar, B. "Genetic resources of Curcuma: Diversity, characterization and utilization." Plant Genetic Resources 3, no. 2 (2005): 230–251.

Sharifi-Rad, Javad, Youssef El Rayess, Alain Abi Rizk, Carmen Sadaka, Raviella Zgheib, Wissam Zam, Simona Sestito, et al. "Turmeric and its major compound curcumin on health: Bioactive effects and safety profiles for food, pharmaceutical, biotechnological and medicinal applications." Frontiers in Pharmacology 11 (2020a): 550909. https://doi.org/10.3389/fphar.2020.01021

Sharifi-Rad, M., N. V. Anil Kumar, P. Zucca, E. M. Varoni, L. Dini, E. Panzarini, . . . J. Sharifi-Rad. "Lifestyle, oxidative stress, and antioxidants: Back and forth in the pathophysiology of chronic diseases." Frontiers in Physiology 11 (2020b): 552535.

Sharma, G. J., P. Chirangini, and R. Kishor. "Gingers of Manipur: Diversity and potentials as bioresources." Genetic Resources and Crop Evolution 58, no. 5 (2011): 753–767. https://doi.org/10.1007/s10722-011-9678-5

Singh, Subhashree, Bhaskar Chandra Sahoo, Asit Ray, Sudipta Jena, Manaswini Dash, Sanghamitra Nayak, Basudeba Kar, and Suprava Sahoo. "Intraspecific chemical variability of essential oil of Curcuma caesia (black turmeric)." Arabian Journal for Science and Engineering 46 (2021): 191–198. https://doi.org/10.1007/s13369-020-04940-6

Sorrenti, V., G. Contarini, S. Sut, S. Dall'Acqua, F. Confortin, A. Pagetta, . . . M. Zusso. "Curcumin prevents acute neuroinflammation and long-term memory impairment induced by systemic lipopolysaccharide in mice." Frontiers in Pharmacology 9 (2018): 183.

Tag, H., A. K. Das, and H. Loyi. "Anti-inflammatory plants used by the Khamti tribe of Lohit district in eastern Arunachal Pradesh, India." Natural Product Radiance 6, no. 4 (2007): 334–340.

Tamang, Sapan. "Black turmeric-Curcuma Caesia Roxb, and its medicinal properties: A review." International Journal of Agriculture and Environmental Research 8, no. 1 (2022): 168–175. http://dx.doi.org/10.22004/ag.econ.333825

Vairappan, Charles Santhanaraju, Ummul Mardiah Elias, Tulasiramanan R. Ramachandram, and Takashi Kamada. "Secondary metabolites from rhizome of Curcuma caesia Roxb. (Zingiberaceae)." Biochemical Systematics and Ecology 48 (2013): 107–110. https://doi.org/10.1016/j.bse.2012.11.008

Vishwavidyalaya, Ghasidas, and Bilaspur Koni. "Nutritional evaluation of Curcuma species collected from different agro climatic regions of chhattisgarh." American Journal of Ethnomedicine 4, no. 2 (2017): 20.

Wang, P., J. Li, H. Li, B. Li, Y. Jiang, F. Bao, and S. Zhang. "Is emotional memory enhancement preserved in amnestic mild cognitive impairment? Evidence from separating recollection and familiarity." Neuropsychology 27, no. 6 (2013): 691.

16 Asafoetida
Phytochemistry and Pharmacological Applications

Naveen Kumar, Anoop Kumar, Piyush Sharma, Pankaj Kumar Chaurasia, and Sunita Singh

16.1. INTRODUCTION

Asafoetida, called Hing in India, is an important spice of Indian cuisine and is produced in compounded form with Pathani Hing and Irani Hing being the commonly used varieties. Ayurveda and Unani systems of medicine illustrate that asafoetida is known as Hingu in ancient texts. In Persia, it was revered as the "food of gods" (Shah and Zare, 2014). It is known by different names such as 'Devil's dung', Kama, Anguza, A-wei, Asafetida, Hajupihka, Asant, Aza, Hingu, Perunkaya, Rechina fena, Zaz, Sheingho, Anjadana (Amalraj and Gopi, 2017). It is obtained from the roots and stems of the Ferula plant that belongs to the Umbelliferae family native to the Central Asian Mountains. There are more than 170 Ferula species distributed throughout Central Asia, Europe, and North Africa (Divya et al., 2014). It is classified into two most important types: Milky white 'Hing Kabuli Safaid' that is derived from *F. rubicaulis*, and the brown or red 'Hinglal' (Irani and Pathan types). Iran produces all forms of asafoetida, including tears called 'Irani Ras' and mass known as 'Irani Hing' (Ravindran et al., 2004).

Asafoetida mainly contains three fractions: resin, gum, and essential oil. The resin mainly contains terpenoids, coumarins, and sesquiterpene coumarins, while gum contains rhamnose, glucose, larabinose, galactose, and glucuronic acid. The essential oil contains monoterpenes, volatile terpenoids, and sulfur-containing compounds, which provide a pungent odour and bitter taste to it (Iranshahy and Iranshahi, 2011). Its long history of use has made it an attractive spice and reliable phytomedicine worldwide due to the presence of these bioactive compounds. It is used for treatment of many diseases, including amenorrhea, asthma, bronchosis, callus, abdomen cancer, liver cancer, cholera, colic, cold, colitis, constipation, convulsion, corn, cough, cramp, diarrhea, dyspepsia, enterosis, epilepsy, felon, flatulence, fracture, frigidity, fungus, hypoglycemia, indigestion, gas, pain, stomachache, gastrosis, mucososis, nervousness, neurasthenia, neurosis, thrombosis, osteosis, tumour, wart, water retention, whitlow, and worm (Duke, 2002). Over time, many researchers have investigated the efficacy of asafoetida in modern medicines and reported that asafoetida have various pharmacological and biological properties (Figure 16.1) (Amalraj and Gopi, 2017; Alruways, 2023). The present manuscript discusses the phytochemistry and traditional and pharmacological applications of asafoetida.

16.2. ORIGIN AND PROCESSING

Ferula asafoetida, native to central Asia, particularly Iran, Iraq, and Afghanistan, is the main source of asafoetida. It can be grown in the Himalayas. With improved agricultural practices, the plant has been successfully grown in parts of India. Ferula farming has the potential to improve farmer livelihoods in India's Kinnaur, Lahaul, Jammu and Kashmir, Himachal Pradesh, and Bharmour regions, and has the potential to change the economies of India's hill states (Sood, 2020). The *Ferula asafoetida* plant has a thick stem with a coarse surface and can reach a height of up to 2.5

 DOI: 10.1201/9781003559894-16

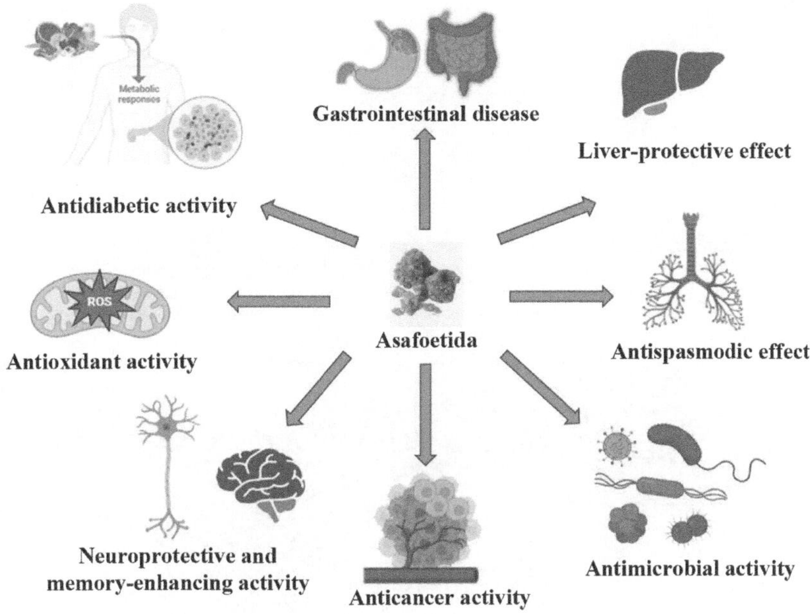

FIGURE 16.1 Biological activities of asafoetida.

m. Its leaves are 60 cm thick, divided into many pieces with a serrated edge, and it produces yellowish flowers at the end of its stem (Hashemi et al., 2019). Figure 16.2 shows an image of the *Ferula asafoetida* plant.

When the plant is 4–5 years old, it has large taproots. Traditionally, asafoetida resin is exuded by cutting the taproot from the top and protecting the cut surface from sunlight. This method gives a good yield of resin, but it has a high mortality rate for the plants. However, different cutting methods such as two-sided cutting, one-sided cutting, horizontal cutting, vertical cutting, concave, and 45-degree cutting methods have been recommended to extract the resin with the lowest plant mortality. The highest amount of resin was obtained through the concave method, while the lowest was obtained through the superficial cutting method (Kumar et al., 2024b).

16.3. TRADITIONAL USES

Asafoetida is widely used in Ayurveda and Unani medicine in various proprietary medicines such as Raj parvartini vati, Heeng Karpur vati, Atisar har vati, and many others. There are also 20 different types of Heeng wadi churna with various compositions available, along with Heeng wadi Ghritam and Kayam churna in the market. It is also a key ingredient in Hingvashtaka churna, a popular Ayurvedic remedy for indigestion containing other herbs such as ginger and long pepper (Shah and Zare, 2014). In Ayurveda, it is classified as an appetizer and restorer of consciousness. Dried gum extract is used to treat whooping cough and asthma in Saudi Arabia while in Afghanistan it is used to treat hysteria and whooping cough. The plant decoction is used as a vermifuge in China, while the gum is chewed for amenorrhea in Malaysia and as an antiepileptic in Morocco. Resin extract is used as an emmenagogue, expectorant, aphrodisiac, and antispasmodic in United States. Dried leaf and stem extract is taken as an aphrodisiac in Brazil (Mahendra and Bisht, 2012). The extract of *F. asafoetida L.* is used in Egypt as an antispasmodic, diuretic, vermifuge, and analgesic, taken orally with warm water (Buddrus et al., 1985). It is a common ingredient in the Nepali diet and is believed to have medicinal properties, such as being a sedative, carminative, antispasmodic, diuretic, anthelmintic, emmenagogue, and expectorant, with aphrodisiac effects (Eigner and Scholz, 1999). In

FIGURE 16.2 *Ferula asafoetida* plant.

India it is used with other ingredients such as rock salt and vinegar for gastrointestinal problems such as flatulence, stomachaches, diuretic, and digestive discomfort (Sood, 2020). People in the US, China, and Fiji also use it for similar purposes. For respiratory problems, it is used for angina pectoris, asthma, whooping cough, and pneumonia in India, Saudi Arabia, Afghanistan, and Fiji. In terms of nervous disorders, it is used to treat hysteria (Mahendra and Bisht, 2012) and as a sedative in Ayurveda, Afghanistan, and Nepal. In ancient Rome, it is used as a replacement for silphion cyrenaicum for tuberculosis while it is used as an anticonvulsant and antiepileptic in Iranian folk medicine (Iranshahy and Iranshahi, 2011).

16.4. PHYTOCHEMISTRY/CHEMICAL COMPOSITION

Asafoetida products consist of alcohol-soluble resin (40–64%), water-soluble gum (25%), and volatile oil (10–17%) (Iranshahy and Iranshahi, 2011). Generally, it has 68% carbohydrate, 16% moisture, 4% protein, 1% fat, 7% minerals, and 4% fibre (Amalraj and Gopi, 2017). Recently, Attarian et al. (2024) found carbohydrates as the main component in mass and tear asafoetida samples (Attarian et al., 2024). Analytical techniques, such as GC-MS, LC-MS, FTIR, and NMR spectroscopy, etc., have become crucial tools to characterize a wide range of components in plants, including *F. asafoetida* oleo-gum-resins and others (Kumar et al., 2024a; Attarian et al., 2024). For example, Singh et al. (2023) identified 141 metabolites, including various vitamins, nucleosides, flavonoids, coumarins, sesquiterpene coumarins, sulfur compounds, and sugars derivatives in *F. asafoetida* using advanced GC–MS/headspace and UHPLC-PDA-QTOFIMS techniques (Singh et al., 2023). The major sulphur-containing compounds, i.e., 2-butyl (E, Z)-1-propenyl disulfide, 1-(methylthio) propyl (E,Z)-1-propenyl disulfide, and 2-butyl 3-(methylthio)-2-propenyl disulfide, have been identified in *Ferula asafoetida* (Takeoka, 2001; Divya et al., 2014; Zomorodian et al., 2018). Further, the chemical composition of asafoetida is influenced by factors such as harvesting dates, collection time of oleo-gum-resin, temperature, soil properties, and geographical location (Karimi et al., 2020; Kavoosi and Rowshan, 2013). According to Karimi et al. (2020), the main constituents of oleo-gum-resin (OGR) essential oil include α-pinene, β-pinene, bicyclogermacrene, and (Z)-β-ocimene and varied throughout the harvesting months (highest amount in September and lowest in July (Karimi et al., 2020). Similarly, Karimian et al. reported the highest OGR amount in September (13.2 g) and the lowest in July (6.3 g), with the total number of components ranged from 43–47, representing 96.8% to 98.6% of the total oil, and α-pinene, (Z)-propenyl sec-butyl disulfide, and β-pinene were the main components of the OGR oils (Karimian et al., 2020). The major constituents of asafoetida with chemical structure that have been reported to date are summarized in Table 16.1.

TABLE 16.1
Major Constituents of Asafoetida

Chemical Constituents	Structure	References
	Coumarins and sesquiterpene coumarins	
Umbelliprenin 5-Hydroxyumbelliprenin 8-Hydroxyumbelliprenin Tadshiferin Asacoumarin A 8-Acetoxy-5-S-hydroxyumbelliprenin	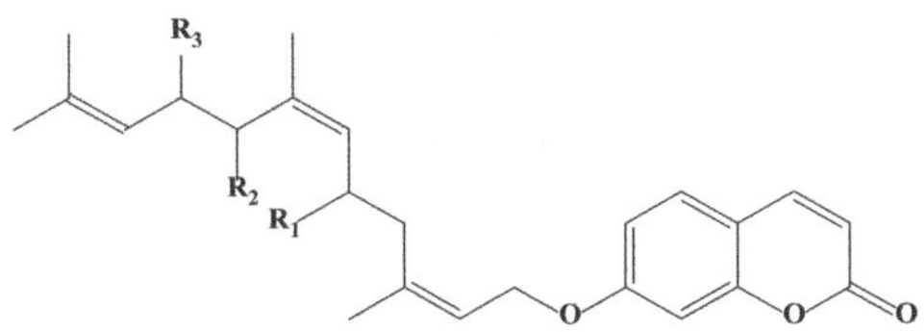 **R1=R2=R3=H - Umbelliprenin** **R1= OH, R2=R3=H -5-Hydroxyumbelliprenin** **R1=R3= H, R2=OH - 8-Hydroxyumbelliprenin** **R1=R2= H, R3=OH - Tadshiferin** **R1=R2= OH, R3=H - Asacoumarin A** **R1= OH, R2= OAc, R3=H - 8-Acetoxy-5-S-hydroxyumbelliprenin**	(Iranshahy and Iranshahi, 2011; Amalraj and Gopi, 2017; Singh et al., 2023; Attarian et al., 2024)
Franesiferol A Gummosin	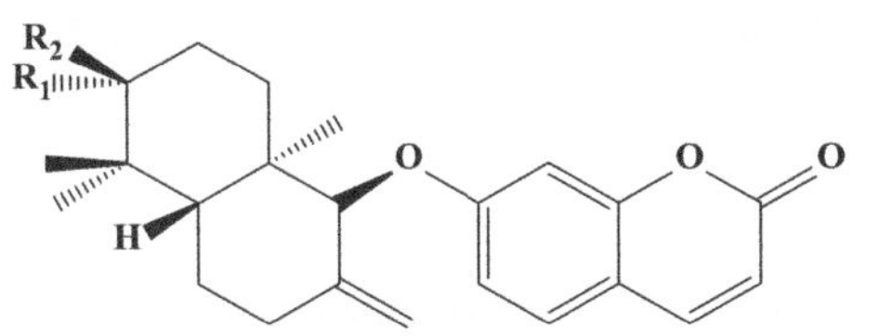 **R1=OH, R2=H - Franesiferol A** **R1= OH, R2=R3=H - Gummosin**	
Assafoetidin	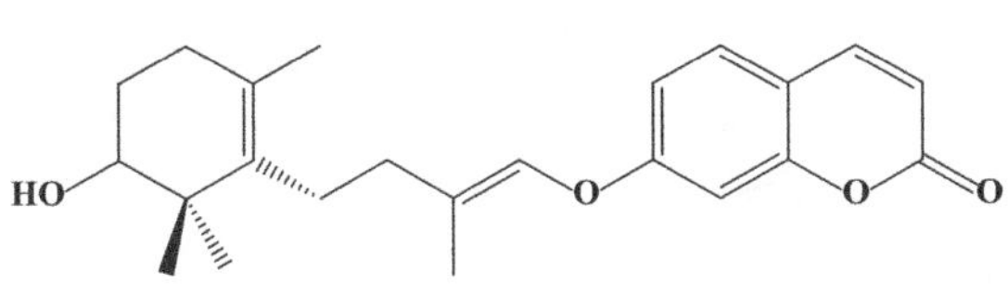	
Franesiferol B	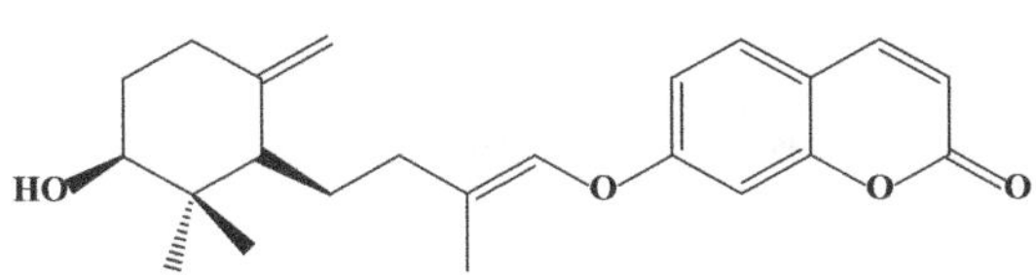	
Franesiferol C	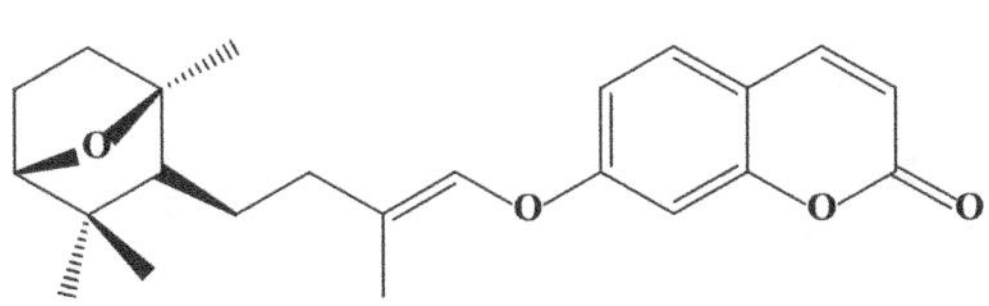	

(Continued)

TABLE 16.1 (*Continued*)
Major Constituents of Asafoetida

Chemical Constituents	Structure	References
	Coumarins and sesquiterpene coumarins	
Galbanic acid	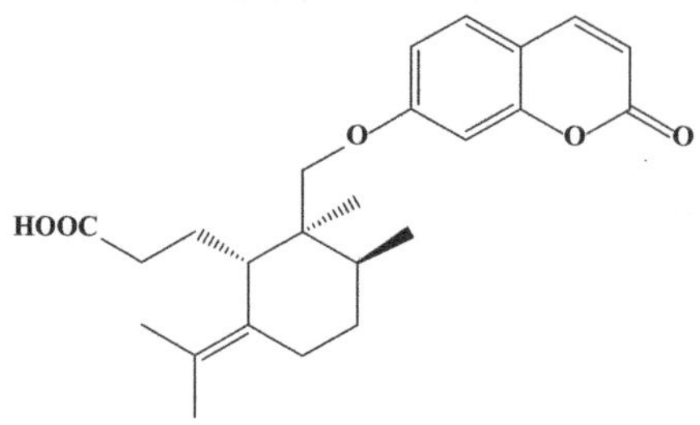	
Conferol	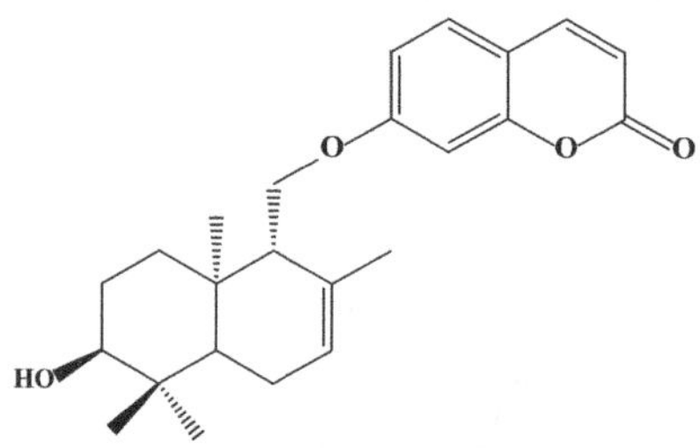	
Assafoetidinol A Assafoetidinol B	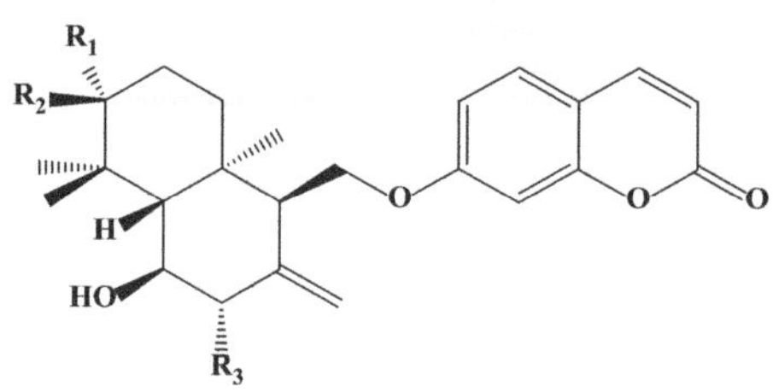 **R1=OH, R2=R3=H - Assafoetidinol A** **R1=H, R2=OAc, R3=OH - Assafoetidinol B**	
Epi-samarcandin Epi-samarcandin acetate	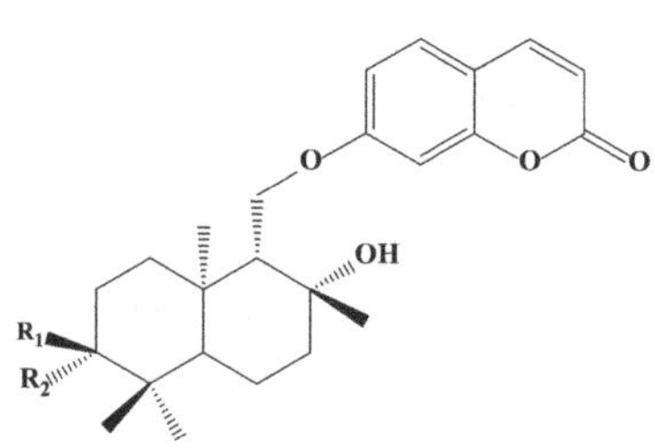 **R1= H, R2= OH - Epi-samarcandin** **R1=H, R2= OAc - Epi-samarcandin acetate**	
Ferocaulicin	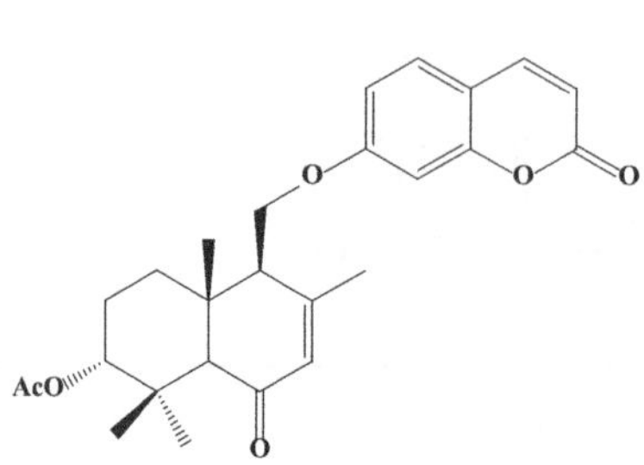	

TABLE 16.1 (*Continued*)
Major Constituents of Asafoetida

Chemical Constituents	Structure	References
	Coumarins and sesquiterpene coumarins	
Foetidine	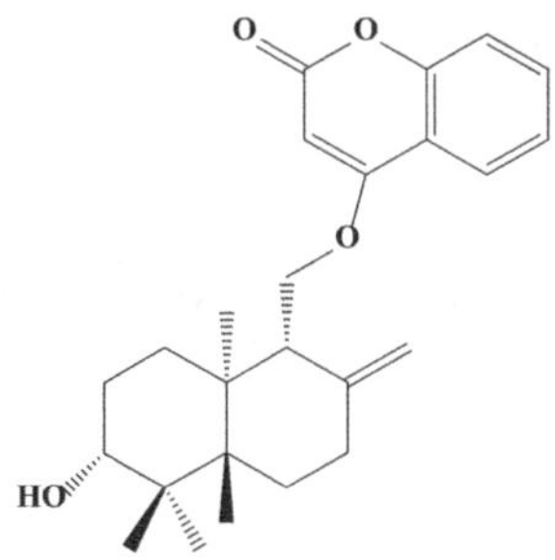	
Karatavicinol A	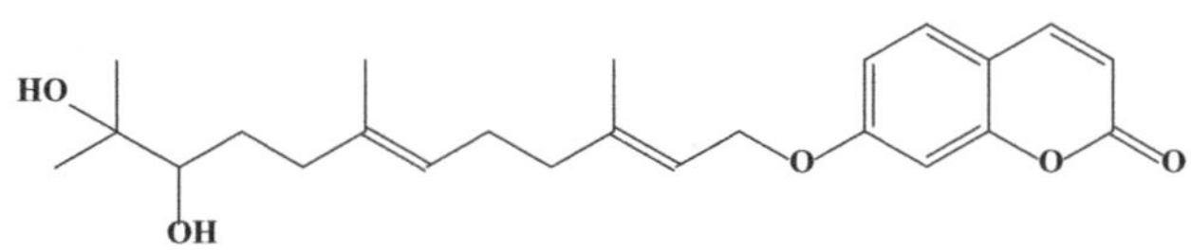	
Feselol	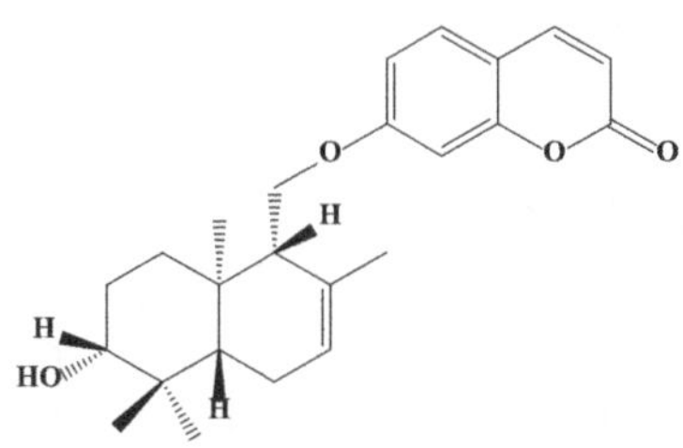	
Epi-conferdione	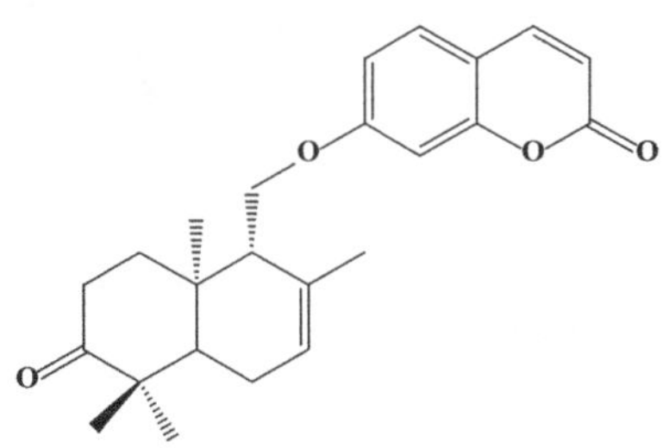	

Sulphur-Containing Compounds

Chemical Constituents	Structure	Chemical Constituents	Structure	
2-Butyl 1-propenyl disulfide	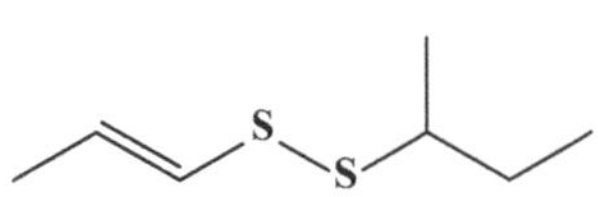	1-(Methylthio) propyl 1-propenyl disulfide	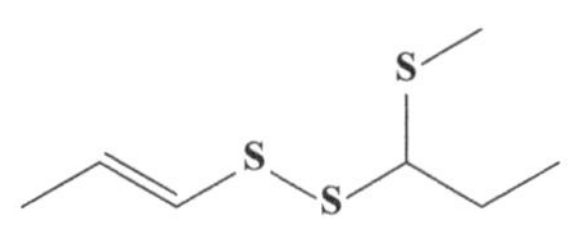	(Iranshahy and Iranshahi, 2011; Amalraj and Gopi, 2017)
2-Butyl 3-(methylthio)-2-propenyl disulfide	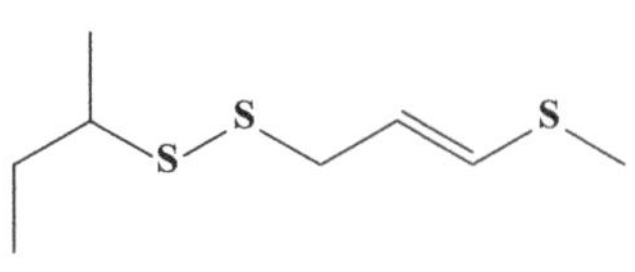	2-Butyl methyl disulfide	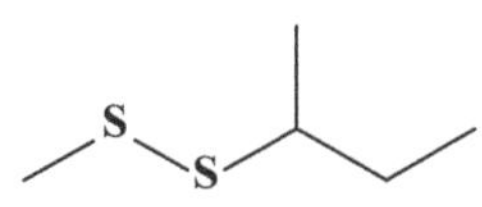	

(*Continued*)

TABLE 16.1 (*Continued*)
Major Constituents of Asafoetida

Sulphur-Containing Compounds

Chemical Constituents	Structure	Chemical Constituents	Structure
Foetisulfide A	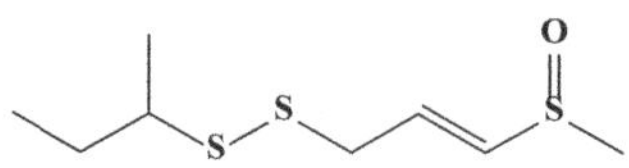	Foetisulfide C	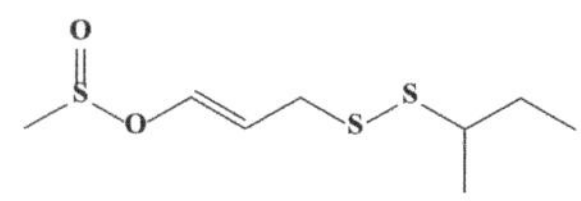
Methyl (Z)-1-propenyl disulfide	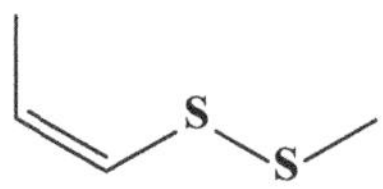	Methyl (E)-1-propenyl disulfide	
Bis (1-methyl propyl) disulfide	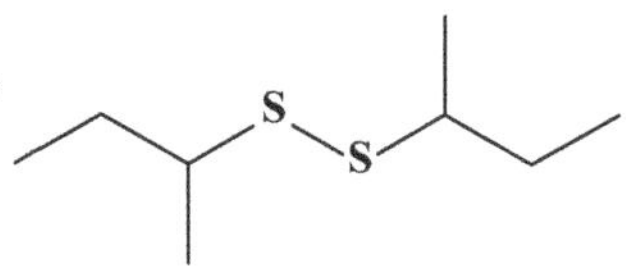	Bis (1-methyl thio) propyl disulfide	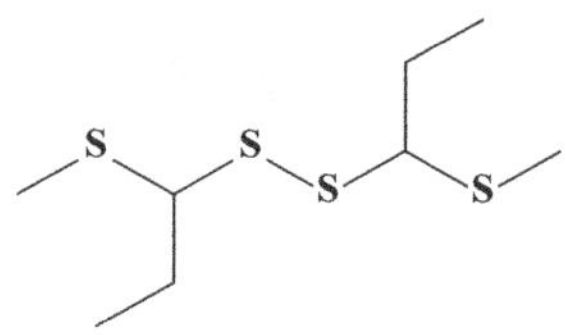
1-methylthio-(Z)-1-propene	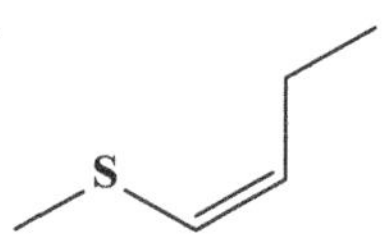	1-methylthio-(E)-1-propene	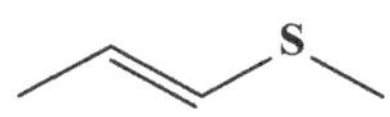
Fetidone A	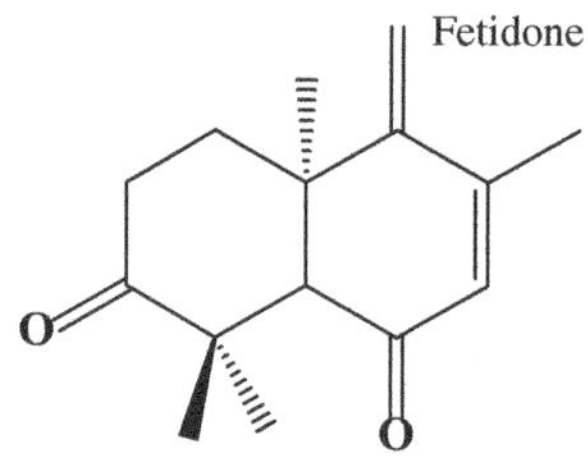	Fetidone B	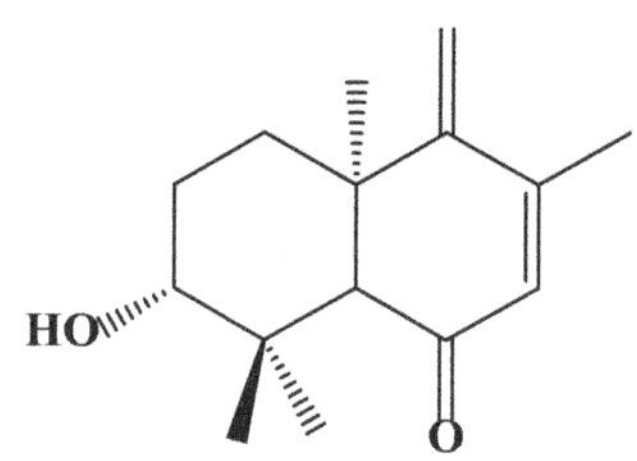

Terpenes/Terpinoids

Chemical Constituents	Structure	Chemical Constituents	Structure	
α-pinene	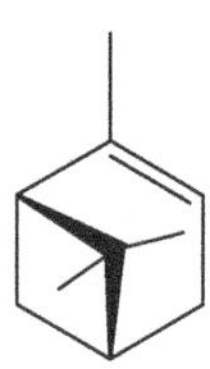	β-pinene	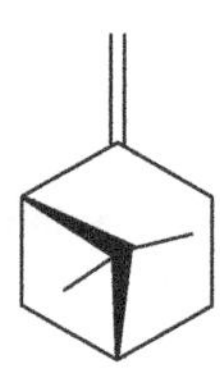	(Singh et al., 2023; Attarian et al., 2024)
Tricyclene	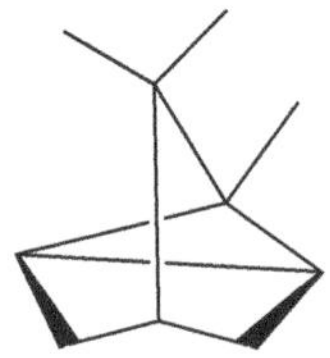	Camphene	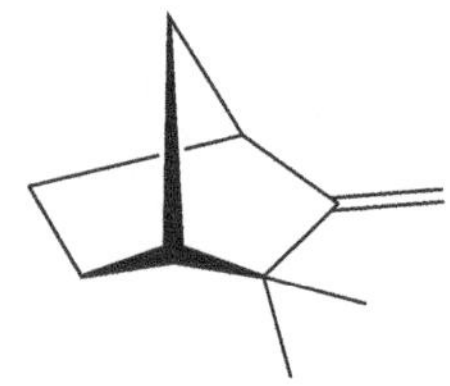	

TABLE 16.1 (*Continued*)
Major Constituents of Asafoetida

		Terpenes/Terpinoids		
Chemical Constituents	**Structure**	**Chemical Constituents**	**Structure**	
Myrcene	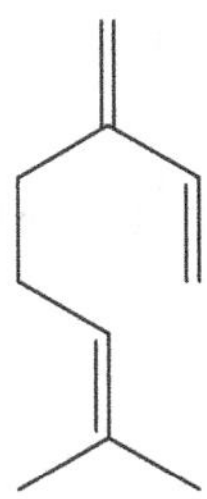	β-cymene	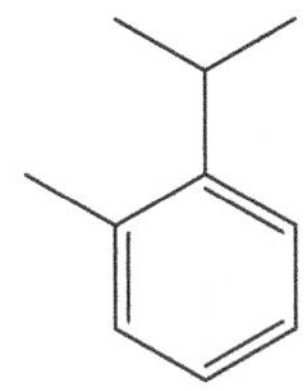	
Limonene	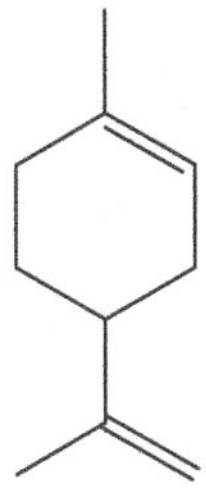	β-ocimene	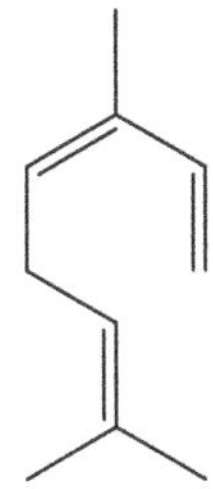	
α-terpinene	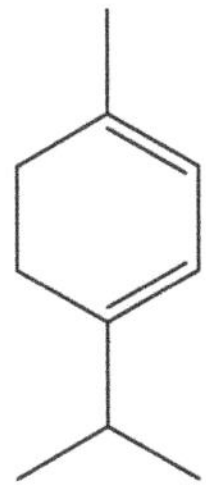	Caryophyllene	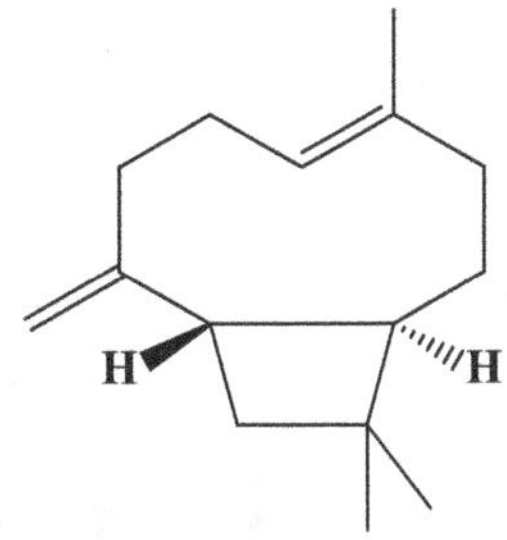	
		Phenolic Acids and Other Compounds		
Chemical Constituents	Structure	Chemical Constituents	Structure	
Gallic acid	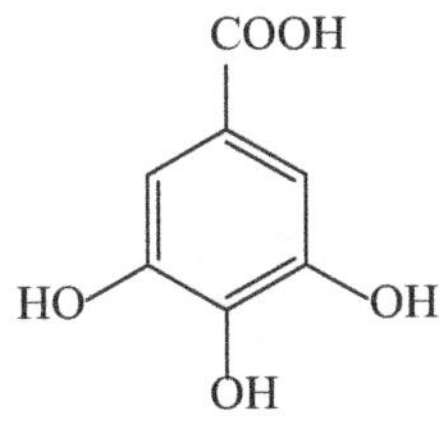	Caffeic acid	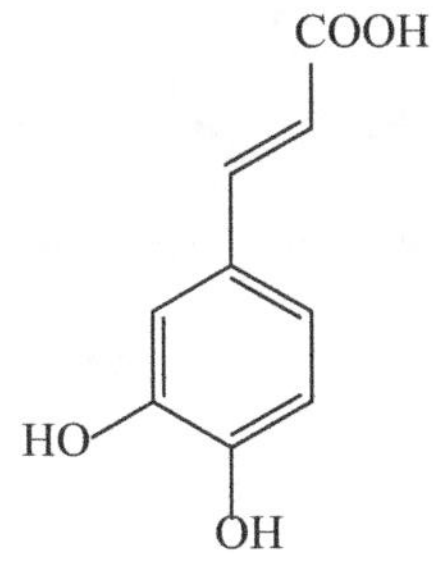	(Iranshahy and Iranshahi, 2011; Amalraj and Gopi, 2017; Singh et al., 2023; Attarian et al., 2024)

(*Continued*)

TABLE 16.1 (*Continued*)
Major Constituents of Asafoetida

Phenolic Acids and Other Compounds			
Chemical Constituents	Structure	Chemical Constituents	Structure
Ferulic acid	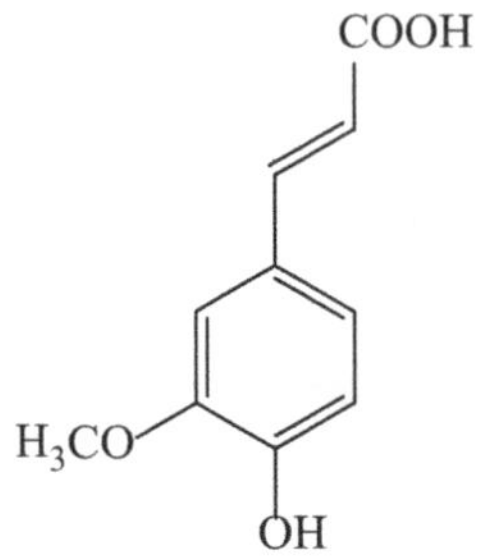	Glucuronic acid	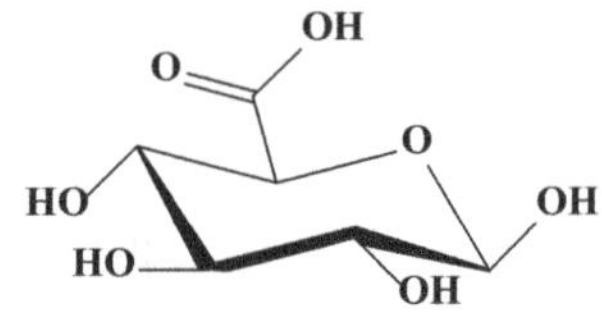
Galactose	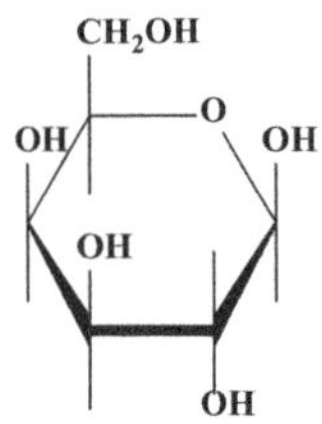	Arabinose	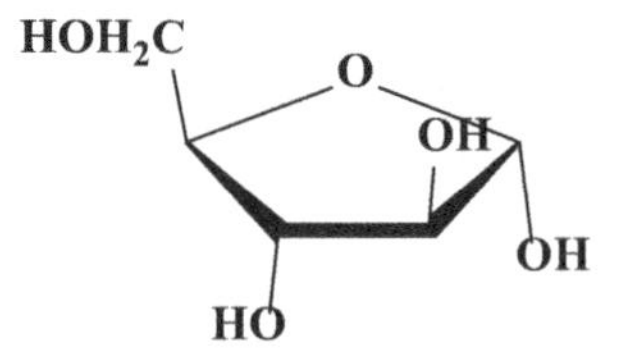
Oleic acid	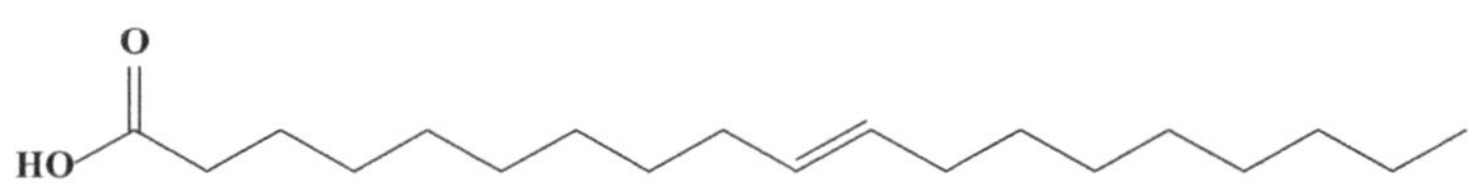		
β-sitosterol	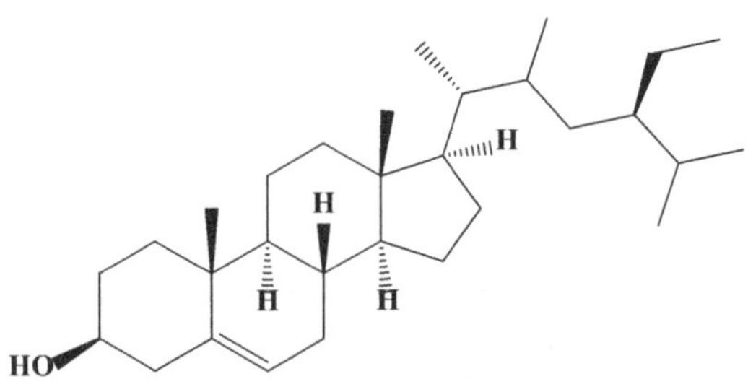		
Luteolin 7-β-D-glucopyranoside	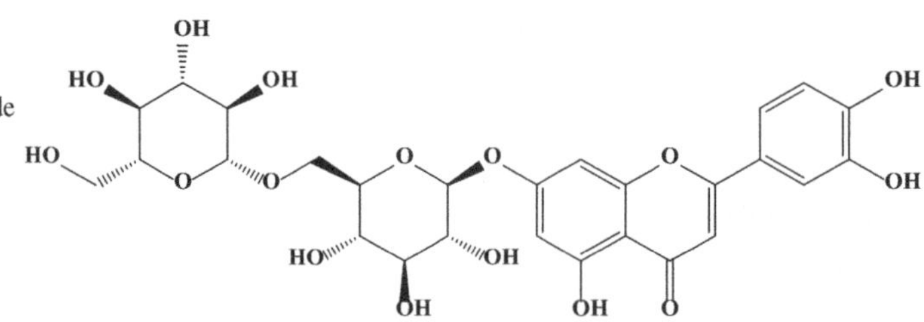		

16.5. PHARMACOLOGICAL APPLICATIONS

Asafoetida have various pharmacological applications, including gastric, antiulcer, antioxidant, antiviral, antifungal, antibacterial, antidiabetic, antispasmodic, neuroprotective, cancer chemopreventive, and many more (Figure 16.1) (Duke, 2002; Amalraj and Gopi, 2017; Alruways, 2023).

16.5.1. Gastrointestinal Diseases

Traditional Iranian medicine recommends asafoetida to treat gastrointestinal ailments, especially for flatulence and dyspepsia. Asafoetida is believed to increase gastric temperature, remove waste from the stomach, relieve pain and bloating, and have sedative effects (Babaeian et al., 2015). Mala et al. studied the effectiveness of asafoetida on 48 patients suffering from functional dyspepsia. The patients were divided into two groups, one receiving two capsules of 250 mg asafoetida and the other receiving placebo for 30 days. The group receiving asafoetida experienced larger reduction in overall severity score than the placebo group. Additionally, 69% of patients were reporting improvement in functional dyspepsia symptoms, such as bloating, indigestion, etc., in the asafoetida group (Mala et al., 2018). Jalilzadeh et al. evaluated the antidiarrheal effect of ethanolic extract of asafoetida in male rats. A 0.1 of LD_{50} therapeutic dose was used, and their intestinal passage of materials was measured after treatment; results showed that 100 and 200 mg/kg asafoetida considerably condensed the intestinal material movement (Jalilzadeh-Amin et al., 2017).

16.5.2. Hepatoprotective or Liver-Protective Effect

The liver is essential for detoxifying the body and performing metabolic functions. Asafoetida gum extract has strong antioxidant properties, particularly in lowering oxidative stress in the liver. The extract's superoxide dismutase plays a vital role in protecting the liver. Asafoetida and its sulphide components exhibit powerful antioxidant activity, restoring antioxidant enzyme and liver enzyme activity and safeguarding against toxic effects of carbon tetra chloride on liver (Mahboubi, 2021). According to Fatima et al., asafoetida extract has liver hepatoprotective effects against sodium arsenate in Swiss albino mice. The administration of 100 mg/kg asafoetida aqueous extract for 4 and 6 weeks reduced the levels of aspartate aminotransferase (AST), alanine aminotransferase (ALT), alkaline phosphatase (ALP), and bilirubin in mice treated with sodium arsenate, indicating the liver-protective effects of the extract (Fatima et al., 2017). Dandagi et al. (2008) formulated the polyherbal suspensions using *M. charantia Linn.*, *N. jatamansi*, and *F. asafoetida* extracts. These formulations demonstrated appreciable hepatoprotective activity evaluated against LIV-52 as standard (Dandagi et al., 2008).

16.5.3. Antispasmodic and Relaxant Effect

In a research, the antispasmodic effect on isolated guinea pig ileum was investigated by Fatehi et al. (2004), and a dose of 3 mg/ml F. asafoetida decreased the average amplitude of spontaneous contractions to 54% (Fatehi et al., 2004). Bagheri et al. (2014) examined the relaxant effect of essential oil and aqueous extract of asafoetida on ileum segments, and 0.3% concentration of asafoetida reduced 43% acetylcholine-induced contraction, indicating its potential as a spasmolytic treatment (Bagheri et al., 2014). Bayrami et al. studied the relaxant effects of asafoetida on guinea pig tracheal chains in two groups with different stimuli (60 mM/L KCl and 10 mM/L methacholine), and the results showed that the asafoetida extract has a strong relaxant effect primarily due to umbelliprenin (Bayrami et al., 2013).

16.5.4. Antiulcer Effect

Gastrointestinal ulcers, such as ulcerative colitis and peptic ulcers, are on the rise globally. These ulcers disrupt the stomach and/or duodenum's mucosal integrity, causing local defects and inflammation. Vijayasteltar et al. developed a green method to create potent powder formulations of asafoetida gum for dietary use and gut health. They used a water-based ultrasound process to disperse the gum on a soluble fibre matrix, resulting in microencapsulated particles (called Asafin) suitable for capsules and tablets. This Asafin product demonstrated potent antiulcerogenic efficacy, and at 250 mg kg^{-1} of Asafin inhibited 94% of ulcer formation, and at 100 mg kg^{-1} of Asafin showed large diminution in ethanol-induced gastric damage (Vijayasteltar et al., 2017). Asafoetida water suspension extensively reduced basal gastric secretion (8.83 ml to 4 ml) and completely inhibited ulceration in Wistar albino rats. It prevented the depletion of nonprotein sulfhydryl groups in the stomach, which are important for protecting the gastric mucosa. Moreover, in the presence of a necrotizing agent, it also increased gastric wall mucus and inhibited stomach lesions (Alqasoumi et al., 2011). Histopathology experiments showed that asafoetida extract had cytoprotective effects with positive effects on liver lesions, reducing inflammation and damage to the gastric mucosa. Bagheri et al. reported that aqueous extract of asafoetida significantly reduced the ulcer index in mice, and macroscopic and microscopic ulcer indexes were 3.2 and 2.1 in the asafoetida group as compared to 13.2 and 11 in the indomethacin group (Bagheri et al., 2018).

16.5.5. Antiparasitic or Antihelminthic Effect

Clotrimazole is usually used to treat *Blastocystis hominis* (intestinal parasite) in humans, despite its adverse effects. El Deeb et al. (2012) reported that the powder and essential oil of asafoetida significantly inhibited the in vitro growth of *Blastocystis* sp. (El Deeb et al., 2012). Gundamaraju studied the antihelminthic effect of asafoetida aqueous extract on *Pheretima postuma*, comparing it to piperazine citrate. It caused paralysis and death in the worms, similar to piperazine, and the aqueous extract of asafoetida at 100 mg/ml had the most significant impact in killing the worms, with higher effectiveness compared to piperazine citrate (Gundamaraju, 2013). These findings indicate that the asafoetida gum aqueous extract has a stronger antihelminthic effect, making it a potential treatment for helminth infections.

16.5.6. Antidiabetic Activity

Latifi et al. examined the impact of alcoholic extract of asafoetida on 42 male Wistar diabetic rats' glucose, lipid levels, and liver and kidney function. After 42 days of treatment, 150 mg/kg body weight ethanolic extract dose showed significant antidiabetic activity, reducing fasting blood sugar levels and increasing serum insulin levels. Moreover, it also had a protective effect on liver and kidney function markers (Latifi et al., 2019). A 50 mg/kg body weight dose of asafoetida extract in streptozotocin-induced diabetic rats showed excellent hypoglycemic activity. Ferulic acid in the extract potentially increased plasma insulin levels and inhibited alpha glucosidase activity, while tannins have also shown antidiabetic activity (Iranshahi and Alizadeh, 2012). In a study, asafoetida gum water suspension was tested on obese diabetic rats for eight weeks, and the suspension reduced fasting blood sugar as well as body weight and abdominal fat, indicating its antiobesity effects, which may protect the liver against fatty liver disease (Bagheri et al., 2012).

16.5.7. Antimicrobial Activity

Essential oil obtained from *F. asafoetida* showed excellent antibacterial and antifungal activity against foodborne bacteria and fungi. In a study, essential oil obtained from Pathani and Irani *F. asafoetida* and Pathani oil inhibited up to 49% growth of bacteria, whereas Irani volatile oil inhibited 70% to 75% (Divya et al., 2014). Bhatnagar et al. reported that red and white *F. asafoetida* extracts showed strong antibacterial activity against *Shigella flexneri* and *S. aureus*. Extracts in organic solvents such as hexane showed stronger activity as compared to the aqueous extracts, and both the extracts had comparable antibacterial activities, indicating similar chemical composition of red and white *F. asafoetida* (Bhatnager et al., 2015). Zomorodian et al. (2018) evaluated the antimicrobial and antifungal activity of *F. asafoetida* essential oil. It showed antibacterial activity against gram-positive and gram-negative bacteria at concentrations of 0.5–8 μL/mL and 16–128 μL/mL, respectively, and antifungal activity against filamentous fungi at concentrations of 0.06–2 μL/mL (Zomorodian et al., 2018). Due to the bioactive sesquiterpene cumarines present in *F. asafoetida*, it has shown strong antiviral properties, with novel chemical scaffolds such as Farnesiferol C and Farnesiferol b inhibiting HRV-2. Additionally, compounds such as 5′S-hydroxyumbelliprenin, 8′-acetoxy-5′S-hydroxyumbelliprenin, and others demonstrated greater strength against influenza A virus (H1N1) than amantadine (Dissanayake and Perera, 2020). According to Lee et al., sesquiterpene coumarins, i.e., galbanic acid, farnesiferol C, and epi-conferdione from asafoetida, showed antiviral effect against influenza H_1N_1 with IC_{50} values of 0.26 μg/ml, 0.29 μg/ml, and 0.32 μg/ml, respectively, comparable to amantadine (Lee et al., 2009). Kellerin, sesquiterpene coumarins present in *F. asafoetida* gum resin, showed potential antiviral activity against herpes virus with a significant inhibition of cytopathic effects (Ghannadi et al., 2014).

16.5.8. Antioxidant Activity

The Ferula plant contains antioxidants such as ferulic acid and cadinene, which can prevent oxidative damage and preserve cellular health. These antioxidants play a crucial role in preventing disease by inhibiting the production of free radicals and reducing inflammation. They can be easily extracted from the plant and used in the food and pharmaceutical industries, making Ferula a significant source of natural antioxidants for maintaining health and preventing diseases (Sangchooli et al., 2024). Hydroalcoholic extract (FLHE) and essential oil (FLEO) of *F. asafoetida* showed good antioxidant activities compared to synthetic antioxidant BHT (butylated hydroxytoluene), with IC_{50} values of 787.13 and 2375.66 μg/mL for FLHE and FLEO, respectively (Ahmadvand et al., 2013). Methanolic extract of *F. asafoetida* showed good peroxidation inhibition, Fe^{2+} chelating, DPPH radical–scavenging, and nitric oxide–scavenging activity. The extract has values of 82% and 88% for peroxidation inhibition at 24 hours and 72 hours, respectively. The Fe^{2+} chelating ability was 0.57 ± 0.02 $mgml^{-1}$, and the IC_{50} values for DPPH radical–scavenging and nitric oxide–scavenging activity was reported 380 ± 12 and 270 ± 3 $mgml^{-1}$, respectively (Dehpour et al., 2009).

16.5.9. Anticancer Activity

The use of *F. asafoetida* could help lower cancer incidence and improve cancer treatment outcomes. Saleem et al. reported that pretreatment of Swiss albino mice with asafoetida at certain doses (300, 400, and 500μg/200μl) could reverse early stages of carcinogenesis (Saleem et al., 2001). Nigam et al. (2013) researched the anticancer properties of *F. asafoetida* in different countries. They collected data from various sources such as the WHO website and reported that in countries where asafoetida is not commonly used, cancer rates are higher (Nigam and Sachan, 2013). Panwar et al. (2015) used *F. asafoetida* against 1, 2-dimethylhydrazine-induced rat colon carcinogenesis and reported that the 10 mg/100 g body weight supplementation dose of asafoetida reduced tumour size, multiplicity, and incidence. It also reduced serum total sialic acid levels and restored the normal histoarchitecture (Panwar et al., 2015). Lee et al. treated various human cancer cell lines with *F. asafoetida* fractions, incubated with MTT, solubilized in DMSO, and their absorbance was measured at 550nm using an ELISA reader. *F. asafoetida* showed the best potency against the HepG2, Hep3B, and MCF-7 cancer cell lines, with IC_{50} values of 0.51, 2.6, and 3.4 μg/mL, respectively (Lee et al., 2009).

16.5.10. Neuroprotective and Memory Enhancing Activity

Tayeboon et al. examined the effects of *F. asafoetida* at a concentration of 100 mg/mL on glutamate-induced cell damage, and they found that the asafoetida extract increased cellular viability and reduced apoptotic/necrotic cell death induced by glutamate (Tayeboon et al., 2013). Moghadam et al. conducted in vitro and in vivo studies to check the effects of asafoetida on isolated sciatic nerves and peripheral neuropathy in mice. In vitro studies showed that asafoetida increased the nerve action potential amplitude, which indicated asafoetida's potential as a nerve stimulant and neuroprotective agent (Moghadam et al., 2014). The effects of asafoetida extract on neurons was examined and reported that 0.01–1 μg/ml concentrations of asafoetida improved neuron survival, whereas a higher concentration (10μg/ml) was toxic (Moghadam et al., 2013). *F. asafoetida* extract has memory-enhancing potential due to its acetyl cholinesterase-inhibiting and antioxidant properties. This can be beneficial for those who are suffering from Alzheimer's disease and could be used as an adjuvant to existing antidementia therapies (Adiga et al., 2012). Bagheri et al. examined the impact of asafoetida on preventing dementia induced by D-galactose and $NaNO_2$ in mice. The groups that received asafoetida had greater memory retention compared to the dementia control group. Thus, Asafoetida may prevent and treat amnesia (Bagheri and Dashti-R, 2015).

16.6. CONCLUSION

Asafoetida is derived from the roots of *Ferula asafoetida*. It contains important chemical ingredients, including carbohydrates, proteins, minerals, fibre, vitamins, and polysulphides. Traditionally, it has been used in Ayurvedic and Unani medicine to treat various diseases, such as gastrointestinal, intestinal parasites, respiratory, and neurological disorders. Recent research confirmed that asafoetida showed numerous medicinal benefits, including liver-protective, neuroprotective, antiparasitic, antiulcer, antidiabetic, antibacterial, antifungal, antiviral, anticancer, and antioxidant properties. These benefits are due to its bioactive phytochemicals, such as coumarins, sesquiterpene coumarins, terpenes, and sulphur-containing compounds. However, further research is needed to fully understand its mechanism of action and potential integration into modern medicine.

REFERENCES

Adiga, S., Bhat, P., Chaturvedi, A., Bairy, K., and Kamath, S. (2012). Evaluation of the effect of Ferula asafoetida Linn. gum extract on learning and memory in Wistar rats. *Indian Journal of Pharmacology* 44(1), 82–87.

Ahmadvand, H., Amiri, H., Dehghani Elmi, Z., and Bagheri, S. (2013). Chemical composition and antioxidant properties of Ferula-assa-foetida leaves essential oil. *IJPT* 12(2), 52–57.

Alqasoumi, S., Al-Dosari, M., Al-Howiriny, T., Al-Yahya, M., Al-Mofleh, I., and Rafatullah, S. (2011). Gastric antiulcer activity of a pungent spice Ferula assafoetida L. in rats. *Farmacia* 59(6), 750–759.

Alruways, M.W. (2023). Antimicrobial effects of Ferula species-an herbal tactic for management of infectious diseases. *Journal of King Saud University-Science*, 102806.

Amalraj, A., and Gopi, S. (2017). Biological activities and medicinal properties of Asafoetida: A review. *Journal of Traditional and Complementary Medicine* 7(3), 347–359.

Attarian, B., Kavoosi, G., Bordbar, Z., and Sadeghi, H. (2024). Proximate composition, physico-chemical properties, techno-functional properties, nutritional quality, and functional activity of Ferula assafoetida oleo-gum-resin. *Journal of Food Composition and Analysis* 129, 106073.

Babaeian, M., Naseri, M., Kamalinejad, M., Ghaffari, F., Emadi, F., Feizi, A., et al. (2015). Herbal remedies for functional dyspepsia and traditional Iranian medicine perspective. *Iranian Red Crescent Medical Journal* 17(11).

Bagheri, S.M., Azizian, H., Rezvani, M.E., and Esmaeili, D.M. (2012). Anti-Obesity, Fat Lowering and Liver Steatosis Protective Effects of Ferula Asafoetida Gum in Type 2 Diabetic Rats: Possible Involvement of Leptin.

Bagheri, S.M., and Dashti-R, M.H. (2015). Influence of asafoetida on prevention and treatment of memory impairment induced by d-galactose and NaNO2 in mice. *American Journal of Alzheimer's Disease & Other Dementias®* 30(6), 607–612.

Bagheri, S.M., Hejazian, S.H., and Dashti-R, M.H. (2014). The relaxant effect of seed fs essential oil and oleo-gum-resin of Ferula assa-foetida on isolated rat's ileum. *Annals of Medical and Health Sciences Research* 4(2), 238–241.

Bagheri, S.M., Yadegari, M., Zare-Mohazabiye, F., Momeni-Asl, H., Mirjalili, A., Anvari, M., et al. (2018). Effect of Ferula assa foetida oleo-gum-resin on gastric ulcer in indomethacin-ulcerated rats. *Journal of Current Research in Scientific Medicine* 4(1), 42–46.

Bayrami, G., Boskabady, M.H., Iranshahi, M., and Gholamnezhad, Z. (2013). Relaxant effects of asafoetida extract and its constituent umbelliprenin on guinea-pig tracheal smooth muscle. *Chinese Journal of Integrative Medicine*, 1–6.

Bhatnager, R., Rani, R., and Dang, A.S. (2015). Antibacterial activity of Ferula asafoetida: A comparison of red and white type. *Journal of Applied Biology & Biotechnology* 3(2), 18–21.

Buddrus, J., Bauer, H., Abu-Mustafa, E., Khattab, A., Mishaal, S., El-Khrisy, E., et al. (1985). Foetidin, a sesquiterpenoid coumarin from Ferula assa-foetida. *Phytochemistry* 24(4), 869–870.

Dandagi, P., Patil, M., Mastiholimath, V., Gadad, A., and Dhumansure, R. (2008). Development and evaluation of hepatoprotective polyherbal formulation containing some indigenous medicinal plants. *Indian Journal of Pharmaceutical Sciences* 70(2), 265.

Dehpour, A.A., Ebrahimzadeh, M.A., Fazel, N.S., and Mohammad, N.S. (2009). Antioxidant activity of the methanol extract of Ferula assafoetida and its essential oil composition. *Grasas y aceites* 60(4), 405–412.

Dissanayake, K., and Perera, W. (2020). Medicinal importance of Ferula asafetida oligo-gum resins against infective diseases. *Journal of Medicinal Plants Studies* 8(2), 135–139.

Divya, K., Ramalakshmi, K., Murthy, P.S., and Rao, L.J.M. (2014). Volatile oils from Ferula asafoetida varieties and their antimicrobial activity. *LWT-Food Science and Technology* 59(2), 774–779.

Duke, J.A. (2002). *Handbook of Medicinal Herbs* (2nd ed.). CRC Press. https://doi.org/10.1201/9781420040463

Eigner, D., and Scholz, D. (1999). Ferula asa-foetida and Curcuma longa in traditional medical treatment and diet in Nepal. *Journal of Ethnopharmacology* 67(1), 1–6.

El Deeb, H.K., Al Khadrawy, F.M., and El-Hameid, A.K.A. (2012). Inhibitory effect of Ferula asafoetida L.(Umbelliferae) on Blastocystis sp. subtype 3 growth in vitro. *Parasitology Research* 111, 1213–1221.

Fatehi, M., Farifteh, F., and Fatehi-Hassanabad, Z. (2004). Antispasmodic and hypotensive effects of Ferula asafoetida gum extract. *Journal of Ethnopharmacology* 91(2–3), 321–324.

Fatima, N., Fatmi, N., Shahzada, M., Sharma, S., Kumar, R., and Ali, M. (2017). Hepatoprotective effect of Ferula assafoetida against arsenic induced toxicity in Swiss albino mice. *Journal of Drug Discovery, Development and Delivery* 4(1), 1030.

Ghannadi, A., Fattahian, K., Shokoohinia, Y., Behbahani, M., and Shahnoush, A. (2014). Anti-viral evaluation of sesquiterpene coumarins from Ferula assa-foetida against HSV−1. *Iranian Journal of Pharmaceutical Research: IJPR* 13(2), 523.

Gundamaraju, R. (2013). Evaluation of anti-helmintic activity of Ferula foetida "Hing-A natural Indian spice" aqueous extract. *Asian Pacific Journal of Tropical Disease* 3(3), 189–191.

Hashemi, M.S., Hashempur, M.H., Lotfi, M.H., Hemat, H., Mousavi, Z., Emtiazy, M., et al. (2019). The efficacy of asafoetida (Ferula assa-foetida oleo-gum resin) versus chlorhexidine gluconate mouthwash on dental plaque and gingivitis: A randomized double-blind controlled trial. *European Journal of Integrative Medicine* 29, 100929.

Iranshahi, M., and Alizadeh, M. (2012). Antihyperglycemic effect of Asafoetida (Ferula assafoetida Oleo-Gum-Resin) in streptozotocin-induced diabetic rats. *World Applied Sciences Journal* 17(2), 157–162.

Iranshahy, M., and Iranshahi, M. (2011). Traditional uses, phytochemistry and pharmacology of asafoetida (Ferula assa-foetida oleo-gum-resin)—A review. *Journal of Ethnopharmacology* 134(1), 1–10.

Jalilzadeh-Amin, G., Yousefi, A., and Abasipour-Dalivand, S. (2017). Antidiarrheal effect of hydroalcoholic extract of Ferula assa foetida in rat. *Journal of Gorgan University of Medical Sciences* 19(1), 7–13.

Karimi, A., Krahmer, A., Herwig, N., Hadian, J., Schulz, H., and Meiners, T. (2020). Metabolomics approaches for analyzing effects of geographic and environmental factors on the variation of root essential oils of Ferula assa-foetida L. *Journal of Agricultural and Food Chemistry* 68(37), 9940–9952.

Karimian, V., Sepehry, A., Barani, H., Ebrahimi, S.N., and Mirjalili, M.H. (2020). Productivity, essential oil variability and antioxidant activity of Ferula assa-foetida L. oleo-gum-resin during the plant exploitation period. *Journal of Essential Oil Research* 32(6), 545–555.

Kavoosi, G., and Rowshan, V. (2013). Chemical composition, antioxidant and antimicrobial activities of essential oil obtained from Ferula assa-foetida oleo-gum-resin: Effect of collection time. *Food Chemistry* 138(4), 2180–2187.

Kumar, N., Kumar, P., and Sharma, P. (2024a) "Technological advancement for the chemical characterization of essential oils," in *Pharmacological Aspects of Essential Oils*. CRC Press, 23–40.

Kumar, R., Yadav, N., Rana, A., Chauhan, R., Singh, S., Kumar, D., et al. (2024b). Ferula assa-foetida L., an important Central and South Asian traditional spice and medicinal herb: A comprehensive review. *Journal of Applied Research on Medicinal and Aromatic Plants*, 100548.

Latifi, E., Mohammadpour, A.A., Fathi, B., and Nourani, H. (2019). Antidiabetic and antihyperlipidemic effects of ethanolic Ferula assa-foetida oleo-gum-resin extract in streptozotocin-induced diabetic wistar rats. *Biomedicine & Pharmacotherapy* 110, 197–202.

Lee, C.-L., Chiang, L.-C., Cheng, L.-H., Liaw, C.-C., Abd El-Razek, M.H., Chang, F.-R., et al. (2009). Influenza A (H1N1) antiviral and cytotoxic agents from Ferula assa-foetida. *Journal of Natural Products* 72(9), 1568–1572.

Mahboubi, M. (2021). The beneficial effects of Ferula asafoetida oleo-gum resin in gastrointestinal disorders. *Bulletin of Faculty of Pharmacy Cairo University* 59(1), 50–63.

Mahendra, P., and Bisht, S. (2012). Ferula asafoetida: Traditional uses and pharmacological activity. *Pharmacognosy Reviews* 6(12), 141.

Mala, K., Thomas, J., Syam, D.S., Maliakel, B., and Krishnakumar, I. (2018). Safety and efficacy of Ferula asafoetida in functional dyspepsia: A randomized, double-blinded, placebo-controlled study. *Evidence-Based Complementary and Alternative Medicine* 2018(1), 4813601.

Moghadam, F.H., Dehghan, M., Zarepur, E., Dehlavi, R., Ghaseminia, F., Ehsani, S., et al. (2014). Oleo gum resin of Ferula assa-foetida L. ameliorates peripheral neuropathy in mice. *Journal of Ethnopharmacology* 154(1), 183–189.

Moghadam, F.H., Zarch, B.V., and Shafiei, M. (2013). Double edged effect of gum-resin of ferula assa-foetida on lifespan of neurons. *Iranian Journal of Basic Medical Sciences* 16(4), 668.

Nigam, U., and Sachan, S. (2013). Evaluation of Ferula asafoetida for its anticancerous activity in different countries. *Journal of Pharmacognosy and Phytochemistry* 2(4), 74–76.

Panwar, R., Rana, S., Dhawan, D.K., and Prasad, K.K. (2015). Chemopreventive efficacy of different doses of Ferula asafoetida oleo-gum-resin against 1, 2-dimethylhydrazine (DMH) induced rat colon carcinogenesis. *The Journal of Phytopharmacology* 4(6), 282–286.

Ravindran, P., Pillai, G.S., and Babu, K.N. (2004). Under-Utilized Herbs and Spices.

Saleem, M., Alam, A., and Sultana, S. (2001). Asafoetida inhibits early events of carcinogenesis: A chemopreventive study. *Life Sciences* 68(16), 1913–1921.

Sangchooli, T., Aboulhassanzadeh, S., Aghazadeh, H., Paeizi, M., Shokri, D., and Malekzadeh, M. (2024). Chemical Composition, Antioxidant Properties, and Antimicrobial Activity of Ferula assa-foetida L. Essential Oil against Pathogenic Bacteria.

Shah, N., and Zare, A. (2014). Asafoetida (Heeng): The well-known medicinal-condiment of India & Iran. *The Scitech Journal* 1(4), 30–36.

Singh, P.P., Joshi, R., Kumar, R., Kumar, A., and Sharma, U. (2023). Comparative phytochemical analysis of Ferula assa-foetida with Ferula jaeschkeana and commercial oleo-gum resins using GC–MS and UHPLC-PDA-QTOF-IMS. *Food Research International* 164, 112434.

Sood, R. (2020). Asafoetida (Ferula asafoetida): A high-value crop suitable for the cold desert of Himachal Pradesh, India. *Journal of Applied and Natural Science* 12(4), 607–617.

Takeoka, G. (2001). Volatile Constituents of Asafoetida. ACS Symposium Series, pp. 33–44. https://doi.org/10.1021/bk-2001-0794.ch004

Tayeboon, G.S., Tavakoli, F., Hassani, S., Khanavi, M., Sabzevari, O., and Ostad, S.N. (2013). Effects of Cymbopogon citratus and Ferula assa-foetida extracts on glutamate-induced neurotoxicity. *In Vitro Cellular & Developmental Biology-Animal* 49, 706–715.

Vijayasteltar, L., Jismy, I., Joseph, A., Maliakel, B., Kuttan, R., and Krishnakumar, I. (2017). Beyond the flavor: A green formulation of Ferula asafoetida oleo-gum-resin with fenugreek dietary fibre and its gut health potential. *Toxicology Reports* 4, 382–390.

Zomorodian, K., Saharkhiz, J., Pakshir, K., Immeripour, Z., and Sadatsharifi, A. (2018). The composition, antibiofilm and antimicrobial activities of essential oil of Ferula assa-foetida oleo-gum-resin. *Biocatalysis and Agricultural Biotechnology* 14, 300–304.

17 Thyme

Chemistry and Pharmacology

Alka Singh, Shalu Verma, Kiran Dobhal, and Chetan Singh Chauhan

17.1. INTRODUCTION

Traditional medicine has historically relied heavily on plants and their extracts to treat and prevent various infectious diseases. Around the world, medicinal plants have a significant role in conventional healing practices. Recently, phytomedicines and dietary supplements have garnered increasing attention as alternative treatments for contagious ailments because of their high concentration of secondary metabolites such as polyphenols, which include flavonoids, tannins, and alkaloids, as well as steroids and essential oils. There are about 400 species in the genus Thymus, and many of them have been utilized in conventional medicine. *Thymus vulgaris*, or Ajwain, is a fragrant perennial shrub rich in phytoconstituents and nutrients. It is planted all over the world, and it belongs to Lamiaceae family. It has been used for ages because of its culinary, medicinal, and flavouring properties and is also known as garden thyme. Thyme thrives best in dry, sunny areas that are not shaded, particularly in temperate to warm regions (Saad et al., 2005).

It is a woody, bushy shrub that usually reaches an average height of 10–40 cm. It has small, oval, gray-green leaves that are heavily scented and packed with tiny glands. The shrub produces clusters of purple or pink blooms. Small, typically 2.5–5 mm long, the leaves can be linear or linear-lanceolate, ovate, or oblong, with a pointy tip, a tapering base that extends into a petiole, and curled edges (Stahl et al., 2012). The lower part of the leaves is grey, hairy, and speckled with glands, while the top layer is light grey or light brownish grey, occasionally appearing light olive green, and covered with many fine hairs. The calyx is 4 mm long, tubular, bilabiate, and has a glandular texture. The flowers are grouped in axillary whorls. The corolla is bilabiate, purplish, and is approximately twice the length of the calyx. The style has a bilobed stigma, and the stamens are didynamous (Reddy et al., 2014).

Thyme is an extremely valuable edible plant because it may be used in various ways, including in food additives, medications, and cosmetics. Because of its strong perfume, its leaves are often used, either dried or fresh, as a seasoning in various cuisine preparations, such as stews, soups, sauces, fish, and meat dishes, and to flavour liqueurs and herbal teas. Minerals, vitamins, flavonoids, antioxidants, and phytonutrients are all abundant in thyme. Thyme and its essential oils, specifically thymol, and carvacrol, have been demonstrated in numerous studies to have therapeutic effects in treating various illnesses. These advantages contribute to its various pharmacological characteristics, which include anti-inflammatory, anticancer, and antioxidant activities. Furthermore, thyme's antiviral, antibacterial, antifungal, and antiseptic qualities have long been known (Halat et al., 2022). Asthma, bronchitis, and chronic obstructive pulmonary disease are among the respiratory conditions treated using *T. vulgaris* extracts in traditional medicine, along with other medical conditions (Salehi et al., 2018). The plant's antibacterial, antifungal, antioxidant, antiseptic, antispasmodic, antitussive, and antiviral activities are the cause. In phytotherapy, thyme oil is also frequently used to treat cancer, infections, hypertension, and skin disorders including acne (Hosseinzadeh et al., 2015). The oil also enhances the immune system and aids in combating colds, infectious diseases,

DOI: 10.1201/9781003559894-17

flu, and chills. It is an effective urinary antiseptic, particularly beneficial for conditions such as cystitis and urethritis. Additionally, thyme is commonly utilized as a seasoning for fish, vegetables, and poultry and for enriching the flavours of liqueurs and herbal infusions (Jain et al., 2022).

17.2. BOTANICAL DESCRIPTION AND GEOGRAPHIC DISTRIBUTION

The taxonomic classification of the herb is as follows (Dauqan et al., 2017):

> It belongs to the Kingdom Plantae, Subkingdom Tracheobionta, Superdivision Spermatophyta, Division Magnoliophyta, Class Magnoliopsida, Subclass Asteridae, Order Lamiales, Family Lamiaceae, Genus: Thymus L., Species: Thymus vulgaris L.

Its historical significance dates back to the 1st century CE, when Dioscorides' work mentions that it was primarily recognized as a medicinal plant. The term comes from the Greek word *thymos*, which signifies courage or strength. Before expanding worldwide, thyme was mostly used as a condiment in the Mediterranean region, despite its early use in medicine (Halat et al., 2022).

17.3. PHYTOCHEMISTRY

Numerous biochemical substances, namely terpenoids, steroids, flavonoids, tannins, alkaloids, and saponins, are abundant in *T. vulgaris* L. Thymol, carvacrol, p-cymene, α-pinene, linalool, borneol, and 1,8-cineole are among the notable monoterpene phenolic chemicals found in thyme, as shown in Figure 17.1. Using gas chromatography-mass spectrometry (GC-MS) investigation, Heidari et al. discovered other chemicals, including creosol, 2-methoxy-4-methylphenol, thiophenol (benzenethiol), loliolide, 3-methoxy-5-methylphenol, and quinic acid (Heidari et al., 2018; Jain et al., 2022).

Additionally, thyme is a good source of flavonoids: apigenin, 6-hydroxy luteolin, luteolin, and other methyl-flavones; these include curvilineal, 5-demethylnobiletin, 8-methoxycirsilineol, cirsimaritin, 7-methoxy-luteolin, gardening B, thymosin, sideritoflavone, xanthomicrol, and thymosin. In addition to flavones such as quercetin-7-O-glucoside and flavanones such as apigenin and naringenin, Roby et al. reported that the methanolic extract of *T. vulgaris* includes phenolic acids such as caffeic acid, cinnamic acid, p-coumaric acid, rosmarinic acid, ferulic acid, quinic acid, and carnosic acid (Roby et al., 2013).

Oxygenated sesquiterpenes (1.84%), monoterpene hydrocarbons (28.69%), and sesquiterpene hydrocarbons (5.04%) were discovered to be relatively low in *T. vulgaris* essential oil, but oxygenated monoterpenes (56.53%) were found to be highly concentrated (Van et al., 1981). The predominant compound, thymol, was found to make up 51.34% of the essential oil, whereas the remaining constituents made up less than 19% (Soleimani et al., 2022).

On the basis of gas chromatography and GCMS analysis of thyme essential oil, 41 components made up 97.85% of all chemicals found. Among the oil's principal components were camphor

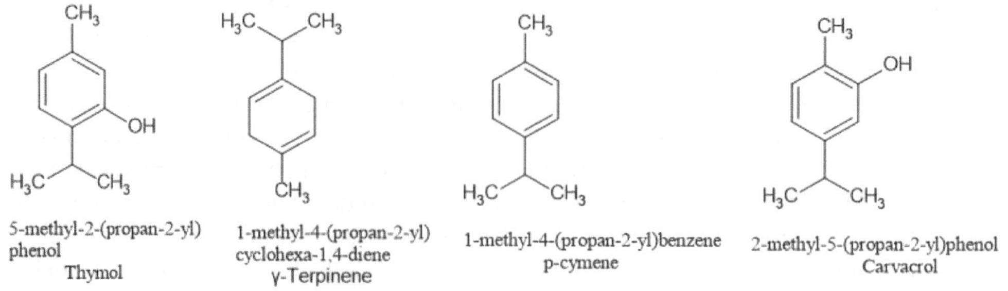

FIGURE 17.1 Major phytoconstituents of thyme.

(39.39%), camphene (17.57%), α-pinene (9.55%), 1,8-cineole (5.57%), borneol (5.03%), and β-pinene (4.32%). The amount of the remaining components was below 2%. The essential oil from thyme was abundant in significant monoterpene hydrocarbons, including myrcene, α-pinene, β-pinene, and camphene (Imelouane et al., 2009).

With 54.82% of the oil's total components, oxygenated monoterpenes comprised the largest group. According to Uhl, there are notable amounts of phenols in thyme, such as p-cymene (9.1–22.2%), linalool (2.2–4.8%), borneol (0.6–7.5%), α-pinene (0.9–6.6%), camphor (0–7.3%), carvacrol (0.4–20.6%), 1,8-cineole (0.2–14.2%), thymol (varying from 12–61%), and camphor (0–7.3%). Thyme's antioxidative qualities are primarily attributed to its main phenolic components, carvacrol, and thymol (Kowalczyk et al., 2020).

An excellent source of vitamins, thyme is exceptionally high in ascorbic acid (vitamin C) and retinol (vitamin A). Ascorbic acid promotes the immune system's ability to fight infections, whereas Vitamin A possesses antioxidant qualities. Furthermore, thyme contains a healthy dose of pyridoxine, also known as vitamin B6, which is believed to help lower stress. This herb also contains trace levels of folic acid, vitamin K, and vitamin E—essential minerals necessary for healthy growth (Halat et al., 2022).

17.4. PHARMACOLOGICAL APPLICATIONS OF *THYMUS*

17.4.1. Antimicrobial Properties

The antibacterial capabilities of *T. vulgaris* were examined by Sartoratto et al. using a bioautographic method. They used a microdilution technique to determine the essential oil's minimum inhibitory concentration (MIC) (Sartoratto et al., 2004). They found that *T. vulgaris* was efficient against *Salmonella cholerasuis* and *Enterococcus fecium.* Wirtu et al. (2024) also investigated the chemical compositions and antimicrobial characteristics of essential oils extracted from seven plants grown in Murcia, Spain. After assessing the vital oils' antibacterial activity against 10 pathogenic species, they concluded that *T. vulgaris* possesses antimicrobial qualities, indicating that it might be a source of antimicrobial agents for the food sector (Rota et al., 2008).

The antibacterial efficacy of *Thymus vulgaris* essential oil was assessed against five critical foodborne bacteria, including *Salmonella enterica* Enteritidis S64, *P. aeruginosa*, *E. coli*, *L. monocytogenes*, and *S. aureus*. According to the research, *T. vulgaris* was more effective against gram-positive bacteria (Galgano et al., 2023). Borugă et al. examined the chemical composition and antibacterial properties of *T. vulgaris* essential oil derived from Romanian plants in a different investigation (Nagoor et al., 2017). Their investigation by utilizing gas chromatography and GCMS revealed that the main ingredients were p-cymene (8.41%), γ-terpinene (30.90%), and thymol (47.59%). The disc diffusion method evaluated the antibacterial effectiveness against bacteria and fungi linked to food rotting. The findings showed that *T. vulgaris* essential oil has powerful antibacterial qualities (Borugă et al., 2014).

17.4.2. Antibacterial Properties

Thymus vulgaris's total polyphenol and flavonoid content was assessed for its antibacterial and antioxidant qualities. Chrysine, kaempferol, apigenin, luteolin, and quercetin were the primary flavonoids found in it (Diniz et al., 2023).

Asbaghian et al. compared the chemical constituents, antibacterial, and antioxidant actions of essential oils extracted from *Thymus vulgaris*, *Thymus kotschyanus*, and *Thymus caucasicus* aerial sections. The broth dilution method was employed to evaluate the antibacterial efficacy. The essential oil of *T. vulgaris* showed vigorous biological activity against *Streptococcus faecalis* at 25 μg/ml and *E. coli* at 12.5 μg/ml. *Staphylococcus aureus* demonstrated the least sensitivity of all the gram-positive bacteria, needing 100 μg/ml to inhibit. Limited efficacy was also shown by *T. vulgaris* oil against *Salmonella typhi*, a gram-negative bacterium (Asbaghian et al., 2011).

Using the disc diffusion method, Kon et al. evaluated the antibacterial qualities of *T. vulgaris* essential oil against *S. aureus* and *E. coli*, both alone and in combination with 34 other essential oils. A fractional inhibitory concentration (FIC) index of 0.26 suggested that the essential oils of *T. vulgaris* and *Cinnamomum zeylonicum* worked in concert (Kon et al., 2012). Al-Balushi et al. (2013) also looked into the cytotoxic and antibacterial effect of petroleum ether, chloroform, and hydroalcoholic extracts from Omani thymus leaves. The disc diffusion method assessed the antibacterial effect against *Klebsiella pneumoniae*, *P. aeruginosa*, *E. coli*, and *S. aureus*. The *T. vulgaris* leaf extracts produced potent effects, with inhibition zones ranging from 7–20 mm (Dobhal et al., 2023).

Fadil et al. used a combination of essential oils from *T. vulgaris* L., *Rosmarinus officinalis* L., and *Myrtus communis* L. to investigate the sensitivity of a strain of *Salmonella typhimurium*. The antibacterial properties of *T. vulgaris* essential oil were more potent than those of other oils. To increase *Salmonella typhimurium* sensitivity, they proposed a formulation that contains 55% essential oils of *T. vulgaris* L. and 45% essential oils of *Myrtus communis* L. (Fadil et al., 2018).

The vulnerability of Enterobacteriaceae that generate bla-extended-spectrum beta-lactamase (bla-ESBL) to *T. vulgaris* essential oil, both alone and in conjunction with cefotaxime, was investigated (Sateriale et al., 2023). Multi-drug resistant (MDR) strains were the ones under investigation (Akermi et al., 2022). The findings showed that thyme oil exhibited strong action against all MDR bacteria, including those that produce blaESBL (Nabavi et al., 2015). The MIC values and inhibition zones ranged from 2.87–11.5 μg/ml and 24–40 mm/10 μl, respectively. Furthermore, thyme oil demonstrated an additive impact against *Enterobacter cloacae* that produces ESBL (with a FICI of 0.987) and a synergistic effect against *E. coli* that produces blaSHV-12 (with a FICI of 0.28), when paired with cefotaxime (Benameur et al., 2019).

17.4.3. Antioxidant Properties

In response to environmental stressors and other situations, the body releases unstable molecules called free radicals. Free-radical scavengers or antioxidants are substances that can stop or lessen the harm that these molecules cause. Due to their ability to neutralize these dangerous chemicals and improve general health, antioxidants are frequently referred to as free-radical scavengers. P-cymene-2,3-diol (2,3-dihydroxy-4-isopropyl-1-methylbenzene), a phenolic molecule obtained from thyme's hexane extract, showed potent antioxidant activity that outperformed that of butylated hydroxyanisole and tocopherol (Vergun et al., 2022).

Lee et al. examined the chemical makeup and free radical–scavenging qualities of the basil and thyme leaves extracts. At a concentration of 5 μg/ml, they discovered that eugenol, thymol, carvacrol, and 4-allylphenol all had potent antioxidant properties, successfully preventing the oxidation of hexanal by almost 100% over 30 days. These compounds' antioxidant properties were on par with well-known antioxidants such as butylated hydroxytoluene and α-tocopherol. The antioxidant properties of essential oils from *T. vulgaris* (thyme) and *T. serpyllum* L. (wild thyme) were also examined by Kulisic et al., who found that both thyme and wild thyme essential oils exhibited strong antioxidant activity in vitro. Additionally, portions of these essential oils demonstrated antioxidant properties in the thiobarbituric acid reactive substance (TBARS) assay (Lee et al., 2005).

The antibacterial and antiradical qualities of essential oils derived from *Origanum vulgare* L. and *T. vulgaris* were examined by Vuković et al. After 30 and 60 minutes, they discovered that the essential oil of *T. vulgaris* had antiradical activity of 9.69 μl/ml and 5.84 μl/ml, respectively. This activity was higher than that of ascorbic acid but lower than that of butylated hydroxytoluene (BHT) (Kačániová et al., 2012).

Chizzola et al. (2008) investigated the antioxidant activity of thyme leaf extracts and commercial thyme that is high in thymol in a variety of extract types. They evaluated antioxidative activity using the rerric-reducing antioxidant power (FRAP) assay, DPPH decolorization, and the total phenolic content as determined by the Folin-Ciocalteu method; 60% ethanol was the extractant that produced the best results. High concentrations of phenolic chemicals, such as carvacrol and thymol, in essential oils showed exceptional antioxidant qualities.

Gedikoğlu et al. used hydro distillation (HD) and microwave-assisted extraction procedures to extract essential oils from *Thymbra spicata* (zahter) and *T. vulgaris* (thyme). They next examined the oils' antibacterial qualities, FRAP value, and free radical–scavenging activity. For both extraction techniques, thyme essential oil had a higher antioxidant capacity than zahter essential oil. Furthermore, it was discovered that the methanol extracts of thyme and zahter had a larger phenolic content than their ethanol counterparts. Two assays were used to assess antioxidant activity: FRAP and DPPH free radical–scavenging activity. Thyme essential oil generated by HD (159.59±12.79 µg/ml) has a substantially lower ($p < 0.05$) free radical–scavenging activity than that obtained by MAE (93.77±13.0 µg/ml). Assiri et al. reported that after 60 minutes of incubation, thyme oil exhibited high antiradical activity, successfully quenching 55% of galvinoxyl radicals and 65% of DPPH radicals (Gedikoğlu et al., 2019).

17.4.4. Antifungal Properties

Thymus oils are potent against fungal diseases because they include phenolic chemicals such as thymol, which can break down fungal cell membranes. According to Numpaque et al. (2011), their investigation into the antifungal effects of thymol showed alterations in the hyphae's structure, leading to hyphal aggregation, reduced hyphal diameters, and lysis of the hyphal wall.

The antifungal properties of essential oils extracted from different chemotypes of *T. vulgaris* against *C. albicans* were investigated by (Giordani et al., 2004). With an MIC 80% of 0.016 µl/ml, the essential oil of the thymol chemotypes of *T. vulgaris* was the most effective; this was mostly because of the high concentration of thymol (63.2%) (Soković et al., 2008).

The antifungal effects of *T. vulgaris* essential oil and its constituents against *Rhizopus oryzae* were investigated by Moghadam et al. (2023), who examined the antifungal properties of *T. vulgaris* L. essential oil and contrasted it with synthetic thymol in a different investigation concerning *Aspergillus niger*. The results showed that thymus oil dilutions at concentrations of 1, 1/2, and 1/4 had more potent antifungal effects than medicines such as gentamicin (8 mg/ml) and streptomycin sulphate (72% SP). At a 10% dilution, however, the synthetic thymol demonstrated strong action. Thymol, a naturally occurring monoterpene phenol, was largely responsible for the oil's exceptional antifungal qualities.

17.5. CYTOTOXICITY

The cytotoxic effects of essential oils from *Rosmarinus officinalis* L. and *T. vulgaris* L. on the human respiratory epithelial cell line A549 were assessed in vitro by (Miladi et al., 2013). They used the MTT colourimetric test to evaluate cytotoxicity. After 72 hours, their dose-dependent investigations showed that the oils had an IC50 of 8.50±0.01 µg/ml and 10.50±0.01 µg/ml, respectively. This suggests that these extracts were not cytotoxic to the A549 cell line at any of the tested dosages. Al-Balushi et al. (2013) examined the cytotoxic and antimicrobial qualities of petroleum ether, chloroform, and hydroalcoholic extracts of Omani thyme leaves in a different study. To evaluate cytotoxic action, they used a brine shrimp test and discovered that, at increasing concentrations, the petroleum ether and chloroform extracts killed almost all of the shrimp larvae.

17.6. ANTISPASMODIC ACTIVITY

T. vulgaris L. (Lamiaceae) is a recognized medicinal herb with secretomotor and broncholytic properties. Because of their low quantities, phenolic compounds may only have a limited impact on the antispasmodic actions of smooth muscles. The possible spasmolytic effects of flavonoids, such as luteolin, apigenin, luteolin glycosides, cirsilineol, 8-methoxycirsilineol, and thymol, were also examined.

Begrow et al. (2010) investigated the effects of thymol and carvacrol separately, as well as a variety of thyme extract formulations, including those with normal and very low quantities of these compounds. They assessed mucociliary clearance in vivo and performed functional investigations on the ileum and trachea's contraction and relaxation. Regardless of the kind of stimulation (acetylcholine, K+, or Ba++), their results showed that thymol and carvacrol successfully affected the smooth muscles in the rat ileum and trachea. However, when stimulated by Ba++ or endothelin-1, extracts of thyme with low thymol content reduced contractions by around 40%, suggesting that thymol is not the primary component causing these effects.

17.7. ANTIDERMATOPHYTIC ACTIVITY

Soković et al. (2008) examined the effects of thymus oil on human pathogenic fungi, such as *Trichophyton mentagrophytes*, *Trichophyton rubrum*, and *Trichophyton tonsurans*. Together with the brand-name medication bifonazole, they evaluated the therapeutic efficacy of a 1% thymus vulgaris essential oil and thymol solution. Following a 37-day observation period, all of the animals received successful treatment.

Significant antidermatophytic properties of *T. vulgaris* essential oil were shown by Jain et al. against a variety of filamentous fungus, including *Microsporum fulvum*, *Microsporum gypseum*, *Trichophyton mentagrophytes*, *Trichophyton rubrum*, *Trichophyton tonsurans*, and *T. soudanense*. The significant amounts of phenolic chemicals found may be responsible for the noteworthy antifungal activities against the chosen pathogenic fungus.

17.8. CARDIOVASCULAR TREATMENT

Ramchoun et al. (2012) examined the properties of *T. vulgaris* aqueous extracts on hypercholesterolaemia, hypotriglyceridemia, and their antioxidant qualities. The FRAP assay, antioxidant property, and measurements of the inhibition of oxidative erythrocyte hemolysis caused by 2,2′-azobis (2-amidinopropane) hydrochloride were among the techniques used to assess the antioxidant activity of the polyphenol-rich extracts. Triton WR-1339 was injected intraperitoneally at a dose of 200 mg/kg body weight to cause hyperlipidemia in rats. The plasma levels of triglycerides and total cholesterol did not significantly alter following a 24-hour treatment with the polyphenol-rich extract of *T. vulgaris.*

17.9. ANTIVIRAL ACTION

Thyme and its bioactive constituents have the potential to defend against the harmful effects of COVID-19 and avoid its complications. At the end of 2021, a double-blind, randomized clinical research in the U.S. found that inhaling essential oils from frankincense, clove bud, orange, and thyme increased levels of lymphocytes in female patients recuperating from COVID-19. In vitro research on the efficacy of thymol essential oil (TEO) against different RNA viruses, including the coronavirus, concentrated on how well it worked in preventing feline infectious peritonitis in cats. TEO at 27 μg/ml and 270 μg/ml effectively inhibited the coronavirus's transcription. Carvacrol is the main ingredient that gives TEO its antiviral properties in solutions and the air, and it also demonstrated viricidal efficacy against airborne coronaviruses. According to the study, TEO could render 99.99% of airborne viruses inactive, indicating great promise for usage in subsequent applications.

17.10. ANTICANCER ACTIVITY

Worldwide, cancer is observed as a major threat. With around 10 million fatalities in 2020, it is one of the major causes of death globally. Its incidence is predicted to raise by 47% by 2040, making it a growing hazard. High cytotoxicity to both the tumour and healthy tissues was demonstrated by the chemotherapeutic chemotherapy drugs that were previously utilized to treat cancer patients. Because medicinal plants and their phytochemicals have a lesser risk of side effects than conventional chemotherapy drugs, investigations for new treatment possibilities have been suggested. Numerous studies have shown the advantages of many medicinal plant extracts, as well as their capacity to damage cancer cells by a variety of molecular pathways, such as necroptosis, autophagy, and intrinsic or extrinsic caspase-dependent apoptosis. The monoterpenoid phenols carvacrol and thymol, which are among the many chemicals found in thyme, have pharmacological characteristics that make them highly promising for use in cancer treatment. In lung, breast, and prostate cancer cell lines, thymol showed proapoptotic and antiproliferative characteristics; as a result, it could potentially serve as a therapeutic agent (Elbe et al., 2020). The effectiveness of thymol in bladder cancer cells was examined by Li et al. (2017), who found that it suppressed cell proliferation in a dose- and time-dependent manner. Through the intrinsic route, caspase-3/9, JNK, and p38 activation, cytochrome C release, and downregulation of antiapoptotic Bcl-2 family proteins, thymol caused cell cycle arrest at the G2/M phase, ROS production, and death. According to Jung et al. (2018), carvacrol can prevent cell migration and proliferation in nonsmall cell lung cancer (NSCLC) by suppressing the expression of the tyrosine kinase receptor (AXL) and by preventing AXL from becoming phosphorylated in response to ligand stimulation.

17.11. FUTURE PERSPECTIVES

A reliable method of getting them to the disease's target cells must be discovered for both natural and synthetic drugs to have their therapeutic effects. Due to the irritative impacts, TEOs and their constituents may be lost while interacting with physiological enzymes. By mixing with natural and synthetic polymers, microfibers, etc., TEO can be consumed. For instance, thymol and lignosulfonate were found to have outstanding biocompatibility. Like thymol, lignosulfonate is an antioxidant, which produces a synergistic effect. Both sustained-release and tailored drug delivery methods for high lipophilic medicines exploit this combination in microvesicles. Thymol exhibits notable suppression against *S. aureus* and *E. coli* when incorporated into a porous membrane composed of cellulose acetate. Future materials for wound healing may benefit from this method. Propylene glycol vesicles and glycerol liposomes that contain TEO have also shown promise against *S. mutans*. Methylcellulose microspheres may mask obnoxious taste and improve poor absorption. Thymol has been shown in previous studies to be a successful chronic wound healer. Treatment results for chronic wounds may be enhanced by the addition of targeted-release antibacterial drugs to the dressing. A novel dressing technique has been developed that incorporates electrospun polycaprolactone microfibers with mesoporous silicon dioxide loaded with thymol. Consequently, more studies and developments in the advancement of therapeutically active microvesicles are anticipated.

17.12. CONCLUSION

It has been established that *T. vulgaris* contains many biological substances with various pharmacological actions, such as antiviral, antibacterial, antioxidant, antifungal, anti-inflammatory, antispasmodic, cytotoxic, and anticarcinogenic qualities. The oil also helps fight off some illnesses and strengthen the immune system. Often used to flavour liqueurs and herbal teas, thyme is also used to season seafood, poultry, vegetables, and soups. There is a lot of room for more research because of its many medical uses, especially in examining the cytotoxic and anticancer effects of essential oil and aqueous extracts. It's critical to precisely calculate the right dosages of various plant fractions and evaluate their toxicological effects. Effective therapeutic agents may be developed by using *T. vulgaris* L.'s chemical constituents and their combinations to treat a range of ailments.

REFERENCES

Abdi-Moghadam, Zohreh, Yeganeh Mazaheri, Alieh Rezagholizade-Shirvan, Maryam Mahmoudzadeh, Mansour Sarafraz, Mahnaz Mohtashami, Samira Shokri et al. "The significance of essential oils and their antifungal properties in the food industry: A systematic review." *Heliyon* (2023). https://doi.org/10.1016/j.heliyon.2023.e21386

Akermi, Sarra, Slim Smaoui, Mariam Fourati, Khaoula Elhadef, Moufida Chaari, Ahlem Chakchouk Mtibaa, and Lotfi Mellouli. "In-depth study of thymus vulgaris essential oil: Towards understanding the antibacterial target mechanism and toxicological and pharmacological aspects." *BioMed Research International* 2022, no. 1 (2022): 3368883. https://doi.org/10.1155/2022/3368883

Al-Balushi, Aisha Haroon, Maryam Aljabri, Md Sohail Akhtar, Sadri Said, Afaf Weli, Qasim Al-Riyami, and Ahlam A. AlAbri. "Antibacterial and cytotoxic activities of thymus vulgaris leaves grown in Oman." *International Journal of Pharmaceutical Sciences and Research* 4, no. 11 (2013): 4253. http://doi.org/10.13040/IJPSR.0975-8232.4(11).4253-57

Asbaghian, Shiva, Ali Shafaghat, Khalil Zarea, Fakhraddin Kasimov, and Farshid Salimi. "Comparison of volatile constituents, and antioxidant and antibacterial activities of the essential oils of thymus caucasicus, T. kotschyanus and T. vulgaris." *Natural Product Communications* 6, no. 1 (2011): 1934578X1100600133.

Begrow, Frank, Jonas Engelbertz, Björn Feistel, Romanus Lehnfeld, Katrin Bauer, and Eugen J. Verspohl. "Impact of thymol in thyme extracts on their antispasmodic action and ciliary clearance." *Planta Medica* 76, no. 4 (2010): 311–318. https://doi.org/10.1055/s-0029-1186179

Benameur, Qada, Teresa Gervasi, Vito Pellizzeri, Mária Pľuchtová, Hassiba Tali-Maama, Farida Assaous, Badia Guettou et al. "Antibacterial activity of thymus vulgaris essential oil alone and in combination with cefotaxime against bla ESBL producing multidrug resistant enterobacteriaceae isolates." *Natural Product Research* 33, no. 18 (2019): 2647–2654. https://doi.org/10.1080/14786419.2018.1466124

Borugă, O., C. Jianu, C. Mişcă, I. Goleţ, A. T. Gruia, and F. G. Horhat. "Thymus vulgaris essential oil: Chemical composition and antimicrobial activity." *Journal of Medicine and Life* 7, no. Spec Iss 3 (2014): 56. PMCID: PMC4391421 PMID: 25870697

Chizzola, Remigius, Hanneliese Michitsch, and Chlodwig Franz. "Antioxidative properties of thymus vulgaris leaves: Comparison of different extracts and essential oil chemotypes." *Journal of Agricultural and Food Chemistry* 56, no. 16 (2008): 6897–6904. https://pubs.acs.org/doi/10.1021/jf800617g

Dauqan, Eqbal M. A., and Aminah Abdullah. "Medicinal and functional values of thyme (thymus vulgaris L.) herb." *Journal of Applied Biology & Biotechnology* 5, no. 2 (2017): 17–22. http://www.jabonline.in/admin/php/uploads/188_pdf.pdf

Diniz, A. F., B. Santos, L. M. M. O. Nóbrega, V. R. L. Santos, W. S. Mariz, P. S. C. Cruz, R. O. Nóbrega et al. "Antibacterial activity of thymus vulgaris (thyme) essential oil against strains of pseudomonas aeruginosa, Klebsiella pneumoniae and staphylococcus saprophyticus isolated from meat product." *Brazilian Journal of Biology* 83 (2023): e275306. https://doi.org/10.1590/1519-6984.275306

Dobhal, Kiran, Shalu Verma, Alka Singh, Ruchika Garg, and Vikash Jakhmola. "Exploration of the pharmacological aspects of Thymus Vulgaris Essential Oil (TEO)." In *Pharmacological Aspects of Essential Oils*, pp. 187–198, 2023. https://doi.org/10.1201/9781003389774

Elbe, H., G. Yigitturk, T. Cavusoglu, Y. Uyanikgil, and F. Ozturk. "Apoptotic effects of thymol, a novel monoterpene phenol, on different types of cancer." *Bratislava Medical Journal/Bratislavske Lekarske Listy* 121, no. 2 (2020). https://doi.org/10.4149/BLL_2020_016

Fadil, Mouhcine, Kawtar Fikri-Benbrahim, Saad Rachiq, Bouchaib Ihssane, Sara Lebrazi, Marwa Chraibi, Taoufik Haloui, and Abdellah Farah. "Combined treatment of thymus vulgaris L., Rosmarinus officinalis L. and Myrtus communis L. essential oils against Salmonella typhimurium: Optimization of antibacterial activity by mixture design methodology." *European Journal of Pharmaceutics and Biopharmaceutics* 126 (2018): 211–220. https://doi.org/10.1016/j.ejpb.2017.06.002

Galgano, Michela, Francesco Pellegrini, Daniela Mrenoshki, Paolo Capozza, Ahmed Hassan Omar, Anna Salvaggiulo, Michele Camero et al. "Assessing contact time and concentration of thymus vulgaris essential oil on antibacterial efficacy in vitro." *Antibiotics* 12, no. 7 (2023): 1129. https://doi.org/10.3390/antibiotics12071129

Gedikoğlu, Ayça, Münevver Sökmen, and Ayşe Çivit. "Evaluation of thymus vulgaris and thymbra spicata essential oils and plant extracts for chemical composition, antioxidant, and antimicrobial properties." *Food Science & Nutrition* 7, no. 5 (2019): 1704–1714. PMCID: PMC6526640 PMID: 31139383

Giordani, Regli, P. Regli, J. Kaloustian, C. Mikail, L. Abou, and H. Portugal. "Antifungal effect of various essential oils against Candidaalbicans: Potentiation of antifungal action of amphotericin B by essential oil from thymus vulgaris." *Phytotherapy Research* 18, no. 12 (2004): 990–995. https://doi.org/10.1002/ptr.1594.

Hammoudi Halat, Dalal, Maha Krayem, Sanaa Khaled, and Samar Younes. "A focused insight into thyme: Biological, chemical, and therapeutic properties of an indigenous mediterranean herb." *Nutrients* 14, no. 10 (2022): 2104. https://doi.org/10.3390/nu14102104

Heidari, Zahra, Ali Salehzadeh, Seyed Ataollah Sadat Shandiz, and Sara Tajdoost. "Anti-cancer and anti-oxidant properties of ethanolic leaf extract of thymus vulgaris and its bio-functionalized silver nanoparticles." *3 Biotech* 8 (2018): 1–14. https://doi.org/10.1007/s13205-018-1199-x

Hosseinzadeh, Saleh, Azizollah Jafarikukhdan, Ahmadreza Hosseini, and Raham Armand. "The application of medicinal plants in traditional and modern medicine: A review of thymus vulgaris." *International Journal of Clinical Medicine* 6, no. 9 (2015): 635–642. http://doi.org/10.4236/ijcm.2015.69084

Imelouane, B., H. Amhamdi, Jean-Paul Wathelet, M. Ankit, Khadija Khedid, and A. El Bachiri. "Chemical composition and antimicrobial activity of essential oil of thyme (thymus vulgaris) from Eastern Morocco." *International Journal of Agriculture and Biology* 11, no. 2 (2009): 205–208. https://hdl.handle.net/2268/16021

Jain, Neetu, and Poonam Choudhary. "Phytochemistry, traditional uses and pharmacological aspect of thymus vulgaris: A review." *Indian Journal of Pharmaceutical Sciences* 84, no. 6 (2022): 1369–1379. https://doi.org/10.36468/pharmaceutical-sciences.1035

Jung, Chi Young, So-Young Kim, and Chuhee Lee. "Carvacrol targets AXL to inhibit cell proliferation and migration in non-small cell lung cancer cells." *Anticancer Research* 38, no. 1 (2018): 279–286. https://doi.org/10.21873/anticanres.12219

Kačániová M, Vukovič N, Hleba L, Bobková A, Pavelková A, Rovná K, Arpášová H. "Antimicrobial and antiradicals activity of origanum vulgare L. and thymus vulgaris essential oils." *Journal of Microbiology, Biotechnology and Food Sciences.* 2, no. 1 (2012): 263–271. https://office2.jmbfs.org/index.php/JMBFS/article/view/7204

Kon, K., and M. Rai. "Antibacterial activity of thymus vulgaris essential oil alone and in combination with other essential oils". (2012): 50–56. https://repo.knmu.edu.ua/handle/123456789/867

Kowalczyk, Adam, Martyna Przychodna, Sylwia Sopata, Agnieszka Bodalska, and Izabela Fecka. "Thymol and thyme essential oil—New insights into selected therapeutic applications." *Molecules* 25, no. 18 (2020): 4125. https://doi.org/10.3390/molecules25184125

Lee, Seung-Joo, Katumi Umano, Takayuki Shibamoto, and Kwang-Geun Lee. "Identification of volatile components in basil (Ocimum basilicum L.) and thyme leaves (thymus vulgaris L.) and their antioxidant properties." *Food Chemistry* 91, no. 1 (2005): 131–137. https://doi.org/10.1016/j.foodchem.2004.05.056

Li, Jing, Jia Li, Guowei Cai, Lin Shen, and Furong Lu. "Proapoptotic and growth-inhibitory effects of plumbagin on human gastric cancer cells via suppression of signal transducer and activator of transcription 3 and protein kinase B." *Alternative Therapies in Health & Medicine* 23, no. 7 (2017).

Miladi, H., R. Ben Slama, D. Mili, S. Zouari, A. Bakhrouf, and E. Ammar. "Essential oil of thymus vulgaris L. and rosmarinus officinalis L.: Gas chromatography-mass spectrometry analysis, cytotoxicity and antioxidant properties and antibacterial activities against foodborne pathogens." (2013): 729–739.https://doi.org/10.4236/ns.2013.56090

Nabavi, Seyed Mohammad, Anna Marchese, Morteza Izadi, Valeria Curti, Maria Daglia, and Seyed Fazel Nabavi. "Plants belonging to the genus thymus as antibacterial agents: From farm to pharmacy." *Food Chemistry* 173 (2015): 339–347. https://doi.org/10.1016/j.foodchem.2014.10.042

Nagoor Meeran, Mohamed Fizur, Hayate Javed, Hasan Al Taee, Sheikh Azimullah, and Shreesh K. Ojha. "Pharmacological properties and molecular mechanisms of thymol: Prospects for its therapeutic potential and pharmaceutical development." *Frontiers in Pharmacology* 8 (2017): 380. https://doi.org/10.3389/fphar.2017.00380

Numpaque, Manuel A., Luis A. Oviedo, Jesús H. Gil, Carlos M. García, and Diego L. Durango. "Thymol and carvacrol: Biotransformation and antifungal activity against the plant pathogenic fungi colletotrichum acutatum and botryodiplodia theobromae." *Tropical Plant Pathology* 36 (2011): 3–13. https://doi.org/10.1590/S1982-56762011000100001

Prasanth Reddy, V., K. Ravi Vital, P. V. Varsha, and S. Satyam. "Review on thymus vulgaris traditional uses and pharmacological properties." *Med Aromat Plants* 3, no. 164 (2014): 2167–0412. http://doi.org/10.4172/2167-0412.1000167

Ramchoun, Mhamed, Hicham Harnafi, Chakib Alem, Berthold Büchele, Thomas Simmet, Mustapha Rouis, Fouad Atmani, and Souliman Amrani. "Hypolipidemic and antioxidant effect of polyphenol-rich extracts from moroccan thyme varieties." *e-SPEN Journal* 7, no. 3 (2012): e119–e124. https://doi.org/10.1016/j.clnme.2012.02.005

Roby, Mohamed Hussein Hamdy, Mohamed Atef Sarhan, Khaled Abdel-Hamed Selim, and Khalel Ibrahim Khalel. "Evaluation of antioxidant activity, total phenols and phenolic compounds in thyme (Thymus vulgaris L.), sage (Salvia officinalis L.), and marjoram (Origanum majorana L.) extracts." *Industrial Crops and Products* 43 (2013): 827–831. https://doi.org/10.1016/j.indcrop.2012.08.029

Rota, María C., Antonio Herrera, Rosa M. Martínez, Jose A. Sotomayor, and María J. Jordán. "Antimicrobial activity and chemical composition of thymus vulgaris, thymus zygis and thymus hyemalis essential oils." *Food Control* 19, no. 7 (2008): 681–687. https://doi.org/10.1016/j.foodcont.2007.07.007

Saad, Bashar, Hassan Azaizeh, and Omar Said. "Tradition and perspectives of Arab herbal medicine: A review." *Evidence-Based Complementary and Alternative Medicine* 2, no. 4 (2005): 475–479. https://doi.org/10.1093/ecam/neh133

Salehi, Bahare, Abhay Prakash Mishra, Ila Shukla, Mehdi Sharifi-Rad, María del Mar Contreras, Antonio Segura-Carretero, Hannane Fathi, Nafiseh Nasri Nasrabadi, Farzad Kobarfard, and Javad Sharifi-Rad. "Thymol, thyme, and other plant sources: Health and potential uses." *Phytotherapy Research* 32, no. 9 (2018): 1688–1706. https://doi.org/10.1002/ptr.6109

Sateriale, Daniela, Giuseppina Forgione, Giuseppa Anna De Cristofaro, Chiara Pagliuca, Roberta Colicchio, Paola Salvatore, Marina Paolucci, and Caterina Pagliarulo. "Antibacterial and antibiofilm efficacy of thyme (thymus vulgaris L.) essential oil against foodborne illness pathogens, Salmonella enterica subsp. enterica serovar typhimurium and bacillus cereus." *Antibiotics* 12, no. 3 (2023): 485. https://doi.org/10.3390/antibiotics12030485

Sartoratto, Adilson, Ana Lúcia M. Machado, Camila Delarmelina, Glyn Mara Figueira, Marta Cristina T. Duarte, and Vera Lúcia G. Rehder. "Composition and antimicrobial activity of essential oils from aromatic plants used in Brazil." *Brazilian Journal of Microbiology* 35 (2004): 275–280. https://doi.org/10.1590/S1517-83822004000300001

Soković, Marina, J. Glamočlija, Ana Ćirić, D. Kataranovski, Petar D. Marin, Jelena Vukojević, and Dejan Brkić. "Antifungal activity of the essential oil of thymus vulgaris L. and thymol on experimentally induced dermatomycoses." *Drug Development and Industrial Pharmacy* 34, no. 12 (2008): 1388–1393. https://doi.org/10.1080/03639040802130053

Soleimani, Mohsen, Ahmad Arzani, Vida Arzani, and Thomas H. Roberts. "Phenolic compounds and antimicrobial properties of mint and thyme." *Journal of Herbal Medicine* 36 (2022): 100604. https://doi.org/10.1016/j.hermed.2022.100604

Stahl-Biskup, Elisabeth, and Rimantas P. Venskutonis. "Thyme." In *Handbook of Herbs and Spices*, pp. 499–525. Woodhead Publishing, 2012. https://doi.org/10.1533/9780857095671.499

Van Den Broucke, C. O., and J. A. Lemli. "Pharmacological and chemical investigation of thyme liquid extracts." *Planta Medica* 41, no. 02 (1981): 129–135. 10.1055/s-2007-971689

Vergun, Olena, Liudmyla Svydenko, Olga Grygorieva, Vladimíra Horčinová Sedláčková, Katarína Fatrcová Šramková, Eva Ivanišová, and Ján Brindza. "Polyphenol component and antioxidant activity of thymus spp." *Slovak Journal of Food Sciences* 16 (2022). https://doi.org/10.5219/1715

Wirtu, Shuma Fayera, Krishnaraj Ramaswamy, Rahul Maitra, Sidharth Chopra, Ashutosh Kumar Mishra, and Leta Tesfaye Jule. "Isolation, characterization and antimicrobial activity study of thymus vulgaris." *Scientific Reports* 14, no. 1 (2024): 21573. https://doi.org/10.1038/s41598-024-71012-2

18 Unveiling the Secrets of Nutmeg

A Deep Dive into Its Phytochemistry and Pharmacology

Reshma P, Shaik Nazir Basha, and Reddappa JB

18.1. INTRODUCTION

Spices are naturally occurring plant materials, such as seeds, fruits, roots, bark, or other materials, mainly used to flavour, colour, or preserve food. Spice plants have historically been considered essential human survival and social development resources. There is historical evidence that spices were used to prevent diseases (Pandey et al., 2016). Spices are widely used in daily life, both in cooking and for household medical needs. Medical professionals are increasingly turning to natural products due to the rise in lifestyle disorders in the modern era (Wink, 2022). The World Health Organization reports that about 80% of the global population uses herbs for healthcare (Jamshidi-Kia et al., 2018). Although many spices in the kitchen have therapeutic characteristics, herbal remedies are safe and effective (Yadav and Agarwala, 2011). Various spices, such as black cumin (*Nigella sativa*), fennel (*Foeniculum vulgare*), pepper (*Piper nigrum*), ginger (*Zingiber officinale*), and turmeric (*Curcuma longa*), are found to be widely used across the globe to treat numerous health conditions. A highly valued medicinal plant, *Myristica fragrans*, known as nutmeg, has numerous therapeutic benefits for treating many ailments (Nguyen et al., 2021).

Nutmeg, botanically *Myristica fragrans* (Houtt.), is a perennial evergreen tree belonging to the Myristicaceae family. Members of this plant family are indigenous to Asia, Africa, the Pacific Islands, and the Americas (Kitamura and Poonswad, 2013). It is even known as the nutmeg family for one of its notable contributors, *Myristica fragrans*, which offers well-known and vital spices such as nutmeg (seed) and mace (fleshy aril), which might be generally used in culinary traditions. *Myristica fragrans* is known by different names worldwide. It is called *jaiphal* in Hindi, *pala* in Indonesia, *josat at-Tib* in Arabic, *jaifal* in Urdu, *muscade* in French, *moschokarido* in Greek, and *roudoukou* in Chinese (Naeem et al., 2016). With the ever-growing thirst for natural products, nutmeg is quite interesting because it is a precious source of bioactive compounds in various medical and pharmaceutical applications.

18.2. TAXONOMY AND BOTANY

The nutmeg tree originates from the Banda Islands, located in the Moluccas region of eastern Indonesia (Peter, 2001), and it was discovered by the Portuguese in 1512. The Dutch played a significant role in spreading the value of nutmeg. The name 'nutmeg' is derived from the Latin term *nux muscatus*, which means 'musky nut' (Gupta and Rajpurohit, 2011; Naeem et al., 2016). Well-drained soil rich in organic matter and high in fertility is necessary for growing nutmeg (Orwa et al., 2009). It flourishes in well-drained soils, including sandy loam, clayey loam, and laterite-rich soils, with an optimal pH range of 6.5–7.5. The best growing condition of this crop has been found between

DOI: 10.1201/9781003559894-18

2000–3500 mm of rainfall (Naeem et al., 2016). *M. fragrans* grows in warm weather, preferring 18–34°C, although it can tolerate 12–38°C (Saputro et al., 2016).

As the name implies, they are dominated by evergreen trees of the Myristicaceae family, particularly the *Myristica* genus, which account for 32% of the total number of trees in the ecosystem (Chandran and Mesta, 2001). The Myristicaceae family is distributed across tropical regions, mainly inhabiting the dense forests of Asia, Africa, Madagascar, South America, and Polynesia. *Myristica fragrans* is widely cultivated in China, Indonesia, Taiwan, Malaysia, India, Grenada, South America, and Sri Lanka (Naeem et al., 2016). The classification details of *Myristica fragrans* are presented in Table 18.1 (Kumari et al., 2021; Sultan et al., 2023).

M. fragrans is a fragrant evergreen tree that can reach a height of 5–13 meters and occasionally 20 meters. Plants are distinguished by their smooth, greyish brown bark that emits a watery red or pink sap (Sultana et al., 2018). The branches are distributed in whorls. The pointed, alternating dark green leaves (5–15 cm x 2–7 cm) have a glossy upper surface and are distributed along the branches on 1-cm-long stems. The flowers are single-sexed, bell-shaped, waxy, and pale yellow. Male flowers are 5–7 mm long and occur in groups of 1 to 10, while female flowers are 1 cm long and occur in bunches of 1 to 3 flowers per group. Nutmeg cross-pollinates due to insufficient blossoms of both sexes on a single tree (Naeem et al., 2016). Nutmeg trees exhibit five distinct sex forms: exclusively male, exclusively female, bisexual female, bisexual male, and hermaphroditic (Krishnamoorthy, 2000).

Nutmeg fruit typically measures between 6–9 cm in length and has a smooth, drooping, fleshy, yellow exterior (Asgarpanah and Kazemivash, 2012). The pericarp flesh splits into two halves at maturity (Tripathi and Dwivedi, 2015). Nutmeg seeds are fleshy, firm, dark brown, and rich in oil. Young nutmeg seeds have wrinkles and are tender when pressed. After harvesting, they reveal greenish white mace and seeds with tender shells. They have high moisture and shrink and wrinkle after drying (Siregar and Murtiningsih, 1996). The mace, a smooth-surfaced aril, comprehensively covers the nut. The mace of nutmeg is reddish brown in colour, irregularly shaped, delicate, and wrinkled upon touch (Saputro et al., 2016). The botany of nutmeg is presented in Figure 18.1.

18.3. USES AND NUTRITIONAL COMPOSITION

Nutmeg, a versatile spice, provides farmers with a stable income and has been integral to cooking and medicine for centuries. It enhances soups, meats, and vegetables and pairs with spices such as white pepper, cloves, ginger, and cinnamon. Once discarded, its pericarp is now used in value-added products for dishes such as custards, pies, soups, sauces, egg-based meals, jams, pickles, syrups, and squash. Nutmeg kernels, either milled or freshly grated, are added to food at the final stage, while mace, the seed's lacy covering, adds a saffron-like hue to light-coloured dishes. Beyond culinary use, nutmeg features in cosmetics, balms, and medicinal syrups (Naeem et al., 2016).

TABLE 18.1
Taxonomic Description of *Myristica fragrans*

Taxonomic level	Taxon
Kingdom	Plantae
Sub kingdom	Viridiplantae
Division	Tracheophyta
Class	Magnoliopsida
Order	Magnoliales
Family	Myristicaceae
Genus	Myristica
Species	Fragrans
Common name	Nutmeg
Hindi/Urdu name	Jaiphal

FIGURE 18.1 Morphology of *M. fragrans* Houtt.: (a) nutmeg tree; (b) fruits; (c) split fruit showing mace (aril) and kernel (nutmeg); (d) mace; (e) kernel.

Nutmeg contains chemical compounds such as alkaloids, phlorotannins, anthraquinones, cardiac glycosides, and flavonoids, which contribute to its antioxidant, health-promoting, and disease-preventing properties (Ginting et al., 2017). Anthraquinones act as natural laxatives, aiding constipation, slow digestion, and piles treatment (Lawless, 1995; Suvachananonda, 2007; Kumari et al., 2021). Its antimicrobial flavonoids help treat skin issues such as eczema, scabies, and blotches (Osemwegie et al., 2017) and may alleviate tuberculosis, colds, fever, and respiratory problems. Cardiac glycosides enhance heart function, aiding conditions such as heart failure, paralysis, poor circulation, fatigue, muscle aches, and neuralgia (Agbogidi and Azagbaekwe, 2013).

Nutmeg essential oil, used in products such as camphor, perfumes, and synthetic pine oil, contains curative compounds with antidepressant, antifungal, digestive, and aphrodisiac properties. Elemicin and myristicin provide calming effects, eugenol relieves toothaches, and nutmeg oil soothes pain. Decoctions with honey treat gastritis and nausea, while whole kernels enhance flavour. Mace lignin has the ability to function as a skin-lightening component (Naeem et al., 2016). Nutmeg butter finds application in fragrances, tobacco products, toothpaste, and digestive treatments. *M. fragrans* seeds aid in treating diarrhea, mouth sores, and insomnia, with Malabaricon C inhibiting microorganisms (Shinohara et al., 1999). Nutmeg seed oil reduces *Listeria monocytogenes* infection by inhibiting listeriolysin O (Smith-Palmer et al., 2002). Myristicin, its main aromatic component, supports medicinal uses as an analgesic, stomachic, digestive, hypnotic, aphrodisiac, and amenorrheal agent (Chatterjee et al., 2015).

Nutmeg contains essential macronutrients, including carbohydrates, fibre, and energy-yielding compounds. It provides vital fat-soluble vitamins such as A and E, and minerals such as magnesium, copper, iron, zinc, calcium and potassium (Agbogidi and Azagbaekwe, 2013). Its composition includes 26.73% fat, 18.67% protein, and 28.93% carbohydrates, with a high energy value of 3938.33 Kcal/kg (Okiki et al., 2023). Nutmeg's bioactive compounds, such as alkaloids, phlobatannins, anthraquinones, cardiac glycosides, and flavonoids, enhance its antioxidant and medicinal properties (Akinduko et al., 2024). It is rich in magnesium (403.6 ppm) and low in lead (95.3 ppm).

18.4. PHYTOCHEMICAL PROFILE

The primary metabolites in nutmeg, including carbohydrates, proteins, and fatty acids (or nutmeg butter), make up about 80% of the dried nutmeg kernel weight. The remaining 20% comprises secondary metabolites, such as essential oils and phenolic compounds. The distinct scent and flavour of nutmeg come from the essential oil found in both the kernel and the mace, which makes it a unique culinary ingredient. Nutmeg contains nearly 16% essential oil and is rich in monoterpenes and phenylpropanoids. Mace, processed separately, has lower fat and carbohydrate content than nutmeg (Abourashed and Alfy, 2016). Phytochemical analysis revealed a higher abundance of steroids, while terpenoids and saponins were in moderate amounts. In contrast, tannins, alkaloids, and phenols were identified in smaller quantities. Notably, anthraquinones, anthocyanins, and glycosides were absent. Quantitative results indicated terpenoid content at 19.00 ± 3.18%, saponins at 12.04 ± 3.33%, and steroids at 32.75 ± 5.42%. Meanwhile, phenols and alkaloids were detected in lower concentrations, measured at 0.46 ± 0.10% and 2.75 ± 0.82%, respectively (Nkwocha et al., 2018).

The essential oil extracted from *M. fragrans* (MFEO), derived from its leaves, mace, kernels, and seeds, is extensively utilized in Ayurveda, aromatherapy, and perfumery. This oil is a clear to pale yellow liquid with a distinct spicy fragrance (Nikolic et al., 2021). Various extraction methods, including hydrodistillation, steam distillation, supercritical fluid extraction, microwave-assisted method, and ultrasound-assisted approaches, have been employed to extract MFEO (Azwanida, 2015; Ashokkumar et al., 2022). The extraction yield of MFEO varied depending on the extraction method and plant part used, ranging from 0.7–3.2% in leaves, 8.1–10.3% in mace, 0.3–12.5% in seeds, and 6.2–7.6% in kernels (Ashokkumar et al., 2022). Among these techniques, classical hydrodistillation using the Clevenger apparatus remains the most commonly used method for estimating volatile oils.

18.4.1. MFEO Chemical Composition

Nutmeg's distinct fragrance comes from its essential oil, which is composed of various terpenes, including α-pinene, sabinene, p-cymene, myrcene, camphene, and γ-terpinene, along with terpene derivatives, such as terpinol, geraniol, and linalool. Additionally, it contains phenylpropanoids such as myristicin, safrole, and elemicin (Widelski and Kukula-Koch, 2017). The essential oil is typically analyzed using gas chromatography-mass spectrometry (GC-MS) techniques (Al-Maskri et al., 2011; Ahmad et al., 2008; Arshad et al., 2014; Hanif et al., 2011). International organizations such as the European Pharmacopoeia (EP), British Pharmacopoeia (BP), the International Organization for Standardization (ISO), and the fragrance and flavour industry oversee the quality standards for nutmeg essential oil. The standard composition comprises nine essential chemical markers: α-pinene, β-pinene, sabinene, limonene, 3-carene, safrole, γ-terpinene, 4-terpineol, and myristicin. The flavour and fragrance industry also specifies limits for eugenol, methyl eugenol, isoeugenol, and elemicin. Notably, methyl eugenol and safrole are recognized as carcinogenic compounds, making them critical parameters for assessing nutmeg oil quality (Riyadi et al., 2014). Furthermore, eugenol, limonene, and isoeugenol are known allergens (IFRA, 2007).

The primary components of MFEO include monoterpenes such as sabinene, β-pinene, β-terpineol, p-menth-8-en-1-ol, and terpinen-4-ol; phenylpropenes such as eugenol, methyl eugenol, and myristicin; and sesquiterpenes including germacrene D and β-bergamotene, along with other minor compounds (Atta-ur-Rahman et al., 2000; Dupuy et al., 2013; Francis et al., 2019). Key constituents of the leaves include eugenol, sabinene, myristicin, caryophyllene, and β-myrcene. Mace primarily contains 3-carene, sabinene, d-limonene, α-pinene, and β-pinene. The core chemical components of the kernel and seed include α-pinene, sabinene, d-limonene, β-pinene, and β-myrcene (Ashokkumar et al., 2022). In total, 40 compounds have been identified by GC-MS. The physicochemical characteristics of young nutmeg seed oil closely align with the quality standards set by the EP and the fragrance industry. Mace oil contained significant amounts of γ-terpinene, myristicin, safrole, α-pinene, terpinen-4-ol, and sabinene (Muhammad et al., 2016). Essential oil from nutmeg seeds cultivated in the Andaman and Nicobar Islands, India, obtained via hydrodistillation, exhibited elevated levels of myristicin, elemicin, α-thujene, sabinene, α-pinene, limonene, 4-terpineol, and γ-terpinene (Waman, 2020). A recent study reported that nutmeg seeds grown in the Western Ghats (South India), including the rind and shell, contained 27.7% sabinene in their essential oil (Ashokkumar et al., 2022). In comparison, nutmeg seed oil from Grenada exhibited a significantly higher sabinene concentration of 52.8% (Mickus et al., 2021). The molecular structures of major essential oil components extracted from various parts of *M. fragrans* were visualized using ChemDraw software, as shown in Figure 18.2.

The essential oil extracted from the mace of nutmeg grown in South India contains several minor constituents, including γ-asarone (1.4%), copaene (0.4%), α-phellandrene (1.9%), methyl eugenol (0.6%), terpineol-4-ol (1.7%), caryophyllene (0.5%), p-cymene (1.1%), eugenol (0.4%), α-thujene (1.0%), β-linalool (0.3%), β-terpineol (0.2%), germacrene-D (0.3%), γ-elemen (0.2%), and α-terpineol (0.4%). The kernel oil consists of 3-carene (1.8%), copaene (1.7%), α-thujene (1.5%), isoterpinolene (1.3%), γ-terpinene (1.2%), β-terpineol (0.8%), γ-asarone (0.8%), α-phellandrene (0.8%), γ-elemen (0.7%), myristicin (0.7%), p-cymene (0.7%), 4-carene (0.7%), caryophyllene (0.5%), p-menth-8-en-1-ol (0.5%), safrole (0.5%), α-terpinyl acetate (0.4%), β-phellandrene (0.3%), camphene (0.3%), and α-cubebene (0.3%) (Ashokkumar et al., 2022). Additionally, nutmeg seed essential oil contains minor compounds such as copaene (2.4%), β-phellandrene (2.2%), γ-asarone (2.3%), terpineol-4-ol (2.4%), α-terpinyl acetate (0.4%), myristicin (1.9%), camphene (0.4%), α-bergamotene (0.6%), α-phellandrene (0.5%), α-thujene (1.3%), 3-carene (1.2%), 4-carene (0.7%), *p*-cymene (0.7%), isoterpinolene (0.9%), β-terpineol (1.1%), γ-elemene (0.8%), methyleugenol (0.7%), *p*-menth-8-en-1-ol (0.9%), γ-terpinene (1.0%), and safrole (0.7%) (Ashokkumar et al., 2022; Mickus et al., 2021; Waman, 2020).

18.4.2. Nutmeg Oleoresin

Mace and nutmeg seeds are well-known for their high oleoresin content, significantly enhancing their flavour and aromatic properties. Compared to essential oils, oleoresins demonstrate excellent

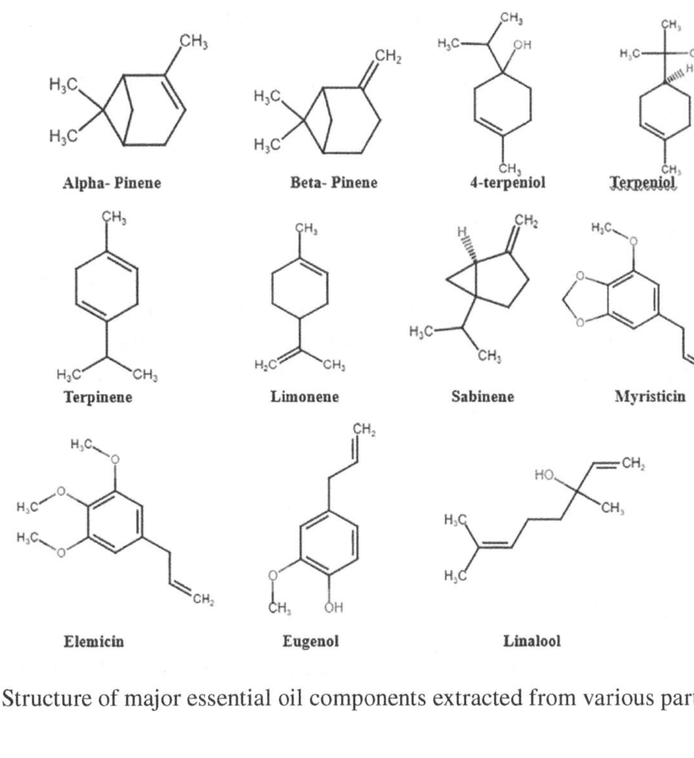

FIGURE 18.2 Structure of major essential oil components extracted from various parts of *M. fragrans*.

thermal stability, making them suitable for a broader range of processing applications (El-Shemy, 2018). The key constituents found in nutmeg oleoresin samples are sabinene, myristicin, elemicin, α-pinene, β-pinene, limonene, terpinen-4-ol, and myristic acid (Morsy, 2016). Oleoresins extracted using ethanol, ethyl acetate, and iso-propyl alcohol demonstrated varying profiles of volatile components, accounting for 84.0%, 95.3%, and 95.7% of the total constituents, respectively. Ethanol-extracted oleoresin contained 40 components, with essential compounds including elemicin (9.3%), myristic acid (7.3%), 2–(4-allyl-2,6-dimethoxyphenoxy)-1–(3,4,5-trimethoxyphenyl)-1-propanol (7.0%), dehydrodiisoeugenol combined with licarin E (6.4%), and myristicin (4.2%). Ethyl acetate-extracted oleoresin comprised 51 components, representing 95.3% of the total, with significant constituents including glycerine-1,3-myristate (29.6%), elemicin (11.5%), myristic acid (5.0%), dehydrodiisoeugenol combined with alizarin E (5.9%), and myristicin (5.0%). Iso-propyl alcohol-extracted oleoresin, containing 37 components and accounting for 95.7% of the total, featured elemicin (17.8%), dehydrodiisoeugenol + licarin E (9.4%), 2–(4-allyl-2,6-dimethoxyphenoxy)-1–(3,4,5-trimethoxyphenyl)-1-propanol (7.0%), myristic acid (5.3%), and myristicin (8.1%) as major constituents. Elemicin emerged as the predominant compound across all the oleoresins (Kapoor et al., 2013).

18.5. PHARMACOLOGICAL ACTIVITIES

18.5.1. Antioxidant Activity

Nutmeg exhibits antioxidant activity due to compounds such as eugenol and β-caryophyllene, which contain hydrogen atoms located at benzylic or allylic sites. These compounds have high antioxidant activity because peroxy radicals easily abstract atomic hydrogen under oxidative stress. Eugenol also enhances antioxidant properties by promoting the activity of enzymes such as catalase, superoxide dismutase, glutathione peroxidase, glucose-6-phosphate dehydrogenase, and glutamine transferase. Compounds having structural similarity to catechols, such as caffeic acid, are considered good antioxidants due to their ability to readily transfer phenolic hydrogen or electrons to suitable acceptors, such as reactive oxygen species or lipid peroxyl radicals. Calliste et al. (2010) reported that lignan-based compounds play a role in enhancing the antioxidant capacity of nutmeg seeds. Upon absorption, lignans and their glycosides undergo metabolic conversion into biologically active molecules with catechol frameworks. Nutmeg seed oil demonstrated potent antioxidant activity using the DPPH assay (Nikolic et al., 2021; Orabi et al., 2022). The shade-dried nutmeg fruit flesh had a higher free radical–scavenging ability than the sun-dried and fresh ones (Sipahelut et al., 2020).

18.5.2. Immune-Regulating and Radioprotective Effects

Fresh nutmeg and mace contain lignans that possess radioprotective and immune-enhancing properties. These lignans in the aqueous extract of fresh nutmeg mace PUC18 plasmid counteract DNA damage caused by radiation within cells. In mammalian splenocytes, they inhibit proliferation following stimulation by the T-cell mitogen concanavalin A (Con A), primarily by arresting the G1 stage of cell division while enhancing apoptosis, indicated by a rise in pre-G1 phase cells. Mace lignans trigger programmed cell death depending on dosage. Mace lignans protect splenocytes from radiation-induced damage by reducing reactive oxygen species without being cytotoxic to lymphocytes. Additionally, they inhibit DNA fragmentation in splenocytes exposed to radiation (Checker et al., 2008).

18.5.3. Antimicrobial Activity

Lignans from nutmeg seeds exhibit microbial inhibitory effects against pathogenic bacteria such as *Staphylococcus aureus*, *Shigella dysenteriae*, and *Bacillus subtilis* (Abourashed & El-Alfy, 2016).

Ethyl acetate extracts derived from nutmeg fruit flesh demonstrate inhibitory activity against gram-positive and gram-negative bacteria, with minimum inhibitory concentration (MIC) values ranging from 0.625–1.25 mg/mL (Shafiei et al., 2012). Analysis by Cho et al. (2007) identified myristic acid and trimyristin as the primary antibacterial compounds present in nutmeg seeds. The essential oil of nutmeg seeds is reported to contain several antimicrobial compounds, including carvacrol, α-pinene, γ-cymene, β-pinene, and β-caryophyllene (Dorman and Deans, 2004). Nutmeg has shown significant antimicrobial activity against various phytopathogenic fungi and oomycetes, including *Fusarium culmorum*, *Botrytis cinerea*, and *Phytophthora cinnamomi*. The hydroethanolic extract of nutmeg, primarily composed of fatty acids, methoxyeugenol, and elemicin, inhibited mycelial growth and induced necrotic cell death by targeting ergosterol biosynthesis. This highlights nutmeg's potential as a source of novel antimicrobial agents (Cruz et al., 2024).

18.5.4. Cancer-Preventive and Liver-Protective Effects

The spice nutmeg resists harmful cancer-causing agents and also has hepatoprotective properties. Hussain and Rao (1991) found that oral mace prevented 3-methylcholanthrene-induced cervical carcinogenesis in Swiss albino mice. Nutmeg also protected liver-damaged rats when included in their diet. Kyriakis et al. (1994) showed that nutmeg essential oil inhibits hepatic enzymes such as cytochrome P450, aryl hydrocarbon hydroxylase, and glutathione-S-transferase, reducing xenobiotic detoxification and activation.

18.5.5. Anti-Inflammatory Activity

Nutmeg and its oil exhibit notable anti-inflammatory properties (Mueller et al., 2010), with oil showing medicinal properties resembling the effects of nonsteroidal anti-inflammatory drugs (NSAIDs) (Olajide et al., 2000). However, this activity is specific to petroleum ether extracts. The total nutmeg extract also activates AMPK enzyme, a therapeutic target for metabolic syndrome. Active bioactive molecules such as specific lignan derivatives, including tetrahydrofuroguaiacin B and 2,5-bis-aryl-3,4-dimethyltetrahydrofuran lignans, fragrancing C1, nectarine B, saucernetindiol, verrucose, Gallatin, and nectarine A, were identified, with some showing strong AMPK activation in C_2C1_2 cells at 5 μM concentration. These findings support the potential of nutmeg and its components in treating metabolic disorders (Nguyen et al., 2010).

18.5.6. Memory Enhancer

Research findings indicate that administering a 5 mg/kg dose of n-hexane extract from nutmeg seeds significantly enhances cognitive function, particularly in young and aged mice, while mitigating cognitive deficits induced by scopolamine and diazepam. The memory-enhancing effects may be attributed to the plant's antioxidant, anti-inflammatory, or procholinergic properties, although the exact mechanism remains unclear (Parle et al., 2004).

18.6. TOXICITY

Overconsumption of nutmeg can result in intoxication, affecting the cardiovascular and central nervous systems. This is primarily due to the psychoactive compounds myristicin, safrole, and elemicin, which are present in nutmeg. At high doses, these compounds may lead to symptoms such as rapid heartbeat, nausea, vomiting, restlessness, hallucinations, and seizures. However, such adverse effects typically result from excessive consumption of nutmeg and do not occur when used in regular culinary amounts (Ehrenpreis et al., 2014). Nutmeg exposure has also been linked to dose- and duration-dependent liver damage. Increased doses and prolonged exposure lead to inflammatory cell infiltration and hepatocyte necrosis. Additionally, increased serum concentrations of liver enzymes, including aspartate aminotransferase (AST) and alanine aminotransferase (ALT), indicate liver stress and damage. Nutmeg exposure has been shown to alter the levels of metabolites associated with liver function, underscoring its impact on hepatic metabolism (Cao et al., 2020).

18.7. CONCLUSION

Myristica fragrans (nutmeg) holds significant pharmacological importance, providing various health benefits, including antioxidant, antimicrobial, anti-inflammatory, and anticancer activities. The essential oil, abundant in terpenes and phenylpropanoids, contributes largely to its distinctive aroma and medicinal properties. Additionally, nutmeg's oleoresins enhance its therapeutic potential. Nutmeg's oleoresins also contribute to its flavour and stability, making it useful in food processing and medicinal applications. Its lignans and other bioactive compounds exhibit immunomodulatory and radioprotective effects, further supporting its diverse therapeutic roles. Nutmeg's pharmacological properties highlight its potential as a valuable natural resource, warranting further research to explore its mechanisms of action and clinical applications.

REFERENCES

Abourashed, E. A., and A. T. El-Alfy. 2016. Chemical diversity and pharmacological significance of the secondary metabolites of nutmeg (*Myristica fragrans* Houtt.). *Phytochemistry Reviews* 15, no. 6: 1035–1056. https://doi.org/10.1007/s11101-016-9469-x

Agbogidi, O., and O. Azagbaekwe. 2013. Health and nutritional benefits of nutmeg (*Myristica fragrans* Houtt.). *Scientia Agriculturae* 1, no. 2: 40–44.

Ahmad, I., M. A. Hanif, R. Nadeem, M. S. Jamil, and M. S. Zafar. 2008. Nutritive evaluation of medicinal plants being used as condiments in the south asian region. *Journal of the Chemical Society of Pakistan* 30, no. 3: 400–405.

Akinduko, A. A., A. A. Akindahunsi, U. H. Uzoka, and C. O. Nwonuma. 2024. Nutritional and safety assessment of aqueous extract of nutmeg seeds, *Myristica fragrans*. *African Journal of Food, Agriculture, Nutrition & Development* 24, no. 8. https://doi.org/10.18697/ajfand.133.23905

Al-Maskri, A. Y., M. A. Hanif, M. Y. Al-Maskari, A. S. Abraham, J. N. Al-Sabahi, and O. Al-Mantheri. 2011. Essential oil from *Ocimum basilicum* (Omani Basil): A desert crop. *Natural Product Communications* 6, no. 10: 1934578X1100601020.

Arshad, Z., M. A. Hanif, R. W. K. Qadri, and M. M. Khan. 2014. role of essential oils in plant diseases protection: A review. *International Journal of Chemical and Biochemical Sciences* 6: 11–17.

Asgarpanah, J., and N. Kazemivash. 2012. Phytochemistry and pharmacologic properties of *Myristica fragrans* Hoyutt: A review. *African Journal of Biotechnology* 11, no. 65: 12787–12793. https://doi.org/10.5897/AJB12.1043

Ashokkumar, K., S. Vellaikumar, M. Murugan, M. K. Dhanya, and S. Aiswarya. 2022. Compositional variation in the leaf, mace, kernel, and seed essential oil of nutmeg (*Myristica fragrans* Houtt.) from the Western Ghats, India. *Natural Product Research* 36, no. 1: 432–435. https://doi.org/10.1080/14786419.2020.17 7171.

Atta-ur-Rahman, Choudhary, M. L., A. Farooq, A. Ahmed, M. I. Zafar, B. Demirci, and K. H. C. Baser. 2000. Antifungal activities and essential oil constituents of some spices from Pakistan. *Journal of the Chemical Society of Pakistan* 22: 60–65.

Azwanida, N. N. 2015. A review on the extraction methods use in medicinal plants, principle strength and limitation. *Medicinal & Aromatic Plants* 4: 3–8. https://doi.org/10.4172/2167-0412.1000196

Calliste, C., D. Kozlowski, J. Duroux, Y. Champavier, A. Chulia, and P. Trouillas. 2010. A new antioxidant from wild nutmeg. *Food Chemistry* 118, no. 3: 489–496. https://doi.org/10.1016/j.foodchem.2009.05.010

Cao, Z., W. Xia, X. Zhang, H. Yuan, D. Guan, and L. Gao. 2020. Hepatotoxicity of nutmeg: A pilot study based on metabolomics. *Biomedicine & Pharmacotherapy* 131: 110780. https://doi.org/10.1016/j.biopha.2020.110780

Chandran, M. D. S., and D. K. Mesta. 2001. On the conservation of the *myristica* swamps of the western ghats. In *Forest Genetic Resources: Status, Threats, and Conservation Strategies*, edited by U. R. Shaanker, K. N. Ganeshaiah, and K. S. Bawa, 1–19. New Delhi: Oxford and IBH. https://doi.org/10.1007/s11273-021-09825-5

Chatterjee, S., S. Gupta, and P. S. Variyar. 2015. Comparison of essential oils obtained from different extraction techniques as an aid in identifying aroma significant compounds of nutmeg (*Myristica fragrans*). *Natural Product Communications* 10, no. 8: 1934578X1501000833.

Checker, S., D. Chatterjee, S. Sharma, S. Gupta, P. Variyar, A. Sharma, and T. Poduval. 2008. Immunomodulatory and radioprotective effects of lignans derived from fresh nutmeg mace (*Myristica fragrans*) in Mammalian Splenocytes. *International Immunopharmacology* 8, no. 5: 661–669. https://doi.org/10.1016/j.intimp.2008.01.012

Chinelo, N. C., F. N. Nworah, I. U. Okagu, and O. R. Nwagwe. 2018. Proximate and phytochemical analysis of *Monodora myristica* (African nutmeg) from Nsukka, Enugu State, Nigeria. *Journal of Food and Nutrition Research* 6, no. 9: 597–601. https://doi.org/10.12691/jfnr-6-9-9.

Cho, J. Y., G. J. Choi, S. W. Son, K. S. Jang, H. K. Lim, S. O. Lee, N. D. Sung, K. Y. Cho, and J. C. Kim. 2007. Isolation and antifungal activity of lignans from *Myristica fragrans* against various plant pathogenic fungi. *Pest Management Science* 63, no. 9: 935–940. https://doi.org/10.1002/ps.1420

Cruz, A., E. Sánchez-Hernández, A. Teixeira, R. Oliveira, A. Cunha, and P. Martín-Ramos. 2024. Phytoconstituents and ergosterol biosynthesis-targeting antimicrobial activity of nutmeg (*Myristica fragrans* Houtt.) against phytopathogens. *Molecules* 29, no. 2: 471. https://doi.org/10.3390/molecules29020471

Dorman, H. D., and S. G. Deans. 2004. Chemical composition, antimicrobial and in vitro antioxidant properties of *Monarda citriodora* var. *citriodora*, *Myristica fragrans*, *Origanum vulgare* ssp. *hirtum*, *Pelargonium* sp., and *Thymus zygis* oils. *Journal of Essential Oil Research* 16, no. 2: 145–150. https://doi.org/10.1080/10412905.2004.9698679

Dupuy, N., J. Molinet, F. Mehl, F. Nanlohy, Y. Le Dréau, and J. Kister. 2013. Chemometric analysis of mid infrared and gas chromatography data of Indonesian nutmeg essential oils. *Industrial Crops and Products* 43: 596–601. https://doi.org/10.1016/j.indcrop.2012.07.073

Ehrenpreis, J. E., C. DesLauriers, P. Lank, P. K. Armstrong, and J. B. Leikin. 2014. Nutmeg poisonings: A retrospective review of 10 years experience from the illinois poison center, 2001–2011. *Journal of Medical Toxicology* 10, no. 2: 148–151. https://doi.org/10.1007/s13181-013-0379-7

El-Shemy, H. 2018. *Potential of Essential Oils*, 196 p. BoD—Books on Demand.

Francis, S. K., B. James, S. Varughese, and M. S. Nair. 2019. Phytochemical investigation on *Myristica fragrans* stems, bark. *Natural Product Research* 33: 1204–1208. https://doi.org/10.1080/14786419.2018.1457670

Ginting, B., M. Mustanir, H. Helwati, L. S. Desiyana, E. Eralisa, and R. Mujahid. 2017. Antioxidant activity of n-hexane extract of nutmeg plants from south aceh province. *Journal Natural* 17: 39–44. https://doi.org/10.24815/jn.v17i1.6969

Gupta, A. D., and D. Rajpurohit. 2011. Antioxidant and antimicrobial activity of nutmeg (*Myristica fragrans*). In *Nuts and Seeds in Health and Disease Prevention*, edited by V. R. Preedy, R. R. Watson, and V. B. Patel, 831–839. London: Academic Press.

Hanif, J. M. A., M. Y. Al-Maskari, A. Al-Maskari, A. Al-Shukaili, A. Y. Al-Maskari, and J. N. Al-Sabahi. 2011. Essential oil composition, antimicrobial and antioxidant activities of unexplored Omani Basil. *Journal of Medicinal Plants Research* 5, no. 5: 751–757.

Hussain, S., and A. Rao. 1991. Chemopreventive action of mace (*Myristica fragrans*, Houtt) on methylcholanthrene-induced carcinogenesis in the uterine cervix in mice. *Cancer Letters* 56, no. 3: 231–234. https://doi.org/10.1016/0304-3835(91)90007-5

International Fragrance Association (IFRA). 2007. *Analytical Procedure: GC/MS Quantitation of Potential Fragrance Allergens in Fragrance Compounds*. Version 3. Accessed September 20, 2024. http://www.ifraorg.org.\

Jamshidi-Kia, F., Z. Lorigooini, and H. Amini-Khoei. 2018. Medicinal plants: History and future perspective. *Journal of Herbmed Pharmacology* 7, no. 1: 1–7. https://doi.org/10.15171/jhp.2018.01

Kapoor, I. P. S., B. Singh, G. Singh, C. S. De Heluani, M. P. De Lampasona, and C. A. Catalan. 2013. Chemical composition and antioxidant activity of essential oil and oleoresins of nutmeg (*Myristica fragrans* Houtt.) fruits. *International Journal of Food Properties* 16, no. 5: 1059–1070.

Kitamura, S., and P. Poonswad. 2013. Nutmeg–vertebrate interactions in the Asia-Pacific region: Importance of frugivores for seed dispersal in *Myristicaceae*. *Tropical Conservation Science* 6, no. 5: 608–636. https://doi.org/10.1177/194008291300600503

Krishnamoorthy, B. 2000. Sex conversion in nutmeg. *Spice India* 13: 11–12.

Kumari, I., H. Kaurav, and G. Chaudhary. 2021. Myristica fragrans (jaiphal): A significant medicinal herbal plant. *International Journal for Research in Applied Sciences and Biotechnology* 8, no. 2: 213–224. https://doi.org/10.31033/ijrasb.8.2.27

Kyriakis, J. M., P. Banerjee, E. Nikolakaki, T. Dai, E. A. Rubie, M. F. Ahmad, J. Avruch, and J. R. Woodgett. 1994. The stress-activated protein kinase subfamily of c-jun kinases. *Nature* 369, no. 6476: 156–160. https://doi.org/10.1038/369156a0

Lawless, J. 1995. *The Illustrated Encyclopedia of Essential Oils: The Complete Illustrated Guide to the Use of Oils in Aromatherapy and Herbalism*. Shaftesbury, Dorset, UK: Element.

Mickus, R., G. Jančiukė, V. Raškevičius, V. Mikalayeva, I. Matulytė, M. Marksa, and V. A. Skeberdis. 2021. The effect of nutmeg essential oil constituents on novikoff hepatoma xell viability and communication through Cx43 gap junctions. *Biomedicine & Pharmacotherapy* 135: 111229. https://doi.org/10.1016/j.biopha.2021.111229

Morsy, N. F. 2016. A comparative study of nutmeg (*Myristica fragrans* Houtt.) oleoresins obtained by conventional and green extraction techniques. *Journal of Food Science and Technology* 53: 3770–3777. https://doi.org/10.1007/s13197-016-2363-0

Mueller, M., S. Hobiger, and A. Jungbauer. 2010. Anti-inflammatory activity of extracts from fruits, herbs and spices. *Food Chemistry* 122, no. 4: 987–996. https://doi.org/10.1016/j.foodchem.2010.03.041

Muhammad, I. S., A. Mahmood, R. Ayesha, Q. S. Zahoor, A. Q. Muhammad, M. Sania, and A. Amir. 2016. Chemical composition of the essential oils of nutmeg and mace by GC-FID/MS indigenous to Pakistan and evaluation of their biological activities. *Latin American Journal of Pharmacy* 35, no. 10: 2176–2184.

Naeem, N., R. Rehman, A. Mushtaq, and J. B. Ghania. 2016. Nutmeg: A review on uses and biological properties. *International Journal of Chemical and Biochemical Sciences* 9: 107–110.

Nguyen, P. H., V. De Tran, D. T. Pham, T. N. P. Dao, and R. S. Dewey. 2021. Use of and attitudes towards herbal medicine during the COVID-19 pandemic: A cross-sectional study in Vietnam. *European Journal of Integrative Medicine* 44: 101328. https://doi.org/10.1016/j.eujim.2021.101328

Nguyen, P. H., T. V. T. Le, H. W. Kang, J. Chae, S. K. Kim, K.-i. Kwon, D. B. Seo, S. J. Lee, and W. K. Oh. 2010. AMP-Activated Protein Kinase (AMPK) activators from *Myristica fragrans* (Nutmeg) and their anti-obesity effect. *Bioorganic & Medicinal Chemistry Letters* 20, no. 14: 4128–4131. https://doi.org/10.1016/j.bmcl.2010.05.067

Nikolic, V., L. Nikolic, A. Dinic, I. Gajic, M. Urosevic, L. Stanojevic, J. Stanojevic, and B. Danilovic. 2021. Chemical composition, antioxidant and antimicrobial activity of nutmeg (*Myristica fragrans* Houtt.) seed essential oil. *Journal of Essential Oil Bearing Plants* 24, no. 2: 218–227. https://doi.org/10.1080/0972060X.2021.1907230

Nkwocha, C. C., F. N. Nworah, I. U. Okagu, and O. R. Nwagwe. 2018. Proximate and phytochemical analysis of monodora myristica (African Nutmeg) from Nsukka, Enugu State, Nigeria. *Journal of Food and Nutrition Research* 6, no. 9: 597–601. https://doi.org/10.12691/jfnr-6-9-9

Okiki, P. A., C. P. Nwobi, O. B. Akpor, E. Adewole, and R. D. Agbana. 2023. Assessment of nutritional and medicinal properties of nutmeg. *Scientific African* 19: e01548. https://doi.org/10.1016/j.sciaf.2023.e01548

Olajide, O. A., J. M. Makinde, and S. O. Awe. 2000. Evaluation of the pharmacological properties of nutmeg oil in rats and mice. *Pharmaceutical Biology* 38, no. 5: 385–390. https://doi.org/10.1076/phbi.38.5.385.5976

Orabi, M., J. O. Abdulsattar, and Z. O. Nasi. 2022. Phytochemical profile, antimicrobial, antioxidant activity and cyclooxygenase 2 inhibitory properties of nutmeg (*Myristica fragrans)* seeds extract. *Egyptian Journal of Chemistry* 65, no. 1: 317–326. https://doi.org/10.21608/ejchem.2021.78192.3831

Orwa, C., A. Mutua, R. Kindt, R. Jamnadass, and A. Simons. 2009. *Agroforestree Database: A Tree Reference and Selection Guide*. Version 4.

Osemwegie, O. O., C. O. Nwonuma, A. P. Oluyori, P. O. Abraham, A. A. Akanbi, D. O. Opaleke, and O. O. Alejolowo. 2017. In vitro antimicrobial and in vivo lead acetate poison abatement study of *Garcinia kola* Heckel. *Journal of Taibah University for Science* 11, no. 6: 883–894. https://doi.org/10.1016/j.jtusci.2017.06.001

Pandey, R., R. K. Tiwari, and S. S. Shukla. 2016. Omics: A newer technique in herbal drug standardization and quantification. *Journal of Young Pharmacists* 8: 76–81. https://doi.org/10.5530/jyp.2016.2.4

Parle, M., D. Dhingra, and S. K. Kulkarni. 2004. Improvement of mouse memory by *Myristica fragrans* seeds. *Journal of Medicinal Food* 7, no. 2: 157–161. https://doi.org/10.1089/1096620041224193

Peter, K. V. 2001. *Handbook of Herbs and Spices*, Vol. 2, 238–248. Cambridge: Woodhead Publishing Ltd.

Riyadi, E., N. Andarwulan, and D. N. Faridah. 2014. The volatile compounds profile as identity for nutmeg oil, patchouli oil, and fresh ginger oil from Indonesia. *Journal Mutu Pangan* 1, no. 1: 19–25.

Saputro, M. A., N. Andarwulan, and D. N. Faridah. 2016. Physical characterization and essential oil properties of west sumatra mace and nutmeg seed (*Myristica fragrans* Houtt) at different ages at harvest. *Journal of Pharmacognosy & Phytochemistry* 5, no. 6: 371–376.

Shafiei, Z., N. N. Shuhairi, N. Md Fazly Shah Yap, C. A. Harry Sibungkil, and J. Latip. 2012. Antibacterial activity of *Myristica fragrans* against oral pathogens. *Evidence-Based Complementary and Alternative Medicine*, 2012. https://doi.org/10.1155/2012/825362

Shinohara, C., S. Mori, T. Ando, and T. Tsuji. 1999. Arg-Gingipain inhibition and anti-bacterial activity selective for *Porphyromonas gingivalis* by Malabaricone C. *Bioscience, Biotechnology, and Biochemistry* 63, no. 8: 1475–1477. https://doi.org/10.1271/bbb.63.1475.

Sipahelut, S. G., A. Y. Kastanja, and Z. Patty. 2020. Antioxidant activity of nutmeg fruit flesh-derived essential oil obtained through multiple drying methods. *EurAsian Journal of BioSciences* 14, no. 1: 21–26.

Siregar, H. M., and T. Murtiningsih. 1996. Nutmeg development effect on essential oil content. In *Proceedings of the 1st National Symposium on Medicinal and Aromatic Plants*. Asian Pacific Information Network on Medicinal and Aromatic Plants (APINMAP). Bogor.

Smith-Palmer, A., J. Stewart, and L. Fyfe. 2002. Inhibition of listeriolysin O and phosphatidylcholine-specific production in *Listeria monocytogenes* by subinhibitory concentrations of plant essential oils. *Journal of Medical Microbiology* 51: 567–574. https://doi.org/10.1099/0022-1317-51-7-567

Sultan, M. T., F. Saeed, H. Raza, A. Ilyas, F. Sadiq, A. Musarrat, M. Afzaal, M. Hussain, M. A. Raza, and E. Al JBawi. 2023. Nutritional and therapeutic potential of nutmeg (*Myristica fragrans*): A concurrent review. *Cogent Food & Agriculture* 9, no. 2: 2279701. https://doi.org/10.1080/23311932.2023.2279701

Sultana, A., A. G. Najeeya, and A. Anjum. 2018. Traditional unani uses with multiple pharmacological activities of aril of *Myristica fragrans* (mace). *CELLMED* 8, no. 2: 6–1.

Suvachananonda, T. 2007. *Potential in Inhibiting the Vibrio Parahaemolyticus by the Medicinal Plant Extracts*. Student's Report (B.S.), Assumption University.

Tripathi, I. P., and N. Dwivedi. 2015. Pharmacognostical standardization of nutmeg seeds (*Myristica fragrans* Houtt.)—A traditional medicine. *International Journal of Pharmaceutical Sciences and Research* 6, no. 7: 3096–3102. https://doi.org/10.13040/IJPSR.0975-8232.6(7).3096-02

Waman, A. A. 2020. Essential oil composition of clove and nutmeg from Andaman and Nicobar Islands, India. *Indian Journal of Biochemistry & Biophysics* 57: 95–100.

Widelski, W., and K. Kukula-Koch. 2017. Psychoactive drugs. *Pharmacognosy*, 363–374. Elsevier. https://doi.org/10.1016/B978-0-12-802104-0.00017-2

Wink, M. 2022. Current understanding of modes of action of multicomponent bioactive phytochemicals: Potential for nutraceuticals and antimicrobials. *Annual Review of Food Science and Technology* 13, no. 1: 337–359. https://doi.org/10.1146/annurev-food-052720-100326

Yadav, R. N. S., and M. Agarwala. 2011. Phytochemical analysis of some medicinal plants. *Journal of Phytology* 3, no. 12: 10–14.

19 Spices That Heal

From Kitchen to Medicine

Nikita Patel, Stephen Oyedeji, and Ramar Krishnamurthy

19.1. INTRODUCTION

The culinary and therapeutic traditions of spices have been woven together to create a complex tapestry that has developed over ages. In addition to being necessary for flavour, spices—from the soothing scent of cinnamon to the strong spiciness of cayenne pepper—are also vital for the age-old fight against diseases and imbalance. As we continue to explore the connection between diet and health, one thing becomes abundantly evident that: spices have remarkable therapeutic properties (Duke et al., 2003). Spices are always used extensively in cooking and in medicine for a variety of purposes, including expectorant, eccoprotic, diuretic, and carminative aperients. Since ancient times, spices were traditionally utilized for health purposes. Spices such as fenugreek, coriander, turmeric, cinnamon, cumin, clove, and others are often used as remedies for a variety of illnesses (Figure 19.1). Spices have historically had a comprehensive impact on human health when included in meals. All spices have been incorporated into Indian cooking from ancient times in both traditional and everyday dishes (Sachan et al., 2018). Around 50 of the 80 varieties of spices cultivated worldwide are cultivated in India, making it a major producer of a wide variety of spices. In addition to adding flavour to food, spices are an excellent source of iron, calcium, vitamins B and C, and other antioxidants (Rathore and Shekhawat, 2008). Numerous medical sectors, including the pharmaceutical, cosmetic, and aromatic fragrance industries, employ spices. The varied Indian culinary tastes specifically oppose the multidrug-safe clinical exclusion of enterococci with distinct lethal genetic components (Vasanthi et al., 2010).

Without flavors, the culinary world would be lifeless. Similar to the herbs of its natural verdant companion, fragrances give various foods across the world a variety of flavors, shades, and tastes. They also provide a wide range of amazing phytonutrients that can improve human health and well-being (Figure 19.2). Although culinary tastes have long been used for their many health advantages, extensive study conducted in the last 20 years has the potential to explore and elucidate the hidden wonders they contain (Sharangi and Guha, 2013). In fact, they may be able to combat persistent conditions, such as cancerous development, cardiovascular infections, and several severe pneumonic, neurological, and immune system disorders. In large quantities of food, such as soups, dinners, stews, and sauces derived from cooking foundations, flavors are the main source of spore-forming microscopic organisms; in extreme circumstances, they grow and expand to contagious and potentially hazardous levels. For both vegetarians and nonvegans, flavors are a crucial component of Indian cuisine. These are the common ingredients included in food to impart it with taste and aroma (Banerjee and Sarkar, 2003). A normal Indian kitchen, including ginger, garlic, turmeric, pepper, coriander, onion, ajwain, jeera, tea, tulsi, neem leaves, and other ingredients, actually serves as a sort of homegrown pharmacy. Clove buds, cinnamon bark, ginger roots, peppercorn berries, cumin seeds and saffron, all have flavors. Carnosic corrosive, carnosol, rosmarinic corrosive, thymol, carvacrol, 6-gingerol, 6-shogaol, zingerone, curcumin, capsaicin, vanillin, eugenol, caffeic corrosive, and ferulic corrosive represent a few components of the dynamic cancer prevention agent that are included in tastes (Gupta et al., 2013).

Foods that support a healthy lifestyle in several ways are sometimes referred to as functional foods. They provide energy and building blocks to countless substances that are essential for

 DOI: 10.1201/9781003559894-19

anyone's development and resilience. Nutrients can be regarded as pharmaceuticals if they have an effect on bodily components that go beyond dietary effects, such as well-being and disease prevention. Both traditional and ethnic functional food-based medications not only prevent and manage ongoing medical conditions without causing adverse drug responses in humans, but they also manage weight, neurological problems, or cardiovascular infections (Amiri et al., 2020). The area of culinary medicine combines knowledge of food, culinary skills, and nutrition to assist patients in maintaining their health, preventing disease, and curing illnesses linked to nutrition. The goal of culinary medicine is to assist people in making excellent clinical decisions about how to access and consume delicious meals that prevent and treat ailments and restore prosperity. With food and drink, culinary medicine aims to help the patient think about themselves safely, sustainably, and delightfully (Hauser, 2020; La Puma, 2020). Therefore, the current chapter focuses on the unique qualities, active ingredients, and uses of certain important spices, together with their immunomodulatory and functional properties.

19.1.1 The Ancient Wisdom of Spices

Spices were considered as one of the most precious commodities traded in antiquity and the Middle Ages. The word 'aroma' from ancient Greek is frequently used to describe spices. Various spices were utilized by the ancient Egyptians to flavour meals, make cosmetics, and embalm their dead. The earliest concrete proof of spice use may be found in the texts and artwork of ancient cultures. Only members of the higher class had access to spices throughout the Roman era, and they were prized as much as gold. Rare spices were utilized in Roman cookery as a sign of status, and subsequently in the Middle Ages and the Renaissance, the wealthy became overly fond of spicy meals (Rathore and Shekhawat, 2008). India has a long history of employing spices in medicinal and cookery, as evidenced by the Indus Valley civilization. Archaeological evidence suggested that the inhabitants of the Indus Valley employed a wide range of spices, such as coriander, cumin, and mustard. These results imply that ancient India had a highly developed culinary culture and an understanding of the therapeutic benefits of spices. Cloves were mentioned in Roman Empire documents as early as the 1st century CE, and they played a vital role in India's external commerce with civilizations such as Egypt, Arabia, and China (Gupta, 2010; Shukla and Nagendra, 2018).

Throughout history, spices have been used in religious rites and rituals in various civilizations, and they have played a significant part in spiritual and religious traditions. They have been used by people to represent fortune, blessings, purity, and a closer relationship with God. Spices have a strong connection to India's religious and spiritual traditions. As seen by the use of holy turmeric in Hindu ceremonies and strong incense in Buddhist rites, spices have long been valued for their purifying and symbolic qualities. Their inclusion in temples, shrines, and family altars emphasizes how significant they are in both the spiritual and material worlds. Spices are commonly offered to the gods in Hinduism since it is thought that they have purifying qualities (Flatt, 2016; Yadav et al., 2020). People use spices such as cloves and cardamom in incense and other ceremonial gifts because they believe that their scents are pleasant to the gods. Many Indian homes and temples use incense sticks, which are often produced with a mixture of spices and herbs, to cleanse the air and provide a calm space for prayer and meditation (Flood, 2005).

A healthy, balanced diet is the most crucial component of attaining health. The medicinal properties of each spice are thought to be determined by a distinct combination of *rasa* (taste), *guna* (quality), *virya* (potency), and *vipaka* (post-digestive effect). For example, people believe that turmeric, with its strong, bitter flavour, has heating qualities and can be used to balance *Kapha dosha*. It is also well-known for its anti-inflammatory, antioxidant, and antibacterial qualities, which make it a useful treatment for a number of illnesses, including skin disorders, digestive issues, and arthritis. Another powerful spice with warming and digestive qualities is ginger. It is used to heal respiratory conditions, indigestion, and nausea. Ginger is also used in Ayurveda to boost immunity and improve circulation (Padakatti and Meti, 2020).

19.2. TURMERIC: THE GOLDEN HEALER

Curcuma longa, known as turmeric, is found in tropical and subtropical areas of the world. It is extensively grown throughout Asia, mostly in China and India. It has a short stem and can reach a height of 1 meter. All around the world, turmeric is a vital spice that has many uses, especially in the East. Because of its advantageous qualities, it is utilized as traditional medicine in Asian regions, including Bangladesh, Pakistan, and India, in addition to being used as a spice (Ravindran et al., 2007). Turmeric has been proven in recent studies to have anticancer, antidiabetic, antioxidant, hypolipidemic, anti-inflammatory, antibacterial, antifertility, antivenom, hepatoprotective, nephroprotective, and anticoagulant properties. Additionally, this plant has demonstrated anti-HIV properties to prevent AIDS. Because of these therapeutic benefits, turmeric is now regarded as a spice with an extensive list of therapeutic uses. Curcumin, the yellow-colored substance found in turmeric, is a vital part of this plant (Nasri et al., 2014; Verma et al., 2018).

19.2.1. Bioactives and Therapeutic Potential

Turmeric possess 69% carbohydrates, 6% protein, 5% fat, 3.5% minerals, and 13% moisture. The steam distillation process yields the essential oil, which contains 53% sesquiterpenes, 25% zingiberene, 1% a-phellandrene, 0.6% sabinene, 1% cineol, and 0.5% borneol. The yellow colour is caused by curcumin (3.1–4.0%), which is made up of curcumin I, curcumin II, and curcumin III in high, moderate, and low levels, respectively (Ammon et al., 1991; Jyotirmayee et al., 2022).

Turmeric has also yielded demethoxy and bisdemethoxy compounds of curcumin. Curcumin is soluble in glacial acetic acid, alcohol, alkali, ketone, and chloroform, and it melts at 176–177°C to produce a reddish brown salt with alkali. The rhizomes have been shown to contain mono- and di-demethoxycurcumin, curzerenone, curdione, tumerone a, and tumerone b (Chanda et al., 2019).

It has been demonstrated that curcumin is a major scavenger of oxygen free radicals. It has antioxidant properties similar to those of vitamin C. It can delay oxidation of hemoglobin or lipids and has a strong ability to prevent activated macrophages from producing reactive oxygen species (ROS), including H_2O_2, superoxide anions, and nitrite radicals. Additionally, its compounds demethoxycurcumin and bisdemethoxycurcumin have antioxidant properties (Jayaprakasha et al., 2006). Curcumin pretreatment reduces ischemia-induced oxidative stress and cardiac alterations. Curcumin's impact on an inducible stress protein was measured in vitro, and the results showed that cells were more resistant to oxidative damage (Akter et al., 2019; Memarzia et al., 2021).

Turmeric protects the gastrointestinal system in a number of ways. Additionally, turmeric prevents the development of ulcers in rats exposed to gastrointestinal insults, such as stress, alcohol, indomethacin, reserpine, and pyloric ligation, while also promoting the production of mucus in the stomach wall. It also promotes the release of digestive enzymes, bicarbonate, gastrin, and secretin while inhibiting intestinal spasm. When 25 patients with stomach ulcers received 600 mg of powdered turmeric five times a day, the study revealed that 48% of the patients had fully recovered. No negative responses or anomalies in the blood were noted (Memarzia et al., 2021).

19.3. GINGER: THE ROOT OF VITALITY

Zingiber officinale Rosc., also known as ginger, originated from Southeast Asia. It is utilized as a spice to enhance the flavour of cuisine in many nations. In addition, traditional herbal medicine has made use of ginger rhizomes. Because of its rich phytochemistry, ginger is thought to deliver health benefits (Jung and John, 2002). The two broad categories of volatiles and nonvolatiles were used by Jolad et al. (2004) to classify fresh ginger. Monoterpenoid hydrocarbons and sesquiterpene are the compounds responsible for the unique flavour and scent of ginger. Numerous conditions, such as degenerative diseases including rheumatism and arthritis, digestive issues such as indigestion, constipation, and ulcers, cardiovascular diseases, vomiting, diabetes mellitus, and cancer, may be treated with ginger. Additionally, it possesses antioxidant and anti-inflammatory attributes that slows down the aging process (Nicoll and Michael, 2009).

19.3.1. Bioactives and Therapeutic Potential

Depending on the region of source and whether the rhizomes are dry or fresh, ginger has a variety of components. The primary source of scent from ginger is due to its volatile oil, which has a yield in the range from 1–3%. More than 50 constituents have been identified, primarily from the oil, which consist of monoterpenoids (i.e., b-phellandrene, (+)-camphene, cineole, geraniol, curcumene, citral, terpineol, and borneol) and sesquiterpenoids (i.e., a-zingiberene (30–70%), b-sesquiphellandrene (15–20%), b-bisabolene (10–15%), (E-E)-a-farnesene, arcurcumene, and zingiberol). After drying, some of the constituents change into less odour-defining substances (Shoaib et al., 2016). Many compounds found in fresh ginger are recognized to have anti-inflammatory, antibacterial, antioxidant, and gastroprotective qualities.

Ginger rhizome is a great source of dietary fibers, which include essential oils, moisture, protein, fat, minerals, vitamins, and carbohydrates, along with other health advantages. The nutritional profile and content vary according to its particular type, variation, drying and storage conditions, extraction, and curing technique. It is used to cure rheumatism, cough, colic, heart palpitations, swellings, dyspepsia, lack of appetite, lower blood sugar, lessen convulsions, and strengthen bones (Fahmi et al., 2019).

The antioxidant characteristics, glycemic index, and carbohydrate-hydrolyzing enzyme activities of ginger-based fruit beverages were assessed in a recent research study by Ademosun et al. (2021). Ferric-reducing antioxidant power tests, 2,2-diphenyl-1-picrylhydrazyl and 2,2′-azino-bis (3-ethylbenzothiazoline-6-sulfonic acid), radicals were used to evaluate the in vitro antioxidant activity. In comparison to other commercial and ginger-based drinks, the results demonstrated that sample beverages exhibited better antioxidant qualities against DPPH, FRAP, and ABTS radicals. All beverages, nevertheless, exhibited modest glycemic indices. The use of ginger to reduce postoperative nausea and vomiting was the subject of several research studies. An investigation by Kamali et al. (2020) explored the efficacy of ginger and dexmedmoidine in preventing nausea and vomiting during abdominal hysterectomy. Although the thyroid gland was improved in both nonsmokers and smokers, Mahassni and Bukhari (2019) reported that the extracted form of ginger rhizome had distinct effects on the immune cells and antibodies. According to their findings, ginger might assist smokers who suffer from anemia, and it may boost humoral immunity or the antibody response in nonsmokers who are battling infections. According to Kim et al. (2022), ginger and Novafen are both useful in reducing discomfort in females with primary dysmenorrhea. For primary dysmenorrhea, therapy with natural herbs and nonsynthetic drugs may be helpful.

19.4. CINNAMON: THE SWEET SPICE OF HEALTH

The tropical evergreen cinnamon tree (*Cinnamomum zeylanicum*) is indigenous to Sri Lanka and the Malabar coast of India. In many languages, it is referred to by different names, including *yookgway* in Chinese, *kaneel* in German, *cannelle* in French, *canella* in Spanish, and *dallchini* in Hindi. *Cinnamomum* is a botanical name that comes from the Arabic and Hebraic word *amomon*, which means "fragrant spice plant". From ancient times, the Egyptians and Chinese used cinnamon in their food preparations and traditional medicine. Today, there are about 250 species of the cinnamon genus known worldwide, which are used not only as a flavouring but also for medicinal, antimicrobial, and antioxidant purposes. Additionally, cinnamon has natural food preservation properties (Hameed et al., 2016). Essential oils and significant elements, including cinnamaldehyde, cinnamic acid, eugenol, and cinnamate, are mostly found in cinnamon. Initially, it was used to cure colds, toothaches, and foul breath. According to Jakhetia et al. (2010), the volatile oil found in tree bark has several therapeutic benefits, including antibacterial, antioxidant, antiulcer, antidiabetic, and antifungal effects.

19.4.1. Bioactives and Therapeutic Potential

The main components of cinnamon include caffeic, gallic, vanillic, protocatechuic, ferulic, and p-coumaric acids. The plant portion from which the volatile components are harvested determines the chemistry of its essential oil. The most prevalent component isolated from bark essential oil is cinnamonaldehyde, which ranges in quantity from 60–90%. The remaining minor volatile chemicals include oxygenated compounds and hydrocarbons, such as benzyl benzoate, b-caryophyllene, cinnamonyl acetate, linalool, and eugenyl acetate (Jayaprakasha et al., 1997). The primary ingredients of cinnamon leaf oil, blossoms, and fruits include eugenol (with a concentration of >80%), (E)-cinnamyl acetate, and caryophyllene (Wong et al., 2014).

Oxidation of fatty acid and lipid peroxidation may be considerably inhibited in vitro by the aqueous and ethanolic extract of cinnamon. Numerous flavonoids that have been extracted from cinnamon have antioxidant and free radical–scavenging properties. Cinnamaldehyde has the capacity to block the generation of nitric oxide along with the expression of inducible nitric oxide. At 0.1, 0.5, and 1 μg/μL, the potent inhibitory activity was recorded (Shobana and Naidu, 2000). With a notably high therapeutic window, in vivo investigations in animals have also demonstrated a lack of notable toxic effects on the liver and kidney. The effects on the number of nuclei, the distribution of embryos based on nucleus number, and the survivability of mouse embryos were assessed after giving cinnamon extract to female mice for two weeks. The antiproliferative properties of cinnamon aldehyde were responsible for the considerable reduction in nucleus count and the major alteration in embryo distribution based on nucleus count (Domarացký et al., 2007). Samarasekera et al. (2005) evaluated the ability of leaf and bark essential oils to repel *Aedes aegypti*, *Culex quinquefasciatus*, and *Anopheles tessellatus* mosquitoes. In comparison to leaf oil, bark oil demonstrated mortality against *A. tessellatus* and *C. quinquefasciatus*. According to Yang et al. (2005), bark oil was less efficient than dphenothrin or pyrethrum against the eggs and adult females of *Pediculus humanus capitis*. The menstrual pattern of polycystic ovarian syndrome (PCOS) patients significantly improved while using *Cinnamomum zeylanicum*. Ovarian size showed a noticeable improvement, and patients claimed full PCOS amelioration. Most patients responded well to the treatment, and it was determined that *Cinnamomum zeylanicum* might be an alternate treatment for PCOS (Khan and Begum, 2019).

19.5. BLACK PEPPER: THE PIPERINE POWERHOUSE

Pepper, also known as *Piper nigrum* L., is regarded as the "king of spices" due to its enormous market share worldwide. The term "pepper" comes from the Sanskrit word *pippali*, meaning berry. This woody, pungent, perennial climber can reach up to a height of 50–60 cm. Black and white peppers are processed differently and are harvested at different stages. White pepper is obtained by removing the pulp from ripe fruit, whereas black pepper is prepared by drying unripe fruit until wrinkled, resulting in black pepper with the pulp. There are many uses for both of the peppers, including as pesticides, preservatives, spices, and in herbal medicine (Kumar et al., 2011). *P. nigrum* is mostly grown in hot, humid climates. It originated primarily from the South Indian Peninsula's Western Ghats, although it was later brought to other South and Southeast Asian nations. Vietnam is currently the top producer, with the southern part of the country being the primary location for black pepper cultivation. The most popular preparation methods by using seeds/fruits were paste, pills, tabs, and powdered form. It was also discovered that *Piper nigrum* and its chemical components have significant pharmacological characteristics (Takooree et al., 2019).

19.5.1. Bioactives and Therapeutic Potential

More than 50 chemicals have been observed after profiling the fruits, seeds, and roots of *P. nigrum*. Several secondary metabolites, such as alkaloids, glycosides, terpenoids, steroids, flavonoids, tannins, and anthraquinones, were found in *P. nigrum* fruit extracts after phytochemical screening demonstrated their existence. A- and b-pinene, a-phellandrene, methylpropanal, myrcene, limonene, linalool, 3-methylbutyric acid, butyric acid, and 2- and 3-methylbutanal were the principal odourants found from *P. nigrum* fruit (Singh *et al.*, 2013). In contrast, piperine was identified as the primary component in alcohol and ethyl acetate oleoresins, with many additional components present in lower amounts. Other phytochemical investigations of ethanolic extracts of *P. nigrum* fruits likewise revealed piperine to be the main active component (Zarai et al., 2013).

Numerous in vitro tests have been conducted to evaluate the antioxidant capacity of *P. nigrum* using various mechanisms. Using the thiocyanate technique, Gülçin (2005) discovered that the ethanol extract had a greater ferric-reducing power and a lower scavenging activity against DPPH, superoxide anion, hydrogen peroxide, and overall antioxidant activity than the aqueous extract of seeds. It was also noted that the antioxidant capacities of various fruit extracts and fractions varied respectively. Researchers are investigating *P. nigrum* as a therapeutic pain reliever due to the interest in finding new and safe natural compounds that alleviate pain. Tasleem et al. (2014) employed the tail immersion, hot-plate, analgesy-meter, and acetic acid–induced writhing test to assess the analgesic efficacy of *P. nigrum* and its constituent piperine in hexane and alcoholic extracts. In the analgesy-meter test, the ethanol extract was most effective after a 60-minute dosage of 10 mg/kg, but piperine had the maximal analgesic effect after 120 minutes at 5 mg/kg dose.

Male Wistar rats treated with a methanolic extract of *P. nigrum* fruit showed anxiolytic and antidepressant-like effects (Hritcu et al., 2015). Through the reduction of oxidative stress in the rat hippocampal region, the extract mitigated the impairment of spatial memory caused by amyloid beta. Piperine was shown to mitigate the abnormalities in motor coordination and cognitive functioning caused by 1-methyl-4-phenyl-1,2,3,6-tetrahydropyridine (MPTP) (Yang et al., 2015). Furthermore, piperine completely restored the motor deficit, partially reversed the decrease in locomotor activity, and reversed the higher frequency of apomorphine-induced contralateral rotations seen in the 6-OHDA (right striatum) group, (Correia et al., 2015).

19.6. CUMIN: THE DIGESTIVE DYNAMO

Cuminum cyminum L., also known as cumin, is commonly used as a spice due to its volatile oil, which gives the seeds their distinct scent. In addition to spices, it also helps with a variety of conditions, including jaundice, dyspepsia, and toothache. Many regions of India grow cumin, which is sent to markets abroad where it is in high demand in a variety of forms, including seeds, powder form, essential oil, and oleoresin (Chen et al., 2018). An important spice in Indian kitchens, cumin contains essential oil that has antibacterial properties. It has been used as an astringent in broncopulmonary spasms, as a cough remedy, as a pain reliever, and as a carminative and eupeptic to treat mild stomach issues (Romagnoli et al., 2010).

A member of the Apiaceae family, cumin is a flowering plant native to the Middle East and Eastern India. It has favorable qualities in relation to health, hence being used globally. Cumin seed and its refined, sweet-smelling water are used for stimulant, antispasmodic, carminative, antibacterial, anti-inflammatory, and wound-mending agent. Cumin is often used to treat simple stomach problems, loose stools, and farts. It is also thought to be an appetite stimulator (Mehdizadeh et al., 2017).

19.6.1. Bioactives and Therapeutic Potential

Alkaloids, coumarins, anthraquinones, flavonoids, glycosides, proteins, steroids, resins, saponins, tannins, dietary fibers, minerals, fats, vitamins A, B, C, and E, and other nutrients are abundant in cumin seeds. Cumin seeds also include organic acids such aspartic, citric, malic, tartaric, propionic, ascorbic, oxalic, maleic, and fumaric acids. The fruits include luteolin, apigenin, and estrogenic isoflavonoids (Kang et al., 2019). The seeds have a significant amount of flavonoids, with quercetin, luteolin, and apigenin being the main flavonoids. The main bioactive component of *C. cyminum* is cuminaldehyde. In addition, it includes cymene, limonene, eugenol, safranal, γ-terpinene, γ-Phellandrene, β-myrcene, α-Phellandrene, α- and β-pinene, cuminic alcohol (cuminol), and limonene (Nirmala et al., 2020).

Cumin hydrosols had encouraging antifungal properties against a variety of phytopathogenic fungi in an in vitro investigation. In comparison to other commercial items, O'Riordan and Wilkinson (2008) observed that market samples of cumin were free of aflatoxin contamination, a toxin generated by *Aspergillus flavus*. Cumin oil, especially high in a-pinene, had strong antifungal effects on the growth of *Aspergillus* strains (Mnif and Sami, 2015).

Eugenol, a secondary metabolite of *C. cyminum*, has been shown to have antibiofilm and quorum-sensing inhibitory capacity against negative bacterial pathogens. For example, Packiavathy et al. (2012) demonstrated that the methanolic extract from *C. cyminum* hindered *Chromobacterium violacems'* ability to produce violacein. Additionally, it was observed that *C. cyminum* extracts significantly disrupt acyl homoserine lactone (AHL) and control physiological processes linked to biofilm formation, such as exopolysaccharide (EPS) synthesis and flagellar motility. Additionally, they were shown to effectively suppress the in vitro biofilm development in strains of *Pseudomonas*

FIGURE 19.1 Therapeutic potential of spices.

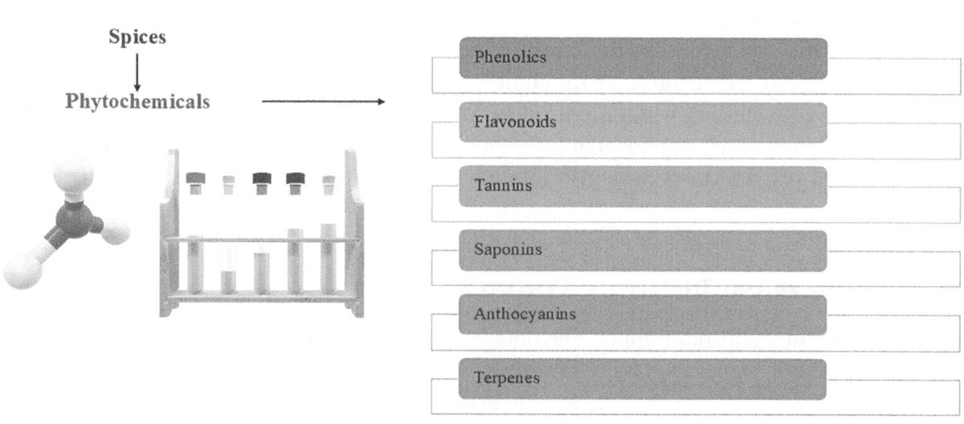

FIGURE 19.2 Phytochemistry of spices.

aeruginosa, *Proteus mirabilis*, and *Serratia marceescens* to eliminate the biofilm architecture. Using the α-amylase inhibition test, Tahir et al. (2016) assessed the antidiabetic potential of the cumin-based emulsions and discovered that the B5 emulsion had the highest suppression of α-amylase activity. Additionally, they observed that the antidiabetic action of cumin was dosage dependent, with the peak activity occurring at a level of 100 μl/ml. The immunomodulatory impact of cumin in streptozotocin (STZ)-induced diabetic rats was also assessed by Moubarz et al. (2016), which showed potent results.

REFERENCES

Ademosun, Mary T., Olufunmilayo S. Omoba, and Aderonke I. Olagunju. "Antioxidant properties, glycemic indices, and carbohydrate hydrolyzing enzymes activities of formulated ginger-based fruit drinks." Journal of Food Biochemistry 45, no. 3 (2021): e13324.

Akter, Jesmin, Md Amzad Hossain, Kensaku Takara, Md Zahorul Islam, and De-Xing Hou. "Antioxidant activity of different species and varieties of turmeric (Curcuma spp): Isolation of active compounds." Comparative Biochemistry and Physiology Part C: Toxicology & Pharmacology 215 (2019): 9–17.

Amiri Ardekani, Ehsan, Hossein Askari, and Abdolali Mohagheghzadeh. "Memorial functional foods: A new concept from Bavi tribe." Journal of Ethnic Foods 7 (2020): 1–10.

Ammon, Hermann PT, and Martin A. Wahl. "Pharmacology of Curcuma longa." Planta Medica 57, no. 01 (1991): 1–7.

Banerjee, Mousumi, and Prabir K. Sarkar. "Microbiological quality of some retail spices in India." *Food Research International* 36, no. 5 (2003): 469–474.

Chanda, Sayantani, and T. V. Ramachandra. "Phytochemical and pharmacological importance of turmeric (Curcuma longa): A review." Research & Reviews: A Journal of Pharmacology 9, no. 1 (2019): 16–23.

Chen, Jingwang, Taihua Mu, Miao Zhang, Dorothée Goffin, Hongnan Sun, Mengmei Ma, Xingli Liu, and Duqin Zhang. "Structure, physicochemical, and functional properties of protein isolates and major fractions from cumin (Cuminum cyminum) seeds." International Journal of Food Properties 21, no. 1 (2018): 685–701.

Correia, Alyne Oliveira, Abílio Augusto Pimentel Cruz, Arôdo Tenório Ribeiro de Aquino, Joanisson Rubens Gomes Diniz, Karizia Bianca Ferreira Santana, Pedro Ivo Martins Cidade, Jaine Dantas Peixoto et al. "Neuroprotective effects of piperine, an alkaloid from the Piper genus, on the Parkinson's disease model in rats." (2015).

Domarацký, M., P. Rehak, Š0 Juhás, and J. Koppel. "Effects of selected plant essential oils on the growth and development of mouse preimplantation embryos in vivo." Physiological Research 56, no. 1 (2007).

Duke, J. A., M. J. Bogenschutz-Godwin, J. deCellier, and P. K. Duke. "Syzygium aromaticum (L.) Merr. and LM Perry (Myrtaceae) clavos, clove, clovetree." CRC Handbook of Medicinal Spices (2003): 281.

Fahmi, Abdelgawad, Naglaa Hassanen, Mariam Abdur-Rahman, and Engy Shams-Eldin. "Phytochemicals, antioxidant activity and hepatoprotective effect of ginger (Zingiber officinale) on diethylnitrosamine toxicity in rats." Biomarkers 24, no. 5 (2019): 436–447.

Flatt, Emma J. "Spices, smells and spells: The use of olfactory substances in the conjuring of spirits." South Asian Studies 32, no. 1 (2016): 3–21.

Flood, Gavin. The tantric body: The secret tradition of Hindu religion. Bloomsbury Publishing, 2005.

Gülçin, İlhami. "The antioxidant and radical scavenging activities of black pepper (Piper nigrum) seeds." International Journal of Food Sciences and Nutrition 56, no. 7 (2005): 491–499.

Gupta, Mradu. "Pharmacological properties and traditional therapeutic uses of important Indian spices: A review." International Journal of Food Properties 13, no. 5 (2010): 1092–1116.

Gupta, Subash C., Bokyung Sung, Ji Hye Kim, Sahdeo Prasad, Shiyou Li, and Bharat B. Aggarwal. "Multitargeting by turmeric, the golden spice: From kitchen to clinic." Molecular Nutrition & Food Research 57, no. 9 (2013): 1510–1528.

Hameed, Imad Hadi, Huda Jasim Altameme, and Ghaidaa Jihadi Mohammed. "Evaluation of antifungal and antibacterial activity and analysis of bioactive phytochemical compounds of Cinnamomum zeylanicum (Cinnamon bark) using gas chromatography-mass spectrometry." Oriental Journal of Chemistry 32, no. 4 (2016): 1769.

Hauser, Michelle E. "Culinary medicine basics and applications in medical education in the United States." In Nutrition education: Strategies for improving nutrition and healthy eating in individuals and communities, vol. 92, pp. 161–170. Karger Publishers, 2020.

Hritcu, Lucian, Jaurès A. Noumedem, Oana Cioanca, Monica Hancianu, Paula Postu, and Marius Mihasan. "Anxiolytic and antidepressant profile of the methanolic extract of piper nigrum fruits in beta-amyloid (1–42) rat model of Alzheimer's disease." Behavioral and Brain Functions 11 (2015): 1–13.

Jakhetia, Vaibhavi, Rakesh Patel, Pankaj Khatri, Neeraj Pahuja, Sunil Garg, Anupriya Pandey, and Sonu Sharma. "Cinnamon: A pharmacological review." Journal of Advanced Scientific Research 1, no. 2 (2010): 19–23.

Jayaprakasha, Guddadarangavvanahally Krishanareddy, L. Jaganmohan Rao, and Kunnumpurath K. Sakariah. "Antioxidant activities of curcumin, demethoxycurcumin and bisdemethoxycurcumin." Food Chemistry 98, no. 4 (2006): 720–724.

Jayaprakasha, G. K., R. P. Singh, J. Pereira, and K. K. Sakariah. "Limonoids from citrus reticulata and their moult inhibiting activity in mosquito culex quinquefasciatus larvae." Phytochemistry 44, no. 5 (1997): 843–846.

Jolad, Shivanand D., R. Clark Lantz, Aniko M. Solyom, Guan Jie Chen, Robert B. Bates, and Barbara N. Timmermann. "Fresh organically grown ginger (Zingiber officinale): Composition and effects on LPS-induced PGE2 production." Phytochemistry 65, no. 13 (2004): 1937–1954.

Jung Park, Eun, and John M. Pezzuto. "Botanicals in cancer chemoprevention." Cancer and Metastasis Reviews 21 (2002): 231–255.

Jyotirmayee, B., and Gyanranjan Mahalik. "A review on selected pharmacological activities of Curcuma longa L." International Journal of Food Properties 25, no. 1 (2022): 1377–1398.

Kamali, Alireza, Shabnam Beigi, Maryam Shokrpour, and Shirin Pazuki. "The efficacy of ginger and doxedetomidine in reducing postoperative nausea and vomiting in patients undergoing abdominal hysterectomy." Alternative Therapies in Health and Medicine 26, no. 2 (2020): 28–33.

Kang, Naixin, Renyikun Yuan, Liting Huang, Zhenjie Liu, Dan Huang, Lu Huang, Hongwei Gao, Yanli Liu, Qiong-Ming Xu, and Shilin Yang. "Atypical nitrogen-containing flavonoid in the fruits of cumin (Cuminum cyminum L.) with anti-inflammatory activity." Journal of Agricultural and Food Chemistry 67, no. 30 (2019): 8339–8347.

Khan, Abiha Ahmad, and Begum Wajeeha. "Efficacy of Darchini in the management of polycystic ovarian syndrome: A randomized clinical study." Journal of Herbal Medicine 15 (2019): 100249.

Kim, Suyong, Chunhoo Cheon, Bonglee Kim, and Woojin Kim. "The effect of ginger and its sub-components on pain." Plants 11, no. 17 (2022): 2296.

Kumar, Suresh, Jitpal Kamboj, and Sunil Sharma. "Overview for various aspects of the health benefits of piper longum linn. fruit." Journal of Acupuncture and Meridian Studies 4, no. 2 (2011): 134–140.

La Puma, John. "Culinary medicine and nature: Foods that work together." American Journal of Lifestyle Medicine 14, no. 2 (2020): 143–146.

Mahassni, Sawsan Hassan, and Oroob Abid Bukhari. "Beneficial effects of an aqueous ginger extract on the immune system cells and antibodies, hematology, and thyroid hormones in male smokers and non-smokers." Journal of Nutrition & Intermediary Metabolism 15 (2019): 10–17.

Mehdizadeh, Leila, Abdollah Ghasemi Pirbalouti, and Mohammad Moghaddam. "Storage stability of essential oil of cumin (Cuminum cyminum L.) as a function of temperature." International Journal of Food Properties 20, no. sup2 (2017): 1742–1750.

Memarzia, Arghavan, Mohammad R. Khazdair, Sepideh Behrouz, Zahra Gholamnezhad, Maryam Jafarnezhad, Saeideh Saadat, and Mohammad H. Boskabady. "Experimental and clinical reports on anti-inflammatory, antioxidant, and immunomodulatory effects of curcuma longa and curcumin, an updated and comprehensive review." BioFactors 47, no. 3 (2021): 311–350.

Mnif, Sami, and Sami Aifa. "Cumin (Cuminum cyminum L.) from traditional uses to potential biomedical applications." Chemistry & Biodiversity 12, no. 5 (2015): 733–742.

Moubarz, Gehan, Mohamed A. Embaby, Nada M. Doleib, and Mona M. Taha. "Effect of dietary antioxidant supplementation (Cuminum cyminum) on bacterial susceptibility of diabetes-induced rats." Central European Journal of Immunology 41, no. 2 (2016): 123–137.

Nasri, Hamid, Najmeh Sahinfard, Mortaza Rafieian, Samira Rafieian, Maryam Shirzad, and Mahmoud Rafieian-Kopaei. "Turmeric: A spice with multifunctional medicinal properties." Journal of HerbMed Pharmacology 3, no. 1 (2014): 5–8.

Nicoll, Rachel, and Michael Y. Henein. "Ginger (Zingiber officinale Roscoe): A hot remedy for cardiovascular disease?" International Journal of Cardiology 131, no. 3 (2009): 408–409.

Nirmala, M. Joyce, Latha Durai, Kagitala Anvesh Rao, and Ramamurthy Nagarajan. "Ultrasonic nanoemulsification of cuminum cyminum essential oil and its applications in medicine." International Journal of Nanomedicine (2020): 795–807.

O'Riordan, Michael J., and Martin G. Wilkinson. "A survey of the incidence and level of aflatoxin contamination in a range of imported spice preparations on the Irish retail market." Food Chemistry 107, no. 4 (2008): 1429–1435.

Packiavathy, Issac Abraham Sybiya Vasantha, Palani Agilandeswari, Khadar Syed Musthafa, Shunmugiah Karutha Pandian, and Arumugam Veera Ravi. "Antibiofilm and quorum sensing inhibitory potential of cuminum cyminum and its secondary metabolite methyl eugenol against gram negative bacterial pathogens." Food Research International 45, no. 1 (2012): 85–92.

Padakatti, Tejashwini, and Meti, Renuka. "Indian spices: Traditional and medicinal use." International Journal of Home Science 6, no. 2 (2020): 42–44.

Rathore, M. S., and N. S. Shekhawat. "Incredible spices of India: From traditions to cuisine." American-Eurasian Journal of Botany 1, no. 3 (2008): 85–89.

Ravindran, P. N., K. Nirmal Babu, and K. Sivaraman. "Turmeric: The golden spice of life." Turmeric: The Genus Curcuma. CRC Press, Boca Raton, FL, USA (2007): 1–14.

Romagnoli, Carlo, Elisa Andreotti, Silvia Maietti, Rai Mahendra, and Donatella Mares. "Antifungal activity of essential oil from fruits of Indian cuminum cyminum." Pharmaceutical Biology 48, no. 7 (2010): 834–838.

Sachan, A. K., Sunil Kumar, Kiran Kumari, and Deepti Singh. "Medicinal uses of spices used in our traditional culture: Worldwide." Journal of Medicinal Plants Studies 6, no. 3 (2018): 116–122.

Samarasekera, Radhika, Kosmulalage S. Kalhari, and Indira S. Weerasinghe. "Mosquitocidal activity of leaf and bark essential oils of ceylon cinnamomum zeylanicum." Journal of Essential Oil Research 17, no. 3 (2005): 301–303.

Sharangi, A. B., and S. Guha. "Wonders of leafy spices: Medicinal properties ensuring Human Health." Science International 1, no. 9 (2013): 312–317.

Shoaib, Muhammad, Aamir Shehzad, Masood Sadiq Butt, Muhammad Saeed, H. Raza, S. Niazi, I. M. Khan, and A. Shakeel. "An overview: Ginger, a tremendous herb." Journal of Global Innovations in Agricultural and Social Sciences 4, no. 4 (2016): 172–187.

Shobana, S., and K. Akhilender Naidu. "Antioxidant activity of selected Indian spices." Prostaglandins, Leukotrienes and Essential Fatty Acids (PLEFA) 62, no. 2 (2000): 107–110.

Shukla, Ankit, and Nagendra Yadav. "Role of Indian spices in Indian history." International Journal of Management Research and Reviews 8, no. 11 (2018): 1.

Singh, Sunita, I. P. S. Kapoor, Gurdip Singh, Carola Schuff, M. P. De Lampasona, and Cesar A. N. Catalan. "Chemistry, antioxidant and antimicrobial potentials of white pepper (Piper nigrum L.) essential oil and oleoresins." Proceedings of the National Academy of Sciences, India Section B: Biological Sciences 83 (2013): 357–366.

Tahir, Hafizah Umaira, Raja Adil Sarfraz, Aisha Ashraf, and Shazia Adil. "Chemical composition and antidiabetic activity of essential oils obtained from two spices (syzygium aromaticum and cuminum cyminum)." International Journal of Food Properties 19, no. 10 (2016): 2156–2164.

Takooree, Heerasing, Muhammad Z. Aumeeruddy, Kannan RR Rengasamy, Katharigatta N. Venugopala, Rajesh Jeewon, Gokhan Zengin, and Mohamad F. Mahomoodally. "A systematic review on black pepper (piper nigrum L.): From folk uses to pharmacological applications." Critical Reviews in Food Science and Nutrition 59, no. sup1 (2019): S210–S243.

Tasleem, Farhana, Iqbal Azhar, Syed Nawazish Ali, Shaista Perveen, and Zafar Alam Mahmood. "Analgesic and anti-inflammatory activities of piper nigrum L." Asian Pacific Journal of Tropical Medicine 7 (2014): S461–S468.

Vasanthi. R., Hannah, and R. P Parameswari. "Indian spices for healthy heart-an overview." Current Cardiology Reviews 6, no. 4 (2010): 274–279.

Verma, Rahul Kumar, Preeti Kumari, Rohit Kumar Maurya, Vijay Kumar, R. B. Verma, and Rahul Kumar Singh. "Medicinal properties of turmeric (curcuma longa L.): A review." International Journal of Chemical Studies 6, no. 4 (2018): 1354–1357.

Wong, Y. C., M. Y. Ahmad-Mudzaqqir, and W. A. Wan-Nurdiyana. "Extraction of essential oil from cinnamon (cinnamomum zeylanicum)." Oriental Journal of Chemistry 30, no. 1 (2014): 37.

Yadav, Virendra Kumar, Nisha Choudhary, S. Heena Khan, Areeba Khayal, Raman Kumar Ravi, Pankaj Kumar, Shreya Modi, and G. Gnanamoorthy. "Incense and incense sticks: Types, components, origin and their religious beliefs and importance among different religions." Journal of Bio Innovation 9 (2020): 1420–1439.

Yang, Wei, Yu-Hua Chen, Hao Liu, and Hong-Dang Qu. "Neuroprotective effects of piperine on the 1-methyl-4-phenyl-1, 2, 3, 6-tetrahydropyridine-induced Parkinson's disease mouse model." International Journal of Molecular Medicine 36, no. 5 (2015): 1369–1376.

Yang, Young-Cheol, Hoi-Seon Lee, Si Hyeock Lee, J. Marshall Clark, and Young-Joon Ahn. "Ovicidal and adulticidal activities of cinnamomum zeylanicum bark essential oil compounds and related compounds against pediculus humanus capitis (Anoplura: Pediculicidae)." International Journal for Parasitology 35, no. 14 (2005): 1595–1600.

Zarai, Zied, Emna Boujelbene, Nadia Ben Salem, Youssef Gargouri, and Adel Sayari. "Antioxidant and antimicrobial activities of various solvent extracts, piperine and piperic acid from piper nigrum." Lwt-Food Science and Technology 50, no. 2 (2013): 634–641.

20 Pharmacological Chemistry of Essential Oils from Spices

Rama Satya Sri Kotipalli, Mani Sharma, Nemala Siva Kumar, and Neeraj Kumar Chouhan

20.1. INTRODUCTION

Essential oils from spices have been valued for centuries for their medicinal, aromatic, and therapeutic properties. Extracted from seeds, bark, and roots of spices such as clove, cinnamon, and black pepper, these oils have been integral to traditional medicine and cultural practices worldwide.

Ancient civilizations used these oils for wound healing, pain relief, and well-being, while modern research highlights their bioactive compounds, such as eugenol and cinnamaldehyde, which offer antimicrobial, anti-inflammatory, and antioxidant benefits. Today, spice-derived essential oils find applications in pharmaceuticals, aromatherapy, cosmetics, and food preservation (Sulieman et al., 2023).

20.1.1. Historical Uses of Essential Oils from Spices

Essential oils extracted from spices have been an integral part of traditional medicine and cultural practices for centuries. Their therapeutic properties were widely recognized across ancient civilizations, where they were used for a variety of health, spiritual, and aromatic purposes. In traditional systems of medicine, essential oils from spices such as clove, cinnamon, nutmeg, and black pepper played a crucial role. In Ayurveda, clove oil was used for its analgesic and antiseptic properties to treat toothaches and infections, while cinnamon oil was valued for its warming effects and ability to enhance circulation. Similarly, Traditional Chinese Medicine (TCM) utilized star anise oil for its antimicrobial and digestive benefits. Native American and Middle Eastern cultures also incorporated spice-derived essential oils for wound healing, respiratory ailments, and general wellness (Sulieman et al., 2023). The use of essential oils from spices varied significantly across cultures and regions. In ancient Egypt, cinnamon and nutmeg oils were highly prized for embalming rituals and as offerings to deities. The Greeks and Romans valued these oils for their aromatic and medicinal properties, often using them in baths and perfumes. In Southeast Asia, oils such as cardamom and turmeric were staples in traditional healing for their anti-inflammatory and digestive benefits. The Middle Eastern cultures frequently employed spice oils in culinary, medicinal, and religious practices, while indigenous tribes of the Americas relied on such oils for their natural healing capabilities (Erichsen-Brown, 2013).

20.1.2. Overview of Essential Oils in Traditional Medicine

Essential oils derived from spices have long been an integral part of traditional medicine systems worldwide. Known for their aromatic and therapeutic properties, these oils were used in various forms to treat ailments, enhance well-being, and support spiritual practices. Their applications ranged from antimicrobial and anti-inflammatory remedies to digestive aids and pain relievers. Traditional systems such as Ayurveda, TCM, and ancient Middle Eastern practices extensively utilized these

DOI: 10.1201/9781003559894-20

oils for their potent health benefits (Shi et al., 2022). Historical texts and records provide a wealth of information on the use of essential oils in traditional medicine. Ancient Egyptian papyri describe the use of spice oils such as cinnamon and nutmeg for embalming and healing wounds. Ayurvedic and Chinese medical texts detail the therapeutic roles of clove, cardamom, and pepper oils in maintaining health and treating specific conditions. Middle Eastern and Greco-Roman records highlight the incorporation of these oils in rituals, perfumery, and medical treatments.

The historical documentation underscores the enduring significance of spice-derived essential oils and their pivotal role in shaping both traditional and modern medicinal practices. This chapter explores the diverse applications of these oils, emphasizing their cultural and historical context, and their transition into contemporary medicine and industry (Schnaubelt, 1999).

20.2. CHEMICAL COMPOSITION OF ESSENTIAL OILS

20.2.1. Major Constituents: Terpenes, Phenolics, and Aldehydes

The major constituents of essential oils include terpenes, phenolics, and aldehydes, each playing a crucial role in their biological and aromatic properties. Terpenes, the most abundant class, are hydrocarbons derived from isoprene units, categorized into monoterpenes like limonene and sesquiterpenes like β-caryophyllene (Figure 20.1).

These compounds contribute to the characteristic aroma and therapeutic effects of essential oils. Phenolics, such as eugenol and thymol, are aromatic compounds with hydroxyl groups that enhance

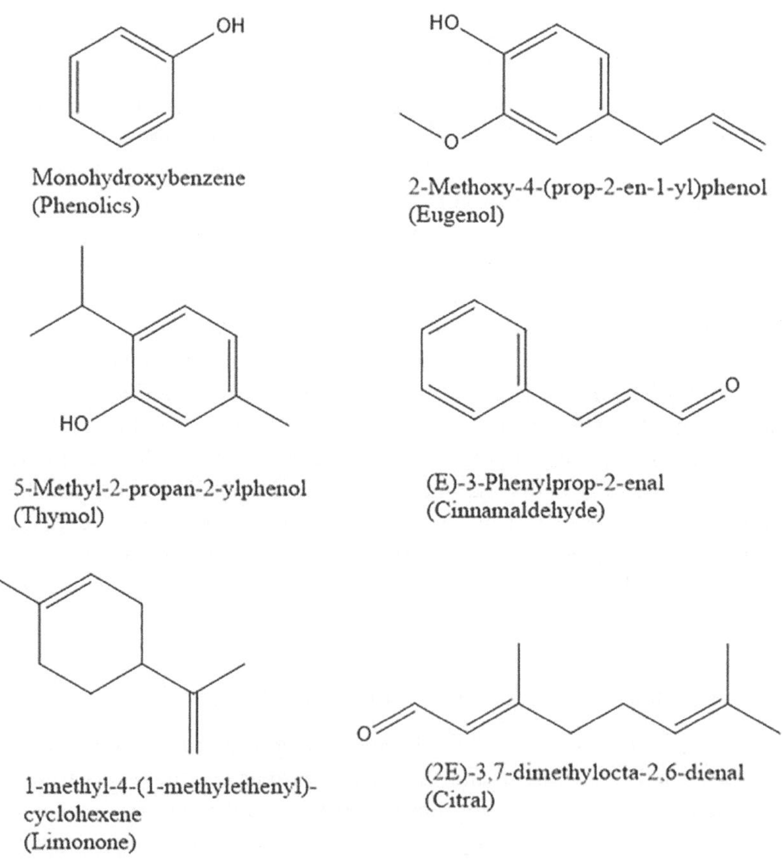

FIGURE 20.1 Structure and IUPAC names of major constituents used in essential oils.

their antioxidant and antimicrobial properties. Aldehydes, known for their sharp and often citrusy aromas, such as citral and cinnamaldehyde, exhibit high reactivity and potent antimicrobial activities. Together, these constituents define the chemical complexity and functionality of essential oils (Dhifi et al., 2016).

20.2.1.1. Structural Characteristics

The structural diversity of essential oil constituents influences their physical, chemical, and biological properties. Terpenes are highly volatile and aromatic due to their small molecular size, with monoterpenes often having cyclic or linear structures and sesquiterpenes being less volatile but more bioactive. Phenolics, characterized by hydroxyl groups on aromatic rings, enhance solubility and antioxidant potential. Aldehydes, with their reactive carbonyl groups, contribute to their distinct aromas and antimicrobial efficacy. This structural variation determines their roles in therapeutic and industrial applications, highlighting the importance of understanding their chemistry (Dhifi et al., 2016).

20.2.1.2. Biosynthetic Pathways

The biosynthesis of essential oil components is driven by specialized metabolic pathways in plants. Terpenes are synthesized through the mevalonate (MVA) pathway in the cytosol and the methylerythritol phosphate (MEP) pathway in plastids, where the precursors isopentenyl diphosphate (IPP) and dimethylallyl diphosphate (DMAPP) undergo condensation to form monoterpenes and sesquiterpenes. Phenolics originate from the shikimic acid pathway, with phenylalanine serving as a key precursor that is converted into cinnamic acid, leading to the formation of various phenolic compounds. Aldehydes are often produced via the fatty acid metabolism pathway, where oxidative cleavage of unsaturated fatty acids or modifications of cinnamic acid derivatives generate reactive aldehydes such as citral and cinnamaldehyde. These biosynthetic routes reflect the enzymatic complexity underlying the chemical and functional diversity of essential oils (Rehman et al., 2016).

20.2.2. Methods for Chemical Analysis and Identification

20.2.2.1. Gas Chromatography-Mass Spectrometry

Gas chromatography-mass spectrometry (GC-MS) is a widely used analytical technique for the separation, identification, and quantification of volatile and semi-volatile compounds in essential oils. The method combines the high-resolution separation capability of gas chromatography with the precise molecular identification provided by mass spectrometry. In GC-MS analysis, the sample is vapourized and carried by an inert gas through a chromatographic column, where components are separated based on their boiling points and polarity. Each separated compound is then ionized and detected by the mass spectrometer, producing a unique mass spectrum that serves as a molecular fingerprint. GC-MS is particularly effective for analyzing complex mixtures such as essential oils, allowing for the detection of individual constituents such as terpenes, phenolics, and aldehydes. Its sensitivity, reproducibility, and ability to provide qualitative and quantitative data make it an indispensable tool in essential oil research (Do et al., 2015).

20.2.2.2. High-Performance Liquid Chromatography

High-performance liquid chromatography (HPLC) is a powerful technique for analyzing nonvolatile or thermally labile components in essential oils. HPLC works by passing a liquid sample through a column packed with a stationary phase, where compounds are separated based on their interactions with the stationary phase and their solubility in the mobile phase. Detection methods such as ultraviolet (UV) or fluorescence spectroscopy are commonly coupled with HPLC to identify and quantify components. HPLC is particularly useful for analyzing phenolics, flavonoids, and other polar compounds in essential oils, providing high sensitivity and precision. This method allows for the detailed profiling of essential oil components, making it a valuable tool for quality control and standardization in industrial and pharmaceutical applications (Milá Cáceres, 2010).

20.2.2.3 Nuclear Magnetic Resonance Spectroscopy

Nuclear magnetic resonance (NMR) spectroscopy is a nondestructive analytical technique used for elucidating the molecular structures of essential oil components. NMR works by exploiting the magnetic properties of atomic nuclei, such as hydrogen (^1H) and carbon (^13C), in a magnetic field. When exposed to radiofrequency energy, nuclei resonate at specific frequencies, producing a spectrum that reveals information about the chemical environment, bonding, and molecular interactions. NMR is particularly effective for structural characterization, stereochemistry determination, and quantification of both volatile and nonvolatile components. Unlike GC-MS and HPLC, NMR does not require separation of the sample, allowing for the direct analysis of essential oils in their natural mixture. Its precision and ability to provide comprehensive structural information make NMR a crucial method in advanced essential oil research and quality assessment (Milá Cáceres, 2010).

20.3. PHARMACOLOGICAL PROPERTIES OF ESSENTIAL OILS

20.3.1. Antimicrobial Activities

Essential oils are renowned for their potent antimicrobial properties, which make them valuable in combating a wide range of pathogenic microorganisms. Their efficacy against bacteria, fungi, and viruses has been extensively documented, highlighting their potential as natural antimicrobial agents in the pharmaceutical, food, and cosmetics industries. The bioactive compounds in essential oils, such as terpenes, phenolics, and aldehydes, contribute to their broad-spectrum antimicrobial effects. These properties have positioned essential oils as promising alternatives to synthetic antimicrobials, especially in addressing drug resistance issues (Wińska et al., 2019).

20.3.1.1. Spectrum of Activity Against Bacteria, Fungi, and Viruses

Essential oils exhibit a broad spectrum of activity against various bacterial, fungal, and viral pathogens. Against bacteria, essential oils are effective against both gram-positive species, such as *Staphylococcus aureus*, and gram-negative species, such as *Escherichia coli*, despite the latter's

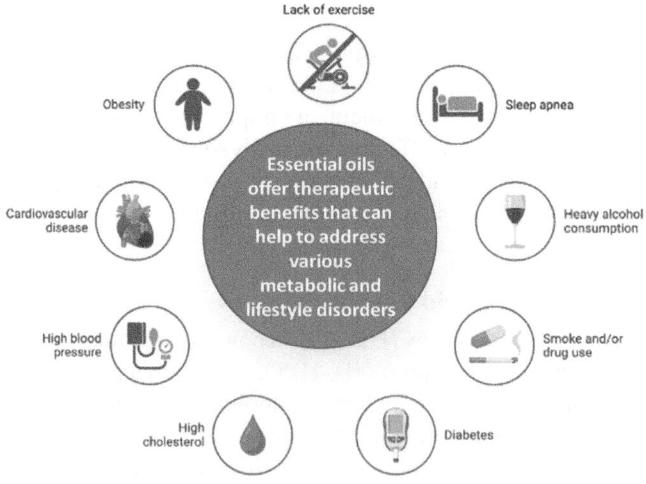

FIGURE 20.2 Therapeutic advantage of essential oils.

more resistant outer membrane. They also display strong antifungal activity, particularly against species such as *Candida albicans* and *Aspergillus niger*, which are known for causing severe infections. Furthermore, certain essential oils demonstrate antiviral properties by inhibiting viral replication or disrupting viral envelopes, as observed with viruses such as herpes simplex virus (HSV) and influenza virus. The diverse activity spectrum of essential oils underscores their utility in managing a variety of infections (Figure 20.2) (Visan and Negut, 2024).

20.3.1.2. Mechanisms of Antimicrobial Action

The antimicrobial action of essential oils involves multiple mechanisms, targeting both structural and functional aspects of microbial cells. Their hydrophobic nature allows them to interact with and disrupt microbial cell membranes, leading to leakage of intracellular contents and cell death. Compounds like phenolics can interfere with the integrity of the lipid bilayer, causing increased permeability. Additionally, essential oils can inhibit enzyme activities and disrupt metabolic pathways essential for microbial growth. For example, aldehydes and terpenoids interfere with ATP synthesis and protein folding. In viral infections, essential oils can disrupt the viral envelope or inhibit viral attachment and replication processes. The ability of essential oils to target multiple cellular pathways makes them effective against resistant strains, offering a multifaceted approach to antimicrobial therapy (Saha, 2023).

20.3.2. Anti-Inflammatory Properties

Essential oils exhibit significant anti-inflammatory properties, which have been widely studied for their potential therapeutic applications. These properties are attributed to the presence of bioactive compounds such as terpenes, phenolics, and aldehydes, which modulate inflammatory responses. Essential oils act by suppressing the activation of inflammatory pathways and reducing the production of mediators responsible for inflammation. This makes them valuable in managing inflammatory conditions, including arthritis, dermatitis, and other chronic inflammatory disorders, with fewer side effects compared to synthetic anti-inflammatory drugs.

20.3.2.1. Inhibition of Pro-Inflammatory Cytokines and Enzymes

The anti-inflammatory effects of essential oils are primarily mediated by their ability to inhibit pro-inflammatory cytokines and enzymes. Essential oil constituents, such as linalool, eugenol, and carvacrol, downregulate the expression of cytokines such as tumour necrosis factor-alpha (TNF-α), interleukin-6 (IL-6), and interleukin-1 beta (IL-1β), which play critical roles in propagating inflammation. Additionally, essential oils inhibit key enzymes, such as cyclooxygenase (COX) and lipoxygenase (LOX), reducing the synthesis of prostaglandins and leukotrienes, which are potent inflammatory mediators. These actions collectively alleviate inflammation and prevent tissue damage, demonstrating the therapeutic potential of essential oils in controlling inflammatory processes (Baptista-Silva et al., 2020).

20.3.2.2. Experimental Models and Clinical Evidence

The anti-inflammatory properties of essential oils have been validated through numerous experimental models and clinical studies. In animal models, essential oils have shown significant efficacy in reducing inflammation induced by carrageenan, lipopolysaccharides (LPS), and other pro-inflammatory agents. For instance, the administration of lavender or tea tree oil in rodent models demonstrated a marked reduction in paw swelling and inflammatory markers. Clinical evidence further supports these findings, with studies indicating the effectiveness of essential oils in alleviating symptoms of inflammatory diseases such as rheumatoid arthritis and atopic dermatitis. Patients treated with essential oil formulations reported decreased pain, swelling, and improved quality of life, highlighting their potential as complementary therapies in clinical settings (Talati et al., 2023).

20.3.3. Antioxidant Effects

20.3.3.1. Free Radical–Scavenging Activity

The antioxidant activity of essential oils is primarily due to their ability to scavenge free radicals, such as reactive oxygen species (ROS) and reactive nitrogen species (RNS). Compounds such as eugenol, thymol, and carvacrol donate electrons or hydrogen atoms to stabilize free radicals, thereby preventing chain reactions that lead to oxidative damage. Essential oils have been shown to exhibit strong radical-scavenging activity in in vitro assays, such as the DPPH (2,2-diphenyl-1-picrylhydrazyl) and ABTS (2,2′-azino-bis(3-ethylbenzothiazoline-6-sulfonic acid)) assays. This activity underscores their potential as natural agents for combating oxidative stress and its associated complications (Muscolo et al., 2024).

20.3.3.2. Protection Against Oxidative Stress-Induced Damage

Essential oils also protect against damage induced by oxidative stress in biological systems. By reducing lipid peroxidation, protein oxidation, and DNA damage, they help maintain cellular integrity and function. Studies using both in vitro and in vivo models have demonstrated the protective effects of essential oils against oxidative damage caused by environmental factors, toxins, and aging. For example, rosemary and clove essential oils have shown protective effects in models of neurodegenerative diseases by reducing ROS levels and enhancing endogenous antioxidant enzymes, such as superoxide dismutase (SOD) and catalase (CAT). This protective mechanism highlights the therapeutic potential of essential oils in mitigating oxidative stress-related disorders.

20.3.4. Anticancer Potential

Essential oils exhibit promising anticancer properties, making them a focus of research for natural cancer therapeutics. Their bioactive constituents, including terpenes, phenolics, and aldehydes, demonstrate the ability to target cancer cells selectively while sparing normal cells. These oils exert antiproliferative effects, inhibit metastasis, and enhance the efficacy of conventional chemotherapy. The diverse anticancer mechanisms of essential oils underline their potential as complementary or standalone agents in cancer treatment.

20.3.4.1. Cytotoxicity Against Cancer Cells

The cytotoxic effects of essential oils on cancer cells have been demonstrated in numerous in vitro and in vivo studies. Essential oils, such as those derived from oregano, thyme, and frankincense, exhibit potent cytotoxicity against a variety of cancer cell lines, including breast, lung, and colorectal cancers. Bioactive compounds, such as thymol, eugenol, and limonene, induce cytotoxic effects by disrupting the cancer cell membrane, altering mitochondrial function, and reducing cell viability. These oils often exhibit dose-dependent effects, with higher concentrations leading to significant cancer cell death while showing minimal toxicity to normal cells, highlighting their therapeutic potential (Orrenius, Nicotera, and Zhivotovsky, 2011) (Sharma et al., 2024).

20.3.4.2. Mechanisms of Apoptosis Induction and Cell Cycle Arrest

The anticancer effects of essential oils are largely mediated by their ability to induce apoptosis and disrupt the cell cycle in cancer cells. Essential oil components, such as linalool and β-caryophyllene, trigger intrinsic and extrinsic apoptotic pathways by activating caspases, increasing pro-apoptotic proteins like Bax, and downregulating antiapoptotic proteins such as Bcl-2 (Chouhan et al., 2024). Additionally, these compounds arrest the cell cycle at critical checkpoints, such as the G1/S or G2/M phases, by modulating cyclins and cyclin-dependent kinases. By inducing programmed cell death and halting uncontrolled cell division, essential oils effectively inhibit tumour growth and progression, positioning them as valuable candidates in the fight against cancer (Sitarek et al., 2020).

20.4. MECHANISMS OF ACTION

20.4.1. Modulation of Signaling Pathways

Essential oils influence critical signaling pathways, such as NF-κB, MAPK, and PI3K/Akt, which regulate processes such as inflammation, oxidative stress, apoptosis, and cell proliferation. These interactions support immune regulation, tumour suppression, and tissue repair, positioning essential oils as promising agents for complex conditions, such as cancer, neurodegeneration, and inflammatory disorders.

20.4.1.1. Key Pathways Affected

Essential oils modulate pathways such as NF-κB to reduce pro-inflammatory cytokines, MAPK to regulate stress responses, and PI3K/Akt to inhibit tumour growth and promote apoptosis. Compounds such as eugenol and thymol showcase these versatile effects (de Sousa et al., 2023).

20.4.1.2. Molecular Targets

By acting on molecular targets, such as GPCRs, COX-2, LOX, and transcription factors, essential oils mediate anti-inflammatory, antioxidant, and apoptotic effects. These targeted actions enhance their potential for treating diverse diseases while maintaining physiological balance.

20.4.2. Interaction with Cellular Targets

Essential oils interact with cellular structures such as receptors, enzymes, and organelles to regulate signaling pathways, metabolism, and defence mechanisms, contributing to their broad therapeutic efficacy.

20.4.2.1. Receptor Binding and Enzyme Inhibition

Essential oils bind to receptors such as GPCRs and TRP channels for pain relief and immune modulation. They also inhibit key enzymes, including COX-2, LOX, and AChE, delivering anti-inflammatory, antimicrobial, and neuroprotective effects.

20.4.2.2. Cellular Uptake and Distribution

The lipophilic nature of essential oils facilitates efficient cellular uptake through passive diffusion, allowing them to cross membranes and target organelles like mitochondria. This enhances their bioavailability, enabling them to regulate energy production, oxidative stress, and other cellular functions effectively (Aqil et al., 2013).

20.4.3. Synergistic Effects with Other Compounds

Essential oils enhance therapeutic potential when combined with natural or synthetic compounds, boosting effects such as anti-inflammatory, anticancer, and antimicrobial activity. This synergy targets multiple pathways, creating versatile and effective treatment strategies, especially for conditions where single-agent therapies are insufficient (Vaou et al., 2022).

20.4.3.1. Combination Therapies

Essential oils enhance therapeutic efficacy when paired with natural products or conventional drugs. These combinations amplify pharmacological effects, improving outcomes in cancer, antimicrobial treatments, and reducing drug resistance. For example, combining essential oils with chemotherapy agents or antibiotics boosts effectiveness while lowering required drug doses.

20.4.3.2. Enhanced Efficacy and Reduced Toxicity

Essential oils improve drug efficacy and reduce toxicity by enhancing drug absorption and targeting while minimizing side effects. Oils such as lavender and rosemary enhance the anticancer activity of chemotherapy agents while mitigating their harmful effects, offering safer and more effective treatment options (Aljaafari et al., 2021).

20.5. EXTRACTION METHODS

20.5.1. Traditional Extraction Techniques

Traditional methods are widely used for essential oil production due to their simplicity, cost-effectiveness, and ability to preserve oil quality. Though modern methods such as supercritical CO_2 extraction are gaining popularity, traditional techniques remain valuable for their efficiency and accessibility.

20.5.1.1. Steam Distillation

Steam distillation involves passing steam through plant materials to vaporize essential oils, which are then condensed and separated. Ideal for flowers, leaves, and stems, this method operates at low temperatures, preserving heat-sensitive compounds and ensuring high-quality oils (Machado et al., 2022).

20.5.1.2. Cold Pressing

Cold pressing mechanically extracts oils from citrus fruit peels without heat, preserving their purity and natural fragrance. This simple, cost-effective method is ideal for citrus oils, protecting delicate compounds from heat degradation.

20.5.1.3. Solvent Extraction

Solvent extraction uses solvents such as ethanol or hexane to dissolve oils from delicate or resinous plant materials, such as jasmine and tuberose. Although it yields high quantities, trace solvents may remain, requiring further purification (Stratakos and Koidis, 2016).

20.5.2. Modern Extraction Technologies

20.5.2.1. Supercritical Fluid Extraction (SFE)

SFE uses supercritical CO_2 to efficiently extract essential oils without high temperatures or harmful solvents. This method ensures high selectivity, faster processing, and the preservation of sensitive compounds. It is ideal for high-value oils from herbs, spices, and medicinal plants, offering pure, high-quality oils with minimal environmental impact.

20.5.2.2. Microwave-Assisted Extraction (MAE)

MAE utilizes microwave energy to rapidly heat plant materials, causing cell rupture and releasing essential oils. It significantly reduces extraction time and energy consumption while preserving oil quality by operating at lower temperatures. This method is effective for low-yield plants or oils located deep within the plant matrix.

20.5.2.3. Ultrasound-Assisted Extraction (UAE)

UAE uses high-frequency sound waves to break plant cell walls, enhancing oil diffusion into solvents. This energy-efficient, nonthermal method offers quick and efficient extraction, particularly for tough or fibrous plant tissues. UAE minimizes heat exposure, preserving the therapeutic properties of essential oils (Ramachandran).

20.5.3. Comparison of Methods and Yield Efficiency

Traditional methods, such as steam distillation and cold pressing, are cost-effective but often yield less and take longer. Modern methods, such as supercritical fluid, microwave-assisted, and ultrasound-assisted extraction, offer higher efficiency and better preservation of bioactive compounds but are costlier and require specialized equipment.

20.5.3.1. Advantages and Disadvantages

Traditional methods are affordable and simple, ideal for large-scale use, but less efficient for delicate materials. Modern methods provide higher yields and better compound preservation but involve higher costs and complex setups. Selection depends on cost, quality, and plant material.

20.5.3.2. Yield and Purity Considerations

Yield and purity depend on the method used. Traditional techniques may lose some volatile compounds, resulting in lower yields. Modern methods achieve higher yields and retain bioactive compounds but may vary in purity due to solvents or plant matrix complexity. The choice balances yield, purity, and application needs (Belwal et al., 2018).

20.6. QUALITY CONTROL AND STANDARDIZATION

20.6.1. Importance of Purity and Consistency

The purity and consistency of essential oils are vital for their therapeutic effectiveness and safety. Small variations in purity can affect their performance, while consistency ensures therapeutic properties and regulatory compliance. Impurities can reduce benefits or cause harm.

20.6.1.1. Ensuring Therapeutic Efficacy

Purity and consistency directly impact the therapeutic effectiveness of essential oils. Impurities or inconsistent production can reduce their medicinal properties. Standardization ensures optimal active compound concentrations and reliable therapeutic outcomes.

20.6.1.2. Preventing Adverse Reactions

Ensuring purity and consistency is essential to avoid adverse reactions, such as allergies, skin irritation, or toxicity. Contaminants, such as synthetic additives, can cause harm. Strict quality control and standardization reduce risks and ensure safety, especially for sensitive populations (Salgueiro, Martins, and Correia, 2010).

20.6.2. Analytical Techniques for Quality Assurance

Analytical methods, such as chromatography and spectroscopy, are used to verify the quality, purity, and consistency of essential oils. These techniques help identify active compounds and detect contaminants or adulteration.

20.6.2.1. Chromatographic Methods

Techniques such as gas chromatography (GC) and high-performance liquid chromatography (HPLC) separate and identify components of essential oils, ensuring purity and authenticity. GC is used for volatile compounds, while HPLC is ideal for nonvolatile substances.

20.6.2.2. Spectroscopic Methods

Spectroscopy methods, such as nuclear magnetic resonance (NMR) and Fourier transform infrared (FTIR), provide molecular insights into the composition of essential oils, verifying their identity and detecting contaminants. These techniques confirm the authenticity and therapeutic properties of the oils (Sadgrove, Padilla-González, and Phumthum, 2022).

20.6.3. Regulatory Standards and Guidelines

Regulatory standards ensure the quality, safety, and efficacy of essential oils. Compliance with these guidelines is necessary for legal sale and global trade, ensuring products are consistently produced, accurately labeled, and safe for consumers.

20.6.3.1. International Standards

International bodies such as the International Organization for Standardization (ISO) and the European Pharmacopoeia set guidelines for essential oil quality, ensuring consistency across markets. These standards help prevent fraud and ensure products are free from contamination.

20.6.3.2. Compliance and Certification

Compliance with regulatory standards, such as ISO or organic certifications, assures consumers of product safety and authenticity. Certification also enables manufacturers to access global markets and boosts credibility (Calixto, 2000).

20.7. SAFETY AND TOXICOLOGY

20.7.1. Dosage and Administration

Proper dosage and administration are crucial for the safe therapeutic use of essential oils. Excessive doses can lead to adverse effects. Dosages depend on factors such as age, health, and the condition being treated. Essential oils are commonly administered topically, via inhalation, or orally, with appropriate dosage monitored for efficacy and safety.

20.7.1.1. Therapeutic Dosages

Essential oils require precise dosing based on specific oils and their intended effects. Dilution is often necessary, especially for topical or oral use. Standardized dosing ensures safety, balancing therapeutic benefits with avoiding adverse reactions.

20.7.1.2. Routes of Administration

Essential oils are applied through various methods, including topical use (diluted in carrier oils), inhalation (for respiratory or mood effects), and oral ingestion (under professional supervision). The route of administration impacts absorption and overall effectiveness.

20.7.2. Potential Adverse Effects

Essential oils can cause side effects such as skin irritation, allergic reactions, or gastrointestinal distress, especially with improper use. Sensitive populations (e.g., children, pregnant women) require extra caution. Monitoring is essential to mitigate adverse effects.

20.7.2.1. Toxicity Studies

Toxicity studies assess the harmful effects of essential oils in high doses. These studies identify toxic compounds, determine safe exposure levels, and establish safety standards for essential oil use.

20.7.2.2. Risk Assessment

Risk assessment evaluates potential hazards of essential oils based on dosage, method of use, and individual health conditions. It identifies safety margins and guides safe usage practices, minimizing harm while optimizing therapeutic effects.

20.7.3. Safety Assessment Protocols

Safety protocols include in vivo (on living organisms) and in vitro (on cell cultures) testing, risk assessments, and adherence to regulatory guidelines to ensure safe use of essential oils.

20.7.3.1. In Vivo and in Vitro Testing

In vivo testing studies the physiological effects on organisms, while in vitro testing evaluates cellular and biochemical effects. Both methods provide data on safety and risks.

20.7.3.2. Regulatory Requirements

Regulatory bodies such as the FDA and EMA set safety standards for essential oils. They ensure proper labeling, manufacturing, and testing, protecting consumers from unsafe or adulterated products (Aziz et al., 2018).

20.8. APPLICATIONS IN THERAPEUTICS

20.8.1. Current Medicinal Uses

Essential oils are used for treating various conditions, with benefits such as antimicrobial, anti-inflammatory, and antioxidant effects. They are effective for respiratory issues, anxiety, inflammation, and infections, making them valuable in both traditional and modern medicine.

20.8.1.1. Approved Therapeutic Applications

Essential oils such as lavender, peppermint, and tea tree are approved for medical use. Lavender aids anxiety and sleep, peppermint helps with digestion, and tea tree oil treats skin infections, with studies supporting their safety and efficacy.

20.8.1.2. Over-the-Counter (OTC) Products

Essential oils are available in OTC products such as creams and capsules for treating muscle pain, skin issues, and insomnia. Their accessibility makes them popular, but proper usage education is crucial to avoid misuse.

20.8.2. Potential for New Drug Development

Research into essential oils suggests potential for new drug development. Compounds from these oils may offer novel treatments for conditions such as cancer, pain, and neurodegenerative diseases, supported by advancements in extraction and formulation techniques.

20.8.2.1. Novel Compounds and Formulations

Active compounds from essential oils are being isolated to create targeted therapies with improved efficacy and fewer side effects. New formulations aim to enhance bioavailability and stability for better therapeutic outcomes.

20.8.2.2. Preclinical and Clinical Development

Essential oils undergo preclinical and clinical testing to ensure safety and efficacy. Preclinical research involves lab and animal studies, while clinical trials assess their effects in humans to confirm their suitability for medical use.

20.8.3. Case Studies and Clinical Trials

Case studies and clinical trials demonstrate the effectiveness of essential oils in treating conditions such as anxiety and pain, providing evidence-based recommendations for their use in medical practice.

20.8.3.1. Evidence from Clinical Research

Clinical research supports the use of essential oils for treating anxiety, sleep disorders, and pain. These studies provide scientific backing for incorporating essential oils into medical treatments.

20.8.3.2. Success Stories and Challenges

Essential oils have shown success in treating conditions such as anxiety and insomnia, but challenges such as inconsistent dosing and potential drug interactions must be addressed for broader clinical use (Hou et al., 2022).

TABLE 20.1
Summary of Key Findings and Future Directions in Essential Oils Research

Section	Summary
Summary of Key Findings	Essential oils show broad bioactivity (antimicrobial, anti-inflammatory, antioxidant, anticancer). They modulate biological systems, making them valuable in modern therapeutics for health and medicine (Ni et al., 2021).
Major Insights and Discoveries	Essential oils are effective for chronic conditions such as cancer and inflammation. Active compounds (terpenes, phenolics) are key for targeted effects. Advances in extraction and molecular understanding enable precise medical applications (de Lavor et al., 2018).
Future Directions in Research	Research is focused on improving efficacy, bioavailability (e.g., nanoparticles), and clinical applications. Synergy with other compounds offers potential for integrated therapies (Bunse et al., 2022).
Emerging Trends and Innovations	Nanotechnology, microencapsulation, and advanced extraction methods (supercritical fluid, microwave-assisted) are enhancing yields, precision, and therapeutic development (Albuquerque et al., 2022).
Unanswered Questions and Research Gaps	Gaps include long-term safety, pharmacokinetics, and clinical trial data for human applications. Addressing these will unlock the full therapeutic potential of essential oils (Pezantes-Orellana et al., 2024).

20.9. CONCLUSION

Essential oils exhibit a broad range of therapeutic properties, including antimicrobial, anti-inflammatory, antioxidant, and anticancer effects. Their active compounds, such as terpenes and phenolics, modulate biological systems, making them valuable in medicine. Advances in extraction and molecular research have enhanced understanding, while innovations in delivery systems, such as nanoparticles, are improving efficacy. However, gaps remain in long-term safety, interactions, and clinical application, emphasizing the need for further research to fully harness their potential in healthcare. A summary of key findings, insights, and future directions in essential oils research is presented in Table 20.1.

REFERENCES

Albuquerque, P. M., Azevedo, S. G., de Andrade, C. P., D'Ambros, N. C. D. S., Pérez, M. T. M., & Manzato, L., 2022. Biotechnological applications of nanoencapsulated essential oils: A review. *Polymers*, *14*(24), p. 5495.

Aljaafari, M. N., AlAli, A. O., Baqais, L., Alqubaisy, M., AlAli, M., Molouki, A., Ong-Abdullah, J., Abushelaibi, A., Lai, K. S., & Lim, S. H. E., 2021. An overview of the potential therapeutic applications of essential oils. *Molecules*, *26*(3), p. 628.

Aqil, F., Munagala, R., Jeyabalan, J., & Vadhanam, M. V., 2013. Bioavailability of phytochemicals and its enhancement by drug delivery systems. *Cancer Letters*, *334*(1), pp. 133–141.

Aziz, Z. A., Ahmad, A., Setapar, S. H. M., Karakucuk, A., Azim, M. M., Lokhat, D., Rafatullah, M., Ganash, M., Kamal, M. A., & Ashraf, G. M., 2018. Essential oils: Extraction techniques, pharmaceutical and therapeutic potential-a review. *Current Drug Metabolism*, *19*(13), pp. 1100–1110.

Baptista-Silva, S., Borges, S., Ramos, O. L., Pintado, M., & Sarmento, B., 2020. The progress of essential oils as potential therapeutic agents: A review. *Journal of Essential Oil Research*, *32*(4), pp. 279–295.

Belwal, T., Ezzat, S. M., Rastrelli, L., Bhatt, I. D., Daglia, M., Baldi, A., Devkota, H. P., Orhan, I. E., Patra, J. K., Das, G., & Atanasov, A. G., 2018. A critical analysis of extraction techniques used for botanicals: Trends, priorities, industrial uses and optimization strategies. *TrAC Trends in Analytical Chemistry*, *100*, pp. 82–102.

Bunse, M., Daniels, R., Gründemann, C., Heilmann, J., Kammerer, D. R., Keusgen, M., Lindequist, U., Melzig, M. F., Morlock, G. E., Schulz, H., & Wink, M., 2022. Essential oils as multicomponent mixtures and their potential for human health and well-being. *Frontiers in Pharmacology*, *13*, p. 956541.

Calixto, J. B., 2000. Efficacy, safety, quality control, marketing and regulatory guidelines for herbal medicines (phytotherapeutic agents). *Brazilian Journal of Medical and Biological Research*, *33*, pp. 179–189.

Chouhan, N. K., Eedara, A., Talati, M. N., Ambadipudi, S. S., Andugulapati, S. B., & Pabbaraja, S., 2024. Glucosyltriazole amphiphile treatment attenuates breast cancer by modulating the AMPK signaling. *Drug Development Research*, *85*(4), p. e22215.

de Lavor, É. M., Fernandes, A. W. C., de Andrade Teles, R. B., Leal, A. E. B. P., de Oliveira Júnior, R. G., Gama e Silva, M., de Oliveira, A. P., Silva, J. C., de Moura Fontes Araújo, M. T., Coutinho, H. D. M., & da Silva Almeida, J. R. G., 2018. Essential oils and their major compounds in the treatment of chronic inflammation: A review of antioxidant potential in preclinical studies and molecular mechanisms. *Oxidative Medicine and Cellular Longevity*, *2018*(1), p. 6468593.

de Sousa, D. P., Damasceno, R. O. S., Amorati, R., Elshabrawy, H. A., de Castro, R. D., Bezerra, D. P., Nunes, V. R. V., Gomes, R. C., & Lima, T. C., 2023. Essential oils: Chemistry and pharmacological activities. *Biomolecules*, *13*(7), p. 1144.

Dhifi, W., Bellili, S., Jazi, S., Bahloul, N., & Mnif, W., 2016. Essential oils' chemical characterization and investigation of some biological activities: A critical review. *Medicines*, *3*(4), p. 25.

Do, T. K. T., Hadji-Minaglou, F., Antoniotti, S., & Fernandez, X., 2015. Authenticity of essential oils. *TrAC Trends in Analytical Chemistry*, *66*, pp. 146–157.

Erichsen-Brown, C., 2013. *Medicinal and other uses of North American plants: A historical survey with special reference to the eastern Indian tribes*. Courier Corporation.

Hou, T., Sana, S. S., Li, H., Xing, Y., Nanda, A., Netala, V. R., & Zhang, Z., 2022. Essential oils and its antibacterial, antifungal and anti-oxidant activity applications: A review. *Food Bioscience*, *47*, p. 101716.

Machado, C. A., Oliveira, F. O., de Andrade, M. A., Hodel, K. V. S., Lepikson, H., & Machado, B. A. S., 2022. Steam distillation for essential oil extraction: An evaluation of technological advances based on an analysis of patent documents. *Sustainability*, *14*(12), p. 7119.

Milá Cáceres, L., 2010. Quality risk management application review in pharmaceutical and biopharmaceutical industries. *Bioprocessing Journal*, *9*(1).

Muscolo, A., Mariateresa, O., Giulio, T., & Mariateresa, R., 2024. Oxidative stress: The role of antioxidant phytochemicals in the prevention and treatment of diseases. *International Journal of Molecular Sciences*, *25*(6), p. 3264.

Ni, Z. J., Wang, X., Shen, Y., Thakur, K., Han, J., Zhang, J. G., Hu, F., & Wei, Z. J., 2021. Recent updates on the chemistry, bioactivities, mode of action, and industrial applications of plant essential oils. *Trends in Food Science & Technology*, *110*, pp. 78–89.

Orrenius, S., Nicotera, P., & Zhivotovsky, B., 2011. Cell death mechanisms and their implications in toxicology. *Toxicological Sciences*, *119*(1), pp. 3–19.

Pezantes-Orellana, C., German Bermúdez, F., Matías De la Cruz, C., Montalvo, J. L., & Orellana-Manzano, A., 2024. Essential oils: A systematic review on revolutionizing health, nutrition, and omics for optimal well-being. *Frontiers in Medicine*, *11*, p. 1337785.

Ramachandran, K., 2021. Advancements in ultrasound-assisted drug delivery systems in pharmaceutics. *Journal of Pharmacy and Bioallied Sciences*, *13*(5), pp. 798–805.

Rehman, R., Hanif, M. A., Mushtaq, Z., & Al-Sadi, A. M., 2016. Biosynthesis of essential oils in aromatic plants: A review. *Food Reviews International*, *32*(2), pp. 117–160.

Sadgrove, N. J., Padilla-González, G. F., & Phumthum, M., 2022. Fundamental chemistry of essential oils and volatile organic compounds, methods of analysis and authentication. *Plants*, *11*(6), p. 789.

Saha, S., 2023. Chapter-12 plants derived bioactive compounds: A new era in antimicrobial therapeutics. *Multidisciplinary*, p. 151.

Salgueiro, L., Martins, A. P., & Correia, H., 2010. Raw materials: The importance of quality and safety. A review. *Flavour and Fragrance Journal*, *25*(5), pp. 253–271.

Schnaubelt, K., 1999. *Medical aromatherapy: Healing with essential oils*. Frog Books.

Sharma, M., Ambadipudi, S. S., Chouhan, N. K., Nayak, V. L., Pabbaraja, S., Andugulapati, S. B., & Sistla, R., 2024. Design, synthesis and biological evaluation of novel cationic liposomes loaded with melphalan for the treatment of cancer. *Bioorganic & Medicinal Chemistry Letters*, *97*, p. 129549.

Shi, J., Yang, Y., Zhou, X., Zhao, L., Li, X., Yusuf, A., Hosseini, M. S., Sefidkon, F., & Hu, X., 2022. The current status of old traditional medicine introduced from Persia to China. *Frontiers in Pharmacology*, *13*, p. 953352.

Sitarek, P., Merecz-Sadowska, A., Śliwiński, T., Zajdel, R., & Kowalczyk, T., 2020. An in vitro evaluation of the molecular mechanisms of action of medical plants from the Lamiaceae family as effective sources of active compounds against human cancer cell lines. *Cancers*, *12*(10), p. 2957.

Stratakos, A. C., & Koidis, A., 2016. Methods for extracting essential oils. In Preedy, V. (ed.) *Essential oils in food preservation, flavor and safety*. Academic Press, London, pp. 31–38.

Sulieman, A. M. E., Abdallah, E. M., Alanazi, N. A., Ed-Dra, A., Jamal, A., Idriss, H., Alshammari, A. S., & Shommo, S. A., 2023. Spices as sustainable food preservatives: A comprehensive review of their antimicrobial potential. *Pharmaceuticals*, *16*(10), p. 1451.

Talati, M. N., Vemireddy, S., Seelam, S. D., & Halmuthur Mahabalarao, S. K., 2023. Synthesis and immunomodulatory activity of novel amino acid analogues of fluoxetine. *Synthetic Communications*, *53*(10), pp. 731–743.

Vaou, N., Stavropoulou, E., Voidarou, C. C., Tsakris, Z., Rozos, G., Tsigalou, C., & Bezirtzoglou, E., 2022. Interactions between medical plant-derived bioactive compounds: Focus on antimicrobial combination effects. *Antibiotics*, *11*(8), p. 1014.

Visan, A. I., & Negut, I., 2024. Coatings based on essential oils for combating antibiotic resistance. *Antibiotics*, *13*(7), p. 625.

Wińska, K., Mączka, W., Łyczko, J., Grabarczyk, M., Czubaszek, A., & Szumny, A., 2019. Essential oils as antimicrobial agents—myth or real alternative? *Molecules*, *24*(11), p. 2130.

21 Clinical Studies and Assessments of Functional Spices

Sagar and Jeena Gupta

21.1. INTRODUCTION

Throughout human history, culinary items have been used for therapeutic purposes. Ancient societies such as Egypt, India, China, and Greece understood that food served as both medicine and sustenance. Kitchen essentials such as garlic, ginger, and turmeric were essential to therapeutic procedures in systems such as Ayurveda and Traditional Chinese Medicine (TCM). Hippocrates' dictum, 'Let food be thy medicine and medicine be thy food', perfectly captures the conventional wisdom on the health advantages of everyday foods. The groundwork for investigating the medicinal potential of common components has been established by these traditions (Prasad et al., 2011; Rivlin, 2001). Modern science has confirmed many ancient beliefs by identifying bioactive chemicals in everyday kitchen ingredients. For example, garlic contains allicin, a powerful antibacterial agent, while turmeric has curcumin, known for its anti-inflammatory properties. Research in phytochemistry and pharmacology has revealed how these natural compounds help combat chronic illnesses such as diabetes, heart disease, and arthritis (Shang et al., 2019; Bhatwalkar et al., 2021). Using culinary ingredients for preventive healthcare offers significant benefits. Foods such as green tea, honey, and cinnamon enhance immunity, reduce inflammation, and promote overall health. Including them in daily diets helps build resilience against diseases, supporting a prevention-over-cure approach (Chacko et al., 2010; Stephen et al., 2023).

Household products are often used medicinally, with practices differing across cultures. For instance, in Japan, green tea is prized for its antioxidant properties, while in India, turmeric milk is a go-to remedy for colds (Prasad et al., 2011; Prasanth et al., 2019). The Mediterranean diet, rich in fresh produce, garlic, and olive oil, is also associated with heart health and longevity, showcasing diverse global traditions in kitchen-based remedies (Martinez-Gonzalez et al., 2016).

Recent studies underscore growing interest in the medicinal potential of culinary ingredients. Research suggests turmeric can reduce inflammation as effectively as ibuprofen, while garlic demonstrates antibacterial properties and cholesterol-lowering effects. Such findings pave the way for these ingredients to be incorporated into nutraceuticals and functional foods (El-Saadony et al., 2023; El-Saber Batiha et al., 2020a).

However, challenges remain, variations in active ingredient levels due to cultivation, storage, and preparation can affect consistency, and the absence of standardized dosages complicates their therapeutic use. Addressing these issues is essential for reliable and effective applications (Busia, 2024). By combining traditional knowledge with modern scientific research, we can develop standardized, evidence-based natural remedies that merge the best of both worlds. This approach has the potential to transform healthcare by promoting holistic, natural solutions (Acs et al., 2005; Rizvi et al., 2022). This chapter focuses on exploring the therapeutic potential of common culinary ingredients (Figure 21.1), highlighting their active compounds, health benefits, and scientific validation to advance a comprehensive approach to well-being (Ramalingum et al., 2014).

DOI: 10.1201/9781003559894-21

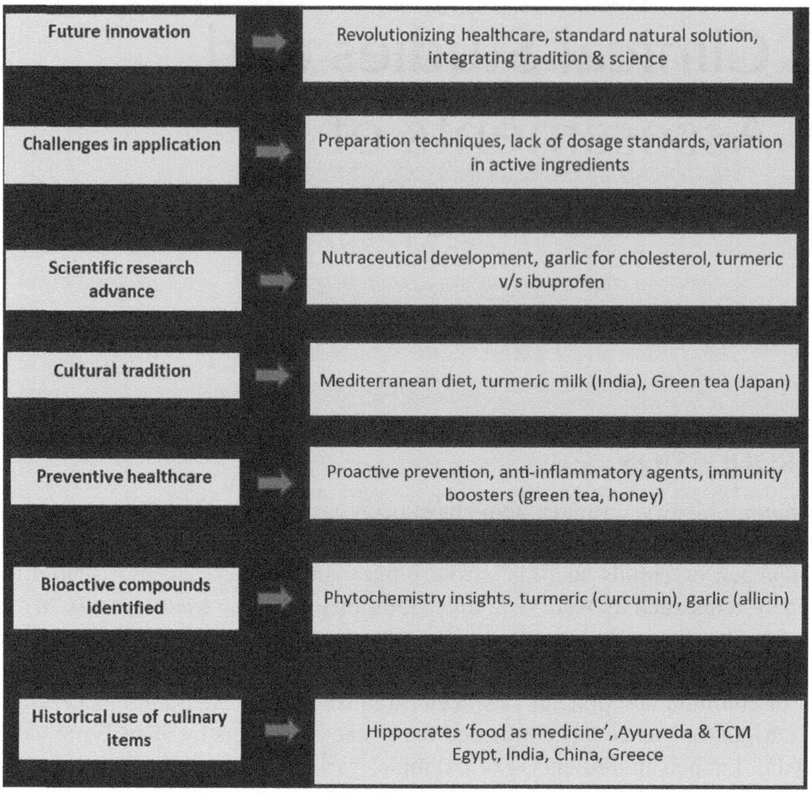

FIGURE 21.1 Therapeutic potential of culinary products.

21.2. GARLIC (*ALLIUM SATIVUM*)

Garlic (*Allium sativum*) is widely valued for its antibacterial, anti-inflammatory, and cardiovascular benefits. Its key bioactive compound, allicin, helps promote heart health by lowering cholesterol and blood pressure (Banerjee et al., 2002). Garlic also boosts the immune system, combats infections naturally, and contains antioxidants that may protect against oxidative stress, reducing the risk of neurological conditions such as Parkinson's and Alzheimer's. These health benefits, supported by both traditional knowledge and modern scientific research, make garlic an exceptional natural remedy (Ried, 2016).

Garlic and its derivatives have demonstrated significant potential for cardiovascular health, particularly in managing hypertension, cholesterol, and metabolic syndrome. Garlic supplements, such as aged garlic extract (AGE), time-released garlic tablets, and raw crushed garlic, were shown to reduce systolic blood pressure (SBP) and diastolic blood pressure (DBP) comparably to standard antihypertensive medications. Studies highlight dose-dependent effects, with higher doses and prolonged use yielding better results. Garlic also positively influenced lipid profiles by reducing triglycerides, LDL cholesterol, and total cholesterol, while increasing HDL cholesterol, along with improving nitric oxide levels, arterial stiffness, and gut microbiota diversity. These effects suggest garlic's utility in mitigating risk factors associated with hypertension, atherosclerosis, and metabolic syndrome. For example, a meta-analysis of 12 studies noted average reductions of 8.3 mmHg in SBP and 5.5 mmHg in DBP, corresponding to a 16–40% lower risk of cardiovascular events. Additionally, specific formulations, like AGE, showed unique benefits for gut health and were particularly effective in hypertensive individuals (Fu et al., 2023; Hutchins et al., 2020; Mirunalini et al., 2011).

21.3. TURMERIC (*CURCUMA LONGA*)

Turmeric (*Curcuma longa*) is renowned for its medicinal benefits, largely attributed to its active compound, curcumin. Curcumin has powerful anti-inflammatory, antioxidant, and anticancer properties. Its anti-inflammatory effects make it effective in managing arthritis by reducing joint pain and swelling, while its antioxidant properties help neutralize free radicals and reduce oxidative stress, which contributes to aging and chronic diseases (Aggarwal et al., 2014).

Curcumin also supports digestive health by stimulating bile production, enhancing digestion, and improving liver function. Its neuroprotective properties may boost brain function and lower the risk of neurodegenerative diseases like Alzheimer's by increasing levels of brain-derived neurotrophic factor (BDNF), a protein critical for neuron survival and function (Gupta et al., 2013).

In addition, curcumin exhibits anticancer potential by inhibiting cancer cell growth, promoting apoptosis (programmed cell death), and suppressing angiogenesis (the formation of blood vessels that feed tumours). These diverse health benefits, validated by traditional use and modern research, make turmeric an invaluable natural remedy (Cole et al., 2007).

Clinical trials investigating turmeric, particularly its active compound curcumin, have highlighted its potential therapeutic effects across various conditions. Numerous studies have focused on its anti-inflammatory, antioxidant, and anticarcinogenic properties. For example, randomized controlled trials (RCTs) have shown that curcumin supplementation can significantly alleviate symptoms of chronic inflammatory diseases such as arthritis, reducing pain and improving joint function (White et al., 2019). In metabolic health, turmeric has demonstrated efficacy in lowering blood glucose, triglycerides, and LDL cholesterol, while increasing HDL cholesterol in patients with metabolic syndrome (Qin et al., 2017). Furthermore, trials in patients with cardiovascular conditions reported reductions in blood pressure and markers of oxidative stress (Hajleh et al., 2022). In neurodegenerative disorders, such as Alzheimer's disease, curcumin has been linked to improved cognitive function and reduced amyloid plaque formation (Voulgaropoulou et al., 2019). However, clinical trials emphasize the importance of bioavailability, as curcumin is poorly absorbed in its natural form. Formulations such as curcumin-piperine combinations or liposomal curcumin have been used to enhance absorption, yielding more consistent results (Khederzadeh et al., 2024). These findings suggest turmeric's promise as a complementary therapy, although further large-scale, long-term studies are necessary to confirm its efficacy and safety.

21.4. GINGER (*ZINGIBER OFFICINALE*)

Ginger (*Zingiber officinale*) is a widely recognized medicinal plant with a rich history in traditional and modern medicine. Known for its anti-inflammatory, antioxidant, and digestive properties, ginger is particularly effective in treating nausea and vomiting, making it a natural remedy for morning sickness during pregnancy and motion sickness. Its active compound, gingerol, plays a key role in these benefits. Ginger's anti-inflammatory effects also help alleviate muscle soreness after exercise and symptoms of osteoarthritis (Bode et al., 2011).

Ginger supports heart health by lowering blood sugar and cholesterol levels, making it beneficial for individuals with type 2 diabetes or those at risk of cardiovascular disease. It also aids digestion by accelerating stomach emptying, which can reduce indigestion and bloating. Scientific studies have validated ginger's medicinal properties, solidifying its reputation as a versatile natural remedy for various conditions (Tapsell et al., 2006). From digestive health to pain relief and cardiovascular support, ginger offers a wide range of health benefits.

In therapeutic settings, ginger has shown significant promise, particularly in managing nausea and vomiting during pregnancy (NVP), with its safety and effectiveness consistently supported by research when compared to conventional antiemetic drugs (Anh et al., 2020). However, evidence regarding its efficacy for chemotherapy-induced nausea and vomiting (CINV) and postoperative nausea and vomiting (PONV) has been inconsistent due to methodological limitations, such as small sample sizes and study design flaws. Beyond its antiemetic properties, ginger has demonstrated potential in improving metabolic markers, including blood sugar and cholesterol profiles, and alleviating pain, particularly in dysmenorrhea. Emerging research also suggests ginger's preventive role in colorectal cancer through its anti-inflammatory and pro-apoptotic mechanisms, although most studies focus on modifiable risk factors rather than direct patient outcomes. Although minor side effects such as heartburn have been reported, these were rare. Nonetheless, the therapeutic potential of ginger is constrained by methodological shortcomings, such as small sample sizes, inconsistent evaluation methods, and dosage variability. These limitations highlight the need for larger, rigorously designed clinical trials to validate its efficacy across a broader range of conditions (Anh et al., 2020).

21.5. CINNAMON (*CINNAMOMUM VERUM*)

Cinnamon (*Cinnamomum verum*), often referred to as "true cinnamon", has been used medicinally for centuries. Its primary active compound, cinnamaldehyde, provides significant health benefits, including antibacterial and anti-inflammatory effects. Cinnamon is particularly effective at lowering blood sugar levels and improving insulin sensitivity, making it a valuable aid for managing type 2 diabetes. Research shows that it can reduce fasting blood sugar and enhance glucose metabolism, supporting its role as a dietary supplement for diabetes management (Spence, 2024).

Cinnamon's anti-inflammatory properties also help combat chronic inflammation, a factor in many diseases, including cardiovascular conditions. Regular consumption has been shown to lower total cholesterol, LDL (bad) cholesterol, and triglycerides while maintaining HDL (good) cholesterol, reducing the risk of heart disease (Gruenwald et al., 2010).

Cinnamon has demonstrated potent antibacterial properties, showing effectiveness against infections caused by *Candida albicans* and *Helicobacter pylori*, which are often linked to gastrointestinal issues (Mohammadabadi et al., 2024). Additionally, it offers a wide range of health benefits, making it a versatile natural remedy. In the context of diabetes management, studies have highlighted cinnamon's potential efficacy. A study administering 1 g of cinnamon daily showed significant improvements in anthropometric parameters (BMI, body fat, visceral fat), glycemic indices (FPG, 2hpp, HbA1c, fasting insulin, insulin resistance), and lipid profiles (total cholesterol, LDL-C, HDL-C), excluding triglycerides. The effects were more pronounced in individuals with a BMI ≥ 27, suggesting that baseline BMI may influence cinnamon's therapeutic potential. Another trial using 3 g/day for 90 days as an adjunct therapy reported significant reductions in HbA1c, fasting glucose, and insulin levels, with notable improvements in HOMA-IR scores among cinnamon users. Importantly, no severe adverse events were reported, supporting cinnamon's safety profile.

However, the Brazilian Diabetes Society has refrained from recommending cinnamon due to inconsistent evidence and modest effects. Limitations of current research include short intervention periods, dietary confounders, and the exclusion of insulin users. Despite these challenges, cinnamon's affordability and safety highlight the need for larger, well-controlled studies involving diverse populations, varied dosages, and longer durations to confirm its role in diabetes management and prevention (Bibi et al., 2024; Lira Neto et al., 2022).

21.6. HONEY

Honey, a natural substance produced by bees, has been valued for its medicinal and health benefits for centuries. Its antibacterial properties, derived from enzymes that produce hydrogen peroxide, make it an effective agent for wound healing and infection prevention. Research has shown that honey is particularly useful in treating burns and surgical wounds by promoting tissue regeneration and reducing inflammation (Bogdanov et al., 2008).

One of honey's most recognized benefits is its ability to soothe sore throats and coughs. Its thick consistency coats the throat, while its antibacterial properties help reduce inflammation and combat infection, making it a go-to remedy for respiratory issues, especially during cold and flu seasons (Mandal et al., 2011).

Honey is not only a natural sweetener but also a potent antioxidant, rich in flavonoids and phenolic acids, which help combat oxidative stress and reduce the risk of chronic diseases (Cortés-Rojas et al., 2014). Regular honey consumption has been linked to improved gut health by promoting beneficial gut bacteria and supporting heart health by lowering LDL cholesterol while increasing HDL cholesterol, thereby reducing cardiovascular risks. Its lipid-modulating properties make honey a heart-healthy alternative to refined sugars (Cortés-Rojas et al., 2014).

Honey's renowned antimicrobial properties stem from hydrogen peroxide (H_2O_2), high osmolarity, low pH, and bioactive compounds such as methylglyoxal (MGO). It demonstrates efficacy against multidrug-resistant bacteria, including *S. aureus*, *P. aeruginosa*, *E. coli*, and *S. typhi*. Manuka honey, noted for its MGO-driven nonperoxide activity, retains antibacterial effects even in catalase-rich environments, unlike peroxide-dependent honeys such as Ulmo. The minimum inhibitory concentrations (MICs) for honey vary widely (3–50%) based on type, concentration, and bacterial strain. Factors such as geographical origin, floral source, and season can influence its antibacterial potency, with effectiveness increasing at concentrations above 25%. Honey's low pH (3.2–4.5) and osmotic pressure further enhance its pathogen-inhibiting properties. Despite variability, honey's broad-spectrum antimicrobial activity, safety, and potential in wound care and infection control underscore its therapeutic value, necessitating further research into local varieties and standardized applications (Almasaudi, 2021).

21.7. CLOVES (*SYZYGIUM AROMATICUM*)

Cloves (*Syzygium aromaticum*) are a highly valued spice known for their therapeutic properties, primarily attributed to eugenol, their active compound with strong anti-inflammatory, antioxidant, and analgesic effects. One of the most notable uses of cloves is in dentistry, where eugenol serves as a natural remedy for tooth pain. Its anesthetic and antibacterial properties help relieve pain, reduce inflammation, and prevent infection, making it a common ingredient in over-the-counter dental products. Beyond dental health, cloves support liver function by reducing oxidative stress. The antioxidants in cloves, particularly eugenol, protect the liver from damage caused by free radicals, promoting detoxification and overall liver health (Batiha et al., 2020).

Cloves provide significant digestive health benefits by stimulating the production of digestive enzymes, alleviating symptoms such as indigestion and gas. They also help protect the stomach lining by enhancing mucus production, thereby reducing the risk of ulcers. For individuals with diabetes, cloves have shown promise in blood sugar management. Compounds in cloves improve insulin activity and help regulate post-meal blood sugar spikes, making them a valuable addition to a diabetes-friendly diet (Ahmed et al., 2018).

The bioactive component of cloves, eugenol, which constitutes 85–92% of clove oil, offers a wide range of medicinal benefits, including antimicrobial, antioxidant, anti-inflammatory, anticancer, analgesic, and anesthetic properties. Eugenol effectively combats bacteria, fungi, and viruses, such as *Candida albicans* and HSV, and protects against oxidative stress, liver damage, and gastric ulcers. Additional compounds such as carvacrol and thymol enhance these effects. Eugenol disrupts microbial cell membranes, provides antioxidant defence, and inhibits inflammatory pathways. In cancer therapy, it induces apoptosis and hinders tumour progression. Cloves also enhance insulin sensitivity, regulate blood sugar levels, and lower cholesterol, providing antidiabetic and antiobesity benefits. They support neurological and sexual health, reduce platelet aggregation, and promote cardiovascular well-being. Although generally safe, excessive consumption may cause hepatotoxicity or allergic reactions. Cloves are widely used in dentistry, for treating inflammation and gastrointestinal disorders, and as a flavouring in food and cosmetics (El-Saber Batiha et al., 2020b).

21.8. LEMON (*CITRUS LIMON*)

Lemon (*Citrus limon*) is a citrus fruit rich in vitamin C and antioxidants, widely valued for its immune-boosting and overall health benefits. Vitamin C plays a critical role in supporting immune function, helping the body fight infections and reducing the severity of cold and flu symptoms. Additionally, the antioxidants in lemon, such as flavonoids and ascorbic acid, protect cells from free radical damage, reduce inflammation, and promote overall well-being (Khanna et al., 2014). Lemon is also known for aiding digestion and detoxification. Its acidity stimulates the production of gastric juices, improving food breakdown and reducing issues such as indigestion and bloating. Furthermore, lemon supports liver function, enhancing the body's natural detoxification processes and aiding in the elimination of toxins. For weight management, lemon is believed to boost metabolism and support fat burning. Drinking lemon water, especially in the morning, is a popular practice to promote hydration and potentially aid weight loss efforts due to its metabolism-enhancing properties. Additionally, lemon serves as a natural remedy for cold and flu symptoms. Its anti-inflammatory and antibacterial properties help soothe sore throats, provide hydration, and reduce inflammation, making it an effective choice for relieving seasonal ailments.

A clinical study assessed fresh lemon juice supplementation in reducing calcium oxalate stone recurrence when combined with a standard diet (normal calcium, reduced salt and animal proteins). Although initial findings at one year showed a protective effect with adherence at 68%, the effect diminished over two years as adherence dropped. Lemon juice supplementation did not increase urinary citrate but decreased urinary sodium excretion, potentially reducing urine saturation and stone risk. However, gastrointestinal side effects, particularly gastric pyrosis, and treatment costs contributed to declining adherence. Both groups maintained good compliance with dietary protein and calcium restrictions. The study's high retention rate and careful monitoring were strengths, but poor adherence limited conclusions. Although results suggest potential benefits of lemon juice supplementation, they highlight the need for strategies to enhance adherence in future trials. Current evidence does not support using lemon juice for stone prevention in clinical practice (Ruggenenti et al., 2022).

21.9. PEPPERMINT (*MENTHA PIPERITA*)

Peppermint (*Mentha piperita*) is a versatile herb valued for its therapeutic properties, primarily due to its active compound, menthol, which provides cooling, relaxing, and pain-relieving effects. Peppermint is particularly effective for addressing digestive and respiratory issues. One of its most well-documented benefits is its ability to alleviate symptoms of irritable bowel syndrome (IBS). The antispasmodic properties of peppermint oil help relax the muscles of the gastrointestinal tract, reducing bloating, gas, and abdominal discomfort associated with IBS (Cortés-Rojas et al., 2014).

In addition to aiding digestion, peppermint is beneficial for headaches and migraines. Topical application of peppermint oil, especially on the temples, has been shown to relieve tension headaches and migraines by improving blood flow and providing a soothing, cooling sensation (Göbel et al., 2016). Peppermint also supports respiratory health. Menthol acts as a natural decongestant, helping to open nasal passages and relieve respiratory congestion, making it an effective remedy for colds and other respiratory conditions. Furthermore, peppermint oil can be used to alleviate muscle soreness. Its cooling and anti-inflammatory effects make it a popular natural solution for soothing muscle aches and reducing inflammation.

A randomized, double-blind, placebo-controlled trial investigated the efficacy of peppermint oil in Rome IV-defined IBS patients. Small-intestinal-release peppermint oil showed statistically significant improvements in secondary outcomes such as IBS symptom severity and abdominal pain, while ileocolonic release did not demonstrate benefits. Despite mild and transient adverse events, peppermint oil's therapeutic gain over placebo was modest (12.4%). Discrepancies with previous studies may be due to stricter criteria and a more diverse patient population. Although not statistically significant, peppermint oil's moderate efficacy in small intestine suggests it remains a valuable option for IBS management. However, the lack of long-term benefits and mild adverse events like belching highlight the need for ongoing evaluation. The study's limitations include a young, predominantly female, and homogenous population, short treatment duration, and potential blinding issues, but the rigorous design provides strong insights into peppermint oil's potential as an IBS treatment (Weerts et al., 2020).

21.10. BASIL (*OCIMUM BASILICUM*)

Basil (*Ocimum basilicum*) is a fragrant herb renowned for its therapeutic properties, which are attributed to its antioxidants, essential oils, and vitamins. One of its key benefits is its ability to reduce tension and anxiety. Basil is considered an adaptogen, meaning it helps the body adapt to stress and improves mental balance. Research has shown that basil can lower cortisol levels, reducing stress and enhancing overall well-being (Jamshidi et al., 2017).

Basil is also beneficial for digestion, helping to alleviate bloating and gas. The essential oils in basil, such as eugenol, have carminative effects, which stimulate digestive enzymes and promote better digestion, reducing discomfort associated with indigestion. Another significant advantage of basil is its anti-inflammatory properties. Compounds such as eugenol, linalool, and citronellol in basil help inhibit pro-inflammatory enzymes, providing relief from conditions such as arthritis and inflammatory bowel disease (Singh et al., 2005). Additionally, basil has potent antibacterial, antiviral, and antifungal effects. Its essential oils have been shown to combat various infections, supporting the immune system and helping prevent illnesses (Hammer et al., 1999).

A study demonstrated that *Ocimum gratissimum* (OG) inhibits breast cancer progression by reducing MMP-2 and MMP-9 enzymatic activities, which are linked to chemotaxis and chemoinvasion. The HB and HL fractions of OG contributed primarily to inhibiting MMP activity and chemotaxis, respectively. Treatment with OG delayed tumour progression in MCF xenografts, with longer preservation of basement membrane integrity. The study suggests that OG's inhibitory effects are due to its impact on multiple processes, including MMP activity, angiogenesis, and cell invasion. OG shows potential as a nontoxic MMP inhibitor for breast cancer treatment, with further purification efforts underway to identify the active compounds (Nangia-Makker et al., 2013).

21.11. APPLE CIDER VINEGAR (ACV)

Apple cider vinegar (ACV), made from fermented apple juice, has been used for centuries for its health benefits. Its primary active ingredient, acetic acid, is responsible for its distinctive sour taste and medicinal properties. One of the most well-known benefits of ACV is its role in weight management. Studies have shown that acetic acid can boost metabolism, reduce fat storage, and suppress appetite, leading to modest weight loss when combined with a healthy diet and regular exercise (Kondo et al., 2009a).

ACV is also beneficial for regulating blood sugar and improving insulin sensitivity. Research indicates that drinking ACV before meals can help reduce blood sugar spikes and enhance insulin sensitivity, making it a useful aid for people with type 2 diabetes or insulin resistance (Johnston et al., 2004). In addition to its effects on weight and blood sugar, ACV promotes digestive health by encouraging the growth of beneficial gut bacteria and aiding digestion. Its antibacterial properties help combat harmful bacteria in the digestive tract, reducing symptoms of indigestion and bloating. ACV is also a popular natural remedy for soothing sore throats and treating respiratory infections. Its acidity and antibacterial properties help inhibit the growth of germs responsible for infections, making it an effective tool for respiratory health (Yagnik et al., 2018).

Clinical research is now being conducted on ACV because of its possible health advantages, especially in the areas of weight management, cholesterol reduction, and blood glucose control. According to studies, ACV can improve fasting glucose levels and lessen post-meal blood sugar surges, which may assist people with type 2 diabetes to control their blood sugar levels, particularly after meals (Johnston, 2009; Kondo et al., 2009b). ACV may also help decrease cholesterol, especially in those who have type 2 diabetes. According to research, consuming ACV in moderation each day might lower total and LDL (bad) cholesterol levels, which may improve cardiovascular health (Kondo et al., 2009b; Fushimi et al., 2002). Human studies have yielded conflicting results, many with small sample sizes or short durations, necessitating additional research to confirm its efficacy, despite some suggesting that ACV may help with weight loss by promoting feelings of fullness and reducing fat accumulation (Sadeghi, 2016). More extensive, long-term research is required to completely comprehend ACV's impact in various health domains, even if the preliminary data is encouraging.

21.12. CARDAMOM (*ELETTARIA CARDAMOMUM*)

Cardamom, also known as *Elachi*, is a spice widely used in Ayurveda and Traditional Chinese Medicine for its therapeutic benefits. Rich in bioactive compounds such as flavonoids, terpenes, and essential oils, cardamom offers numerous health-promoting properties (Ashokkumar et al., 2020). One of cardamom's key benefits is its support for digestive health. It helps alleviate indigestion, bloating, and gas by increasing digestive enzyme production. Additionally, its essential oils reduce stomach acidity and promote mucus production, which can prevent ulcers and ease conditions like heartburn. For respiratory health, cardamom acts as an expectorant, aiding in the removal of mucus from the respiratory tract. This is particularly beneficial for individuals with asthma, bronchitis, and chronic coughs. It also supports lung function by enhancing airflow, offering antiasthmatic benefits.

Cardamom has strong antimicrobial properties, making it effective in combating bad breath and preventing oral infections. It is also effective against bacterial and fungal infections. As a mild diuretic, cardamom helps manage blood pressure by eliminating excess fluid and salt from the body. Its antioxidant properties are beneficial for cardiovascular health, reducing oxidative stress and inflammation. Preliminary studies suggest that cardamom may lower cholesterol levels and enhance metabolism, contributing to weight loss (Souissi et al., 2020).

In addition to its physical health benefits, cardamom has positive effects on mental health. Its essential oil provides calming effects, helping to reduce stress and anxiety. The spice's antioxidants may also help protect against cognitive decline and potentially inhibit tumour growth. Although cardamom is generally safe when used in culinary amounts, excessive use for medicinal purposes should be approached cautiously, and it is advisable to consult a healthcare professional (Korikanthimath et al., 2023).

A study explored the potential of *Elettaria cardamomum* (EC) extract in preventing Alzheimer's disease (AD) in diabetic rats, suggesting it could alleviate brain-specific insulin resistance and improve insulin signaling. Insulin resistance is linked to AD, with abnormal GSK3β activation promoting amyloid-β (Aβ) and tau accumulation. After eight weeks of EC administration, peripheral insulin resistance was reduced, and improvements in glucose and cholesterol levels were observed. In the brain, EC reduced GSK3β activation, thus reducing Aβ and tau accumulation. EC also mitigated oxidative stress by restoring the oxidant/antioxidant balance, increasing glutathione (GSH) and superoxide dismutase (SOD) levels while lowering malondialdehyde (MDA) levels in the brain. Furthermore, EC exhibited anti-inflammatory properties, lowering pro-inflammatory cytokines TNF-α and IL1β, and reducing neuronal degeneration. The extract also modulated the glutamatergic system by correcting the reduced expression of specific glutamate receptors, potentially improving memory and learning. These findings suggest EC could offer neuroprotective benefits, but further studies, including pharmacokinetics and broader animal models, are necessary before clinical application (Gomaa et al., 2019).

21.13. NUTMEG (*MYRISTICA*)

Nutmeg, derived from the seeds of the *Myristica fragrans* tree, is a spice renowned for its rich flavour and therapeutic properties. It contains bioactive compounds such as myristicin, eugenol, and macelignan, which contribute to its wide range of health benefits (Ashokkumar et al., 2020). One of nutmeg's primary uses is as a carminative, which helps relieve nausea, bloating, and indigestion. It also possesses astringent qualities that make it effective in alleviating diarrhea. By stimulating digestive enzymes, nutmeg aids overall digestion. In addition to its digestive benefits, nutmeg has potent anti-inflammatory effects, with compounds such as myristicin and eugenol contributing to the relief of joint and muscle pain, particularly in conditions such as arthritis (Zhang et al., 2016).

Nutmeg is also prized for its mild sedative properties, which can promote relaxation and improve sleep quality, especially when consumed in small doses with warm milk. Its high antioxidant content helps fight oxidative stress, potentially reducing the risk of chronic diseases such as cancer and heart disease. Nutmeg's antibacterial properties also support dental health by combating oral bacteria and freshening breath. Furthermore, nutmeg may enhance cognitive function, including memory and focus, and its mild psychoactive effects are thought to reduce symptoms of anxiety and depression. Additionally, nutmeg supports kidney and liver detoxification, aiding the body's natural ability to eliminate toxins (Ebrahim et al., 2022). This versatile spice, with its array of health benefits, is a valuable addition to both the kitchen and the medicine cabinet.

The study highlights the toxic effects of prolonged, high-dose nutmeg consumption on the medial geniculate body (MGB) of adult Wistar rats. Histological analysis revealed cellular degenerative changes, including hypertrophy, sparse cellular populations, pyknotic nuclei, and vacuolation in the stroma. These findings suggest that nutmeg, known for its neurotoxic effects, may lead to neuronal degeneration and cell death, particularly affecting dopaminergic neurons. Additionally, cytotoxic edema, characterized by intracellular swelling and neuronal vacuolation, was observed, indicating potential electrolyte imbalance due to nutmeg toxicity. The study emphasizes the need for further investigation into the microanatomical effects of nutmeg abuse, which may not be evident through conventional clinical tests. It proposes the development of novel laboratory tests to better assess nutmeg toxicity and its impact on MGB pathology and auditory dysfunction, advocating for further studies to confirm these findings (Adjene et al., 2010).

21.14. BLACK HIMEJ (*TERMINALIA CHEBULA*)

Black Himej (also known as Black Myrobalan or *Terminalia chebula*) is a powerful plant used extensively in Ayurveda, often referred to as Haritaki. This plant is renowned for its wide range of medicinal properties, including healing, revitalizing, and cleansing effects. It is a key ingredient in the traditional Triphala formulation, which is highly valued in Ayurvedic medicine.

One of Black Himej's primary benefits is its ability to act as a natural laxative. It helps regulate bowel movements, alleviating issues such as constipation, bloating, and indigestion, while also enhancing digestion by stimulating digestive enzymes and reducing gas. In addition to its digestive benefits, Black Himej is known for its liver-detoxifying properties. It helps cleanse the blood, supports liver and colon health, and aids in the elimination of toxins from the body (Bag et al., 2013)

Rich in antioxidants and vitamin C, Black Himej boosts the immune system, combats infections caused by bacteria, fungi, and viruses, and reduces inflammation. Its ability to promote metabolism and detoxify the body makes it an aid in weight management and appetite control. Furthermore, its antioxidant properties offer protection against aging and chronic diseases, such as heart disease and cancer (Sadhupati et al., 2024).

Black Himej also offers skin and hair benefits. It is used to treat acne, skin infections, and scalp conditions, promoting healthier skin and hair. Additionally, its positive impact extends to mental health, as it has been shown to enhance memory, reduce stress and anxiety, and improve mental clarity. Some studies also suggest that it may help regulate blood sugar levels, offering potential benefits in managing diabetes (Chakkalakal et al., 2023). This versatile plant has a long history of use and continues to be a valuable natural remedy in traditional and modern healthcare practices.

A study investigated the antimicrobial activity of extracts from *Terminalia chebula* (*T. chebula*) against common microorganisms implicated in dental caries, including *S. mutans*, *L. acidophilus*, and *C. albicans*. Additionally, *S. aureus* and *S. cerevisiae* were tested due to their association with oral infections. The results showed that all extracts of *T. chebula* exhibited growth inhibition against the tested strains, with acetone extracts proving the most potent against both *S. mutans* and *S. aureus*. Interestingly, acetone and ethanol extracts demonstrated greater antimicrobial activity than water and methanol extracts, suggesting that organic solvents may better extract bioactive compounds. The study also found that the antimicrobial activity of the plant was not affected by boiling, indicating the presence of thermo-stable compounds. These findings highlight *T. chebula* as a promising source for developing phytomedicines to combat oral infections and emphasize the need for further phytochemical studies (Rai et al., 2009).

21.15. AJWAIN (*TRACHYSPERMUM AMMI*)

Ajwain (*Trachyspermum ammi*), also known as carom seeds or bishop's weed, is a highly versatile spice commonly used in both cooking and traditional medicine, especially in Ayurveda and Unani systems. The seeds are particularly valued for their strong digestive, antimicrobial, and anti-inflammatory properties, which are largely attributed to thymol, a bioactive compound with multiple health benefits. Ajwain is renowned for its ability to support digestive health by promoting the synthesis of digestive enzymes. It helps alleviate common digestive issues such as bloating, gas, acidity, and indigestion, making it an excellent natural remedy for gastrointestinal discomfort. It is also commonly used as a natural antacid and is especially effective for relieving colic in both adults and babies, particularly after heavy meals.

In addition to digestive benefits, ajwain has bronchodilator properties, which help relax the airways and reduce congestion. This makes it effective in treating respiratory conditions such as coughing, bronchitis, and asthma. Gargling with water infused with ajwain can also help soothe a sore throat, enhancing its respiratory benefits. Ajwain is well-known for its potent anti-inflammatory and antibacterial effects. It is commonly used to reduce joint pain and swelling, particularly in conditions like arthritis. Due to its antibacterial and antifungal properties, ajwain is also effective in treating a variety of skin conditions and infections. It is a natural analgesic, often used to relieve toothaches and menstrual cramps by applying it as an oil or paste.

Furthermore, ajwain is beneficial for weight management. By increasing metabolism, reducing bloating, and lowering water retention, it supports weight control. Its diuretic properties help regulate blood pressure, and its antioxidant content helps protect against free radical damage, which may reduce the risk of chronic illnesses. Ajwain has also been traditionally used as an insect repellent and to treat intestinal worms.

Although ajwain is generally safe, it should be used in moderation. Pregnant or breastfeeding women should consult with a healthcare professional before using it, as with any potent herb (Bairwa et al., 2012). A clinical trial demonstrated that ajwain cream was safe and showed potential in reducing symptoms such as burning, tingling, numbness, and pain compared to a placebo. However, differences in study results may stem from variations in the type of neuropathy and measurement tools used. Previous research on similar treatments, like menthol ointment, also showed significant pain reduction, with mechanisms linked to blocking sodium channels. The study highlights Ajwain's hot and dry temperament in traditional medicine, potentially offering symptom relief for peripheral neuropathy. Limitations include small sample sizes and factors such as chemotherapy dose and treatment duration. Despite showing promising results, further, larger trials are needed to validate these findings and identify effective treatments for CIPN (Heydarirad et al., 2024).

21.16. CONCLUSION

Natural remedies derived from herbs, spices, and other natural products have been valued for their therapeutic properties in traditional and modern medicine. Garlic, turmeric, ginger, cinnamon, and cloves are renowned for their anti-inflammatory, antioxidant, and antimicrobial benefits,

TABLE 21.1
Health Benefits of Common Medicinal Herbs and Spices

	Herb/Spice	Health Benefits	References
1.	**Garlic (*Allium sativum*)**	Antibacterial, anti-inflammatory, boosts heart health, lowers cholesterol and blood pressure, combats infections, antioxidant, supports brain health.	(Banerjee et al., 2002)
2.	**Turmeric (*Curcuma longa*)**	Anti-inflammatory, antioxidant, anticancer, supports digestive and liver health, boosts brain function, reduces oxidative stress.	(Gupta et al., 2013; Cole et al., 2007)
3.	**Ginger (*Zingiber officinale*)**	Anti-inflammatory, antioxidant, aids digestion, treats nausea, alleviates muscle soreness, lowers blood sugar and cholesterol, supports heart health.	(Bode et al., 2011; Tapsell et al., 2006)
4.	**Cinnamon (*Cinnamomum verum*)**	Anti-inflammatory, antibacterial, lowers blood sugar, improves insulin sensitivity, supports heart health, reduces cholesterol and triglycerides.	(Spence, 2024; Gruenwald et al., 2010; Mohammadabadi et al., 2024)
4.	**Honey**	Antibacterial, wound healing, soothes sore throats, combats oxidative stress, improves gut health, supports heart health by lowering LDL and raising HDL cholesterol.	(Bogdanov et al., 2008; Mandal et al., 2011; Cortés-Rojas et al., 2014)
5.	**Cloves (*Syzygium aromaticum*)**	Anti-inflammatory, antioxidant, analgesic, alleviates tooth pain, supports liver health, aids digestion, regulates blood sugar, combats infections.	(Ahmed et al., 2018)
6.	**Lemon (*Citrus limon*)**	Boosts immune function, antioxidant, aids digestion, detoxifies the liver, helps with weight management, relieves cold and flu symptoms.	(Khanna et al., 2014)
7.	**Peppermint (*Mentha piperita*)**	Relieves digestive issues such as IBS, alleviates headaches and migraines, supports respiratory health, reduces muscle soreness.	(Göbel et al., 2016; Cortés-Rojas et al., 2014)
8.	**Basil (*Ocimum basilicum*)**	Reduces stress, alleviates digestive discomfort, anti-inflammatory, combats infections, supports mental health, improves digestion.	(Jamshidi et al., 2017; Singh et al., 2005)
9.	**Apple Cider Vinegar (ACV)**	Aids weight management, regulates blood sugar, improves digestion, combats bacterial infections, soothes sore throats.	(Kondo et al., 2009b; Johnston et al., 2004)
10.	**Cardamom (*Elettaria cardamomum*)**	Supports digestion, alleviates respiratory issues, combats infections, regulates blood sugar, supports heart health, reduces stress and anxiety.	(Korikanthimath et al., 2023; Kochhar, 1981)
11.	**Nutmeg (Myristica fragrans)**	Relieves nausea, indigestion, pain relief, anti-inflammatory, promotes relaxation, supports cognitive function, detoxifies kidneys and liver.	(Ashokkumar et al., 2020; Lounès et al., 2018)
12.	**Black Himej (Terminalia chebula)**	Aids digestion, detoxifies liver and blood, boosts immune system, reduces inflammation, promotes metabolism and weight management, supports mental health.	(Sadhupati et al., 2024)
13.	**Ajwain (Trachyspermum ammi)**	Aids digestion, alleviates bloating, gas, acidity, respiratory support, anti-inflammatory, antibacterial, pain relief for joint pain and toothaches, supports weight loss.	(Bairwa et al., 2012; Heydarirad et al., 2024)

TABLE 21.2
Active Ingredients in Different Spices

Name of Species	Essential Chemical Molecules	Chemical Formula
Garlic	Allicin	$C_6H_{10}S_2$
Turmeric	Curcumin	$C_{21}H_{20}O_6$
Ginger	Gingerol	$C_{17}H_{26}O_4$
Cinnamon	Cinnamaldehyde	C_9H_8O
Honey	Flavonoids and hydrogen peroxide	
Cloves	Eugenol	$C_{10}H_{12}O_2$
Lemon	Vitamin C	$C_6H_8O_6$
Peppermint	Menthol, 1,8-Cineole (Eucalyptol)	$C_{10}H_{18}O$
Basil	Eugenol	$C_{10}H_{12}O_2$
Apple Cider Vinegar	Acetic acid	CH_3COOH
Cardamom	Cineole, Terpinyl acetate, Linalool	$C_{10}H_{18}O$
Nutmeg	Myristicin, Elemicin, Eugenol, Safrole	$C_{11}H_{12}O_3$, $C_{12}H_{16}O_3$, $C_{10}H_{10}O_2$
Black Himej	Tannins (chebulagic acid, chebulinic acid), Gallic acid, Ellagic acid	$C_{76}H_{52}O_{46}$, $C_7H_6O_5$, $C_{14}H_6O_8$
Ajwain	Thymol, Paracymene, Terpenoids	$C_{10}H_{14}O$, $C_{10}H_{14}$

supporting heart, immune, and digestive health. Honey, lemon, and peppermint are celebrated for their soothing effects on respiratory and digestive systems, while basil and apple cider vinegar offer stress relief and metabolic advantages. Cardamom and nutmeg improve digestion, respiratory function, and mental health, with nutmeg also known for its sedative effects. Black Himej promotes detoxification, immunity, and skin health, and Ajwain excels in digestive support and pain relief (Tables 21.1 and 21.2). Each remedy provides a unique spectrum of benefits, validated by both traditional practices and scientific studies. These natural solutions present holistic, accessible ways to support health, although moderation and professional advice are essential for safe use, especially in medical conditions or pregnancy.

REFERENCES

Acs, Z. J., & Audretsch, D. B. (2005). Entrepreneurship, innovation and technological change. *Foundations and Trends® in Entrepreneurship*, *1*(4), 149–195.

Adjene, J. O., & Igbigbi, P. S. (2010). Histological effects of chronic consumption of nutmeg on the lateral geniculate body of adult wistar rats. *Revista Electrónica de Biomedicina*, (1), 33–38.

Aggarwal, B. B., Deb, L., & Prasad, S. (2014). Curcumin differs from tetrahydrocurcumin for molecular targets, signaling pathways and cellular responses. *Molecules*, *20*(1), 185–205.

Ahmed, S., Sulaiman, S. A., Baig, A. A., Ibrahim, M., Liaqat, S., Fatima, S., . . . & Othman, N. H. (2018). Honey as a potential natural antioxidant medicine: An insight into its molecular mechanisms of action. *Oxidative Medicine and Cellular Longevity*, *2018*(1), 8367846.

Anh, N. H., Kim, S. J., Long, N. P., Min, J. E., Yoon, Y. C., Lee, E. G., . . . & Kwon, S. W. (2020). Ginger on human health: A comprehensive systematic review of 109 randomized controlled trials. *Nutrients*, *12*(1), 157.

Almasaudi, S. (2021). The antibacterial activities of honey. *Saudi Journal of Biological Sciences*, *28*(4), 2188–2196.

Ashokkumar, K., Murugan, M., Dhanya, M. K., & Warkentin, T. D. (2020). Botany, traditional uses, phytochemistry and biological activities of cardamom [Elettaria cardamomum (L.) Maton] – A critical review. *Journal of ethnopharmacology*, *246*, 112244.

Bag, A., Bhattacharyya, S. K., & Chattopadhyay, R. R. (2013). The development of Terminalia chebula Retz. (Combretaceae) in clinical research. *Asian Pacific Journal of Tropical Biomedicine*, *3*(3), 244–252.

Bairwa, R., Sodha, R. S., & Rajawat, B. S. (2012). Trachyspermum ammi. *Pharmacognosy Reviews*, *6*(11), 56.

Banerjee, S. K., & Maulik, S. K. (2002). Effect of garlic on cardiovascular disorders: A review. *Nutrition Journal*, *1*(1), 4.

Batiha, G. E., Alkazmi, L. M., Wasef, L. G., Beshbishy, A. M., Nadwa, E. H., & Rashwan, E. K. (2020). Syzygium aromaticum L. (Myrtaceae): Traditional uses, bioactive chemical constituents, pharmacological and toxicological activities. *Biomolecules*, *10*(2), 202. https://doi.org/10.3390/biom10020202

Bhatwalkar, S. B., Mondal, R., Krishna, S. B. N., Adam, J. K., Govender, P., & Anupam, R. (2021). Antibacterial properties of organosulfur compounds of garlic (Allium sativum). *Frontiers in Microbiology*, *12*, 613077.

Bibi, T., Altemimi, A. B., Rabail, R., Munir, S., Shahbaz, M. U., Rizvi, M. K., . . . & Aadil, R. M. (2024). The therapeutic perspective of cinnamon (Cinnamomum verum) consumption against metabolic syndrome. *Journal of Functional Foods*, *122*, 106545.

Bode, A. M., & Dong, Z. (2011). The amazing and mighty ginger. In *Herbal Medicine: Biomolecular and Clinical Aspects* (2nd ed., pp. 125–143). Boca Raton, FL: CRC Press.

Bogdanov, S., Jurendic, T., Sieber, R., & Gallmann, P. (2008). Honey for nutrition and health: A review. *Journal of the American college of Nutrition*, *27*(6), 677–689.

Busia, K. (2024). Herbal medicine dosage standardisation. *Journal of Herbal Medicine*, *46*, 100889.

Chacko, S. M., Thambi, P. T., Kuttan, R., & Nishigaki, I. (2010). Beneficial effects of green tea: A literature review. *Chinese Medicine*, *5*(1), 13.

Chakkalakal, M., Pan, A., Nadora, D., Gahoonia, N., Chaudhuri, R. K., Burney, W., ... & Sivamani, R. K. (2023). Randomized double-blind placebo-controlled supplementation with standardized Terminalia chebula fruit extracts reduces facial sebum excretion, erythema, and wrinkle severity. Journal of Clinical Medicine, 12(4), 1591.

Chandil, S., Pansare, T. A., & Bamoriya, H. (2020). Medicinal importance of traditional herbal plant haritaki (terminalia chebula): An overview.

Cole, G. M., Teter, B., & Frautschy, S. A. (2007). Neuroprotective effects of curcumin. In *The Molecular Targets and Therapeutic Uses of Curcumin in Health and Disease* (pp. 197–212) New York: Springer.

Cortés-Rojas, D. F., de Souza, C. R. F., & Oliveira, W. P. (2014). Clove (Syzygium aromaticum): A precious spice. *Asian Pacific Journal of Tropical Biomedicine*, *4*(2), 90–96.

Ebrahim, R. R., Mostafa, M. H., & El-Sayed, S. R. (2022). Antibacterial Effect of Nutmeg Extract on Cariogenic Microflora in a Group of Egyptian Children. *Al-Azhar Dental Journal for Girls*, *9*(2), 365–371.

El-Saadony, M. T., Yang, T., Korma, S. A., Sitohy, M., Abd El-Mageed, T. A., Selim, S., . . . & Saad, A. M. (2023). Impacts of turmeric and its principal bioactive curcumin on human health: Pharmaceutical, medicinal, and food applications: A comprehensive review. *Frontiers in Nutrition*, *9*, 1040259.

El-Saber Batiha, G., Magdy Beshbishy, A., G. Wasef, L., Elewa, Y. H., A. Al-Sagan, A., Abd El-Hack, M. E., . . . & Prasad Devkota, H. (2020a). Chemical constituents and pharmacological activities of garlic (Allium sativum L.): A review. *Nutrients*, *12*(3), 872.

El-Saber Batiha, G., Alkazmi, L. M., Wasef, L. G., Beshbishy, A. M., Nadwa, E. H., & Rashwan, E. K. (2020b). Syzygium aromaticum L.(Myrtaceae): Traditional uses, bioactive chemical constituents, pharmacological and toxicological activities. *Biomolecules*, *10*(2), 202.

Fu, Z., Lv, J., Gao, X., Zheng, H., Shi, S., Xu, X., . . . & Song, Q. (2023). Effects of garlic supplementation on components of metabolic syndrome: A systematic review, meta-analysis, and meta-regression of randomized controlled trials. *BMC Complementary Medicine and Therapies*, *23*(1), 260.

Fushimi, T., Tayama, K., Fukaya, M., Kitakoshi, K., Nakai, N., Tsukamoto, Y., & Sato, Y. (2002). The efficacy of acetic acid for glycogen repletion in rat skeletal muscle after exercise. *International Journal of Sports Medicine*, *23*(3), 218–222.

Göbel, H., Heinze, A., Heinze-Kuhn, K., Göbel, A., & Göbel, C. (2016). Peppermint oil in the acute treatment of tension-type headache. *Schmerz (Berlin, Germany)*, *30*(3), 295–310.

Gomaa, A. A., Makboul, R. M., El-Mokhtar, M. A., Abdel-Rahman, E. A., Ahmed, I. A., & Nicola, M. A. (2019). Terpenoid-rich Elettaria cardamomum extract prevents Alzheimer-like alterations induced in diabetic rats via inhibition of GSK3β activity, oxidative stress and pro-inflammatory cytokines. *Cytokine*, *113*, 405–416.

Gruenwald, J., Freder, J., & Armbruester, N. (2010). Cinnamon and health. *Critical Reviews in Food Science and Nutrition*, *50*(9), 822–834.

Gupta, S. C., Patchva, S., & Aggarwal, B. B. (2013). Therapeutic roles of curcumin: Lessons learned from clinical trials. *The AAPS Journal*, *15*(1), 195–218.

Hajleh, M. N. A., & Al-Dujaili, E. A. S. (2022). Effects of turmeric concentrate on cardiovascular risk factors and exercise-induced oxidative stress in healthy volunteers; an exploratory study. *Advanced Pharmaceutical Bulletin, 13*(3), 601.

Hammer, K. A., Carson, C. F., & Riley, T. V. (1999). Antimicrobial activity of essential oils and other plant extracts. *Journal of Applied Microbiology, 86*(6), 985–990.

Heydarirad, G., Mirzaei, H. R., Gharehgozlou, R., Buentzel, J., Namazi, M. R., & Pasalar, M. (2024). Exploring the effectiveness of Ajwain cream in treating taxane-induced peripheral neuropathy in cancer patients: A pilot, randomised and double-blind clinical trial. *Indian Journal of Palliative Care, 30*(1), 71.

Jamshidi, N., & Cohen, M. M. (2017). The clinical efficacy and safety of Tulsi in humans: A systematic review of the literature. *Evidence-Based Complementary and Alternative Medicine, 2017*(1), 9217567.

Johnston, C. S. (2009). Medicinal uses of vinegar. In *Complementary and alternative therapies and the aging population* (pp. 433–443). Academic Press.

Johnston, C. S., Kim, C. M., & Buller, A. J. (2004). Vinegar improves insulin sensitivity to a high-carbohydrate meal in subjects with insulin resistance or type 2 diabetes. *Diabetes Care, 27*(1), 281–283.

Khanna, R., MacDonald, J. K., & Levesque, B. G. (2014). Peppermint oil for the treatment of irritable bowel syndrome: A systematic review and meta-analysis. *Journal of Clinical Gastroenterology, 48*(6), 505–512.

Khederzadeh, A., Ebrahimnejad, P., Seyedabadi, M., Babaei, A., Amiri, F. T., Aslani, N., . . . & Mohammadi, H. (2024). Synergistic effect of curcumin and Piperine loaded Niosomal nanoparticles on acute pulmonary toxicity induced by Paraquat in mice. *Toxicology Research, 13*(6), tfae181.

Kochhar, S. L. (1981). *Economic Botany in the Tropics.*

Kondo, T., Kishi, M., Fushimi, T., & Kaga, T. (2009a). Acetic acid upregulates the expression of genes for fatty acid oxidation enzymes in liver to suppress body fat accumulation. *Journal of Agricultural and Food Chemistry, 57*(13), 5982–5986.

Kondo, T., Kishi, M., Fushimi, T., Ugajin, S., & Kaga, T. (2009b). Vinegar intake reduces body weight, body fat mass, and serum triglyceride levels in obese Japanese subjects. *Bioscience, Biotechnology, and Biochemistry, 73*(8), 1837–1843.

Korikanthimath, V. S., Ankegowda, S. J., & Akshitha, H. J. (2023). Cardamom. In *Handbook of Spices in India: 75 Years of Research and Development* (pp. 1577–1659). Singapore: Springer Nature Singapore.

Lira Neto, J. C. G., Damasceno, M. M. C., Ciol, M. A., de Freitas, R. W. J. F., de Araujo, M. F. M., Teixeira, C. R. D. S., . . . & Zanetti, M. L. (2022). Efficacy of cinnamon as an adjuvant in reducing the glycemic biomarkers of type 2 diabetes mellitus: A three-month, randomized, triple-blind, placebo-controlled clinical trial. *Journal of the American Nutrition Association, 41*(3), 266–274.

Mandal, M. D., & Mandal, S. (2011). Honey: Its medicinal property and antibacterial activity. *Asian Pacific Journal of Tropical Biomedicine, 1*(2), 154–160.

Martinez-Gonzalez, M. A., & Martin-Calvo, N. (2016). Mediterranean diet and life expectancy; beyond olive oil, fruits, and vegetables. *Current Opinion in Clinical Nutrition & Metabolic Care, 19*(6), 401–407.

Mirunalini, S., Krishnaveni, M., & Ambily, V. (2011). Effects of raw garlic (Allium sativum) on hyperglycemia in patients with type 2 diabetes mellitus. *Pharmacologyonline, 2*, 968–974.

Mohammadabadi, T., & Jain, R. (2024). Cinnamon: A nutraceutical supplement for the cardiovascular system. *Archives of Medical sciences. Atherosclerotic Diseases, 9*, e72.

Nangia-Makker, P., Raz, T., Tait, L., Shekhar, M. P., Li, H., Balan, V., . . . & Raz, A. (2013). Ocimum gratissimum retards breast cancer growth and progression and is a natural inhibitor of matrix metalloproteases. *Cancer Biology & Therapy, 14*(5), 417–427.

Prasad, S., & Aggarwal, B. B. (2011). Turmeric, the golden spice. In *Herbal Medicine: Biomolecular and Clinical Aspects* (2nd ed., pp. 263–288). Boca Raton, FL: CRC Press.

Prasanth, M. I., Sivamaruthi, B. S., Chaiyasut, C., & Tencomnao, T. (2019). A review of the role of green tea (Camellia sinensis) in antiphotoaging, stress resistance, neuroprotection, and autophagy. *Nutrients, 11*(2), 474.

Qin, S., Huang, L., Gong, J., Shen, S., Huang, J., Ren, H., & Hu, H. (2017). Efficacy and safety of turmeric and curcumin in lowering blood lipid levels in patients with cardiovascular risk factors: A meta-analysis of randomized controlled trials. *Nutrition Journal, 16*(1), 68.

Rai, A. K., & Joshi, R. (2009). Evaluation of antimicrobial properties of fruit extracts of Terminalia chebula against dental caries pathogens. *Jundishapur Journal of Microbiology, 2*(3), 105–111.

Ramalingum, N., & Mahomoodally, M. F. (2014). The therapeutic potential of medicinal foods. *Advances in Pharmacological and Pharmaceutical Sciences, 2014*(1), 354264.

Ried, K. (2016). Garlic lowers blood pressure in hypertensive individuals, regulates serum cholesterol, and stimulates immunity: An updated meta-analysis and review. *The Journal of Nutrition, 146*(2), 389S–396S.

Rivlin, R. S. (2001). Historical perspective on the use of garlic. *The Journal of Nutrition, 131*(3), 951S–954S.

Rizvi, S. A., Einstein, G. P., Tulp, O. L., Sainvil, F., & Branly, R. (2022). Introduction to traditional medicine and their role in prevention and treatment of emerging and re-emerging diseases. *Biomolecules*, *12*(10), 1442.

Ruggenenti, P., Caruso, M. R., Cortinovis, M., Perna, A., Peracchi, T., Giuliano, G. A., . . . & Remuzzi, G. (2022). Fresh lemon juice supplementation for the prevention of recurrent stones in calcium oxalate nephrolithiasis: A pragmatic, prospective, randomised, open, blinded endpoint (PROBE) trial. *EClinicalMedicine*, *43*.

Sadeghi, A. (2016). In vitro assessment of some probiotic properties of Lactobacillus fermentum isolated from pickled garlic. *Journal of Food Quality and Hazards Control*, *3*(2), 67–72.

Sadhupati, D. P., Lakshmisudha, R., Chakravarthy, K. N. K., & Naidana, P. S. (2024). A standardized combination of Terminalia chebula and Withania somnifera extracts enhances immune function in adults: A pilot randomized, double-blind, placebo-controlled clinical study. *Food & Nutrition Research*, *68*. https://doi.org/10.29219/fnr.v68.10297

Shang, A., Cao, S. Y., Xu, X. Y., Gan, R. Y., Tang, G. Y., Corke, H., . . . & Li, H. B. (2019). Bioactive compounds and biological functions of garlic (Allium sativum L.). *Foods*, *8*(7), 246.

Singh, S., Malhotra, M., & Majumdar, D. K. (2005). Antibacterial activity of Ocimum sanctum L. fixed oil. *Indian Journal of Experimental Biology*, *43*(9), 835.

Spence, C. (2024). Cinnamon: The historic spice, medicinal uses, and flavour chemistry. *International Journal of Gastronomy and Food Science*, *35*, 100858.

Stephen, J., Manoharan, D., & Radhakrishnan, M. (2023). Immune boosting functional components of natural foods and its health benefits. *Food Production, Processing and Nutrition*, *5*(1), 61.

Souissi, M., Azelmat, J., Chaieb, K., & Grenier, D. (2020). Antibacterial and anti-inflammatory activities of cardamom (Elettaria cardamomum) extracts: Potential therapeutic benefits for periodontal infections. *Anaerobe*, *61*, 102089.

Tapsell, L., Hemphill, I., Cobiac, L., Sullivan, D. R., Fenech, M., Patch, C., . . . & Inge, K. E. (2006). Health benefits of herbs and spices: The past, the present, the future. *Medical Journal of Australia*, *185*, 4–24.

Voulgaropoulou, S. D., Van Amelsvoort, T. A. M. J., Prickaerts, J., & Vingerhoets, C. (2019). The effect of curcumin on cognition in Alzheimer's disease and healthy aging: A systematic review of pre-clinical and clinical studies. *Brain Research*, *1725*, 146476.

Weerts, Z. Z. R., Masclee, A. A., Witteman, B. J., Clemens, C. H., Winkens, B., Brouwers, J. R., . . . & Keszthelyi, D. (2020). Efficacy and safety of peppermint oil in a randomized, double-blind trial of patients with irritable bowel syndrome. *Gastroenterology*, *158*(1), 123–136.

White, C. M., Pasupuleti, V., Roman, Y. M., Li, Y., & Hernandez, A. V. (2019). Oral turmeric/curcumin effects on inflammatory markers in chronic inflammatory diseases: A systematic review and meta-analysis of randomized controlled trials. *Pharmacological Research*, *146*, 104280.

Yagnik, D., Serafin, V., & J. Shah, A. (2018). Antimicrobial activity of apple cider vinegar against Escherichia coli, Staphylococcus aureus and Candida albicans; downregulating cytokine and microbial protein expression. *Scientific Reports*, *8*(1), 1732.

Zhang, W. K., Tao, S. S., Li, T. T., Li, Y. S., Li, X. J., Tang, H. B., . . . & Wan, C. J. (2016). Nutmeg oil alleviates chronic inflammatory pain through inhibition of COX-2 expression and substance P release in vivo. *Food & Nutrition Research*, *60*(1), 30849.

22 Clinical Studies and Assessment of Functional Food (Vegetables)

*Shubhankar Anand, Bhanu Pratap Singh,
R T Thrisha Lakshmi, and V Dhanyalakshmi*

22.1. INTRODUCTION

Functional foods are a category of foods that go beyond basic nutrition, offering additional health benefits due to their bioactive compounds (Milner 2002). Vegetables, with their abundance of vitamins, minerals, fibre, and phytochemicals, stand out as prime contenders for functional food status (da Silva, Barreira, and Oliveira 2016). The purpose of this review is to examine clinical researches that particularly assess the health benefits of including functional foods made from vegetables in diets. This study aims to shed light on the importance and value of vegetables in boosting general health and well-being by examining the effects of these nutrients on a range of health-related topics, from disease prevention to cardiovascular health. The role of diet in preserving health and averting chronic illnesses has come to be understood more and more in recent years (Organization 2003). As a result, a brand-new food product category called as "functional foods" has emerged (Thompson and Moughan 2008). Functional meals, as opposed to conventional foods, are made to offer more health advantages than just basic nourishment (Hasler 2002). These advantages are usually associated with the management or prevention of a number of medical disorders, such as diabetes, obesity, cardiovascular disease, and some types of cancer (Eyre et al. 2004). Vegetable-related functional foods have drawn the most attention out of all the different kinds of functional meals because of the abundance of bioactive chemicals that are present in vegetables and are thought to offer many health advantages (Davis and Melina 2014).

Functional foods are not a wholly novel idea. Food and health have long been associated historically, and many traditional diets have included items thought to have therapeutic qualities (Goldberg 2012). The phrase "functional food" was initially used in Japan in the 1980s to refer to foods that were fortified with particular nutrients in order to promote health, which is where the contemporary notion of functional foods originated (Howlett 2008). Since then, the market for functional foods has expanded quickly due to advancements in food science and technology, as well as growing consumer awareness of the connection between nutrition and health (Frewer, Scholderer, and Lambert 2003). One very abundant source of useful dietary components is vegetables. They contain a range of bioactive substances with demonstrated health benefits, such as vitamins, minerals, fibre, and phytochemicals (Liu 2013). In particular, phytochemicals are quite interesting because they may lower the chance of developing chronic illnesses (Lila and Raskin 2005). Numerous clinical trials have examined the potential health advantages of these chemicals, which include flavonoids, carotenoids, glucosinolates, and polyphenols (Finley 2005). The health claims made for functional foods need to be supported by clinical trials (Hasler 2002). These studies offer empirical support for the safety and effectiveness of functional foods in fostering wellness and averting illness (Baker et al. 2022). Nonetheless, there are particular difficulties in the planning and execution of clinical trials including functional foods. Functional meals, in contrast to pharmaceutical medications, which

 DOI: 10.1201/9781003559894-22

usually have a single active ingredient, frequently have a complex variety of bioactive substances (Lordan, Ross, and Stanton 2011). Because of its intricacy, it is challenging to identify the ideal dosage and length of intake as well as to isolate the effects of individual components (Dybing et al. 2002). Moreover, a person's diet, lifestyle, genetics, and gut flora can all have an impact on the activities of functional foods (Alkhatib et al. 2017). Because of this, compared to studies on pharmaceutical medications, clinical trials on functional foods sometimes call for bigger sample sizes and longer time frames (Hasler 2002). Notwithstanding these obstacles, the discipline has made great strides, and an increasing amount of research confirms the advantageous effects of functional foods related to vegetables on health.

Due to the high concentration of bioactive chemicals found in vegetables, functional foods related to vegetables are particularly significant among the several categories of functional foods (Shashirekha, Mallikarjuna, and Rajarathnam 2015). Numerous health benefits, such as antioxidant, anti-inflammatory, and anticancer characteristics, have been demonstrated for these substances (Caponio et al. 2022). Among the main bioactive substances included in vegetables are the following.

22.1.1. Flavonoids

Found in a wide range of vegetables, flavonoids are a vast class of polyphenolic chemicals. It has been demonstrated that flavonoids possess antioxidant qualities, which may aid in shielding cells from oxidative damage (Shen et al. 2022). They may also help lower the risk of cardiovascular illnesses and some forms of cancer since they have anti-inflammatory properties (Griffiths et al. 2016).

22.1.2. Carotenoids

These pigments are present in a wide variety of vegetables, especially those with red, orange, or yellow hues (Khoo et al. 2011). Antioxidant-rich carotenoids are building blocks for vitamin A, which is necessary for healthy skin, eyesight, and immune system performance (Kaparapu, Pragada, and Geddada 2020). Carotenoids may also help lower the chance of developing specific cancer types, according to some research (Tapiero, Townsend, and Tew 2004).

22.1.3. Glucosinolates

These sulphur-containing substances are present in Brussels sprouts, broccoli, and cabbage, among other cruciferous vegetables (Sivapalan 2017). The process of food preparation and digestion involves the conversion of glucosinolates into beneficial chemicals known as isothiocyanates (Oliviero, Verkerk, and Dekker 2018). It has been demonstrated that isothiocyanates possess anticancer capabilities, namely in lowering the risk of colorectal cancer (Bao et al. 2014).

22.1.4. Dietary Fibre

Vegetables packed with dietary fibre are great for gut health. Some of the health benefits that come from dietary fibre include regulation in blood sugar, prevention against cholesterol increases, and regular bowel movements (Buttriss and Stokes 2008). It also increases satiety, helping you control weight.

22.1.5. Polyphenols

A set of phytochemicals present in many vegetables are polyphenols (Upadhyay and Dixit 2015). Polyphenols may help decrease the risk of acquiring chronic diseases, such as cancer, diabetes, and cardiovascular disease, due to their proven anti-inflammatory and oxidative characteristics (Rasouli, Farzaei, and Khodarahmi 2017).

There are opportunities as well as problems in the creation of functional foods related to vegetables (Hasler 2002). Variability in the amount of bioactive chemicals in vegetables is one of the key problems, and it can be influenced by a number of factors, including growing circumstances, harvesting techniques, and food processing (Tiwari and Cummins 2013). It is challenging to standardize functional food products and guarantee consistent health benefits because of this diversity (Sanders and Huis in't Veld 1999). The requirement for thorough clinical trials to support health claims presents another difficulty. Although there is increasing evidence on the health advantages of functional foods related to vegetables, further study is required to completely understand the mechanisms of action and to establish the ideal intake levels for various individuals (Liu 2003).

The market for functional foods has a lot of potential despite these obstacles, especially given the rising demand from consumers for natural and healthful products (Van Kleef, Van Trijp, and Luning 2005). Technological developments in food processing, such as fermentation and encapsulation, are also creating new opportunities for the creation of functional foods with improved stability and bioavailability (Reque and Brandelli 2021). Given that studies are still being conducted to determine the health advantages of the bioactive chemicals found in vegetables, the future of functional foods involving vegetables is bright (Patil et al. 2009). More excellent clinical studies are required as the market for functional foods expands in order to offer solid proof of the items' safety and efficacy (Sun-Waterhouse 2011). Furthermore, innovation in food processing and product development is required to improve the nutritional value and consumer appeal of functional foods related to vegetables (Guiné, Ramalhosa, and Valente 2016). In summary, functional foods connected to vegetables have the potential to be very important in avoiding and promoting chronic diseases. We can create functional foods that not only nourish the body but also contribute to a healthier and more sustainable future by utilizing the power of nature's pharmacy.

22.2. BIOACTIVE COMPOUNDS IN VEGETABLES

Supplementing with garlic has been proposed as a cardioprotective agent and is associated with a decrease in atherogenic markers. Aged garlic extract has been shown in clinical trials to reduce blood pressure, suggesting that it could be used as a supplement to more established antihypertensive medications (Jeremic et al. 2020). There is conflicting data supporting the claim that allium vegetables, such as garlic and onions, prevent cancer. Although some meta-analyses indicate no significant connections, others demonstrate a protective benefit (Cairns et al. 2014). Regarding the effect of eating vegetables high in allium on metabolic health, the review showed conflicting results, with some research suggesting benefits and others not (Wan et al. 2019). Using AMSTAR (A Measurement Tool to Assess Systematic Reviews) and GRADE (Grading of Recommendations, Assessment, Development, and Evaluations), the paper evaluated the methodological quality of the included meta-analyses (Shea et al. 2007). It draws attention to the significant variation in research findings and the existence of publication bias in a number of the results.

22.2.1. Polyphenols

Among these, polyphenols have become a class of promising molecules due to their potential health benefits and wide occurrence in several fruits and vegetables (Rasouli, Farzaei, and Khodarahmi 2017). These are a few of the active compounds in them that minimize oxidative stress and possess antioxidant components, which can help avoid persistent conditions (Balsano and Alisi 2009). There are numerous types for polyphenols, but we can divide them into two main categories: flavonoids and phenolic acids (Singla et al. 2019). Flavonoids are one of the subclasses of polyphenols known as anthocyanins, isoflavones, flavones (that include apigenin), flavonols, and flavanones (Tijjani et al. 2020). Each of them has their own individual health benefits.

One frequent flavonoid type that is present in foods such as onions, tea, and apples is quercetin, which is well-known for its antioxidative qualities (Lakhanpal and Rai 2007). Comprising molecules such as epicatechin and epigallocatechin, catechins are members of the flavanol group of flavonoids and display a range of biological activities, including antibacterial, anticarcinogenic, and cardioprotective actions. Eating foods high in polyphenols, such as strawberries and blackberries, can supply a significant amount of these health-promoting substances through diet (Vahapoglu et al. 2021).

Vegetables are rich in polyphenols and flavonoids, which have been demonstrated to have anti-inflammatory, anticancer, and antioxidant qualities (Cuevas-Cianca et al. 2023). The synergistic effects of various bioactive components in vegetables mean that eating a range of vegetables can maximize health benefits. Cruciferous vegetables contain glucosinolates, which are linked to a lower risk of cancer (Soundararajan and Kim 2018). Dietary fibre is essential for preserving gut health and preventing chronic illnesses, such as obesity, diabetes, and cardiovascular disease, in addition to other bioactive chemicals (Gill et al. 2021).

Numerous encapsulation and nano-delivery methods have been developed to improve the stability and bioavailability of polyphenols (Zhang et al. 2021). The efficiency of various methods, such as liposomes, nanoparticles, and nanoemulsions, in delivering polyphenols to specific body locations is discussed in the work. The prospective uses of polyphenols in the food, pharmaceutical, and cosmetic industries are highlighted in the study. The difficulties and potential paths for the creation of nutraceuticals based on polyphenols are also covered. This chapter offers a thorough grasp of the health advantages linked to dietary fibre, polyphenols, bioactive substances found in vegetables, and allium vegetables. Every research study emphasizes the significance of these elements in fostering well-being and averting illnesses via diverse pathways and cooperative impacts (Cohen, Chavez, and Chehimi 2010).

22.2.2. Carotenoids

Plants, some microbes, and carotenogenic fungi contain a variety of pigments called carotenoids, which are yellow, orange, and red in colour (Avalos and Carmen Limón 2015). They coexist with chlorophyll in leaves and are essential to the colour of many fruits and vegetables. The compounds have specific physiological effects on plants and aid in photosynthesis and phototaxis (Tamaki, Shinomura, and Mochida 2023). Furthermore, it has been found that carotenoids not only operate as a shield against reactive oxygen species but also efficiently quench singlet oxygen and scavenge other free radicals (Edge and Truscott 2018). Carotenoids are essential for human and animal diets because they function as precursors to vitamin A. Lycopene is a well-known carotenoid that has been associated with a lower risk of cardiovascular disease and a lower incidence of prostate cancer in men (Przybylska 2020). Among carotenoids, it is considered to be one of the most potent antioxidants due to its high conjugated double-bond count. Eating tomatoes has been linked to increased serum lycopene levels. Tomatoes and tomato-based products are a common source of lycopene (Giovannucci 1999).

Some of the carotenoids that can be obtained from food are lutein, which can be found in spinach, kale, and broccoli; zeaxanthin, which can be found in egg yolks, maize, and spinach; beta-carotene, which can be found in carrots, spinach, and parsley; and lycopene, which can be found in tomatoes and their derivatives, watermelon, and guava. Carotenoids also reduce oxidative stress and scavenge oxygen radicals, demonstrating antioxidant action. Their inclusion in both fruits and vegetables emphasizes how crucial they are to a diet that is well-balanced and healthful (Minocha and Singh 2025).

To address nutritional deficiencies such as vitamin A, for example, genetic engineering techniques have been used to increase the carotenoid content in crops such as rice and tomatoes. Because of their antioxidant qualities, carotenoids are important for human health as well as plant physiology, including photoprotection. Studies have focused on clarifying the mevalonate and nonmevalonate pathways involved in carotenoid production, emphasizing the possibility of using metabolic engineering to produce health-promoting isoprenoids in crops such as tomatoes (Bartley and Scolnik 1995).

Research on the tomato mutation known as the light hyper-responsive high pigment mutant has shown substantial changes in the fruit metabolome, providing insight into the complex regulation of carotenoid levels in plants (Bino et al. 2005). The groundbreaking research on "golden rice", which successfully combated vitamin A deficiency by introducing the beta-carotene production route into rice endosperm through genetic engineering, is also covered in this document. These revelations highlight how carotenoid engineering might improve food's nutritional value while also expanding our knowledge of plant biology (Datta, Majumder, and Datta 2021).

Carotenoids, which are included in spinach, carrots, and other vibrant vegetables, have been linked to a lower chance of developing several malignancies as well as better eye health. Studies on humans have shown that taking supplements containing carotenoids improves skin health and immunity.

22.2.3. Dietary Fibre

High dietary fibre consumption has been linked to a lower risk of obesity and obesity-related disorders, including type 2 diabetes, cardiovascular disease, and several malignancies, according to numerous studies (Waddell and Orfila 2023). The many ways that dietary fibre improves satiety and energy metabolism, as well as how it produces short-chain fatty acids and modulates gut flora, are all covered in this research. Fibres, both soluble and insoluble, have positive effects on health. Insoluble fibres shorten transit times and increase fecal volume, while soluble fibres gel in the gastrointestinal tract to slow down digestion and absorption. High in dietary fibre and found in foods such as broccoli and kale, dietary fibre is essential for intestinal health. The advantages of fibre in controlling weight, enhancing glycemic management, and lowering the incidence of colon cancer have been demonstrated by clinical research (Berndtsson 2019).

The advantageous elements present in the raw and ripe peels of two Indian mango types, illuminating possible uses (Tharanathan, Yashoda, and Prabha 2006). The studies conducted on citrus oil emulsions stabilized by citrus pectin provide information about the mechanisms underpinning emulsion stability (Kaade et al. 2022). The study looks at how acid treatment and citrus variety affect the stabilization process, offering useful insights for the food business. The chemical properties and uses of calcium pectate made from citrus or apple pectin add to our understanding of the gelation features and possible biological applications of these pectin derivatives. Together, these research studies advance our knowledge of the bioactive ingredients found in fruits and how they work in different food applications. Table 22.1 includes classification of different dietary sources.

22.3. CLINICAL STUDIES ON VEGETABLE-RELATED FUNCTIONAL FOODS

22.3.1. Cardiovascular Health

The effect of diets high in vegetables on cardiovascular health has been the subject of several clinical research studies (Lampe 1999). For example, a study on the vegetable-rich Mediterranean diet revealed significant decreases in cardiovascular events in those who followed it. Precursors for nitric oxide generation, nitrate, and nitrate-derived nitrite have an impact on cardiomyocytes and cardiac ischemia/reperfusion. They increase smooth muscle relaxation, lessen arterial stiffness, enhance endothelial function, and lower blood pressure. Beetroot's dietary NO_{3-} may affect metabolic and cardiovascular processes via affecting protein/enzyme activity and gene expression. Bioactive compounds found in beetroot have the potential to amplify the cytoprotective effects of NO, which is produced via the NO_3-NO_2 -/NO pathway (Bangar et al. 2022).

After ischemia/reperfusion, flavonoids, such as luteolin, which is present in fruits, vegetables, and herbs, enhanced systolic and diastolic function in rat hearts (Olabiyi and de Castro Brás 2023). According to some studies, sarcoplasmic reticulum Ca2+-ATPase activity was elevated by p38 MAPK signaling (Wang et al. 2014). In animal models, baicalin, a flavone found in the rhizomes of

TABLE 22.1
Classification and Dietary Source

Phenolic Class	Compound	Dietary Sources
	Flavanoids	
Flavanones	Hesperetin Naringenin	Citrus fruits
Flavonols	Kaempferol Quercetin	Cherry tomato, onion, broccoli, tea, red wine, berries
Flavanols	Catechins	Chocolate, beans, apricot, tea, red wine, cherry, apple
Anthocyanidins	Cyanidin	Aubergine, berries, red wine, red cabbage
Isoflavones	Daidzein Genistein Glycitein	Soy products, peas
Flavones	Apigenin Luteolin	Cereals, parsley, celery
	Phenolic acids	
Hydroxycinnamic acid	Caffeic acid Coumaric acid Ferulic acid Sinapic acid	Kiwi, cherry, plum, apple, pear, potato, garlic
Hydroxybenzoic acid	Gallic acid p-hydroxybenzoic acid Protocatechuic acid	Berries, onion, blackberry, raspberry, strawberry

several herbs, has been demonstrated to provide protection against atherosclerosis and myocardial ischemia injury. Mice given baicalin orally were protected against pressure-overload-induced cardiac hypertrophy. Baicalin decreased apoptosis and fibrosis in heart tissue and ameliorated cardiac dysfunction and myocardial hypertrophy. Flavonoids' hydroxyl groups have the ability to inhibit platelet activation. The heart and brain are two organs where blood flow might be restricted by excessive platelet activity (Kloner, King, and Harrington 2018).

22.3.2. Cancer Prevention

Vegetable consumption has been linked to a lower chance of getting some cancers (Key 2011). Clinical experiments have demonstrated the potential of cruciferous vegetables, such as broccoli and Brussels sprouts, to reduce the risk of cancer due to their high glucosinolate and isothiocyanate content. Among the cruciferous vegetables are cabbage, Brussels sprouts, cauliflower, and broccoli. Dithiolthiones and isothiocyanates, two types of organosulfur compounds that are highly concentrated in these plants, have been shown to enhance the activity of enzymes involved in the detoxification of carcinogens and other foreign chemicals (Egbujor et al. 2022). Sulphorophane is one isothiocyanate that seems to be a very potent inducer of detoxifying enzymes. Indole-3-carbinol, another ingredient found in cruciferous vegetables, has been shown to affect how people metabolize estrogen. More specifically, changes to the estradiol hydroxylation pathway may produce a weaker form of estradiol that may offer protection against cancers associated to estrogen, including endometrial and breast cancers (Mungenaṣt and Thalhammer 2014). Alliums are a vegetable family that includes garlic, onions, scallions, leeks, and chives. They are known to contain chemicals such allyl methyl trisulfide and diallyl sulphide (Upadhyay 2017). It has been shown that allium compounds in vegetable systems promote enzymatic detoxification. Additionally, these compounds' antibacterial properties might prevent bacteria from converting nitrate to nitrite in the stomach, which would lower the amount of nitrite that is available for reaction with secondary amines to form nitrosamines, which could be carcinogenic, especially in the stomach (Duncan et al. 1997).

Epidemiology research has shown a connection between eating cruciferous vegetables, broccoli, cauliflower, Brussels sprouts, and watercress with a decreased risk of acquiring a number of cancer types (Higdon et al. 2007). Glucosinolates and isothiocyanates, which are abundant in cruciferous vegetables, may be the cause of this anticancer action (Abbaoui et al. 2018). Polyphenols, isothiocyanates, and glucosinolates are all present in significant proportions in daikon radish (Barillari et al. 2005). For instance, in *Raphanus sativus*, the most prevalent glucosinolate is 4-methylsulfanyl-3-butenyl glucosinolate, or glucoraphasatin (Jia et al. 2023). It is widely acknowledged that supporting Phase II detoxifying enzyme systems, including NADPH quinone oxidoreductase 1 (NQO_1) is an efficient approach to induce anticancer activity (Preethi et al. 2022). Detoxification is aided by phenol sulfotransferase and glutathione S-transferase. In fact, glucosinolates have been linked to the enhancement of such enzyme activity and have been shown in numerous studies to have anticancer effects (Mithen 2001).

Tomatoes normally contain 11.6–14 mg of lycopene per kg, according to studies (Marques et al. 2015). An unsaturated carotenoid with strong antioxidant qualities is lycopene. Studies have revealed that this substance can affect a number of biological functions, such as cell division, proliferation, and apoptosis. In addition to lycopene, green tomatoes—which have higher tomatine content than red tomatoes—may also possess antitumour properties. Through the destruction of cancer cell membranes and the induction of antigen-specific cellular immunity, tomateine's antitumour effects were demonstrated in vivo. The anticancer drug doxorubicin was shown to be less effective than the commercial tomato glycoalkaloid tomatine (10:1 mixture of α-tomatine and dehydrotomatine), which inhibited HepG2 cancer cells in a dose-dependent manner. These findings suggest that consumers may benefit from consuming green tomatoes with a high tomatine concentration (Friedman 2013).

22.3.3. Gut Health

The significance of vegetable-derived fibre in promoting intestinal health has been extensively studied. Eating more vegetables has been shown in clinical research to enhance the function and diversity of the gut microbiota, hence promoting better digestive health and reducing inflammation (Klinder et al. 2016). Cardiovascular disease, metabolic syndrome, and type 2 diabetes are all significantly influenced by dietary choices. The gut flora plays a major role in the regulation of metabolic health through food (Janssen and Kersten 2015).

The Bifidobacterium species increased 3.8 times after two weeks of consuming vegetables high in ITF (Hiel et al. 2019). In addition, we saw an increase in *B. longum subsp. longum* and, to a lesser extent, *B. bifidum*, *B. adolescentis*, and *B. longum subsp. longum* at the species level (Chichlowski et al. 2020). These findings align with a prior study, which found elevated levels of *B. longum subsp. longum* and *B. adolescentis* in overweight and obese children who received ITF supplements. There hasn't been much research done on how a naturally high-fibre diet with prebiotic activity affects intestinal tolerance and the composition and activity of gut bacteria. Following a three-week course of vegetable injections, consisting of pureed or juiced Jerusalem artichokes, Bifidobacterium levels increased in healthy persons (Campbell 2024).

22.3.4. Metabolic Health

Clinical research has demonstrated that vegetables enhance metabolic health. Studies have demonstrated that diets high in vegetables improve insulin sensitivity, reduce blood pressure, and reduce the incidence of type 2 diabetes (Lampe 1999). The impact of concentrated flavonoid components or plant extracts rich in flavonoids on insulin-glucose homeostasis shows that several of them indicate possible advantages. Through a variety of distinct chemical pathways, flavonoids may affect the metabolism of glucose in the liver, small intestine, muscle, and adipose tissue. It has been demonstrated that certain flavonoids inhibit important enzymes, such as α-amylase, α-glucosidase, and sodium-dependent glucose transporter, which aid in the digestion and absorption of carbohydrates from food in vitro (Sok Yen et al. 2021). This could assist in reducing glycemia after meals. Furthermore, flavonoids may improve the homeostasis of glucose and insulin via a number of signalling pathways. Two main pathways by which certain flavonoids enhance adipocyte function and muscle glucose uptake have been identified by studies conducted on cells and animals: peroxisome proliferator-activated receptor-γ (PPAR-γ) and adenosine 5′-adenosine monophosphate-activated protein kinase (AMPK) (Zygmunt et al. 2010).

22.4. MECHANISMS OF ACTION

22.4.1. Antioxidant Activity

The antioxidant properties of vegetable bioactive components help to prevent oxidative stress, neutralize free radicals, and prevent cellular damage (Balsano and Alisi 2009). Antioxidants delay molecular oxidation and may reduce oxidative damage to the body by inhibiting free radicals. Oxidative damage can make chronic illnesses, such as cancer and cardiovascular disease, worse. Cabbage had better antioxidant activity but a lower total phenolic content than kale (Ismail, Marjan, and Foong 2004). Shallots exhibited lesser antioxidant activity than spinach, but spinach had a higher total phenolic content than other vegetables (Şengül, Yildiz, and Kavaz 2014). The study's findings demonstrate that each type of vegetable has a different level of antioxidant activity, which can be related to a number of antioxidant-rich substances, including phenolic compounds, vitamin C, selenium, b-carotene, and a-tocopherol. Although the high concentrations of a-tocopherol, b-carotene, and ferulic acid in spinach and swamp cabbage were connected with their high antioxidant activity, shallots' great antioxidant activity may be attributed to their flavonoid content (Oboh and Rocha 2007).

22.4.2. Anti-Inflammatory Effects

Vegetables include compounds that can alter inflammatory pathways, reducing chronic inflammation and the diseases it is associated with. It was demonstrated that ginger phenolic compounds reduce the production of NO and PGE2 in LPS-induced RAW 264.7 macrophages (Saanin et al. 2021). Zerumbone and 3-O-methyl kaempferol effectively inhibited the generation of NO and decreased the expression of iNOS in a dose-dependent manner (Chien et al. 2008). However, the anti-inflammatory effects of zerumbone were superior to those of 3-O-methyl kaempferol. It was found that quercetin and luteolin suppressed the expression of iNOS and the generation of TNF-a and NO in LPS-activated macrophages (Aziz, Kim, and Cho 2018). This result was linked to the suppression of the NF-kB pathway. Among the vegetables that contain indole-3-carbinol are broccoli, cabbage, cauliflower, Brussels sprouts, and sprouts. This compound has been associated with a reduction in the production of pro-inflammatory mediators, such as NO, IL-6, and IL-1beta. The green leaves of the Welsh onion plant may have analgesic and anti-inflammatory effects because they have higher concentrations of glutathione peroxidase, catalase, and superoxide dismutase (SOD) and lower concentrations of reactive nitrogen species. Zhu et al. (2012) found that water-soluble polysaccharide from *Chaenomeles speciosa* fruits inhibited the production of TNF-a, IFN-g, and granulocyte colony-stimulating factor gene in RAW 264.7 cells challenged with LPS (Zhu et al. 2012).

Phospholipase A2 (PLA2) releases arachidonic acid from plasma membrane phospholipids during inflammation. Arachidonic acid is processed by enzymes such as COX and LOX to produce inflammatory mediators, such as prostaglandins, thromboxanes, and leukotrienes. Flavonoids can suppress the production of prostaglandins, thromboxanes, and leukotrienes by inhibiting the PLA2 enzyme (Balsinde, Winstead, and Dennis 2002). The numerous characteristics and functions of flavonoids impact immune cell types' activation, maturation, signalling transduction, cytokine production, and secretory processes. For example, flavonoids have been demonstrated to limit dendritic cell (DC) maturation by blocking molecules, such as CD80 and CD86, that are increased during DC maturation and required for CD4 + T cell activation. This would prevent the release of cytokines and the growth of CD4+ T cells. Flavonoids alter the metabolism of iron, which alters DCs' inflammatory responses. Several studies have demonstrated that certain flavonoids can prevent or inhibit mast cells from generating histamine or prostaglandin (Martínez, Mijares, and De Sanctis 2019).

22.4.3. Modulation of Gut Microbiota

Dendritic cell (DC) maturation has been demonstrated to be restricted by flavonoids, for example, by blocking molecules such as CD80 and CD86 that are increased during DC maturation and required for CD4 + T cell activation (Li, Li, and Zhang 2015). This would prevent cytokines from being released and stop CD4+ T cell growth. Flavonoids modify iron metabolism, which in turn modifies the inflammatory responses of DCs. Numerous studies have shown that certain flavonoids can inhibit or stop mast cells from producing prostaglandin or histamine (Amellal et al. 1985). Short-chain fatty acids (SCFAs) are produced when different bacterial species and genera cross-feed one other. SCFAs can be produced by gut bacteria fermenting DFs. According to Den Besten et al. (2013), butyrate, an SCFA produced by bacteria, may have several health benefits. Colonic epithelial cells preferentially employ butyrate to make ketone bodies and carbon dioxide. By lowering luminal pH, SCFA synthesis improves mineral absorption and stops main bile acids from converting into secondary bile acids, which can cause cancer (den Besten et al. 2013).

22.5. CONCLUSION

In order to assess the safety and effectiveness of functional foods and provide scientific proof of their health benefits, clinical trials are essential. By establishing evidence-based assertions, these research studies guarantee that functional meals have a favorable impact on both illness prevention and general well-being. To verify their functional qualities, thorough evaluations are necessary, such as long-term observational studies, biomarker analysis, and randomized controlled trials. To improve dependability, however, issues including individual response variations, study design variability, and regulatory concerns must be resolved. To optimize the potential of functional foods in improving health, future research should concentrate on standardized procedures, individualized nutrition approaches, and the exploration of novel bioactive components. These foods, which are high in bioactive chemicals, help manage and prevent a number of chronic illnesses. To completely understand the mechanisms of action and to pinpoint certain vegetable-based therapies for various medical diseases, more research is required.

REFERENCES

Abbaoui, Besma, Christopher R Lucas, Ken M Riedl, Steven K Clinton, and Amir Mortazavi. 2018. "Cruciferous vegetables, isothiocyanates, and bladder cancer prevention." *Molecular Nutrition & Food Research* 62 (18):1800079.

Alkhatib, Ahmad, Catherine Tsang, Ali Tiss, Theeshan Bahorun, Hossein Arefanian, Roula Barake, Abdelkrim Khadir, and Jaakko Tuomilehto. 2017. "Functional foods and lifestyle approaches for diabetes prevention and management." *Nutrients* 9 (12):1310.

Amellal, M, C Bronner, F Briancon, M Haag, R Anton, and Y Landry. 1985. "Inhibition of mast cell histamine release by flavonoids and biflavonoids." *Planta Medica* 51 (1):16–20.

Avalos, Javier, and M Carmen Limón. 2015. "Biological roles of fungal carotenoids." *Current Genetics* 61:309–324.

Aziz, Nur, Mi-Yeon Kim, and Jae Youl Cho. 2018. "Anti-inflammatory effects of luteolin: A review of in vitro, in vivo, and in silico studies." *Journal of Ethnopharmacology* 225:342–358.

Baker, Mathew T, Peng Lu, Jean A Parrella, and Holli R Leggette. 2022. "Consumer acceptance toward functional foods: A scoping review." *International Journal of Environmental Research and Public Health* 19 (3):1217.

Balsano, Clara, and Anna Alisi. 2009. "Antioxidant effects of natural bioactive compounds." *Current Pharmaceutical Design* 15 (26):3063–3073.

Balsinde, Jesús, Michelle V Winstead, and Edward A Dennis. 2002. "Phospholipase A2 regulation of arachidonic acid mobilization." *FEBS Letters* 531 (1):2–6.

Bangar, Sneh Punia, Nitya Sharma, Nikita Sanwal, Jose M Lorenzo, and JK Sahu. 2022. "Bioactive potential of beetroot (Beta vulgaris)." *Food Research International* 158:111556.

Bao, Yongping, Wei Wang, Zhigang Zhou, and Changhao Sun. 2014. "Benefits and risks of the hormetic effects of dietary isothiocyanates on cancer prevention." *PloS One* 9 (12):e114764.

Barillari, Jessica, Rinaldo Cervellati, Moreno Paolini, Arnaud Tatibouët, Patrick Rollin, and Renato Iori. 2005. "Isolation of 4-methylthio-3-butenyl glucosinolate from Raphanus sativus sprouts (Kaiware Daikon) and its redox properties." *Journal of Agricultural and Food Chemistry* 53 (26):9890–9896.

Bartley, Glenn E, and Pablo A Scolnik. 1995. "Plant carotenoids: Pigments for photoprotection, visual attraction, and human health." *The Plant Cell* 7 (7):1027.

Berndtsson, Emilia. 2019. "Dietary fibre and phenolic compounds in broccoli (Brassica oleracea Italica group) and kale (Brassica oleracea Sabellica group)." In *A Literature Study about the Potential Uses of Side Streams*. Alnarp: Sveriges lantbruksuniversitet.

Bino, Raoul J, CH Ric De Vos, Michal Lieberman, Robert D Hall, Arnaud Bovy, Harry H Jonker, Yury Tikunov, Arjen Lommen, Sofia Moco, and Ilan Levin. 2005. "The light-hyperresponsive high pigment-2dg mutation of tomato: Alterations in the fruit metabolome." *New Phytologist* 166 (2):427–438.

Buttriss, JL, and CS Stokes. 2008. "Dietary fibre and health: An overview." *Nutrition Bulletin* 33 (3):186–200.

Cairns, Kathryn Elizabeth, Marie Bee Hui Yap, Pamela Doreen Pilkington, and Anthony Francis Jorm. 2014. "Risk and protective factors for depression that adolescents can modify: A systematic review and meta-analysis of longitudinal studies." *Journal of Affective Disorders* 169:61–75.

Campbell, Kristina. 2024. *Gut Health for Dummies*. John Wiley & Sons.

Caponio, Giusy Rita, Tamara Lippolis, Valeria Tutino, Isabella Gigante, Valentina De Nunzio, Rosa Anna Milella, Marica Gasparro, and Maria Notarnicola. 2022. "Nutraceuticals: Focus on anti-inflammatory, anti-cancer, antioxidant properties in gastrointestinal tract." *Antioxidants* 11 (7):1274.

Chichlowski, Maciej, Neil Shah, Jennifer L Wampler, Steven S Wu, and Jon A Vanderhoof. 2020. "Bifidobacterium longum subspecies infantis (B. infantis) in pediatric nutrition: Current state of knowledge." *Nutrients* 12 (6):1581.

Chien, TY, LG Chen, CJ Lee, FY Lee, and CC Wang. 2008. "Anti-inflammatory constituents of Zingiber zerumbet." *Food Chemistry* 110 (3):584–589.

Cohen, Larry, Vivian Chavez, and Sana Chehimi. 2010. *Prevention is Primary: Strategies for Community Well Being*. John Wiley & Sons.

Cuevas-Cianca, Sofía Isabel, Cristian Romero-Castillo, José Luis Gálvez-Romero, Zaida Nelly Juárez, and Luis Ricardo Hernández. 2023. "Antioxidant and anti-inflammatory compounds from edible plants with anti-cancer activity and their potential use as drugs." *Molecules* 28 (3):1488.

da Silva, Beatriz Vieira, João CM Barreira, and M Beatriz PP Oliveira. 2016. "Natural phytochemicals and probiotics as bioactive ingredients for functional foods: Extraction, biochemistry and protected-delivery technologies." *Trends in Food Science & Technology* 50:144–158.

Datta, Swapan K, Shuvobrata Majumder, and Karabi Datta. 2021. "Molecular breeding for improved β-carotene synthesis in golden rice: Recent progress and future perspectives." *Molecular Breeding for Rice Abiotic Stress Tolerance and Nutritional Quality*. 287–303. Wiley Online Library.

Davis, Brenda, and Vesanto Melina. 2014. *Becoming Vegan: The Complete Reference to Plant-Base Nutrition*. Book Publishing Company.

den Besten, Gijs, Katja Lange, Rick Havinga, Theo H van Dijk, Albert Gerding, Karen van Eunen, Michael Müller, Albert K Groen, Guido J Hooiveld, and Barbara M Bakker. 2013. "Gut-derived short-chain fatty acids are vividly assimilated into host carbohydrates and lipids." *American Journal of Physiology-Gastrointestinal and Liver Physiology* 305 (12):G900–G910.

Duncan, Callum, Hong Li, Roelf Dykhuizen, Rennie Frazer, Peter Johnston, Gillian MacKnight, Lorna Smith, Kathryn Lamza, Hamish McKenzie, and Les Batt. 1997. "Protection against oral and gastrointestinal diseases: Importance of dietary nitrate intake, oral nitrate reduction and enterosalivary nitrate circulation." *Comparative Biochemistry and Physiology Part A: Physiology* 118 (4):939–948.

Dybing, E, J Doe, J Groten, J Kleiner, J O'Brien, AG Renwick, J Schlatter, Pablo Steinberg, A Tritscher, and R Walker. 2002. "Hazard characterisation of chemicals in food and diet: Dose response, mechanisms and extrapolation issues." *Food and Chemical Toxicology* 40 (2–3):237–282.

Edge, Ruth, and T George Truscott. 2018. "Singlet oxygen and free radical reactions of retinoids and carotenoids—a review." *Antioxidants* 7 (1):5.

Egbujor, Melford Chuka, Maria Petrosino, Karim Zuhra, and Luciano Saso. 2022. "The role of organosulfur compounds as Nrf2 activators and their antioxidant effects." *Antioxidants* 11 (7):1255.

Eyre, Harmon, Richard Kahn, Rose Marie Robertson, ACS/ADA/AHA Collaborative Writing Committee, ACS/ADA/AHA Collaborative Writing Committee Members, Nathaniel G Clark, Colleen Doyle, Yuling Hong, Ted Gansler, and Thomas Glynn. 2004. "Preventing cancer, cardiovascular disease, and diabetes: A common agenda for the American Cancer Society, the American Diabetes Association, and the American Heart Association." *Circulation* 109 (25):3244–3255.

Finley, John W. 2005. "Proposed criteria for assessing the efficacy of cancer reduction by plant foods enriched in carotenoids, glucosinolates, polyphenols and selenocompounds." *Annals of Botany* 95 (7): 1075–1096.

Frewer, Lynn, Joachim Scholderer, and Nigel Lambert. 2003. "Consumer acceptance of functional foods: Issues for the future." *British Food Journal* 105 (10):714–731.

Friedman, Mendel. 2013. "Anticarcinogenic, cardioprotective, and other health benefits of tomato compounds lycopene, α-tomatine, and tomatidine in pure form and in fresh and processed tomatoes." *Journal of Agricultural and Food Chemistry* 61 (40):9534–9550.

Gill, Samantha K, Megan Rossi, Balazs Bajka, and Kevin Whelan. 2021. "Dietary fibre in gastrointestinal health and disease." *Nature Reviews Gastroenterology & Hepatology* 18 (2):101–116.

Giovannucci, Edward. 1999. "Tomatoes, tomato-based products, lycopene, and cancer: Review of the epidemiologic literature." *Journal of the National Cancer Institute* 91 (4):317–331.

Goldberg, Israel. 2012. *Functional Foods: Designer Foods, Pharmafoods, Nutraceuticals*. Springer Science & Business Media.

Griffiths, Keith, Bharat B Aggarwal, Ram B Singh, Harpal S Buttar, Douglas Wilson, and Fabien De Meester. 2016. "Food antioxidants and their anti-inflammatory properties: A potential role in cardiovascular diseases and cancer prevention." *Diseases* 4 (3):28.

Guiné, PF Raquel, Elsa CD Ramalhosa, and Luísa Paula Valente. 2016. "New foods, new consumers: Innovation in food product development." *Current Nutrition & Food Science* 12 (3):175–189.

Hasler, Clare M. 2002. "Functional foods: Benefits, concerns and challenges—a position paper from the American Council on Science and Health." *The Journal of Nutrition* 132 (12):3772–3781.

Hiel, Sophie, Laure B Bindels, Barbara D Pachikian, Gaetan Kalala, Valérie Broers, Giorgia Zamariola, Betty PI Chang, Bienvenu Kambashi, Julie Rodriguez, and Patrice D Cani. 2019. "Effects of a diet based on inulin-rich vegetables on gut health and nutritional behavior in healthy humans." *The American Journal of Clinical Nutrition* 109 (6):1683–1695.

Higdon, Jane V, Barbara Delage, David E Williams, and Roderick H Dashwood. 2007. "Cruciferous vegetables and human cancer risk: Epidemiologic evidence and mechanistic basis." *Pharmacological Research* 55 (3):224–236.

Howlett, John. 2008. *Functional Foods: From Science to Health and Claims,* ILSI Europe Concise Monograph Series. https://scholar.google.com/scholar?hl=en&as_sdt=0%2C5&q=Howlett%2C+John.+2008.+Functional+Foods%3A+From+Science+to+Health+and+Claims%2C+ILSI+Europe+Concise+Monograph++Series.&btnG=

Ismail, Amin, Zamaliah M Marjan, and Chin W Foong. 2004. "Total antioxidant activity and phenolic content in selected vegetables." *Food Chemistry* 87 (4):581–586.

Janssen, Aafke WF, and Sander Kersten. 2015. "The role of the gut microbiota in metabolic health." *The FASEB Journal* 29 (8):3111–3123.

Jeremic, Jovana N, Vladimir Lj Jakovljevic, Vladimir I Zivkovic, Ivan M Srejovic, Jovana V Bradic, Isidora M Milosavljevic, Slobodanka Lj Mitrovic, Nemanja U Jovicic, Sergey B Bolevich, and Andrey A Svistunov. 2020. "Garlic derived diallyl trisulfide in experimental metabolic syndrome: Metabolic effects and cardioprotective role." *International Journal of Molecular Sciences* 21 (23):9100.

Jia, Xiao, Pei Yu, Qi An, Jingnan Ren, Gang Fan, Zelan Wei, Xixiang Li, and Siyi Pan. 2023. "Identification of glucosinolates and volatile odor compounds in microwaved radish (Raphanus sativus L.) seeds and the corresponding oils by UPLC-IMS-QTOF-MS and GC×GC-qMS analysis." *Food Research International* 169:112873.

Kaade, Wael, Carmen Méndez-Sánchez, Carme Güell, Silvia De Lamo-Castellví, Monsterrat Mestres, and Montse Ferrando. 2022. "Complexed biopolymer of whey protein and carboxymethyl cellulose to enhance the chemical stability of lemon oil-in-water emulsions." *ACS Food Science & Technology* 2 (1):41–48.

Kaparapu, Jyothi, Prayaga Murthy Pragada, and Mohana Narasimha Rao Geddada. 2020. "Fruits and vegetables and its nutritional benefits." *Functional Foods and Nutraceuticals: Bioactive Components, Formulations and Innovations*, 241–260. Springer.

Key, TJ. 2011. "Fruit and vegetables and cancer risk." *British Journal of Cancer* 104 (1):6–11.

Khoo, Hock-Eng, K Nagendra Prasad, Kin-Weng Kong, Yueming Jiang, and Amin Ismail. 2011. "Carotenoids and their isomers: Color pigments in fruits and vegetables." *Molecules* 16 (2):1710–1738.

Klinder, Annett, Qing Shen, Susanne Heppel, Julie A Lovegrove, Ian Rowland, and Kieran M Tuohy. 2016. "Impact of increasing fruit and vegetables and flavonoid intake on the human gut microbiota." *Food & Function* 7 (4):1788–1796.

Kloner, Robert A, Kevin S King, and Michael G Harrington. 2018. "No-reflow phenomenon in the heart and brain." *American Journal of Physiology-Heart and Circulatory Physiology* 315 (3):H550–H562.

Lakhanpal, Parul, and Deepak Kumar Rai. 2007. "Quercetin: A versatile flavonoid." *Internet Journal of Medical Update* 2 (2):22–37.

Lampe, Johanna W. 1999. "Health effects of vegetables and fruit: Assessing mechanisms of action in human experimental studies." *The American Journal of Clinical Nutrition* 70 (3):475S–490S.

Li, Jinyao, Jinyu Li, and Fuchun Zhang. 2015. "The immunoregulatory effects of Chinese herbal medicine on the maturation and function of dendritic cells." *Journal of Ethnopharmacology* 171:184–195.

Lila, Mary Ann, and Ilya Raskin. 2005. "Health-related interactions of phytochemicals." *Journal of Food Science* 70 (1):R20–R27.

Liu, Rui Hai. 2003. "Health benefits of fruit and vegetables are from additive and synergistic combinations of phytochemicals." *The American Journal of Clinical Nutrition* 78 (3):517S–520S.

Liu, Rui Hai. 2013. "Dietary bioactive compounds and their health implications." *Journal of Food Science* 78 (S1):A18–A25.

Lordan, Sinéad, R Paul Ross, and Catherine Stanton. 2011. "Marine bioactives as functional food ingredients: Potential to reduce the incidence of chronic diseases." *Marine Drugs* 9 (6):1056–1100.

Marques, CS, Maria João Reis Lima, J Oliveira, and Edite Teixeira de Lemos. 2015. "Tomato lycopene: Functional proprieties and health." *International Scholarly and Scientific Research & Innovation* 10:458–468.

Martínez, Gricelis, Michael R Mijares, and Juan B De Sanctis. 2019. "Effects of flavonoids and its derivatives on immune cell responses." *Recent Patents on Inflammation & Allergy Drug Discovery* 13 (2):84–104.

Milner, JA. 2002. "Functional foods and health: A US perspective." *British Journal of Nutrition* 88 (S2): S152–S158.

Minocha, Neha, and Astha Singh. 2025. "Nutrition essentials: Building a foundation for optimal health through diet." *Impact of Yoga and Proper Diet on Cardiopulmonary Function*, 119–160. IGI Global Scientific Publishing.

Mithen, Richard. 2001. "Glucosinolates–biochemistry, genetics and biological activity." *Plant Growth Regulation* 34:91–103.

Mungenast, Felicitas, and Theresia Thalhammer. 2014. "Estrogen biosynthesis and action in ovarian cancer." *Frontiers in Endocrinology* 5:192.

Oboh, G, and JBT Rocha. 2007. "Antioxidant in foods: A new challenge for food processors." *Leading Edge Antioxidants Research* 1:35–64.

Olabiyi, Ayodeji A, and Lisandra E de Castro Brás. 2023. "Cardiovascular remodeling post-ischemia: Herbs, diet, and drug interventions." *Biomedicines* 11 (6):1697.

Oliviero, Teresa, Ruud Verkerk, and Matthijs Dekker. 2018. "Isothiocyanates from brassica vegetables—effects of processing, cooking, mastication, and digestion." *Molecular Nutrition & Food reSearch* 62 (18):1701069.

Patil, Bhimanagouda S, Guddadarangavvanahally K Jayaprakasha, KN Chidambara Murthy, and Amit Vikram. 2009. "Bioactive compounds: Historical perspectives, opportunities, and challenges." *Journal of Agricultural and Food Chemistry* 57 (18):8142–8160.

Preethi, S, K Arthiga, Amit B Patil, Asha Spandana, and Vikas Jain. 2022. "Review on NAD (P) H dehydrogenase quinone 1 (NQO1) pathway." *Molecular Biology Reports* 49 (9):8907–8924.

Przybylska, Sylwia. 2020. "Lycopene–a bioactive carotenoid offering multiple health benefits: A review." *International Journal of Food Science and Technology* 55 (1):11–32.

Rasouli, Hassan, Mohammad Hosein Farzaei, and Reza Khodarahmi. 2017. "Polyphenols and their benefits: A review." *International Journal of Food Properties* 20 (S2):1700–1741.

Reque, Priscilla Magro, and Adriano Brandelli. 2021. "Encapsulation of probiotics and nutraceuticals: Applications in functional food industry." *Trends in Food Science & Technology* 114:1–10.

Saanin, Sri Nadya, Roro Wahyudianingsih, Merry Afni, Ervi Afifah, Maesaroh Maesaroh, and Wahyu Widowati. 2021. "Suppression of pro-inflammatory cytokines and mediators production by ginger (Zingiber officinale Roscoe) ethanolic extract and gingerol in lipopolysaccharide-induced RAW 264.7 murine macrophage cells." *Indian Journal of Natural Products and Resources (IJNPR) [Formerly Natural Product Radiance (NPR)]* 11(4):260–266.

Sanders, Mary Ellen, and Jos Huis in't Veld. 1999. "Bringing a probiotic-containing functional food to the market: Microbiological, product, regulatory and labeling issues." *Antonie van Leeuwenhoek* 76:293–315.

Şengül, Memnune, Hilal Yildiz, and Arzu Kavaz. 2014. "The effect of cooking on total polyphenolic content and antioxidant activity of selected vegetables." *International Journal of Food Properties* 17 (3):481–490.

Shashirekha, MN, SE Mallikarjuna, and S Rajarathnam. 2015. "Status of bioactive compounds in foods, with focus on fruits and vegetables." *Critical Reviews in Food Science and Nutrition* 55 (10):1324–1339.

Shea, Beverley J, Lex M Bouter, Joan Peterson, Maarten Boers, Neil Andersson, Zulma Ortiz, Tim Ramsay, Annie Bai, Vijay K Shukla, and Jeremy M Grimshaw. 2007. "External validation of a measurement tool to assess systematic reviews (AMSTAR)." *PloS One* 2 (12):e1350.

Shen, Nan, Tongfei Wang, Quan Gan, Sian Liu, Li Wang, and Biao Jin. 2022. "Plant flavonoids: Classification, distribution, biosynthesis, and antioxidant activity." *Food Chemistry* 383:132531.

Singla, Rajeev K, Ashok K Dubey, Arun Garg, Ramesh K Sharma, Marco Fiorino, Sara M Ameen, Moawiya A Haddad, and Masnat Al-Hiary. 2019. "Natural polyphenols: Chemical classification, definition of classes, subcategories, and structures." *Journal of AOAC International* 102(5):1397–1400.

Sivapalan, Tharsini. 2017. *The Bioavailability and Biological Activity of Sulphur-Containing Compounds from Broccoli*. University of East Anglia.

Sok Yen, Foo, Chan Shu Qin, Sharryl Tan Shi Xuan, Puah Jia Ying, Hong Yi Le, Thiviya Darmarajan, Baskaran Gunasekaran, and Shamala Salvamani. 2021. "Hypoglycemic effects of plant flavonoids: A review." *Evidence-Based Complementary and Alternative Medicine* 2021 (1):2057333.

Soundararajan, Prabhakaran, and Jung Sun Kim. 2018. "Anti-carcinogenic glucosinolates in cruciferous vegetables and their antagonistic effects on prevention of cancers." *Molecules* 23 (11):2983.

Sun-Waterhouse, Dongxiao. 2011. "The development of fruit-based functional foods targeting the health and wellness market: A review." *International Journal of Food Science and Technology* 46 (5):899–920.

Tamaki, Shun, Tomoko Shinomura, and Keiichi Mochida. 2023. "Illuminating the diversity of carotenoids in microalgal eyespots and phototaxis." *Plant Signaling & Behavior* 18 (1):2257348.

Tapiero, Haim, Danyelle M Townsend, and Kenneth D Tew. 2004. "The role of carotenoids in the prevention of human pathologies." *Biomedicine & Pharmacotherapy* 58 (2):100–110.

Tharanathan, RN, HM Yashoda, and TN Prabha. 2006. "Mango (Mangifera indica L.), 'The king of fruits'—an overview." *Food Reviews International* 22 (2):95–123.

Thompson, Abby K, and Paul J Moughan. 2008. "Innovation in the foods industry: Functional foods." *Innovation* 10 (1):61–73.

Tijjani, Habibu, Maryam H Zangoma, Zinat S Mohammed, Shakirdeen M Obidola, Chukwuebuka Egbuna, and Suliat I Abdulai. 2020. "Polyphenols: Classifications, biosynthesis and bioactivities." *Functional Foods and Nutraceuticals: Bioactive Components, Formulations and Innovations*. 389–414. Springer International Publishing.

Tiwari, U, and E Cummins. 2013. "Factors influencing levels of phytochemicals in selected fruit and vegetables during pre-and post-harvest food processing operations." *Food Research International* 50 (2):497–506.

Upadhyay, Ravi Kant. 2017. "Nutritional and therapeutic potential of Allium vegetables." *Journal of Nutritional Therapeutics* 6 (1):18–37.

Upadhyay, Swapna, and Madhulika Dixit. 2015. "Role of polyphenols and other phytochemicals on molecular signaling." *Oxidative Medicine and Cellular Longevity* 2015 (1):504253.

Vahapoglu, Beyza, Ezgi Erskine, Busra Gultekin Subasi, and Esra Capanoglu. 2021. "Recent studies on berry bioactives and their health-promoting roles." *Molecules* 27 (1):108.

Van Kleef, Ellen, Hans CM Van Trijp, and Pieternel Luning. 2005. "Functional foods: Health claim-food product compatibility and the impact of health claim framing on consumer evaluation." *Appetite* 44 (3):299–308.

Waddell, Isabella Skye, and Caroline Orfila. 2023. "Dietary fiber in the prevention of obesity and obesity-related chronic diseases: From epidemiological evidence to potential molecular mechanisms." *Critical Reviews in Food Science and Nutrition* 63 (27):8752–8767.

Wan, Qianyi, Ni Li, Liang Du, Rui Zhao, Mengshi Yi, Qiushi Xu, and Yong Zhou. 2019. "Allium vegetable consumption and health: An umbrella review of meta-analyses of multiple health outcomes." *Food Science & Nutrition* 7 (8):2451–2470.

Wang, Lu, Wanshuai Li, Yang Yang, Yamei Hu, Yanhong Gu, Yongqian Shu, Yang Sun, Xuefeng Wu, Yan Shen, and Qiang Xu. 2014. "High expression of sarcoplasmic/endoplasmic reticulum Ca2+-ATPase 2b blocks cell differentiation in human liposarcoma cells." *Life Sciences* 99 (1–2):37–43.

World Health Organization. 2003. *Diet, Nutrition, and the Prevention of Chronic Diseases: Report of a Joint WHO/FAO Expert Consultation*, Vol. 916. World Health Organization.

Zhang, Zhiheng, Chao Qiu, Xiaojing Li, David Julian McClements, Aiquan Jiao, Jinpeng Wang, and Zhengyu Jin. 2021. "Advances in research on interactions between polyphenols and biology-based nano-delivery systems and their applications in improving the bioavailability of polyphenols." *Trends in Food Science & Technology* 116:492–500.

Zhu, Qing, Chaolin Liao, Yongmei Liu, Pengcheng Wang, Wei Guo, Meijun He, and Zebo Huang. 2012. "Ethanolic extract and water-soluble polysaccharide from Chaenomeles speciosa fruit modulate lipopolysaccharide-induced nitric oxide production in RAW264. 7 macrophage cells." *Journal of Ethnopharmacology* 144 (2):441–447.

Zygmunt, Katarzyna, Brandon Faubert, Jordan MacNeil, and Evangelia Tsiani. 2010. "Naringenin, a citrus flavonoid, increases muscle cell glucose uptake via AMPK." *Biochemical and Biophysical Research Communications* 398 (2):178–183.

23 Allergy, Allergens, and Allergenicity Related to Spices

Neetu Yadav

23.1. INTRODUCTION

Many plant materials are used as spices (such as dried seeds, berries, roots, bark) and herbs (such as leaves). Only small amount of spices are generally required to enhance the foods' flavour and taste. Although, most of the spices are considered as safe and healthy, a few may cause allergic properties to it consumers. Up to 2% of people are allergic to fragrances. The most problematic spices for people with allergies are celery, garlic, cinnamon, sesame, turmeric, onion, and mustard. Mustard allergy is the most common type of allergy. Black pepper and vanilla have also been reported to cause allergies. However, allergies can occur due to high concentrations. This means that if you are allergic to nuts or tree pollen, you may also be allergic to certain spices. The symptoms of this type of allergy usually appear gradually. Food allergies are immune reactions that occur immediately after eating certain foods. Even small amounts of food allergies can cause symptoms such as hives, breathing problems, and digestive problems. In some people, food allergies can cause serious and even life-threatening symptoms (called anaphylaxis) in adults. There is no cure, but some children's food allergies may be improved as they get older (Sherman and Billing, 1999).

Products derived from plants are widely used as herbs and spices and are often used in many cuisines all over the world. Spices and herbs are used in small amounts to add flavour to the foods we eat, but they can still cause the adverse reactions. According to 2012 presentations at the American College of Allergy, Asthma and Immunology (ACAAI) Scientific Meeting, an estimated 2–3% of people worldwide are allergic to spices. This number may seem quite low, mainly due to the lack of reliable testing (Sherman and Billing, 1999). Unfortunately, spices are not regulated by the Food and Drug Administration (FDA), so they are not listed on the food labels, making it very difficult for consumers to determine which foods should be avoided. In addition to food, these spices are also used in cosmetics, toothpaste, and various other dental products. Sami Bahana, former president of ACAAI, stated at the conference that as more people in the United States begin using cosmetics and spices, they may soon begin to develop allergies to these substances. Common spice allergies are typically found in pepper, garlic, and vanilla, and some spice blends are much more difficult to control. Symptoms of spice allergies include watery eyes and a runny nose immediately after eating spicy foods such as chili peppers or horseradish. Many people think that these reactions are caused by the immune system, but they are actually caused by irritation of the mucous membranes. When spices in food or cosmetics come into direct contact with the body, skin rashes may appear, and in some cases, irritant contact dermatitis may occur (Paur et al., 2011).

Some life-threatening allergic reactions, such as angioedema and hives, can occur when spices are eaten or inhaled. Inhaling spices can also cause asthma symptoms and angioedema. Skin contact with spices can cause atopic dermatitis, hives, and allergic contact dermatitis.

23.2. SPICE ALLERGY TREATMENT

Spice allergy can be diagnosed simply through allergy tests such as blood tests and patch tests. Once you know which spice you are allergic to, you can avoid using that spice. However, since

 DOI: 10.1201/9781003559894-23

spices are used in many household items and toiletries, it is extremely difficult to completely avoid them.

If symptoms appear, treatment is the same as for food allergies. For severe reactions, using injectable epinephrine and antihistamines can greatly alleviate immediate symptoms. Topical corticosteroids can treat contact dermatitis caused by spices, but ultimately eliminating spices altogether is the best treatment for spice allergies.

23.3. ANISE

Anise is one of the oldest spices used by the historical Greeks, Romans, and Arabs. It was also encouraged with the aid of famous people, including Pliny, Pythagoras, and Hippocrates. Anise has morphological similarities with other species of the Apiaceae family, which may cause severe issues. For instance, misidentification of *Pimpinella anisum* and *Conium maculatum* can result in poisoning and/or death (Colombo et al., 2009). Anise and its essential oil have been utilized in folk medicine drugs for a huge variety of healing purposes, which include as a diuretic, mild expectorant, sedative, laxative, antifungal, antibacterial, antispasmodic, carminative, lacto-inducing, antispasmodic, expectorant, and colon cleaner (Janahmadi et al., 2008). These types and best managed parameters of the herb are described in many reliable compendia around the world, which includes the British Pharmacopoeia, American Pharmacopoeia, and Brazilian Pharmacopoeia. Its use is likewise regulated in Europe. Anise is likewise indexed as "generally recognized as safe" (GRAS) with the aid of the United States Food and Drug Administration (FDA). The usage of anise in folk medication has led many researchers to observe the medicinal quality of the crucial oil of this spice, which include its effective antimicrobial property. Further, many research studies on this essential oil have elucidated its mechanism of action. For instance, anise essential oil can have a broncho dilatory effect, which may be defined by using inhibition of muscarinic receptors. Despite the superb benefits and potential use of important oils, in the case of anise, a study by Janahmadi et al. (2008) reveals that patients with epilepsy need to be cautious while using products containing anise critical oil, as it causes excessive exhilaration (Janahmadi et al., 2008).

23.4. CELERY

Celery is a fibre vegetable that is grown around the world and is widely consumed for the benefits of health, unique taste, and multipurpose. A few data sets explore the development of celery's aroma during maturity. This can help the manufacturer to optimize the harvest time and identify the potential impact of the harvest after commercial maturity. The dried celery of two genotypes selected for biochemical and sensory differences was collected three times and was irradiated using solid micro extracts of gas chromatography-mass spectrometry (SPME GC-MS) and solid micro extract of gas chromatography-sulfa emetri (GC-O) (Pauli et al., 1988). The maturity and the genotype showed significant interactions between the compounds, and the crop collection stage had a greater influence on the quality of the aroma than the plant genotype. This suggests that agricultural practice is important for determining the quality of crops.

The two most frequently consumed celery types are roots (*Apium gravelens Var. Rapaceum*, also known as celery) and stems (*Apium graveolens var. Dulce*). The most common cause of the anaphylaxis for celery is celery roots. Celery roots have considerable potential as "hidden" allergens. The risk to allergic patients is much more complicated by the fact that the ED10 value, the minimum dose required to cause allergic reactions in 10% of allergic patients, is 10 times lower than the ED10 patient with allergies in peanuts (Bauermeister et al., 2009). Celery allergens also show the resistance to food heat treatment. This creates a dangerous food for allergic consumers.

Sensitivity to celery is one of the clinically important causes of food allergies in central Europe, especially in Germany, France, and Switzerland (Marc-Andre et al., 2004; Burney et al., 2014; Ballmer-Weber et al., 2000). More than 70% of patients with pollen allergies are due to the specific fruits, vegetables, and nuts (Breitender et al., 1995; Geroldinger-Simic et al., 2011). Various protein patterns indicate the difference in the content and number of potential allergens. This assumption was confirmed by various IGE recognition models for the serum of the two patients. In addition, the serum blocked by seed extract was reacted in ELISA inhibitory analysis (Afrough et al., 2007).

23.5. CORIANDER

Coriander, also known as cilantro, is a popular herb used in numerous dishes around the world. This is rich source of nutritional benefits due to the valuable phytochemical profiles (Chaurasia and Bharati, 2025; Yadav et al., 2025). Despite its widespread use and various health benefits, some individuals can have an allergic reaction to coriander. To identify these individuals, the Allergen, Individual—Food Coriander test is performed. It measures the presence of specific Immunoglobulin E (IgE) antibodies in the blood that react to proteins in coriander (Sirvent et al., 2014).

The young leaves are popular in most South Asian countries. They are almost always used as fresh because dried leaves loose their flavour. They are often featured in salads, poultry dishes, and soups. They are used essentially in Thailand (where both the root and the leaves are needed for green curry paste), as well as in parts of Vietnam and China, where the chopped leaves are used in many dishes. A famous Mexican dish made with cilantro leaves is guacamole.

Ripe coriander fruits of stronger European varieties contain up to 1.8% essential oil, which is mainly composed of linalool (50–60%). Coriander oil is used to flavour alcoholic beverages, candies, meats, sauces, and tobacco. The fruit and oil are used to correct nausea or burning properties of drugs. This spice was used, among other things, to treat diseases of the digestive tract and joints, allergies, and hay fever, and to stimulate the appetite.

Coriander allergic reaction can occur when the immune system reacts too much to proteins in the herb. This is triggered when the immune system identifies coriander proteins as harmful, even though they aren't, leading to the production of IgE antibodies (Suhonen et al., 1979). When an individual with a coriander allergy consumes the herb, these IgE antibodies cause an allergic reaction.

23.6. CUMIN

Cumin is a potent spice that comes from the dried ripe fruit of *Cuminum cyminum* L. (family: Umbelliferae or Apiaceae). Both the pericarp and seed together constitute a fruit. Cumin that is available in markets is a mixture of fused and separated mericarps (cotyledons), which are yellowish green or yellowish brown, oblong ovoid, and 3–6 mm long. The fruits (seeds) belong to an annual herb which is 15–50 cm in height with long slender and white roots, and branching stem (Boxer et al., 1997).

The spice has a pungent and sweet aroma with some strong bitter notes, especially in the case of ground spices. Ground cumin and whole cumin seeds are frequently employed to enhance the taste of various dishes, particularly in Indian, Middle Eastern, and North African cuisines (Jensen-Jarolim et al., 1997). Allergies to cumin and other spices are uncommon, but they can occur in older children and adults. If any have a cumin allergy, it is possible that they may also have a reaction to other similar foods, such as coriander and dill, or have a pollen allergy.

23.7. FENNEL

Fennel has been utilized in culinary and medicinal applications since ancient time. It is widely cultivated by China, India, Egypt, Germany, Mexico and several other countries. This evergreen perennial herb is approximately 1 meter in height, characterized by pale green stalks, resembling with celery and feathery bright green foliage.

Beyond gastronomy, the plant is employed in therapeutic formulations, gargles, eye-washes, and as a source of natural dyes (yellow/brown pigments from florets and leaves). Essential oils extracted from fennel enhance flavor profiles and fragrances in toothpaste, soaps, perfumes, and air fresheners. Notably, cross-reactivity risks within the *Apiaceae* family necessitate careful allergen evaluation. Carrot, parsley, anise, fennel, and cumin frequently demonstrate shared immune responses due to the structural similarities. Profilin allergens in fennel further enable cross-reactivity with unrelated plant species. Clinical data suggests correlations between spice hypersensitivity and pre-existing sensitivities to birch pollen or wormwood. Among 196 birch pollen–sensitized patients with vegetable-induced oral allergy syndrome (OAS), 103 exhibited *Apiaceae* reactivity, and only one case presented without concurrent apple/hazelnut allergy. Elevated birch-specific IgE levels were observed in *Apiaceae*-reactive cohorts. Fennel consumption may provoke food allergy symptoms or atopic dermatitis in children.

23.8. PARSLEY

Parsley is a biennial garden plant that grows in temperate regions and is an annual in subtropical or tropical regions. It is widely used in Middle Eastern, European, Brazilian, and American cooking (celery) (Sarwar et al., 2016). The former is often used as a garnish, as whole or chopped, and is used in many commercial products, salads, soups and sauces. A group of herbs are used to flavour many dishes. In the UK, parsley is often served with fish or ham. It is the main ingredient in Italian salsa verde and Brazilian cheiro verde. In the Middle East, parsley is a key ingredient in salads such as Lebanese tabbouleh.

Parsley frequently causes allergic symptoms in allergic individuals, particularly those with rhinitis due to pollen allergy. Symptoms include nasal and eye irritation, throat swelling, wheezing, and food allergy symptoms. Contact dermatitis from celery has been reported. Occupational dermatitis has also been reported in the farmers exposed to parsley. Normal foods are safe for pregnant women, but eating too much may cause uterine contractions effect.

23.9. RESULT

The resulting spices are very diverse and are usually used in mixtures. Spice allergies are rare, occurring in 4 to 13 per 10,000 adults (Chen and Bahna, 2011) and are more common in women due to the use of cosmetics. Spice data for children are rare. Allergens present in spice are destroyed during digestion. Therefore, IgE sensitization occurs mainly through inhalation of cross-reactive pollens, especially wormwood and Betula. Respiration disorders are caused by inhaling pollens, while exposure is more likely to cause skin symptoms. Studies on skin testing and specific IgE assays are limited and unreliable. Diagnosis is based on a thorough history and oral examination. The common use of spice mixtures makes it difficult to identify the specific ingredient causing the poisoning.

23.10. CONCLUSION

Fragrance allergies are common as they contain common allergens, but allergy by spice may not be diagnosed. A reliable skin test and determination of specific IgE are required (Bradatan and Sabouraud, 2020). Diagnosis is currently based on a good history and a good titration test. Treatment is strictly prohibited until immunotherapy is developed, which may be difficult due to incomplete or unclear labeling.

REFERENCES

Afrough, B., M. V. Dwek, P. Greenwell. 2007, Identification and elimination of false-positives in an ELISA-based system for qualitative assessment of glycoconjugate binding using a selection of plant lectins. Biotechniques, 43:458–460. https://doi.org/10.2144/000112554.

Ballmer-Weber, B. K., S. Vieths, D. Luttkopf, P. Heuschmann, B. Wuthrich, 2000, Celery allergy confirmed by double-blind, placebo-controlled food challenge: A clinical study in 32 subjects with a history of adverse reactions to celery root. The Journal of Allergy and Clinical Immunology, 106(2):373–378.

Bauermeister, K., B. K. Ballmer-Weber, M. Bublin, P. Fritsche, K. M. Hanschmann, K. Hoffmann-Sommergruber, 2009, Assessment of component-resolved in vitro diagnosis of celeriac allergy. The Journal of Allergy and Clinical Immunology, 124(6):1273–1281. https://doi.org/10.1016/j.jaci.2009.07.033.

Boxer, M. M. D., R. N. Mary Roberts, M. D. Leslie Grammer, 1997, Cumin anaphylaxis: A case report. Journal of Allergy and Clinical Immunology, 99(5):722–723. https://doi.org/10.1016/S0091-6749(97)70039-3.

Bradatan, E., D. Sabouraud, 2020, Spice reactions in children: Allergic or not? Cases reports and literature review. World Allergy Organization Journal, 13(8):100303. https://doi.org/10.1016/j.waojou.2020.100303.

Breitender, H., K. Hoffmann-Sommergruber, G. Riordain, 1995, Molecular characterization of Api g 1, the major allergen of celery (Apium graveolens), and its immunological and structural relationships to a group of 17-kDa tree pollen allergens. European Journal of Biochemistry, 233(2):484–489.

Burney, P. G. J., J. Potts, I. Kummeling, 2014, The prevalence and distribution of food sensitization in European adults. Allergy, 69(3):365–371.

Chaurasia, P. K., S. L. Bharati. 2025. Coriander: A Holistic Outlook on Its Chemistry and Pharmacology. Food Chemistry, 469:142444. https://doi.org/10.1016/j.foodchem.2024.142444

Chen, J. L., S. L. Bahna, 2011, Spice allergy. Annals of Allergy, Asthma and Immunology, 107(3):191–199. https://doi.org/10.1016/j.anai.2011.06.020

Colombo, A. P. V., Susan Bennet, Sean L. Cotton, J. Max Goodson, Ralph Kent, Anne D. Haffajee, Sigmund S. Socransky, Hatice Hasturk, Thomos E. Van Dyke, Floyd Dewhirst, Bruce J. Paster, 2009, The National Center for Biotechnology Information advances science and health by providing access to biomedical and genomic information. Journal of Periodontology, 80(9):1421–1432. https://doi.org/10.1902/jop.2012.110566.

Geroldinger-Simic, M., T. Zelniker, W. Aberer, 2011, Birch pollen-related food allergy: Clinical aspects and the role of allergen-specific IgE and IgG4 antibodies. The Journal of Allergy and Clinical Immunology, 127(3):616–622.

Janahmadi, Mahyar, Sahar Farajnia, Jafar Vatanparast, Habib Abbasipour, Mohammad Kamalinejad, 2008, The fruit essential oil of pimpinella anisum L.(Umblliferae) induces neuronal hyperexcitability in snail party through attentenuation of after hyperpolarization. Journal of Ethnopharmacology, 120(3):360–365. https://doi.org/10.1016/j.jep.2008.09.008.

Jensen-Jarolim, E., A. Leitner, R. Hirschwehr, D. Kraft, B. Wuthrich, O. Scheiner, 1997, Characterization of allergens in Apiaceae spices: Anise, fennel, coriander and cumin. Clinical and Experimental Allergy, 27:1299–1306. https://doi.org/10.1046/j.1365-2222.1997.1580956.x.

Marc-Andre, D., Dominique Penninck, Lilian Cornejo, Philippe Pibarot, 2004, Ultrasonographic diagnosis of portosystemic shunting in dogs and cats. Veterinary Radiology & Ultrasound, 45(5):424–430. https://doi.org/10.1111/.1740-821.2004.04076.x.

Pauli, G., J. C. Bessot, P. A. Braun, A. Dietemann-Molard, M. C. Kopferschmitt-Kubler, R. Thierry, 1988, Celery allergy: Clinical and biological study of 20 cases. Annals of Allergy, 60(3):243–246.

Paur, I., M. H. Carlsen, B. L. Halvorsen, R. Blomhoff, 2011, Chapter 2. Antioxidants in herbs and spices. In I. F. F. Benzie and S. Wachtel-Galor (eds), *Herbal Medicine: Biomolecular and Clinical Aspects* (2nd ed.), CRC Press.

Sarwar, S., M. A. Ayyub, M. Rezgui, S. Nisar, M. I. Jilani, 2016, Parsley: A review of habitat, phytochemistry, ethnopharmacology and biological activities. IJCBS, 9(2016):49–55.

Sherman, P. W., J. Billing, 1999, Darwinian gastronomy: Why we use spices: Spices taste good because they are good for us. BioScience, 49(6):453–463. https://doi.org/10.2307/1313553.

Sirvent, S., B. Canto, F. Gomez, 2014, Detailed characterization of act d 12 and act d 13 from kiwi seeds: Implication in IgE cross-reactivity with peanut and tree nuts. Allergy, 69(11):1481–1488.

Suhonen, R., H. Keskinen, R. Suhonen, F. Bjorksten, E. Vaheri, A. Zitting, 1979, Allergy to coriander: A case report. Allergy, 34(5):327–330. https://doi.org/10.1111/j.1398-9995.1979.tb04374.x.

Yadav, M., S. Kumari, Nagraj, N. Sharma, S. L. Bharati, S. Singh, J. Kumar, P. K. Chaurasia, 2025, Hydrogen peroxide scavenging by methanolic extracts of coriander: An in vitro antioxidant study. Pharmaceutical Drug Regulatory Affair Journal, 8(1):000162. https://doi.org/10.23880/pdraj-16000162.

Index

A

ABTS, 160, 229, 242
acetic acid-induced ulcers, 160
African wonder spice *see* ginger
ajwain, 205, 260
alanine aminotransferase (ALT), 199
alcohol-soluble resin, 192
alkaline phosphatase (ALP), 199
allergen, 281
allergy, 280
alpha-glucosidase enzymes, 145
alpha-glucosidase, 75
Alzheimer's disease, 253
analgesic, 96, 104, 154, 156, 167, 182, 217
anise, 281
anorexia, 156
anthocyanins, 9, 58, 72, 157
anthoxanthins, 157
antibacterial, 14, 16, 40, 44, 74, 91, 154, 205, 207, 229
anticancer, 75, 84, 147, 201, 210, 242
antidermatophytic, 210
antidiabetic activity, 200
anti-helminthic effect, 200
anti-inflammatory, 154, 183, 221, 228, 241, 273
antimicrobial, 200, 207, 220, 240
antioxidant, 183, 201, 205, 208, 220, 273
antipyretic, 128, 154, 182
antispasmodic, 46, 76, 96, 105, 199, 209
antiswelling, 156
apple cider vinegar, 257
asafoetida, 190
aspartate aminotransferase (AST), 199
astaxanthin, 157

B

basil, 58, 208, 257
β-bergamotene, 218
beta-carotene, 2, 7, 25, 157
beta-caryophyllene, 82, 93, 94, 144, 238
black turmeric, 178, 182
brain-derived neurotrophic factor (BDNF), 253
British Pharmacopoeia (BP), 218

C

caffeic acid, 179
cancer, 228, 242, 271
cardamom, 258
cardiovascular, 210, 228, 270
carminative, 156
carotenoid, 156
carotenoids, 267
carvacrol, 211
caryophyllene, 218
catalase (CAT), 242
celery, 281
chemopreventive, 183
cholesterol regulation, 183
cinnamaldehyde, 143, 147, 148, 230
cinnamomum, 143, 148, 229, 254
cinnamon bark oil, 143
cinnamon essential oil (CEO), 148
cinnamon leaf oil, 143
clotrimazole, 200
cloves, 255
copaene, 219
coriander, 282
cough assuaging, 156
coumarin, 116, 120
Curcuma caesia, 178, 181, 182
Curcuma longa, 181, 215, 228, 253
curcumene, 229
curcumin, 228
curcuminoids, 181
cytotoxicity, 209

D

dehydrodiisoeugenol, 220
diabetes mellitus, 19, 61, 75, 108, 145, 228
diaphoresis, 156
diastolic blood pressure (DBP), 252
dietary fibers, 229, 231
dimethylallyl diphosphate (DMAPP), 239
DPPH, 160, 208, 209, 229, 231, 242
dyspepsia, 160

E

elemicin, 217, 220
eugenol, 82, 84, 91, 143, 217
European Pharmacopoeia (EP), 218, 246

F

fennel, 46, 122, 282
fenugreek, 114
Ferula asafoetida, 190, 191, 192
ferulic acid, 179, 206
flavonoids, 109, 116, 157, 181, 206, 255, 267
flavonols, 157
Food and Drug Administration (FDA), 280
FRAP, 208, 209, 229

G

gallic acid, 179
garlic, 252
gastritis, 160
gastrointestinal disorders or disease, 148, 160, 199
GC-MS, 168, 218, 239
geranyl acetate, 160
germacrene D, 218, 219
ginger toxicology, 165, 166, 167

ginger, 154, 155, 156
ginger-citron tea, 154
ginger-hopsu tea, 154
gingerol, 154, 156, 160
gingerone, 160
glucosinolates, 157, 267
glutathione (GSH), 146
gut health, 184

H

HDL cholesterol, 253, 255
hepatoprotective, 199, 221
high-performance liquid chromatography (HPLC), 239
high-performance thin-layer chromatography (HPTLC), 157
Hing, 190
honey, 254
hydrodistillation, 145
hydrogen peroxide, 146
hydroxyl radicals, 146

I

inflammatory bowel diseases, 148
International Organization for Standardization (ISO), 218
isoeugenol, 218
isoflavones, 157
isopentenyl diphosphate (IPP), 239
isothiocyanates, 157

K

kaempferol, 26, 46, 61, 83, 92, 116, 131, 207, 271
kernel oil, 219
king of spices, 230

L

LDL cholesterol, 253, 255
lemon, 256
licarin E, 220
lignans, 82, 157, 220
β-linalool, 219
linalool, 84, 102, 103, 105, 206
lipopolysaccharides (LPS), 241
liver-protective, 221
lutein, 157
lycopene, 157, 269

M

macronutrients, 217
major depressive disorder (MDD), 184
Malabaricon C, 217
malondialdehyde (MDA), 94, 106, 146, 266
Mentha piperita, 256
methylerythritol phosphate (MEP), 239
methylglyoxal, 255
mevalonate (MVA), 239
MFEO, 218
microwave-assisted extraction, 145, 244
minimum inhibitory concentration (MIC), 221, 255
molecular toxicology, 167
β-myrcene, 232
myristic acid, 220
Myristica fragrans, 215, 216, 258
myristicin, 217, 218, 220

N

nature's medicine chest, 156
nerolidol, 102, 103, 105
neurodegenerative diseases, 147
neurological health, 184
neuroprotective, 147, 201
Nigella sativa, 94, 129, 132
NMR, 168, 240
nonsmall cell lung cancer (NSCLC), 211
nutmeg, 215, 216, 217, 258
nutmeg oleoresin, 219

O

Ocimum basilicum, 55, 257
oleo-gum-resin, 192, 193
onion, 72
osteoarthiritis, 37, 84
oxygenated monoterpenes, 207
oxygenated sesquiterpenes, 206

P

parsley, 283
pepper, 230
peppermint, 256
peptic ulcer, 160
phellandrene, 181, 219
phenols, 157
phenylpropanoids, 218
Piper nigrum L., 230
piperine, 80, 83, 260
platelet-activating factor, 146
polycystic ovary syndrome (PCOS), 146, 230
polyphenol, 2, 25, 28, 64, 146, 157, 266
polyphenols, 267
proanthocyanidins, 157
pro-inflammatory, 241
pro-inflammatory cytokines, 146
prostaglandin E2 (PGE2), 146
pulmonary disease, 205

Q

quercetin, 2, 5, 28, 37, 78, 83, 92, 133
quinic acid, 61, 103, 206

R

radioprotective, 220
randomized controlled trials (RCTs), 253
reactive nitrogen species (RNS), 242
reactive oxygen species (ROS), 146, 242

S

saponins, 157, 180, 206, 218
SCFAs (short-chain fatty acids), 148
sesquiterpenes, 228
shogaol, 156, 160
solvent extraction, 145, 244

spice of life *see* ginger
steam distillation, 244
sterols, 157
stress-induced ulcers, 160
sulphur-containing compounds, 192
supercritical fluid extraction, 145, 244
superoxide anions, 146
superoxide dismutase (SOD), 146, 242
systolic blood pressure (SBP), 252
Syzygium aromaticum, 255

T

TAC (total antioxidant capacity), 146
tannins, 2, 9, 10, 22, 65, 74
Terminalia chebula, 259, 261
terpenes, 82, 130
terpenoids, 144, 157, 179, 181
β-terpineol, 219
tetrahydrocurcumin, 160
thiobarbituric acid reactive substance (TBARS), 208
thyme, 58, 94, 205
thymosin, 206
Thymus vulgaris, 205
toxicity, 48, 66, 97, 221
toxicology, 58, 161, 206, 218, 231, 262
Trachyspermum ammi, 260, 261
turmeric, 228, 253

U

ulcers, 184, 199, 228
ultrasound-assisted extraction, 244
UV, 173, 185, 239

V

viral, 117, 240, 241
volatile oil, 192

W

water-soluble gum, 192
wheatgrass, 185

X

xanthophylls, 22, 36, 58, 157

Y

yeast, 95

Z

Zingiber officinale, 156, 215, 228
Zingiberaceae, 154, 156, 178
zingiberene, 228, 229

For Product Safety Concerns and Information please contact our
EU representative GPSR@taylorandfrancis.com Taylor & Francis
Verlag GmbH, Kaufingerstraße 24, 80331 München, Germany